Vegetation Dynamics on the Mountains and Plateaus of the American Southwest

PLANT AND VEGETATION

Volume 8

Series Editor: M.J.A. Werger

For further volumes:
http://www.springer.com/series/7549

John L. Vankat

Vegetation Dynamics on the Mountains and Plateaus of the American Southwest

Springer

John L. Vankat
Professor Emeritus
Department of Botany
Miami University
Oxford, Ohio, USA

Senior Research Ecologist
Merriam-Powell Center for Environmental
Research
Northern Arizona University
Flagstaff, Arizona, USA

Adjunct Professor
School of Forestry
Northern Arizona University
Flagstaff, Arizona, USA

Former Ecologist
Grand Canyon National Park
Arizona, USA

ISSN 1875-1318 ISSN 1875-1326 (electronic)
ISBN 978-94-007-6148-3 ISBN 978-94-007-6149-0 (eBook)
DOI 10.1007/978-94-007-6149-0
Springer Dordrecht Heidelberg New York London

Library of Congress Control Number: 2013935872

Printed on acid-free paper

Springer is part of Springer Science+Business Media (www.springer.com)

Preface

This book provides information essential for anyone interested in the ecology of the American Southwest, including land and resource managers, environmental planners, conservationists, environmentalists, ecologists, land stewards, and students. The book is unique in its coverage of the hows and whys of dynamics (changes) in the major types of vegetation occurring on southwestern mountains and plateaus. The book explains the drivers and processes of change, describes historical changes, and provides conceptual models that diagrammatically illustrate past, present, and potential future changes.

All major types of vegetation are covered: spruce-fir forest, mixed conifer forest, ponderosa pine forest, pinyon-juniper vegetation, subalpine-montane grassland, Gambel oak shrubland, and interior chaparral shrubland. Other types of upper-elevation vegetation, such as alpine tundra, treeline, and most riparian types, are not included. They cover little land area and their dynamics have received little research in the American Southwest. In addition, the focus is on vegetation that is relatively undisturbed, i.e., in natural and near-natural condition, and how it responds to natural disturbances such as fire and drought, as well as to anthropogenic disturbances such as fire exclusion and invasive species. Although intensive land uses such as logging are not included, knowledge of post-disturbance vegetation dynamics is applicable to the restoration and recovery of heavily disturbed areas.

The book has an introductory chapter that is followed by individual chapters on the types of vegetation listed above. Each vegetation chapter begins with an introduction that presents an overall picture of the vegetation. The next section of each chapter covers major drivers, including landscape, climate, soil, and animals, as well as natural disturbances such as fire and anthropogenic disturbances such as livestock grazing. The third section describes key processes of vegetation dynamics, such as succession. The fourth section describes vegetation conditions before Euro-American settlement, evaluates approaches used to determine them, and outlines changes that followed Euro-American settlement. The fifth section of each chapter presents a three-tiered suite of conceptual models unique to the vegetation of that chapter. These models (1) characterize the ecosystem in relation to vegetation and disturbance, (2) describe vegetation dynamics in terms of vegetation states and transitions among them, and (3) illustrate mechanisms driving those dynamics. The sixth section highlights conclusions and key challenges for researchers and managers. Each chapter concludes with a list of the literature references cited. Individual chapters have been written to stand alone; nevertheless, they contain many cross-references.

My interest in southwestern vegetation extends back more than four decades, and I vividly remember the moment it began. I had passed the entrance to the North Rim region of Grand Canyon National Park in north-central Arizona and soon became fascinated by the complexity of forest patterns (I've heard that most park visitors pay more attention to a certain canyon). Over the following years, I had research leaves at Northern Arizona University and recurrent research projects in central and north-central Arizona uplands. After I retired from my long-term position in the Department of Botany, Miami University, Oxford, Ohio, and moved to Flagstaff, Arizona, Grand Canyon National Park offered me a 4-year position to do forest ecology research and to bring research to bear on management issues. I couldn't resist the opportunity. Experiences over those years made me acutely aware of the need for this book.

Flagstaff, Arizona

John L. Vankat

Acknowledgments

I have many people to thank…

My parents exposed me to vegetation dynamics at a young age by living next to a natural area and taking annual camping trips to visit national parks and forests in the western United States. These experiences encouraged a career and also gave me the desire to continue the family tradition and share nature with my children and grandchildren … each an ongoing source of inspiration.

I have had many mentors, including Herbert “Bud” T. Lystrup of Eau Claire Memorial High School in Eau Claire, Wisconsin, William “Bill” H. Muir of Carleton College in Northfield, Minnesota, and Jack Major of the University of California, Davis. They and colleagues and students at Miami University in Oxford, Ohio, were very important in my development.

D. Coleman “Cole” Crocker-Bedford, formerly of Grand Canyon National Park, Arizona, hired me to do forest ecology research and bring research to bear on management issues during 2003–2007, shortly after I had moved to the American Southwest. The research, as well as interactions with him and other staff members, greatly expanded my understanding of southwestern forest dynamics. During a portion of this time, Lisa P. Thomas of the National Park Service’s Southern Colorado Plateau Inventory and Monitoring Network funded me to review literature and develop conceptual models of southwestern montane vegetation, as part of the development of a monitoring program for the Colorado Plateau. This broadened my background. These experiences with the National Park Service provided the impetus for this book.

The following individuals provided many valuable comments on recent versions of chapters: Scott R. Abella of the University of Nevada, Las Vegas (Chap. 4), Leonard F. DeBano retired from the US Forest Service (Chaps. 1 and 8), M. Lisa Floyd-Hanna of Prescott College (Chap. 7), Brian F. Jacobs of Bandelier National Monument (Chaps. 1 and 5), Ellis Q. Margolis of the University of Arizona (Chaps. 1 and 2), Wayne A. Robbie of the US Forest Service (Chaps. 1 and 6), William “Bill” H. Romme of Colorado State University (Chaps. 1 and 3), Charles “Charlie” M. Truettner of Northern Arizona University (Chap. 2), and Thomas “Tom” T. Veblen of the University of Colorado Boulder (Chap. 2). Additional, anonymous reviewers provided valuable suggestions on earlier drafts. Crocker-Bedford, Thomas, Floyd-Hanna, Jacobs, Romme, and Mark A. White of The Nature Conservancy provided useful comments on earlier versions of the conceptual models. The many individuals and organizations that graciously allowed me to use their photographs are acknowledged in the figure legends.

Marinus J. A. Werger of Utrecht University and editor for Springer's Plant and Vegetation book series thoroughly reviewed every chapter, made numerous valuable comments, and facilitated publication of this book.

My progress on this book nearly ended with the abrupt onset of vision problems a few years ago, but Jaime R. Gaitan of Retinal Consultants of Arizona and Dharmendra R. Patel of Mayo Clinic (Scottsdale) surgically restored my vision and enabled completion of this and other projects.

Regardless of the above contributions, I would not have undertaken or completed this project without my wife Betty J. Huffman. She continually encouraged me, patiently served as my sounding board, and gracefully endured periods when my attention was focused on writing. She also provided many of the photographs.

I am a fortunate person.

Contents

1 Introduction ... 2
1.1 Introduction ... 2
1.2 Drivers ... 13
1.2.1 Landscape ... 14
1.2.2 Climate ... 19
1.2.3 Soil ... 24
1.2.4 Animals ... 25
1.2.5 Natural Disturbance ... 25
1.3 Processes ... 35
1.3.1 Succession ... 35
1.3.2 Other ... 37
1.4 Paleoecological History ... 37
1.5 Land-Use History ... 39
1.5.1 Native-American Dominance ... 39
1.5.2 Euro-American Dominance ... 42
1.6 Anthropogenic Disturbance ... 48
1.6.1 Livestock Grazing ... 48
1.6.2 Fire Management ... 49
1.6.3 Modern Climate Change ... 50
1.6.4 Air Pollution ... 51
1.6.5 Invasive Species ... 52
1.6.6 Recreation ... 53
1.6.7 Nearby Land Use ... 54
1.7 Conceptual Models ... 54
1.7.1 Ecosystem-Characterization Models ... 55
1.7.2 Vegetation-Dynamics Models ... 55
1.7.3 Mechanistic Models ... 57
1.8 Conclusions and Challenges ... 57
References ... 58

Part I Forests

2 Spruce-Fir Forest 72
2.1 Introduction 72
2.2 Drivers 78
2.2.1 Landscape 79
2.2.2 Climate 80
2.2.3 Soil 80
2.2.4 Animals 81
2.2.5 Natural Disturbance 81
2.2.6 Anthropogenic Disturbance 91
2.3 Processes 94
2.3.1 Succession 95
2.3.2 Gap Dynamics 97
2.4 Historical Changes 99
2.4.1 Overstory 99
2.4.2 Understory 103
2.5 Conceptual Models 105
2.5.1 Ecosystem-Characterization Model 105
2.5.2 Vegetation-Dynamics Models 110
2.5.3 Mechanistic Model 113
2.6 Conclusions and Challenges 117
References 117

3 Mixed Conifer Forest 124
3.1 Introduction 124
3.2 Drivers 131
3.2.1 Landscape 131
3.2.2 Climate 133
3.2.3 Soil 135
3.2.4 Animals 135
3.2.5 Natural Disturbance 136
3.2.6 Anthropogenic Drivers 148
3.3 Processes 151
3.3.1 Gap Dynamics 151
3.3.2 Succession 152
3.4 Historical Changes 155
3.4.1 Overstory 155
3.4.2 Understory 163
3.5 Conceptual Models 164
3.5.1 Ecosystem-Characterization Model 165
3.5.2 Vegetation-Dynamics Models 165
3.5.3 Mechanistic Model 177
3.6 Conclusions and Challenges 179
References 180

4 Ponderosa Pine Forest 190
4.1 Introduction 190
4.2 Drivers 195
4.2.1 Landscape 195
4.2.2 Climate 197
4.2.3 Soil 198
4.2.4 Animals 199
4.2.5 Natural Disturbance 199
4.2.6 Anthropogenic Disturbance 208
4.3 Processes 216
4.3.1 Regeneration and Thinning 216
4.3.2 Succession 219
4.4 Historical Changes 222
4.4.1 Overstory 222
4.4.2 Understory 233
4.5 Conceptual Models 236
4.5.1 Ecosystem-Characterization Model 236
4.5.2 Vegetation-Dynamics Models 237
4.5.3 Mechanistic Model 251
4.6 Conclusions and Challenges 251
References 252

Part II Woodland, Savanna, Grassland, and Shrublands

5 Pinyon-Juniper Vegetation 268
5.1 Introduction 268
5.2 Drivers 275
5.2.1 Landscape 275
5.2.2 Climate 276
5.2.3 Soil 277
5.2.4 Animals 277
5.2.5 Natural Disturbance 278
5.2.6 Anthropogenic Drivers 286
5.3 Processes 292
5.3.1 Infill and Expansion 293
5.3.2 Succession 297
5.4 Historical Changes 300
5.4.1 Overstory 300
5.4.2 Understory 306
5.5 Conceptual Models 308
5.5.1 Ecosystem-Characterization Model 308
5.5.2 Vegetation-Dynamics Models 308
5.5.3 Mechanistic Models 318
5.6 Conclusions and Challenges 322
References 324

6 Subalpine-Montane Grassland ... 334
6.1 Introduction ... 334
6.2 Drivers ... 341
6.2.1 Landscape ... 341
6.2.2 Climate ... 342
6.2.3 Soil ... 343
6.2.4 Animals ... 343
6.2.5 Natural Disturbance ... 344
6.2.6 Anthropogenic Disturbance ... 347
6.3 Processes ... 350
6.3.1 Rapid Regrowth ... 350
6.3.2 Succession ... 352
6.3.3 Tree Encroachment ... 352
6.4 Historical Changes ... 354
6.4.1 Pre-Euro-American Settlement ... 354
6.4.2 Post-Euro-American Settlement ... 356
6.5 Conceptual Models ... 358
6.5.1 Ecosystem-Characterization Model ... 358
6.5.2 Vegetation-Dynamics Model ... 359
6.5.3 Mechanistic Model ... 366
6.6 Conclusions and Challenges ... 366
References ... 368

7 Gambel Oak Shrubland ... 372
7.1 Introduction ... 372
7.2 Drivers ... 377
7.2.1 Landscape ... 377
7.2.2 Climate ... 379
7.2.3 Soil ... 380
7.2.4 Animals ... 380
7.2.5 Natural Disturbance ... 381
7.2.6 Anthropogenic Disturbance ... 382
7.3 Processes ... 385
7.3.1 Rapid Regrowth ... 385
7.3.2 Succession ... 387
7.4 Historical Changes ... 387
7.4.1 Overstory ... 387
7.4.2 Understory ... 390
7.5 Conceptual Models ... 390
7.5.1 Ecosystem-Characterization Model ... 391
7.5.2 Vegetation-Dynamics Models ... 391
7.5.3 Mechanistic Model ... 398
7.6 Conclusions and Challenges ... 398
References ... 400

8 Interior Chaparral Shrubland 404
8.1 Introduction 404
8.2 Drivers 411
8.2.1 Landscape 412
8.2.2 Climate 413
8.2.3 Soil 414
8.2.4 Animals 414
8.2.5 Natural Disturbance – Fire 414
8.2.6 Anthropogenic Disturbance 419
8.3 Processes 422
8.3.1 Rapid Regrowth 422
8.3.2 Succession 426
8.4 Historical Changes 426
8.4.1 Overstory 426
8.4.2 Understory 431
8.5 Conceptual Models 431
8.5.1 Ecosystem-Characterization Model 431
8.5.2 Vegetation-Dynamics Model 435
8.5.3 Mechanistic Model 435
8.6 Conclusions and Challenges 437
References 439

Brief Professional Biography of John L. Vankat 443

Author Index 445

Subject Index 453

Chapter 1
Introduction

Abstract The vegetation on the mountains and plateaus of the American Southwest includes Spruce-Fir Forest, Mixed Conifer Forest, Ponderosa Pine Forest, Pinyon-Juniper vegetation, Subalpine-Montane Grassland, Gambel Oak Shrubland, and Interior Chaparral Shrubland. These are introduced in relation to gradients of elevation and moisture. Key drivers of vegetation dynamics, i.e., landscape, climate, soil, animals, and natural disturbance, are characterized for the Southwest, with emphasis on the natural disturbances of fire and biotic agents. Processes of vegetation dynamics, such as succession, are outlined. The paleoecological development of today's vegetation and the land-use history and impacts of Native Americans and Euro-Americans are described. This is followed by an overview of anthropogenic drivers affecting vegetation dynamics: livestock grazing, fire management, modern climate change, air pollution, invasive species, recreation, and nearby land use. A nested, three-tiered set of conceptual models is introduced to synthesize information on drivers and processes and diagrammatically illustrate past, present, and future vegetation dynamics. The set consists of ecosystem-characterization, vegetation-dynamics, and mechanistic models. Key conclusions and challenges for researchers and land managers are summarized.

1.1 Introduction

The American Southwest is generally thought of as an arid region with expanses of deserts interrupted only by cities such as Phoenix, Tucson, and Albuquerque. There is some truth to this image of aridity, but the region also has cool, moist mountains and plateaus (Fig. 1.1a, b). These have forests of pines (*Pinus* spp.), Douglas-fir (*Pseudotsuga menziesii*), firs (*Abies* spp.), spruces (*Picea* spp.), and quaking aspen (*Populus tremuloides*). There also are woodlands and savannas of pinyons (*Pinus* spp.) such as Colorado pinyon (*P. edulis*) and junipers (*Juniperus* spp.) such as Utah juniper (*J. osteosperma*), as well as grasslands, shrublands, and – on the highest peaks – alpine tundra.

J.L. Vankat, *Vegetation Dynamics on the Mountains and Plateaus of the American Southwest*, Plant and Vegetation 8, DOI 10.1007/978-94-007-6149-0_1,

Fig. 1.1 (**a, b**) Examples of mountains and plateaus in the American Southwest. (**a**) The San Juan Mountains are in southwestern Colorado (Photograph by Betty J. Huffman). (**b**) The Coconino Plateau (*foreground*) and Kaibab Plateau (*distant background*) are separated by the Grand Canyon in Grand Canyon National Park, north-central Arizona (Photograph by National Park Service)

Although mountains and plateaus cover only a third of the land area of the American Southwest, they are important regionally, nationally, and internationally. Ecologically, they embody a great richness of biodiversity, providing unique habitats in this generally arid region. Economically, they have been major sources of wood, forage, and water, as well as sites for tourism and outdoor recreation that annually bring millions of people from around the world. Aesthetically, they comprise spectacularly scenic landscapes that have inspired humans for millennia.

This book is not a general overview of southwestern vegetation or ecosystems; others have done that (e.g., Lowe 1964; Brown 1982, 1994; Dick-Peddie 1993; Ffolliott and Davis 2008). Instead, this book emphasizes the dynamics (changes) of the major types of vegetation occurring on southwestern mountains and plateaus. It explains the drivers and processes of change, describes historical changes, and provides conceptual models that diagrammatically illustrate past, present, and potential future changes. Understanding of vegetation dynamics is essential to land and resource managers, environmental planners, ecologists, other scientists, conservationists, environmentalists, and others interested in the ecology of the American Southwest. This book focuses on vegetation that is relatively undisturbed, i.e., in natural and near-natural condition, and how it responds to natural disturbances such as fire and drought and to anthropogenic disturbances such as fire exclusion and invasive species. Intensive land uses such as logging are not included, but knowledge of vegetation dynamics is also applicable to the restoration and recovery of heavily disturbed areas.

Each of the following vegetation chapters covers a common vegetation type and begins with an overview of factors influencing it: landscape, climate, soil, animals, and both natural and anthropogenic disturbances. Key processes of vegetation dynamics, such as succession, are described. With this as background, historical vegetation dynamics are illustrated by descriptions of vegetation structure and composition present both before and with the influence of Euro-Americans. This information is then summarized and synthesized in a nested set of conceptual models that (1) characterize the ecosystem in relation to vegetation and disturbance, (2) describe vegetation dynamics in terms of vegetation states and transitions among them, and (3) illustrate mechanisms of those vegetation dynamics. Each chapter ends with a paragraph of conclusions and challenges for researchers and managers.

Geographically, this book covers the mountains and plateaus of Arizona, New Mexico, southwestern Colorado, and southern and central Utah, herein defined as the American Southwest. Most maps in this book include all of these four states, but the area designated as the American Southwest is shown as the study area. There is emphasis on the Colorado Plateau, a region that covers portions of each of the four states and has extensive areas of the vegetation types covered in this book. Findings from elsewhere in the American Southwest are also included. With a focus on the dynamics of relatively undisturbed vegetation, most insight comes from research done in protected areas, especially units of the U.S. National Park System. Pertinent research on lands managed by other federal and state agencies is included as well.

Table 1.1 Common and scientific names of species in this chapter. Primary source: Integrated Taxonomic Information System (2012)

Plants	
Blue spruce	*Picea pungens* Engelm.
Colorado pinyon	*Pinus edulis* Engelm.
Corkbark fir	*Abies lasiocarpa* var. *arizonica* (Merriam) Lemmon
Douglas-fir	*Pseudotsuga menziesii* (Mirbel) Franco
Dwarf mistletoe	*Arceuthobium* Bieb.
Engelmann spruce	*Picea engelmannii* Parry ex Engelm.
Fescue	*Festuca* L.
Fir	*Abies* P. Mill.
Gambel oak	*Quercus gambelii* Nutt.
Juniper	*Juniperus* L.
Limber pine	*Pinus flexilis* James
Mountain mahogany	*Cercocarpus* Kunth
Pine	*Pinus* L.
Pinyon	*Pinus* L.
Ponderosa pine	*Pinus ponderosa* Douglas ex P. Lawson & C. Lawson
Quaking aspen	*Populus tremuloides* Michx.
Serviceberry	*Amelanchier* Medik.
Shrub live oak	*Quercus turbinella* Greene
Southwestern white pine	*Pinus strobiformis* Engelm.
Spruce	*Picea* A. Dietr.
Subalpine fir	*Abies lasiocarpa* var. *lasiocarpa* (Hook.) Nutt.
Utah juniper	*Juniperus osteosperma* (Torr.) Little
White fir	*Abies concolor* (Gord. & Glend.) Lindl. Ex Hildebr.
Animals	
Bark beetle	*Dendroctonus* Erichson, 1836, *Dryocoetes confusus* Swaine, 1912, *Ips* De Geer, 1775, and *Scolytus ventralis* LeConte, 1868
Beaver	*Castor canadensis* Kuhl, 1820
Bobcat	*Lynx rufus* Schreber, 1777
Cattle	*Bos taurus* Linnaeus, 1758
Coyote	*Canis latrans* Say, 1823
Deer	*Odocoileus* Rafinesque, 1832
Elk	*Cervus elaphus* Linnaeus, 1758
Goat	*Capra hircus* Linnaeus, 1758
Mountain lion	*Felis concolor* Linnaeus, 1771
Mule deer	*Odocoileus hemionus* Rafinesque, 1817
Sheep	*Ovis aries* Linnaeus, 1758
Western spruce budworm	*Archips fumiferana* Clemens
Western tent caterpillar	*Malacosoma californicum* Packard, 1864
White-tailed deer	*Odocoileus virginianus* Zimmermann, 1780
Wolf	*Canis lupus* Linnaeus, 1758

Ecological study of the vegetation on southwestern mountains and plateaus began in 1889 with Clinton Hart Merriam's classic research on life zones (Brown et al. 1994). Merriam's first publication on this topic, "Results of a biological survey of the San Francisco Mountain region and desert of the Little Colorado, Arizona" (Merriam and Stejneger 1890), described the elevational distribution of vegetation and proposed the life zone concept, the general principles of which became core ideas in ecology and biogeography (Phillips et al. 1989) and have been applied to upper-elevation ecosystems of the American Southwest for many decades (Brown et al. 1994).

The core of the vegetation classification used in this book has its origins in Merriam's classification of life zones. The classification is also influenced by a series of publications by David E. Brown, usually coauthored by Charles H. Lowe and Charles P. Pase (e.g., Brown et al. 1980; Brown 1994), as well as a revision of the Brown-Lowe-Pase classification by Spence et al. (1995) for the Colorado Plateau. The vegetation types covered in the following chapters are included in the forest and woodland, tall shrubland, and grassland formations proposed by Spence et al. (1995). They are

Spruce-Fir Forest
Mixed Conifer Forest
Ponderosa Pine Forest (*Pinus ponderosa*)
Pinyon-Juniper vegetation
Subalpine-Montane Grassland
Gambel Oak Shrubland (*Quercus gambelii*)
Interior Chaparral Shrubland

Geographic distributions of these vegetation types are shown for the American Southwest (Fig. 1.2) and in more detail for the four states with land included in the American Southwest (Fig. 1.3a–d). Stands of quaking aspen are treated separately in these maps because of the organization of the data used for mapping. However, the rest of this book joins Brown (1994) and others in treating stands of quaking aspen as parts of different types of coniferous forest vegetation. Other types of vegetation on southwestern mountains and plateaus, such as alpine tundra, treeline, and most riparian types, are not included. They cover little land area, and their dynamics have been little researched in the American Southwest.

This book is organized on the above classification of vegetation. Nevertheless, southwestern mountains and plateaus have gradients of vegetation that parallel mostly gradual (occasionally abrupt) changes in various environmental factors. Key environmental factors include elevation and moisture, as presented in Fig. 1.4. The presence of broad areas of transitional vegetation indicates that the vegetation types represent portions of these gradients and are not discrete units. The highest elevations treated in this book have Spruce-Fir Forest (Fig. 1.5), which is characterized by Engelmann spruce (*Picea engelmannii*), subalpine fir (*Abies lasiocarpa* var. *lasiocarpa*) or corkbark fir (*A. lasiocarpa* var. *arizonica*), and quaking aspen. With decreasing elevation, there is a gradual, often patchy transition from Spruce-Fir Forest to Mixed Conifer Forest (Fig. 1.6). This forest

Fig. 1.2 Distribution of the major vegetation types on the mountains and plateaus of the American Southwest. The map shows all of the states of Arizona, New Mexico, Colorado, and Utah, and the American Southwest is illustrated in *red* on the small map (Source: U.S. Geological Survey National Gap Analysis Program (2005). Map prepared by Monica Swihart)

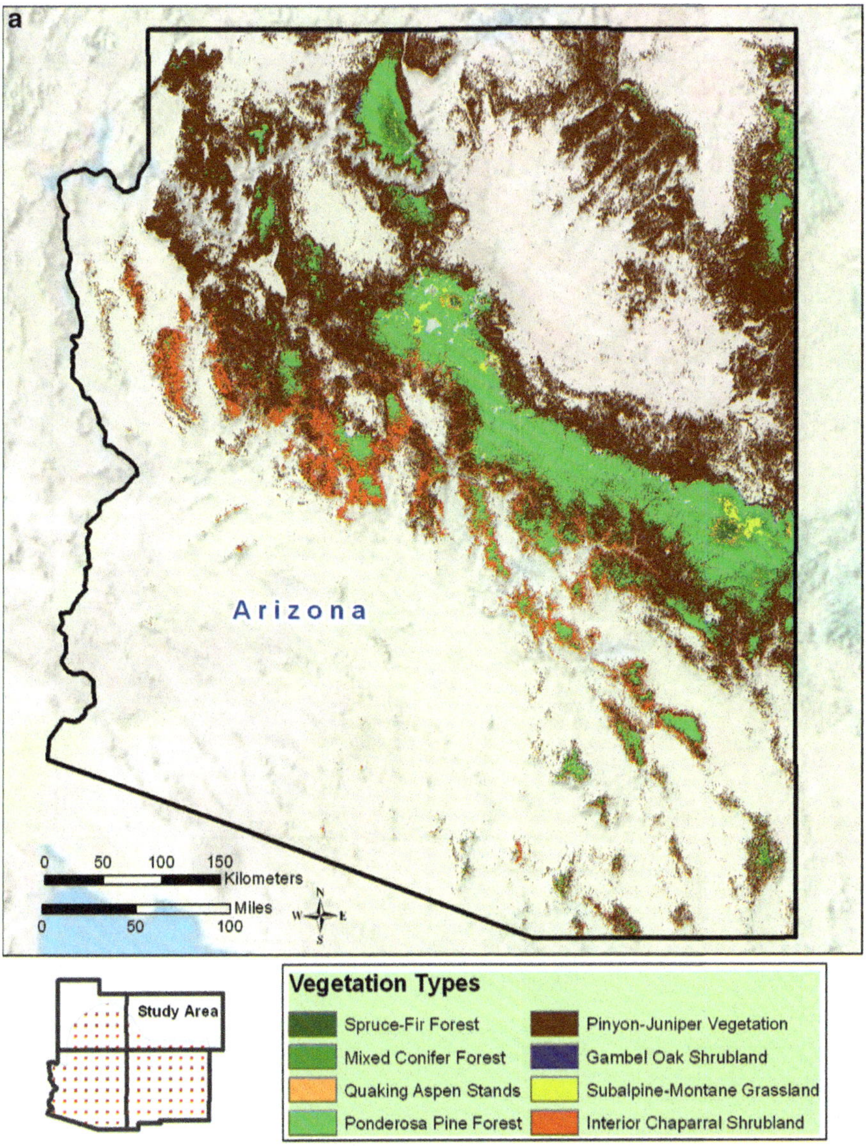

Fig. 1.3 (**a**–**d**) Distribution of the major vegetation types on the mountains and plateaus of (**a**) Arizona, (**b**) New Mexico, (**c**) Colorado, and (**d**) Utah. Each map shows an entire state, and the American Southwest is illustrated in *red* on the small map (Source: U.S. Geological Survey National Gap Analysis Program (2005). Maps prepared by Monica Swihart)

is highly diverse, consisting of a mosaic of topographic- and disturbance-related patches dominated by various combinations of ponderosa pine, Douglas-fir, white fir (*Abies concolor*), blue spruce (*Picea pungens*), southwestern white pine (*Pinus strobiformis*), limber pine (*Pinus flexilis*), quaking aspen, and sometimes

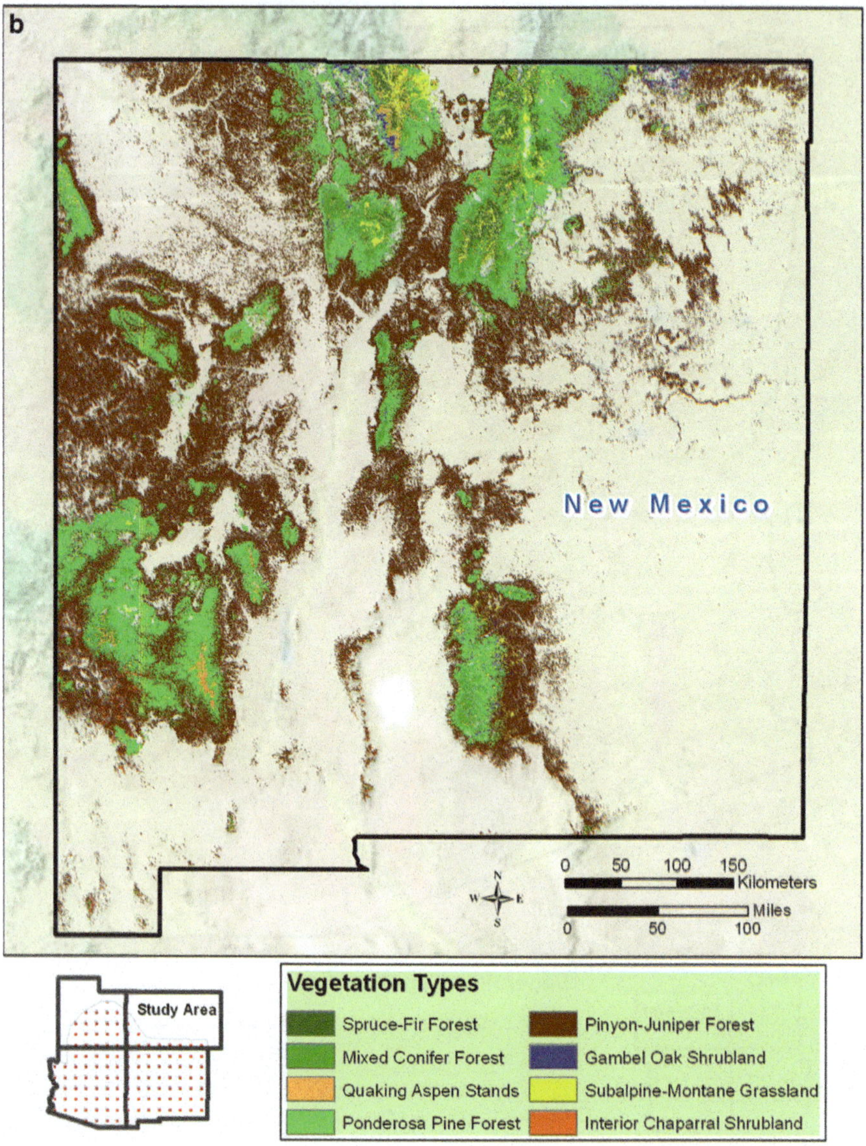

Fig. 1.3 (continued)

other species. With decreasing elevation, ponderosa pine progressively becomes dominant in Ponderosa Pine Forest (Fig. 1.7), first with white fir and/or Douglas-fir as co-dominant tree species and then as the only canopy tree, but often with Gambel oak as a subcanopy tree. At lower elevation, the subcanopy of Ponderosa Pine Forest is dominated by pinyons and junipers, and Ponderosa Pine Forest is

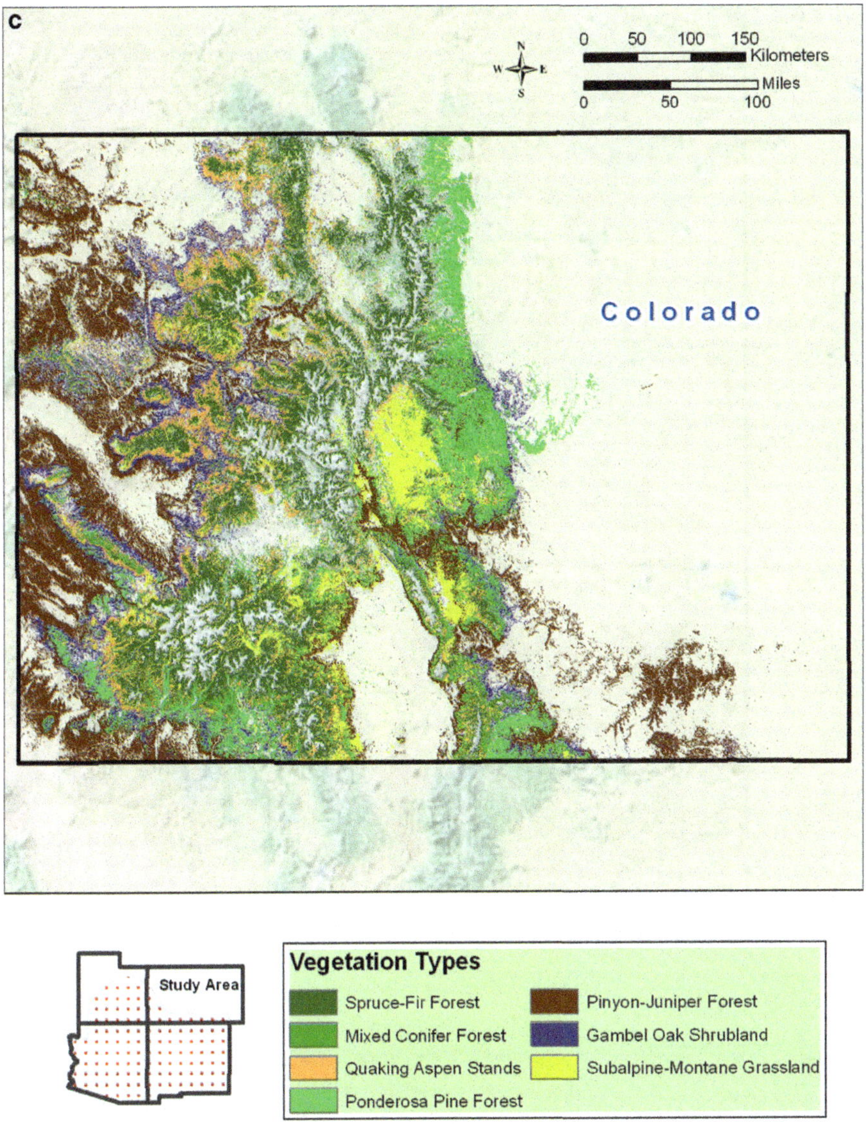

Fig. 1.3 (continued)

intermixed with Pinyon-Juniper vegetation in a mosaic. At the lowest elevations covered in this book, Pinyon-Juniper vegetation (Fig. 1.8) predominates, with pinyons and junipers as canopy dominants.

Three other important types of vegetation are present within the gradient from Spruce-Fir Forest to Pinyon-Juniper vegetation: Subalpine-Montane Grassland,

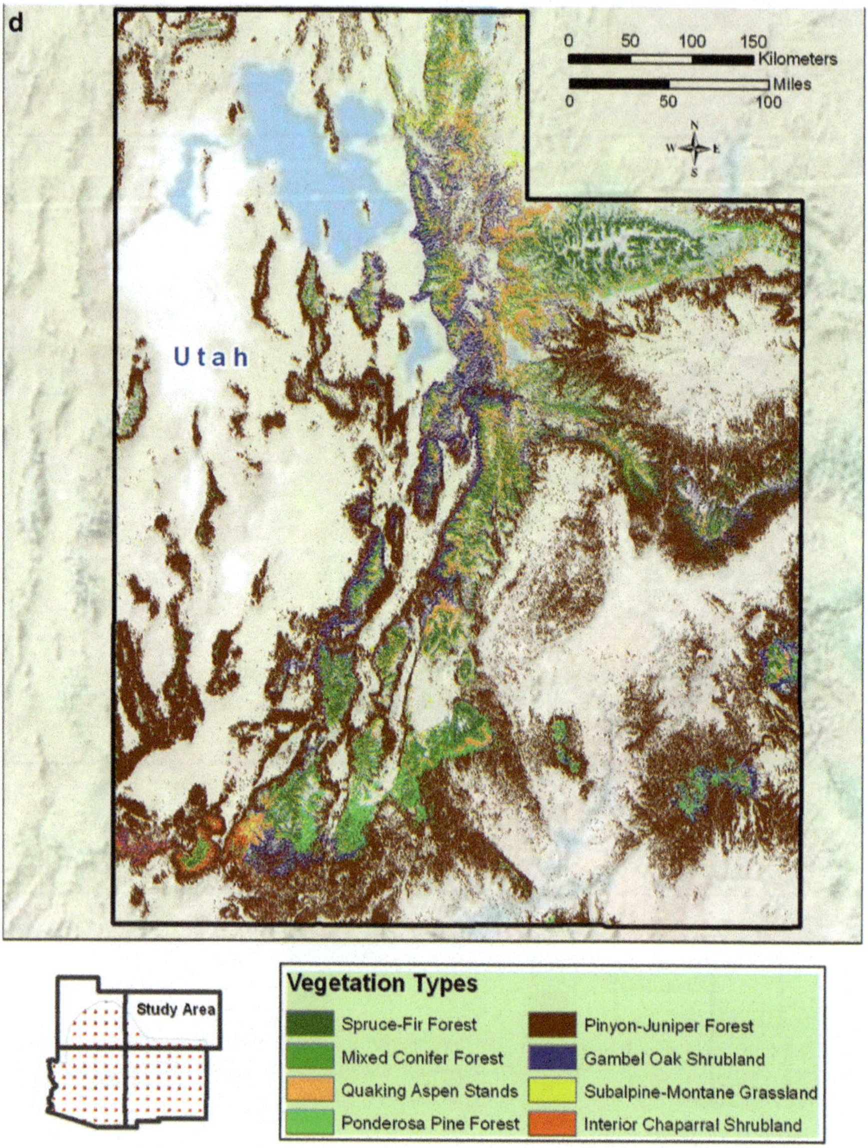

Fig. 1.3 (continued)

Gambel Oak Shrubland, and Interior Chaparral Shrubland. Subalpine-Montane Grassland (Fig. 1.9) occurs in some valley bottoms and on dry, steep slopes across most of the elevational gradient of coniferous forests. Dominants include fescues (*Festuca* spp.), other grasses, and forbs. Gambel Oak Shrubland (Fig. 1.10) occurs on a variety of sites but mostly within or near the elevational range of Ponderosa

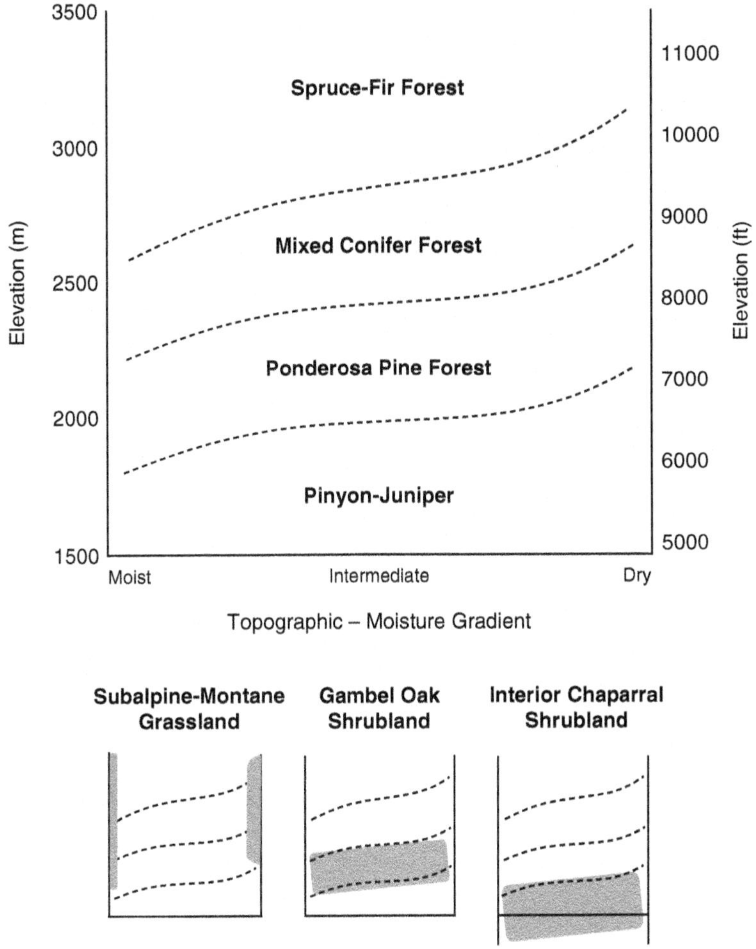

Fig. 1.4 Ecological distribution of vegetation on the mountains and plateaus of the American Southwest along gradients of elevation and moisture. Elevations are approximate and generally representative of mid-latitudes of the region, i.e., landscapes in northern Arizona and northern New Mexico. Elevations above those shown have tree line and alpine tundra vegetation; elevations immediately below those shown have desert shrublands and semi-desert grasslands

Pine Forest. Common shrubs besides Gambel oak include mountain mahoganies (*Cercocarpus* spp.) and serviceberries (*Amelanchier* spp.). Interior Chaparral Shrubland (Fig. 1.11) also occurs on a variety of sites but mostly within or near the elevational range of Pinyon-Juniper vegetation. It is dominated by shrub live oak (*Quercus turbinella*) and other shrubs, many of which are broad-sclerophylls (i.e., have broad, hard leaves).

Fig. 1.5 Spruce-Fir Forest in Fishlake National Forest, central Utah (Photograph by author)

1.2 Drivers

In the context of this book, a driver is any factor, either natural or human-caused, that directly or indirectly affects vegetation dynamics (cf. Nelson et al. 2006). Drivers covered in this section are landscape, climate, soil, animals, and natural disturbance. Anthropogenic disturbance is also a driver of vegetation dynamics and is described in Sect. 1.6 (after background on land-use history in Sect. 1.5). These multiple drivers are incorporated in the conceptual models introduced in Sect. 1.7. The relative importance of these drivers, as well as the information available on them, differs among the types of vegetation on the mountains and plateaus of the American Southwest and is covered in the following chapters. This section provides a general introduction to each major driver.

Fig. 1.6 Mixed Conifer Forest in the North Rim region of Grand Canyon National Park, north-central Arizona (Photograph by author)

1.2.1 Landscape

Topography, which is a key aspect of landscapes, is an important factor affecting the vegetation on southwestern mountains and plateaus. Mountains are uncommon in most areas of the American Southwest, and all have discontinuous distributions, being surrounded by lower elevations (Fig. 1.12). This is particularly evident with the "sky island" mountains of southeastern Arizona and southwestern New Mexico (Fig. 1.13). More continuous areas of mountains include the southern Rocky Mountains of northern New Mexico and adjacent southern Colorado.

Plateaus are more extensive than mountains. The Colorado Plateau, which encompasses much of the American Southwest, is not a single, uniform plateau, but is a physiographic region with many large and small plateaus, such as the Kaibab

Fig. 1.7 Ponderosa Pine Forest in the North Rim region of Grand Canyon National Park, north-central Arizona (Photograph by Betty J. Huffman)

Fig. 1.8 Pinyon-Juniper vegetation on the South Rim of Grand Canyon National Park, north-central Arizona (Photograph by Betty J. Huffman)

Fig. 1.9 Subalpine-Montane Grassland in Coconino National Forest, central Arizona (Photograph by Daniel Barton)

Plateau of north-central Arizona, the Pajarito Plateau of north-central New Mexico, and the Wasatch Plateau of central Utah.

Landscape topography is also a key driver of vegetation on various scales. At a regional scale, the importance of topography is suggested by the correlation of topography with both climate and vegetation (Fig. 1.14). Also, the mass of different mountain ranges is correlated with differences in elevational distributions of vegetation and species (Lowe 1961; Gottfried et al. 1995). At a stand (patch) scale, landscapes within individual mountains and plateaus are topographically diverse. Differences in elevation, slope aspect, slope inclination, and slope position correlate with the distribution of various kinds of vegetation, their structure and composition, and their disturbance regimes (e.g., White and Vankat 1993; Fulé et al. 2003).

Fig. 1.10 Gambel Oak Shrubland in Sugarite Canyon State Park, northeastern New Mexico (Photograph by Wade Patterson)

Fig. 1.11 Interior Chaparral Shrubland in Prescott National Forest, central Arizona (Photograph by Brian Reif)

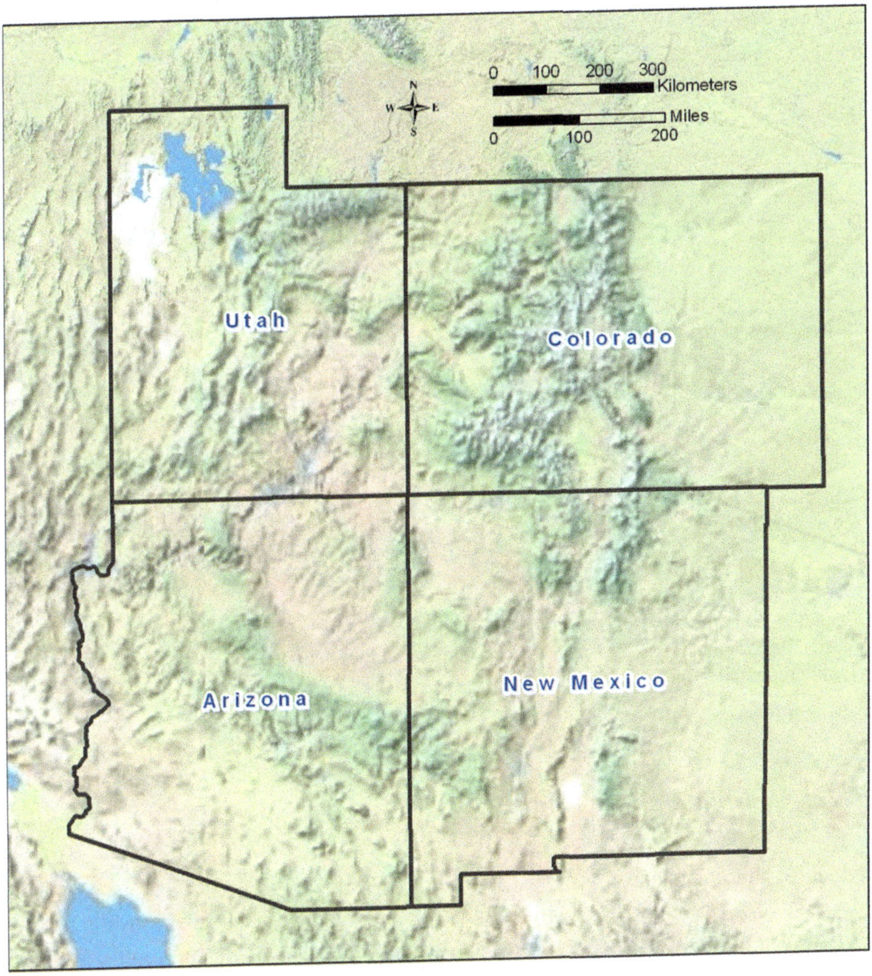

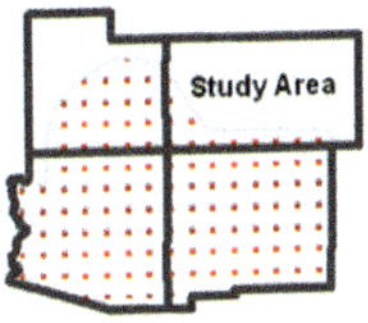

Fig. 1.12 Topography of the American Southwest. The map shows all of the states of Arizona, New Mexico, Colorado, and Utah, and the American Southwest is illustrated in *red* on the small map (Source: U.S. Geological Survey National Gap Analysis Program (2005); see Prior-Magee et al. 2007. Map prepared by Monica Swihart)

Fig. 1.13 Example of a sky island: the Santa Rita Mountains in southeastern Arizona. The tallest peak is Mt. Wrightson at 2,881 m (9,453 ft). Cloud buildup is characteristic of the monsoon season (Photograph by Philip MacAuliffe)

1.2.2 Climate

Much of the American Southwest is dry and warm. The overall weather pattern is dominated by a Hadley Cell in which warm, moist air rises in the tropics, loses much of its moisture, moves north, and descends as dry air in the Southwest. This produces a regional high pressure system with generally little precipitation. Temperatures are high, because solar radiation is usually not reduced by cloud cover and strikes at a more direct angle than at higher latitudes. Moreover, because there is little moisture to be evaporated, most solar radiation is converted to heat.

Climate heterogeneity is characteristic of the American Southwest. It is related to topography and latitude (Fig. 1.15). Topography is important because elevation strongly influences temperature, precipitation, and evapotranspiration (Spence 2001), slope exposure influences temperature and evapotranspiration, and slope steepness influences precipitation runoff. In Utah, mean annual temperature decreases approximately 1.7 °C (3 °F) per 300-m (1,000-ft) increase in elevation (Brown 1960). Latitude is important because it affects temperature and is correlated with the pattern of precipitation. In Utah, mean annual temperature decreases approximately 0.8–1.1 °C (1.5–2.0 °F) per 1° increase in latitude (Brown 1960). The pattern of precipitation shifts from bimodal in Arizona and New Mexico, where there is a peak in winter and a more pronounced peak in mid- to late-summer, to

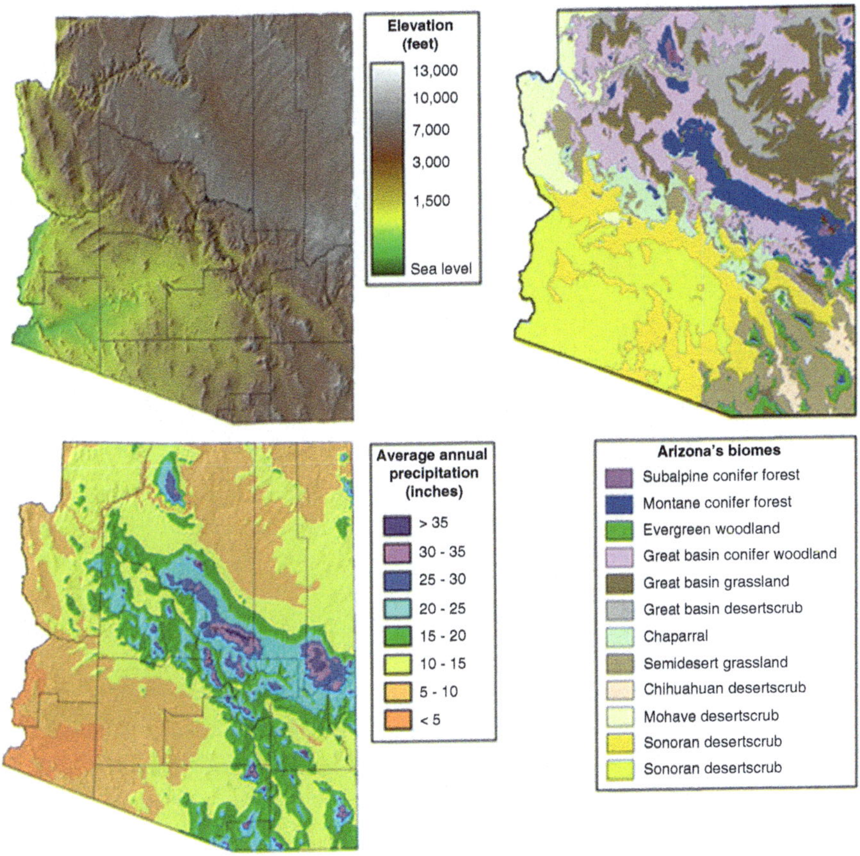

Fig. 1.14 Topography (*upper left*), climate (mean annual precipitation; *lower left*), and biomes (*upper right*) are interrelated in the American Southwest and therefore have similar spatial patterns in Arizona (Map by Joseph Abraham using topographic data from National Geophysical Data Center and precipitation data for 1971–2000 from PRISM Group, Oregon State University. Map provided by Climate Assessment for the Southwest, University of Arizona)

more equitably distributed throughout the year to the north in Utah (cf. Petersen 1994; Higgins et al. 1997; Spence 2001).

In winter, the prevailing high pressure system of the American Southwest sometimes is displaced over the Pacific Ocean. This allows low-pressure storms from the northeastern Pacific Ocean to move southward and affect the Southwest (Petersen 1994; Weng and Jackson 1999), often bringing several successive storms (Petersen 1994). These winter storms are usually larger and longer-lasting than summer storms (see below). Winter precipitation falls mostly as snow at high elevations, where it can form a substantial snowpack. For example, mean annual snowfall is 3.5 m (11.5 ft) at the Bright Angel Ranger Station at 2,560 m (8,400 ft) in the North Rim region of Grand Canyon National Park (Kaibab Plateau) in north-central

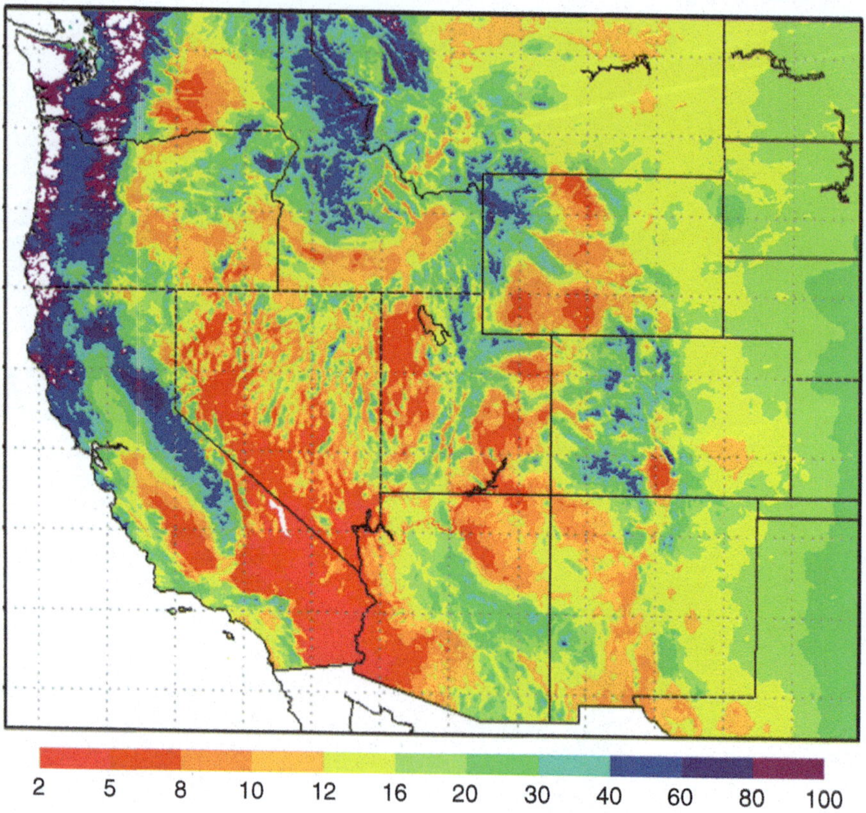

Fig. 1.15 Mean annual precipitation in inches in the western United States (for 1960–1990). Precipitation is generally positively correlated with elevation (Map by Western Regional Climate Center from PRISM data set provided by Oregon Climate Service, Oregon State University)

Arizona, and mean snow depth in February is 76 cm (30 in.; Western Regional Climate Center 2011).

In spring and early summer, temperatures increase and the high pressure system persistently dominates. This results in little precipitation in May and much of June. Beginning in late-June or early-July, moist air masses move into Arizona and New Mexico from the Gulf of California and the Gulf of Mexico (Fig. 1.16) and begin the monsoon season of precipitation that lasts into September. Monsoonal precipitation results from convection off heated surfaces and from convergence and orographic lifting (Petersen 1994; Fig. 1.13). Storms usually show no frontal development but consist of small clusters of convective cells that produce scattered, often-brief precipitation events. Nevertheless, summer precipitation is less variable in timing and amount than winter precipitation. Nearly 50 % of annual precipitation comes during the monsoon season in southern Arizona and New Mexico; this percentage decreases northward, more sharply in Arizona than in New Mexico (Fig. 1.17).

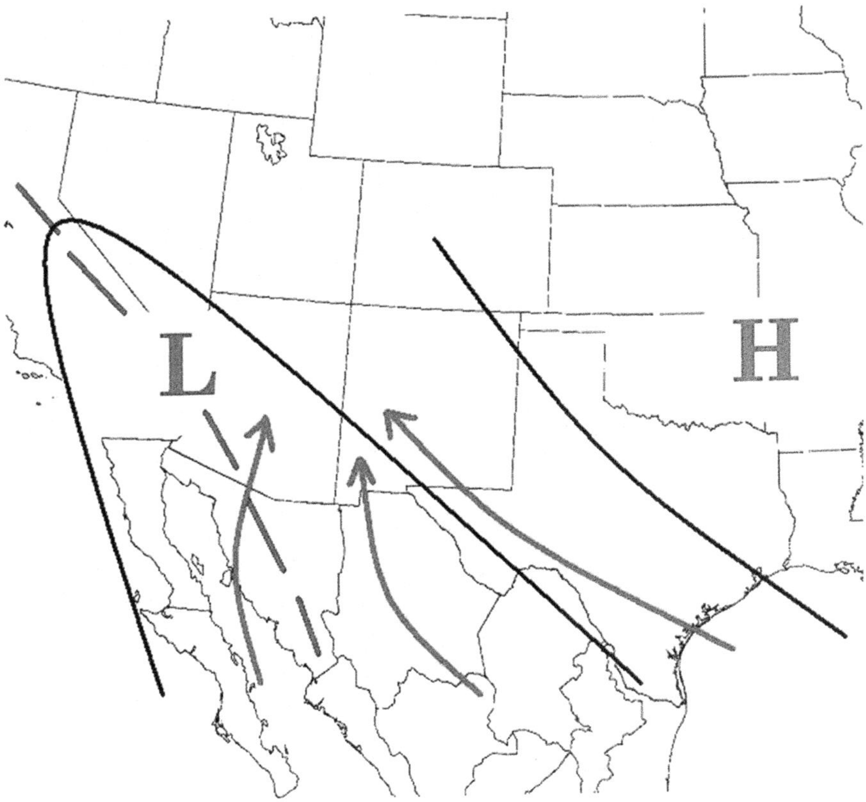

Fig. 1.16 Pattern of regional airflow during the monsoon season. (Map by U.S. National Oceanic and Atmospheric Administration)

A critically important component of southwestern weather-climate is a high frequency of lightning (Fig. 1.18), which provides an abundant ignition source of fires (Sect. 1.2.5.1). The Jemez Mountains of north-central New Mexico averaged 2.1 strikes/km^2/year (5.4 strikes/miles^2/year) during 1985–1994, and a large area in east-central Arizona and adjacent New Mexico averaged 3.1 strikes/km^2/year (8.0 strikes/miles^2/year) during May–September 1990–2005 (calculated from data in Allen (2002) and Evett et al. (2008), respectively). Lightning frequency can be directly proportional to elevation (Hall 2007), although Allen (2002) observed no relationship. Seasonally, lightning increases with the convectional storms of the summer monsoon season (Allen 2001, 2002; Hall 2007; Evett et al. 2008).

Weather patterns of the American Southwest vary on annual and longer time scales. Of particular significance is the El Niño-Southern Oscillation (ENSO). It occurs as sea surface temperatures in the tropical Pacific Ocean affect the latitude of the Pacific jet stream that enters and crosses North America. ENSO climatic variation has occurred in the Southwest for at least the last 2,000 years (Meko et al. 1995; Grissino-Mayer 1996; Grissino-Mayer et al. 1997; Salzer and Kipfmueller 2005). ENSO has a major impact

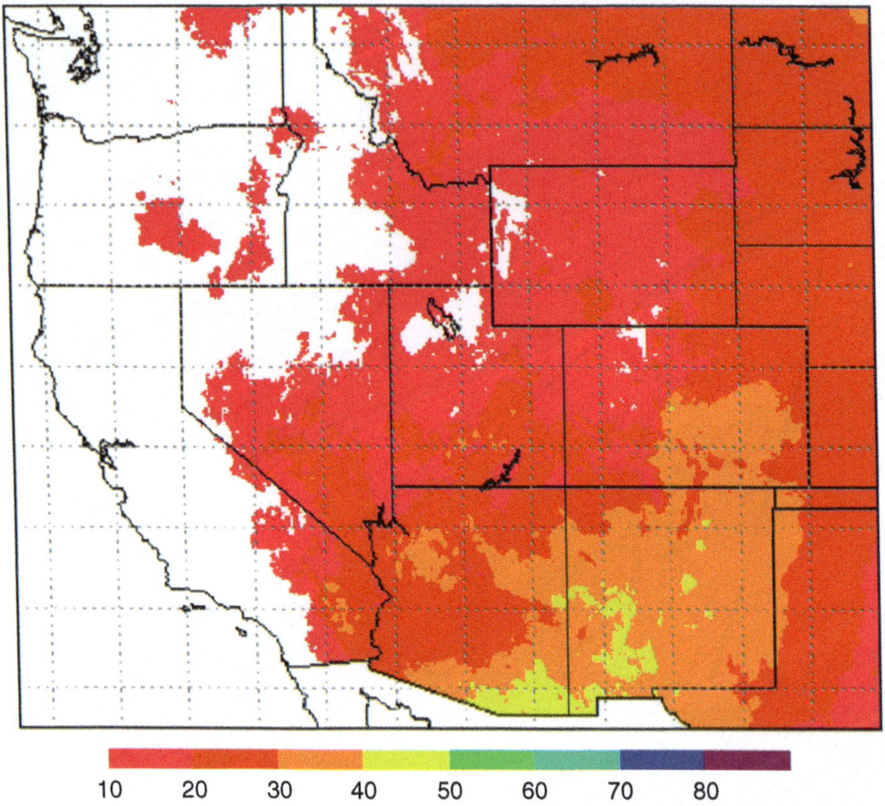

Fig. 1.17 Percentage of mean annual precipitation during July and August (i.e., the monsoon season) in the western United States. Note that the percentage in the American Southwest decreases from 40–50 % in southern Arizona and New Mexico to 10–20 % in central Utah (Map by Western Regional Climate Center from PRISM data set provided by Oregon Climate Service, Oregon State University)

on disturbance in the Southwest, particularly fire frequency (Swetnam and Betancourt 1990, 1998), because it affects precipitation. ENSO entails both El Niño and La Niña episodes. They tend to develop in the spring and peak in the winter when their impacts are greatest. El Niño and La Niña events typically occur every 3–5 years (National Weather Service 2011). El Niño episodes typically last 9–12 months (National Weather Service 2011) and often bring cooler, wetter winters to the Southwest (D'Arrigo and Jacoby 1991), especially in southern Arizona and New Mexico. The effects on monsoonal rainfall are more variable, but precipitation amounts are often normal or above normal (Hereford and Webb 1992). In contrast, La Niña episodes typically last 1–3 years (National Weather Service 2011) and often bring drier winters. La Niña-like conditions are associated with all six severe, multiple-year droughts recorded in western North America since weather instruments have been available (Seager et al. 2007) and generally result in larger wildfires in the Southwest (Haire and McGarigal 2009).

Fig. 1.18 Lightning during the monsoon season, as viewed from the South Rim (Coconino Plateau) across the Grand Canyon toward the North Rim (Kaibab Plateau), Grand Canyon National Park, north-central Arizona (Photograph courtesy of Lewis Wyman)

The location, strength, and effects of El Niño and La Niña are related to a longer time-scaled (usually 20–30 years) phenomenon, the Pacific Decadal Oscillation (Biondi et al. 2001; Mantua and Hare 2002; Brown and Comrie 2004).

1.2.3 Soil

Soils of southwestern mountains and plateaus are as diverse as the topography, climate, and vegetation that influence them. In general, upper-elevation soils of the American Southwest tend to be leached, acidic, and well-developed (Maker and Saugherty 1986), as well as well-drained and ranging from shallow to deep and from fine to moderately coarse textured (Hendricks 1985). Alfisols, Mollisols, and Entisols are common (Hendricks 1985; Maker and Saugherty 1986; DeBano et al. 2008). Soil moisture varies during the year, reflecting patterns of precipitation and snow melt. Soils have formed in residuum (occasionally colluvium or alluvium) from volcanic materials or sandstone, limestone, or igneous rocks (Hendricks 1985). Parent material tends to have less influence on soils at high elevations (Klemmedson and Smith 1979; Maker and Saugherty 1986), although its effects on forest vegetation are likely understudied (Peet 2000, but see Betancourt 1990).

1.2.4 *Animals*

Only a small percentage of the numerous animal species native to the mountains and plateaus of the American Southwest directly affect vegetation dynamics. The majority of these are insects that can cause tree mortality (see Sect. 1.2.5.3 for examples). Other native animals that can be important in vegetation dynamics are large mammals such as mule deer (*Odocoileus hemionus*), white-tailed deer (*O. virginianus*), and elk (*Cervus elaphus*).

1.2.5 *Natural Disturbance*

Disturbances affect many attributes of vegetation on southwestern mountains and plateaus. Broad-scale disturbances unrelated to humans include fire, wind, biotic agents, and climate change, and these are covered below. Disturbances covering little land area are not included, e.g., avalanches (cf. Hebertson and Jenkins 2003). Human-related disturbances, including fire exclusion and modern climate change, are considered in Sect.1.6 as anthropogenic disturbance.

1.2.5.1 Fire

The relatively dry climate and high frequency of lightning result in fire having a major influence on the vegetation of southwestern mountains and plateaus. In fact, the American Southwest has the highest concentration of lightning-initiated forest fires in the conterminous United States (Fig. 1.19). Lightning fire in the American Southwest has several distinctive features (Barrows 1978). These include a long season for lightning fires from April into October, although 82 % occur in June–August. Another distinctive feature is great variation in fuel flammability during the fire season. The fire season can be divided into two periods: spring dry season (April-June) and summer wet season (July–August). Lightning fires tend to be more severe and burn larger area during the spring dry season. Of course there is also spatial variation within the region and temporal variation among years.

Fire appears to have been important in the region throughout most of the Holocene (e.g., Anderson and Shafer 1991; Anderson et al. 2008a,b). Fire frequency and area burned have steadily increased in recent decades (Dickson et al. 2006). Fires in the Southwest are primarily affected by weather-climate and vegetation-fuel (Swetnam and Betancourt 1990, 1998; Swetnam and Baisan 1996; Allen 2001, 2002). Tree-ring studies correlating historical fire with climate frequently have used an index of drought, the Palmer Drought Severity Index (PDSI), which emphasizes winter precipitation in

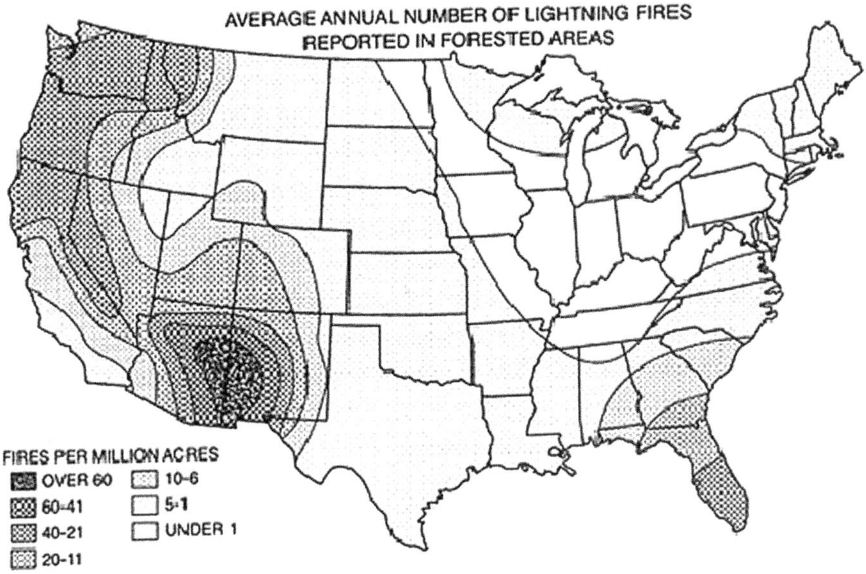

Fig. 1.19 The distribution of lightning-ignited forest fires in the conterminous United States (From Pyne (2001). Courtesy of Stephen Pyne)

the American Southwest. St. George et al. (2010) determined that even summer PDSI reflects the prior winter's precipitation (because of lags in the calculation of PDSI and sensitivity of tree rings to winter precipitation). Therefore, earlier statements in the scientific literature of summer PDSI or summer precipitation have been changed to winter precipitation where reported in this book. As an example of the connection of fire and climate, historical fires in Utah during 1630–1900 were unrecorded when the previous winter was wetter than average, which occurred during El Niño years (Brown et al. 2008). In contrast, fires occurred and were regionally synchronous following drier than average winters, which tended to occur during La Niña years. They also tended to occur in the year following one to three wet winters, which apparently resulted in herbaceous growth that increased the amount and continuity of fine fuels (e.g., Swetnam and Baisan 1996; Allen et al. 2008; Brown et al. 2008; Margolis and Balmat 2009; see Crimmins (2006) for details on the climatology of extreme fire-weather conditions in Arizona and New Mexico).

Although less than 0.5 % of lightning strikes in the Southwest result in wildfires (Hall 2007 and calculated from data in Evett et al. 2008), lightning accounted for nearly 80 % of fires on protected private, state, and federal lands in the region (Barrows 1978, for 1960–1975). Lightning ignitions are most numerous in Arizona with 20/1,000 km^2/year (52/1,000 miles^2/year), followed by New Mexico with 13/1,000 km^2/year (33/1,000 miles^2/year), as calculated from data in Kay (2007) for National Forests during 1970–2002 (see also Vale 2002). In parallel with the pattern of decline in monsoonal storms, lightning ignitions decrease northward into southwestern Colorado and southern and central Utah; both areas have 8/1,000/km^2/year (21/1,000 miles^2/year). Specific areas can have much higher rates of ignitions, e.g., mountains in southern Arizona had rates as high as 2,000/1,000 km^2/year (5,000/1,000 miles^2/year) during the twentieth century (Swetnam and Baisan 2003). Variation within the Southwest is also described in Barrows (1978). Lightning ignitions have outpaced human ignitions in producing more extensive wildland fires in the Southwest (Dickson et al. 2006), although human ignitions have produced the two largest fires in Arizona's recorded history (Wallow Fire of 2011 and Rodeo-Chediski Fire of 2002).

Anecdotal and analytic evidence indicates that the dryness of fuels is important in facilitating lightning ignitions (Hall 2007; Evett et al. 2008). Consequently, lightning ignitions reach a maximum early in the monsoon season, before peaks in lightning strikes and precipitation (Allen 2001, 2002; Hall 2007). This period of late June and early July often has "dry lightning," i.e., lightning accompanied by little or no precipitation. July accounts for 41–45 % of lightning ignitions (Barrows 1978; Evett et al. 2008) and 45 % of the area burned by lightning ignitions (Evett et al. 2008). The probability of ignition and spread of fires decreases with increased humidity and precipitation (Evett et al. 2008). Forests and woodlands as well as elevations above 1,900 m (6,234 ft) have disproportionately high frequencies of lightning ignitions for their land areas (Hall 2007). Barrows (1978) reported lightning fires were (a) most concentrated at ~2,000 to 2,300 m (6,501–7,500 ft) where 40 % of lightning fires occurred, (b) approximately evenly distributed by slope aspect, and (c) concentrated in Ponderosa Pine Forest where nearly 80 % of forest and woodland fires occurred. He also reported that larger fires tended to originate on north and northeast aspects, on steep slopes, and in Ponderosa Pine Forest and Pinyon-Juniper woodland.

Major determinants of different fire regimes include vegetation, elevation, and other aspects of topography (e.g., slope aspect, steepness, and complexity), along with their associated differences in precipitation, evapotranspiration, fuels, etc. The term *fire regime* encompasses patterns of a suite of factors such as frequency, severity, timing, and distribution of fires (the disturbance regime of other factors such as insect outbreaks can be similarly characterized). Basic descriptors of fire regimes include quantitative measures such as mean fire interval (mean time between fires in the study area) and fire turnover time (time to burn an area equivalent in size to the study area; also known as rotation time). Neither descriptor implies that fires burned the entire study area. For review and comparison of different quantitative measures of fire intervals and turnover times, see Baker and Ehle (2001), Fulé et al. (2006), Kou and Baker (2006a,b), Van Horne and Fulé (2006), Farris et al. (2010), and Dugan (2012).

Fire regimes are also characterized as surface, crown, and mixed-severity. A *surface-fire regime* consists of frequent, low-severity fires with flame heights generally <1 m (3.3 ft; Fig. 1.20). Historical mean fire intervals (i.e., before the exclusion of fire in the mid to late nineteenth century) were a few decades or less (see following chapters for more specifics). A *crown-fire regime* is characterized by infrequent, high-severity, stand-replacing (stand-initiating) fires that burn across landscapes (Fig. 1.21). Flame heights exceed the heights of the canopy layer. Historical mean fire intervals were one to a few centuries. A *mixed-severity fire regime* is characterized by fires that burn as low-severity surface fires in some sites of a landscape and occasionally as high-severity crown fires in other sites (Fig. 1.22a, b).

Fig. 1.20 Surface fire in Ponderosa Pine Forest in Lincoln National Forest, south-central New Mexico (Photograph by Kari Greer/U.S. Forest Service)

Fig. 1.21 Crown fire in Mixed Conifer Forest in Apache-Sitgreaves National Forests, east-central Arizona (Photograph by Jayson Coil)

Fig. 1.22 (a) Mixed-severity fire with surface fire in foreground and crown fire in background in Kaibab National Forest, north-central Arizona (Photograph by Carlos Valadez). (b) Area burned by mixed-severity fire in Santa Fe National Forest, north-central New Mexico. Fire crowned in foreground and burned as surface fire most elsewhere (Photograph by Kari Greer/U.S. Forest Service)

Mean fire intervals of the two components of the mixed-severity fire regime are comparable to those of surface-fire and crown-fire regimes. The term mixed-severity fire regime also can be applied to temporal variation in fire severity, i.e., when fires burn as surface fires in some years and include crown fires in other years.

The history of surface fires on southwestern mountains and plateaus has been studied primarily by examination of fire scars in the wood of old living and dead trees, especially ponderosa pines (cf. Swetnam and Baisan 1996; Fig. 1.23a, b). The approach can accurately reconstruct landscape-scale surface-fire history in the Southwest, at least for widespread fires (Farris et al. 2010). Results are reported in terms of mean fire intervals, e.g., 33 years. However, mean intervals are affected by the size of the area studied and the intensity of sampling. In general, the smaller the study area and the lighter the sampling intensity, the fewer fires are observed in tree rings and the longer the mean fire interval. Intervals are usually reported based on fires that had scarred a minimum number or percentage of scarred trees in the study area, e.g., 11 years at ≥10 % scarred, 23 years at ≥25 % scarred, etc. The number or percentage of trees scarred is assumed to be directly proportional to the size of the fire within the study area.

Fire-scar analysis is less useful for crown fires, because remnant, scarred trees are uncommon or absent. Instead, researchers determine the age of cohorts of post-fire initiated stems to determine approximate dates and areas of past fires (e.g., Floyd et al. 2000; Margolis et al. 2007). Other approaches to determining fire history utilize charcoal particles in sedimentary deposits in lakes and bogs (e.g., Allen et al. 2008) or in small alluvial fans (e.g., Frechette and Meyer 2009).

1.2.5.2 Wind

Wind has its greatest impacts on the vegetation of southwestern mountains and plateaus through its effects on fire. Wind augments fire, affecting spread rates and distribution, as well as increasing the probability of crown fire (Fulé et al. 2004). The relationship with fire is synergistic, because fire releases heat that increases local wind through convection.

The direct effects of wind on vegetation constitute a low-frequency, high-severity disturbance regime. This has received little research in the American Southwest. Anecdotal evidence indicates wind disturbance occurs at scales from microbursts affecting a few trees to occasional large storms resulting in blowdowns of trees over multiple square kilometers (Fig. 1.24). A large blowdown occurred in Ponderosa Pine Forest in the Jemez Mountains (Allen 1989), and a series of tornados caused moderately high to severe damage on 2,375 ha (5,868 acres) of mostly Ponderosa Pine Forest in northern Arizona in 2010 (U.S. Forest Service 2010).

Fig. 1.23 (**a**) Cutting a partial cross-section through a fire scar in a ponderosa pine in the North Rim region of Grand Canyon National Park, north-central Arizona. (**b**) Partial cross-section of ponderosa pine, which has been sanded and polished to show tree rings. *Red arrows* point to some of the fire scars (charred areas inside lobes of post-fire growth) (Photographs by author)

Fig. 1.24 Extensive blowdown (*gray area of dead trees*) in Spruce-Fir Forest, Pecos Wilderness, Santa Fe National Forest, north-central New Mexico (Photograph by William M. Ciesla, Forest Health Management International, Bugwood.org)

1.2.5.3 Biotic Agents

The major biotic agents that affect vegetation dynamics on the mountains and plateaus of the American Southwest are native species of bark beetles (*Dendroctonus* spp., *Dryocoetes confusus*, *Ips* spp., and *Scolytus ventralis*), defoliating insects, dwarf mistletoes (*Arceuthobium* spp.), and root-decay fungi. The species have long been present in the region and likely co-evolved with their hosts.

Most bark beetles are restricted to specific tree species (Dahms and Geils 1997). Bark beetles bore small holes into host trees and lay eggs in living tissues. The beetles and their larvae feed on the tissues, and their tunnels reduce the flow of water and nutrients in trees. Bark beetles are usually present in low numbers, persisting in freshly fallen trees (such as windthrows) and less productive living trees where they only occasionally cause tree death. Healthy trees typically produce enough resinous pitch to prevent beetles from successfully boring into tree trunks. Nevertheless, beetle outbreaks occur periodically and can result in high mortality of trees, especially larger, older, stressed trees but also healthy trees. Recent outbreaks have been linked to warmer, drier weather that stresses trees (Breshears et al. 2005; Fig. 1.25). Extensive bark beetle outbreaks can impact fire spread and severity, but the specific effects on fire change through time as fuel characteristics change, first with the fall of needles and later following the fall of snags (Jenkins et al. 2008). Moreover, bark beetles can be associated with fungi that also negatively impact host trees (cf. Paine et al. 1997).

Fig. 1.25 Insect-caused mortality of pinyon in Mesa Verde National Park, southwestern Colorado (Photograph by William M. Ciesla, Forest Health Management International, Bugwood.org)

The major defoliating insect is western spruce budworm (*Archips fumiferana*). It can have different hosts, but primarily affects white fir and Douglas-fir in the American Southwest (Dahms and Geils 1997). When outbreaks last several years, complete defoliation can occur and, if sustained, results in reduced tree vigor and death. Mortality tends to be greater in understory trees (in contrast to bark beetles, which disproportionately affect overstory trees). Outbreaks have been linked to warm fall and winter temperatures and drought (Hebertson and Jenkins 2008). Another important defoliating insect is the western tent caterpillar (*Malacosoma californicum*), which impacts quaking aspen. Annual and biannual defoliation for several consecutive years can minimize tree growth and cause mortality (Allen 1989).

Dwarf mistletoes are semi-parasitic plants, and most southwestern conifers are hosts. Dwarf mistletoes rarely cause mortality, but reduce vigor, making trees more susceptible to insects. They also cause trees to accumulate resins and produce clumps of shoots and needles ("witch's brooms") that can facilitate surface fires laddering into tree canopies (Alexander and Hawksworth 1975; Fig. 1.26) and increase crown scorching during fires (Harrington and Hawksworth 1990; Breece et al. 2008). Prior to Euro-American influence, dwarf mistletoes likely occurred throughout forests and had a distribution similar to their current distribution, but could have been less abundant (Dahms and Geils 1997). They are considered the Southwest's most widespread and damaging forest pathogens (U.S. Forest Service 2011).

Fig. 1.26 Witches broom formed on a ponderosa pine by dwarf mistletoe has branches and resins that facilitate burning. Most witches brooms have many smaller branches (see Fig. 4.12), but here these apparently burned in previous fires (Photograph by Fire and Aviation Program, Grand Canyon National Park)

Root-decay fungi also are widespread. Many species have specific hosts (Dahms and Geils 1997). Root-disease weakens trees, increasing the likelihood of bark-beetle infestation and windthrow. Large canopy trees are more likely to be impacted, especially on mesic sites.

1.2.5.4 Climate Variation

The Southwest has had periodic droughts. Those related to the El Niño-Southern Oscillation have impacted the region periodically for at least 2,000 years (Sect. 1.2.2). Other disturbance agents such as fire and bark beetles interact with drought, as for example occurred with tree mortality in the late-sixteenth, mid-twentieth, and early-twenty-first centuries (e.g., Swetnam and Betancourt 1998; Breshears et al. 2005). Critical aspects of drought include its severity, duration, and linkage to other disturbances. Drought is the most limiting factor for growth of trees and other plants in the Southwest (e.g., Fritts 1976; Swetnam and Betancourt

1998). In addition, southwestern forests appear "particularly sensitive to drought and warmth" (Williams et al. 2010).

Given that disturbance is defined as a "relatively discrete event" (cf. White and Pickett 1985), it can appear questionable that climate variation is included in this section on natural disturbance. However, at least the *onset* of drought and increased temperature can be a natural disturbance in the American Southwest, where vegetation is very sensitive to changes in water balance. This is especially true for forests and woodlands, where onset of drought and warmer temperature triggers increases in fire and insects such as bark beetles (Sects. 1.2.5.1 and 1.2.5.3, respectively). The impacts of this disturbance complex can occur quickly (see Sect. 5.2.5.2 for examples of rapid tree mortality). In fact, the onset of drought is known to encompass the period when drought effects on vegetation are likely to be most pronounced (cf. Swetnam and Baisan 1996; Breshears et al. 2005). Therefore, the onset of drought and elevated temperature is a relatively discrete event, especially on multi-decadal or longer time scales.

1.3 Processes

Disturbances and other changes in the environment are followed by changes in vegetation, i.e., vegetation dynamics. They include everything from regrowth of vegetation following major disturbances to minor changes in species populations following small disturbances or even small fluctuations in the environment. Processes of vegetation dynamics that are important on southwestern mountains and plateaus are succession, gap dynamics, regeneration and thinning, infill and expansion, rapid regrowth, and tree encroachment. These play key roles in the conceptual models of vegetation dynamics (Sect. 1.7).

1.3.1 Succession

Vegetation dynamics after major disturbances typically involve succession, a process important in the dynamics of all vegetation types covered in this book. This ecological process involves the sequential replacement of species through time. It is driven by species characteristics and species-caused changes in site conditions. In addition, stochastic (chance or probabilistic) events such as plant dispersal are important.

A common example of succession follows high-severity crown fire in some high-elevation southwestern coniferous forests. The site is initially dominated by herbaceous plants adapted to open conditions, as well as root sprouts (suckers) of quaking aspen, a shade-intolerant species. The mix of early-successional species apparently depends on many factors, including abundance before fire, severity of the fire, post-fire dispersal, and of course other site conditions. Within a few

Fig. 1.27 Successional stand of quaking aspen (*lighter green* and *rounded crowns*) and Engelmann spruce (*darker green* and *pointed crowns*) in Dixie National Forest, southwestern Utah. Individuals of Engelmann spruce are overtopping the aspen canopy (Photograph by author)

years, aspen sprouts overtop the herbs and dominate the site for several decades or more. The rapidity of this change to aspen dominance also depends on various factors including pre-fire density of aspen, fire severity, intensity of animal browsing of aspen, and other site conditions. Aspen can remain dominant, or conifers can regenerate and decades later overtop the aspens (Fig. 1.27) and form a coniferous forest that dominates the site until the next major disturbance. Factors affecting persistence vs. replacement of aspen include proximity of conifer seed sources, seed dispersal, and site conditions. See Sects. 2.3.1 and 3.3.2 for more details.

Succession is often described as occurring in stages. For example, the succession outlined above could be described as beginning with a herb-dominated stage that changes into an aspen-dominated stage that can change into a conifer-dominated stage. Stages of succession appear in some of the conceptual models (see Sect. 1.7.2). However, the description of successional stages is a simplification done for clarity. Vegetation actually varies more continuously during succession, analogous to how vegetation can vary continuously along spatial environmental gradients (as described in Sect. 1.1).

In the above example of succession, the persistent aspen forest or the replacement coniferous forest can be considered relatively stable, late-successional vegetation. The coniferous forest has been called a "climax" stage, but this classic terminology suggests the vegetation has reached equilibrium and shows no directional change without additional disturbance. In fact, relatively stable, late-successional vegetation rarely appears to be in an equilibrium state. Instead, vegetation continues to change in response to shifts in weather, climate, and other aspects of the environment. Analogously, but at a broader scale, landscapes comprised of different patches of vegetation also do not reach equilibrium in terms of the proportions of patch types. Instead, on-going shifts in the scale, severity, and frequency of disturbances result in continuous changes in those proportions, i.e., non-equilibrium landscapes.

1.3.2 Other

Each of the other processes of vegetation dynamics – i.e., gap dynamics, regeneration and thinning, infill and expansion, rapid regrowth, and tree encroachment – is important in only a few of the vegetation types covered in this book. The process of gap dynamics (in which one or a small cluster of trees dies, opens a gap in the canopy, is replaced, and closes the canopy gap) is important primarily in both Spruce-Fir Forest (Sect. 2.3.2) and Mixed Conifer Forest (Sect. 3.3.1). The process of regeneration followed by thinning (of the regeneration) is important in Ponderosa Pine Forest (Sect. 4.3.1). The processes of infill and expansion involve increases in tree densities, with infill occurring where trees are already present and expansion occurring where trees are absent. Both infill and expansion are important in Pinyon-Juniper vegetation (Sect. 5.3.1). Rapid regrowth following disturbance (especially by resprouting) is a key process in Subalpine-Montane Grassland, Gambel Oak Shrubland, and Interior Chaparral Shrubland (Sects. 6.3.1, 7.3.1, and 8.3.1, respectively). Tree encroachment involves the invasion of trees and is important in stands of Subalpine-Montane Grassland adjacent to forest (Sect. 6.3.3).

1.4 Paleoecological History

The paleoecology of the American Southwest is well-studied (Betancourt et al. 1990; Swetnam et al. 1999). Although most species that characterize today's biotic communities have been in place since the end of the Tertiary (Lowe and Brown 1994), their distributions have changed many times in response to climate. Species migrations were and are critical, because the Southwest had "an unparalleled record of climatic variability" over the last two millennia (Salzer and Kipfmueller 2005) and faces a future with additional variability (Sect. 1.6.3). Southwestern species have been able to respond relatively quickly to past climate variability, because mountains and plateaus provide both refugia and sources for species during climatic fluctuations (Van Devender and Spaulding 1979). This is less likely to occur in regions that are topographically more homogeneous. Species migrate individually rather than as assemblages or communities (Van Devender and Spaulding 1979; Cole 1985), and migrations involve local extinctions and colonizations (Betancourt 1990). Fragmented habitats and the relatively rapid pace of climate change will challenge future migrations.

During the Full-Glacial (ca 21000–13000 years BP), the American Southwest generally had wetter winters and cooler summers with less monsoonal precipitation than today (Betancourt 1990; Petersen 1994; Anderson et al. 2000). Temperature gradients were steeper both latitudinally and elevationally (Petersen 1994). Much of the difference with today was related to a southerly displacement of the jet stream (Van Devender and Spaulding 1979; Petersen 1994; Barlein et al. 1998). This brought Pacific air masses across the Southwest more frequently, accounting for increased precipitation, greater cloud cover, and generally cooler temperatures (Petersen 1994). The southerly displacement of the jet stream also appears to have been one of several factors that inhibited development of a monsoon season (Anderson et al. 2000). Species of today's coniferous forests generally occurred at lower elevations (Cole 1985, 1990; Weng and Jackson 1999; Anderson et al. 2000). Ponderosa pine possibly was absent from the Colorado Plateau because of low summer precipitation (Betancourt and Davis 1984; Betancourt 1990; Cole 1990).

During the Late Glacial and Early Holocene (circa 13000–8500 years BP), the jet stream migrated northward (Van Devender 1977; Thompson et al. 1993; Petersen 1994). Temperatures increased and a summer monsoon season developed. This shifted the precipitation regime from winter-dominated to summer-dominated (Betancourt and Biggar 1985; Weng and Jackson 1999) and resulted in cooler, wetter summers than today (Weng and Jackson 1999). Vegetation composition changed as species migrated toward the higher elevations of modern conditions (Cole 1982, 1985; Weng and Jackson 1999; Anderson et al. 2000). Ponderosa pine spread rapidly across the mid-elevations of the Colorado Plateau (Anderson 1989), but it is unclear whether it migrated from the south or expanded from isolated, small populations (Betancourt and Davis 1984). The spread of ponderosa pine could have been related to increased summer, monsoonal precipitation (Betancourt 1990; Cole 1990; Anderson 1993), although a change in the fire regime with more lightning ignitions in the summer could have triggered the spread (Betancourt and Van Devender 1981; Betancourt 1990).

The Middle and Late Holocene (8500 years BP to present) had increases in temperature and decreases in effective precipitation, as well as climatic variability with wet and cool conditions in the late Holocene (Hasbargen 1994; Weng and Jackson 1999). The vegetation continued to change toward modern conditions, with shifts matching variations in climate. Estimates for the establishment of modern vegetation range from 5000 to 11000 years BP (Van Devender and Spaulding 1979; Cole 1985; Anderson 1993; Thompson et al. 1993; Hasbargen 1994; Murdock 1994; Weng and Jackson 1999; Anderson et al. 2000).

1.5 Land-Use History

Past land use is a key determinant of the present-day structure, composition, and dynamics of vegetation on the mountains and plateaus of the American Southwest. Land uses have included subsistence hunting and gathering, resource exploitation, resource protection, and ecologically based resource management. The history of land use can be divided into two major periods: a long, relatively poorly documented period when Native Americans dominated land use and a period of Euro-American dominance of land use that began in the mid nineteenth century. Specifics of the land-use history across the Southwest are complex; this section focuses on the southern Colorado Plateau for a more concise overview.

1.5.1 Native-American Dominance

1.5.1.1 History

Paleo Indians first entered the Southwest no later than 11000 years BP, but these hunter-gathers likely had little lasting impact on ecosystems of the region (Allen 2002), especially at high elevations. The Paleo Indians were gradually replaced by the Desert Archaic peoples. They were seasonally migratory hunter-gatherers who used high elevations in the summer to hunt deer (*Odocoileus* spp.) and elk and gather wild foods. They appear to have been the first Native Americans with the potential to have impacted high-elevation vegetation on the Colorado Plateau, but their impacts are presumed to have been localized and temporary, as these migratory people would have moved whenever resources became scarce (Allen 2002).

After about 4000 years BP, the importance of agriculture gradually increased. This led to relatively permanent small settlements and villages, which increased the potential for altering landscapes. Population densities varied. An area in northern New Mexico had 1,000–3,500 people in about 100 km^2 (Orcutt 1999 in Allen 2002), but other areas were sparsely populated. Overall, some areas were greatly changed by Native American agriculture, but the overall area was small (Vale 2002). Villages tended to be in areas of Pinyon-Juniper vegetation (Allen 2002; Sect. 5.2.6); therefore, agriculture would seem to have had little effect on landscapes at higher elevations.

The first non-aboriginals known to have entered the Colorado Plateau comprised a Spanish exploring party in New Mexico in 1539. They were followed in 1540 by Spanish explorer Francisco Vásquez de Coronado who brought 500 cattle (*Bos taurus*) and over 5,000 sheep (*Ovis aries*) as food, the first entry of livestock into the present United States (Haskett 1935, 1936; Schlegel 1992). Later, the introduction of livestock as a commercial product (see Schlegel 1992 and Wildeman and Brock

2000 for details) had a large and lasting impact. Some Native American peoples adopted sheepherding into their cultures early in the seventeenth century (Scurlock and Finch 1997). By 1757, New Mexico had 112,000 sheep and goats (*Capra hircus*), over half of which were owned by Native Americans other than Navajos, whose livestock were uncounted (Bailey 1980; Baxter 1987). Details on Native American livestock operations are in Brugge and Gerow (2000). Sheep were more abundant than cattle in New Mexico, at least until the mid nineteenth century, but cattle predominated in Arizona (Wildeman and Brock 2000).

Euro-American fur trappers and traders entered the Colorado Plateau by the 1820s (Peterson 1975; Brown et al. 1994). In the mid nineteenth century, various United States military surveying expeditions visited the Colorado Plateau to consider possible routes across the Southwest and to evaluate the natural resources of the region (Peterson 1975; Wurtz 1991). The first lasting Euro-American settlements on the Colorado Plateau were established by Mormons in the "Arizona Strip" (the portion of Arizona between the Grand Canyon and Utah) in the 1860s (Wurtz 1991). Permanent Euro-American settlements marked the impending end of region-wide dominance of land use by Native Americans, although it continues in large areas of the Colorado Plateau today.

1.5.1.2 Impacts

It is challenging to assess the historical impacts of Native Americans on vegetation of the mountains and plateaus of the American Southwest. Although it is commonly thought that Native Americans had little impact on their environment (except perhaps for their use of fire), Dahms and Geils (1997) suggested that this is a myth for the Southwest. On the broader scale of the North American continent, Vale (1998, 2002) postulated that the historical impacts of Native Americans varied and as a result areas ranged from "pristine" to "humanized." Evidence indicates that the mountain and plateau landscapes of the American Southwest fit this generalization.

Native Americans set fires for hunting, improving yields from wild food crops, and other reasons for thousands of years in many North American landscapes (e.g., Pyne 1982; Kay 2007). Some researchers have concluded that Native American's role in historical fire regimes is often overstated, and its characterization is "a highly speculative venture for ecologists and historians alike" (Barrett et al. 2005). Other researchers have concluded that anthropogenic fires of Native Americans have been understated by fire historians and scientists (Roos et al. 2010).

The question of whether Native Americans used fire as a landscape-scale tool in the American Southwest has been controversial (Allen 2002; Alcoze 2003; Kay 2007). Direct evidence of Native American influence on landscape-scale fire regimes in the Southwest is limited if not lacking, and archaeological, ethnographical, and fire-scar chronological evidence "...indicated that Indians likely had minimal effects on the fire regimes of most upland ecosystems in the Southwest prior to European contact" (Allen 2002). The high incidence of lightning was sufficient to account for the landscape-scale fire regimes prior to Euro-American influence (Swetnam and Baisan 1996; Fulé et al. 2003; Margolis and Balmat 2009). However, it has been argued that lightning-ignited fires were seasonally insufficient to account for the fire regime of quaking aspen stands in the West and of forests of the San Juan Mountains in southwestern Colorado (Kay 2007, but see Grissino-Mayer et al. 2004). Also, Native Americans could have influenced the fire regime in some areas, e.g., the Chiricahua Mountains of southeastern Arizona, but it can be difficult to separate such effects from factors such as climate variation (Seklecki et al. 1996).

Local impacts of Native Americans on the vegetation of the mountains and plateaus of the American Southwest are not disputed. Evidence includes historical accounts such as

> From a high point…we looked down and across the forest to the plain. And as we looked there rose a line of smokes. An Apache was getting ready to hunt deer. And he was setting the woods on fire because the hunter has a better chance under cover of the smoke. (Pinchot 1947 for east-central Arizona in 1900)

Also, an elder of the Southern Paiute near the Arizona-Utah border stated that his people had burned stands of pinyons every 4 years for purposes such as increasing production of edible pinyon "nuts" and reducing damage from insects and disease (B. Pikyavit, in Alcoze (2003), but see Burwell (1999) in Vale (2002)). In addition, the Ute people set fires throughout the San Juan Mountain region into the early twentieth century (Romme et al. 1994, in Baker 2002).

Local impacts have been supported by research, including studies that have shown that Native Americans increased fire frequencies or otherwise altered fire regimes for periods in some areas (e.g., Baisan and Swetnam 1997; Kaye and Swetnam 1999; Roos et al. 2010). Needs for fuel, small shelters, and agriculture were met by cutting trees in Pinyon-Juniper woodland where settlements were concentrated (Betancourt and Van Devender 1981). Construction of large structures for ceremonial and housing purposes at Chaco Canyon in northwestern New Mexico in the tenth to twelfth centuries used thousands of logs of ponderosa pine, spruce, and fir hauled more than 75 km (46 miles; Betancourt et al. 1986; Reynolds et al. 2005). Impacts of the tree cutting for Chaco Canyon appear to have ranged from loss of nearby small forest stands to minor impacts on the dynamics of more-distant, larger forests, where only a narrow size range of trees was cut (Betancourt et al. 1986). Impacts on upper-elevation vegetation by use of fire in food-crop production have been have documented for two areas in Arizona, one dominated by Ponderosa Pine Forest and the other by Pinyon-Juniper vegetation (Roos et al. 2010). Greater concentration of Native American populations in Pinyon-Juniper vegetation presumably resulted in broader-scale impacts than at higher elevations, but landscape-scale effects have not been documented.

Other possible landscape-scale impacts of Native Americans included hunting, which likely altered ecosystems, particularly where elk populations were reduced (Allen 1996; Vale 2002), as well as livestock grazing (see Sects. 1.5.2.2 and 1.6.1).

1.5.2 *Euro-American Dominance*

1.5.2.1 History

The impacts of early Europeans and Euro-Americans on vegetation before the introduction of livestock appear to have been small. Consequently, the transition from Native-American to Euro-American dominance of land use took more than three centuries following Spanish entry. One possible exception is that trappers altered beaver (*Castor canadensis*) populations, possibly causing substantial ecological changes. Impacts of beaver trapping on southwestern vegetation dynamics are unreported, but have been important elsewhere (cf. Naiman et al. 1988; Wright et al. 2004).

The emergence of Euro-American dominance on the Colorado Plateau was facilitated by the development of transportation routes (Wurtz 1991; Scurlock and Finch 1997). For example, completion of the Beale Wagon Road across the southern Colorado Plateau in 1859 impacted Euro-American settlement and land use because it provided a route for driving livestock (Haskett 1936). Railroads followed, and the demand for wood for fuel, ties, trestles, buildings, and mine supports led to widespread logging of Ponderosa Pine Forest beginning in the 1870s (Scurlock and Finch 1997). The railroads also opened new markets for lumber and livestock and accelerated Euro-American settlement (Wurtz 1991; Schlegel 1992; Wildeman and Brock 2000). Logging became a major industry in many areas (Fig. 1.28a, b), occasionally even at high-elevation (deBuys 1985). Livestock grazing became widespread (Fig. 1.29; see Schlegel 1992; Abruzzi 1995; Scurlock and Finch 1997 and Wildeman and Brock 2000 for the history of grazing). For example, over 200,000 sheep reportedly were grazed in Coconino County of north-central Arizona by 1894 (Haskett 1936). But estimates of historical livestock numbers are likely inaccurate (different sources have provided vastly different numbers; see Schlegel 1992 for Arizona). Grazing is now known to have initiated the exclusion of fire as a natural disturbance factor on southwestern mountains and plateaus (Sects. 1.6.1 and 1.6.2).

Concern over increasingly widespread, intensive, and destructive land use led the federal government to establish several Forest Reserves (precursors of today's National Forests) and units of what is now the National Park System on the Colorado Plateau beginning in the 1890s (Scurlock and Finch 1997). This began a period during which land use gradually shifted from resource exploitation to greater focus on resource protection. Although logging and livestock grazing continued, they were more regulated on Forest Reserves/National Forests and were generally absent on units of the National Park System. Also, governmental land-management agencies initiated practices to exclude fires and control predators.

At the time, it was widely accepted that fires were caused primarily by humans (Allen 2002). Fires were therefore viewed as unnatural events from which forests should be protected. Attempts at fire exclusion via fire prevention and suppression became widespread in the early twentieth century (Pyne 1982; Fig. 1.30). Later,

Fig. 1.28 (**a**) Transporting ponderosa pine logs in 1909 using horses and a large-wheeled skidder in Coconino National Forest, north-central Arizona (Photograph by A.G. Varela, Coconino National Forest). (**b**) Logging train in 1928 in the region of Lincoln National Forest and Mescalero Apache Reservation, south-central New Mexico. Many railroad lines were constructed solely to facilitate removal of logs from forests (Photograph by E.S. Shipp, U.S. Forest Service)

Fig. 1.29 Sheep grazing in 1928 in Ponderosa Pine Forest in the Mescalero Apache Reservation in south-central New Mexico (Photograph by E.S. Shipp, U.S. Forest Service)

technological advances, particularly with regard to the use of aircraft (Fig. 1.31), led to effective exclusion by the middle of the twentieth century (Sawyer 1976; Swetnam 1990). Suppression was such a focus of land-management agencies that fire fighters responded to 87 % of lightning-ignited fires in National Forests in Arizona and New Mexico within 3 h of ignition (96 % within 12 h), and 80 % of the fires were controlled within 24 h (Barrows 1978 for 1960–1974). Moreover, from 1940 to 1975, only 3 % of lightning fires grew to an area of 4 ha (10 acres) or more – "a remarkable record of fire control efficiency" (Barrows 1978).

Predators were killed by Euro-American settlers and later by hunters hired by government land managers to protect other wildlife. For example, 7,388 coyotes (*Canis latrans*), 863 bobcats (*Lynx rufus*), 816 mountain lions (*Felis concolor*), and 30 wolves (*Canis lupis*) were killed on the Kaibab Plateau between 1906 and 1939 (Rasmussen 1941). Wolves became extinct on the Kaibab Plateau (Mann and Locke 1931) and elsewhere by 1930.

Logging, especially in Ponderosa Pine Forest, increased through much of the twentieth century, as stimulated by increasing demand, greater access, and mechanized equipment. The average annual cut on National Forest lands in the Southwest increased 10-fold from the late 1900s through the 1980s, but cutting quickly declined thereafter, approaching the level of the late 1900s by 1996 (Dahms

Fig. 1.30 Lookout tree in Kaibab National Forest, north-central Arizona. Trees were used to locate fires for suppression before fire lookout towers were built. Note person on platform at top of tree about 19 m (62 ft) above the ground and person climbing ladder attached to tree. Information in National Park Service (1987) suggested undated photograph is possibly from 1916 (Photograph by U.S. Forest Service, Southwestern Region, Kaibab National Forest)

and Geils 1997). This abrupt decline was related to the loss of most old-growth Ponderosa Pine Forest and was affected by federal environmental legislation (e.g., the National Environmental Policy Act of 1969 and the Endangered Species Act of 1973), federal land-management legislation (e.g., the National Forest Management Act of 1976), and legal challenges by environmental organizations.

Smaller-scale tree cutting, including in protected areas, also occurred. Early Euro-American settlers cut trees for structures and fuel. Tree cutting also took place into the twentieth century – even in National Parks –for construction of administration and tourism infrastructures, as well as for fuel and control of insects such as bark beetles.

Fig. 1.31 Airplane dropping fire-retardant slurry in an attempt to limit the spread of a nearby forest fire in north-central New Mexico (Photograph by Kari Greer/U.S. Forest Service)

Beginning in the 1960s and 1970s, governmental land-management agencies became more ecologically oriented. This led to reduced logging (see above), as well as the reintroduction of fire into landscapes from which it had been excluded for a century or more. Management fires were intended to accomplish specific goals, such as reductions of fuels and small tree densities that accumulated during the period of fire exclusion. Management fires included prescribed fires that are planned and set under specific limits of wind speed, fuel moisture, and other parameters (Fig. 1.32). They also now include selected wildfires that are unplanned, naturally ignited, and have potential to safely accomplish management goals. See Pyne (1982) for a history of cultural fire and Stephens and Ruth (2005) for a review of federal forest-fire policy.

1.5.2.2 Impacts

An important effect of early Euro-American land use on southwestern mountains and plateaus was alteration of fire regimes (e.g., Baisan and Swetnam 1997). There is anecdotal evidence (e.g., Powell 1890) of Euro-Americans causing additional ignitions, and livestock herders in particular have been accused of setting fires. However, fire-history studies do not indicate an abnormally high frequency of fires

Fig. 1.32 Prescribed fire in Ponderosa Pine Forest on the South Rim of Grand Canyon National Park, north-central Arizona (Photograph by Fire and Aviation Program, Grand Canyon National Park)

during the period of Euro-American exploration and settlement, even in locations specifically described as having been burned by herders (Allen 2002). Instead, fire-history studies show an abrupt decrease in fire frequency with the beginning of livestock grazing, which was typically between 1870 and 1900 (e.g., Dieterich 1980; Swetnam and Baisan 2003). Livestock grazing greatly reduced the dense herbaceous layers of open forests and meadows, which formerly had carried surface fires. This was recognized early in the twentieth century:

> During recent years the [ponderosa]-pine type has been heavily grazed by sheep and cattle, and in consequence the grass is kept short, and the damage from fire very much reduced. (Woolsey (1911) for Arizona and New Mexico)

Several other lines of evidence also connect grazing with the change in fire regimes (Allen 2002). For example, areas where Native Americans grazed livestock had earlier declines in fire frequencies (Savage and Swetnam 1990; Touchan et al. 1995). Also, fire frequencies were mostly unchanged in isolated areas likely to have been free of grazing (Grissino-Mayer and Swetnam 1995; Touchan et al. 1995; Fulé et al. 2000). Fire exclusion – first inadvertently by livestock grazing and later by active fire exclusion – caused large changes in forest structure, fuels, and dynamics, as covered in following chapters.

Livestock grazing also directly affected tree recruitment. Some have argued that recruitment was enhanced as sheep reduced competition from the herbaceous understory (Cooper 1960; Belsky and Blumenthal 1997), but sheep also stunted if not killed tree seedlings (e.g., Woolsey 1911). Moreover, grazing altered soils and hydrology by increasing compaction through trampling, which reduced water infiltration rates and increased erosion (Abruzzi 1995; Belsky and Blumenthal 1997).

The ecological impacts of predator control programs are poorly understood, but are thought to have included a role in population increases of deer, as for example on the Kaibab Plateau in the 1910s–1920s (Mitchell and Freeman 1993). Large deer populations in turn affected vegetation in both the short term (Mead 1930) and longer term (Fulé et al. 2003; Binkley et al. 2006).

Commercial logging obviously had major impacts. In fact, little undisturbed Ponderosa Pine Forest remains outside of protected areas such as some units of the National Park System. Higher-elevation forests were generally less impacted by logging. The effects of smaller-scale tree cutting (previous section), including in protected areas, are unstudied.

The reintroduction of fire into ecosystems unburned for many decades has altered vegetation structure, composition, and dynamics. This is described in following chapters.

1.6 Anthropogenic Disturbance

Anthropogenic disturbances (drivers) have had important impacts on the vegetation of southwestern mountains and plateaus. Past and present anthropogenic disturbances include livestock grazing, fire management, modern climate change, air pollution, invasive species, recreation, and nearby land use. These are important in vegetation dynamics and therefore are incorporated in the conceptual models in Sect. 1.7. Intensive land uses that remove near-natural vegetation, such as logging, mining, and suburban development, are outside the scope of this book.

1.6.1 Livestock Grazing

The first broad-scale European and Euro-American impact on the vegetation of southwestern mountains and plateaus was livestock grazing, which initiated the exclusion of fire (previous section). Grazing also had direct effects on vegetation, altering its structure, composition, and dynamics by selective herbivory and possibly

Fig. 1.33 Smoke plume from a landscape-scale crown fire started by recreationists camping in Ponderosa Pine Forest near the San Francisco Peaks, north-central Arizona (Photograph by Betty J. Huffman)

by creation of nutrient-rich patches through urination and defecation. Some of these effects persist today (see following chapters). The number of livestock grazed on public lands in the American Southwest has decreased since the 1920s–1930s (Raish et al. 1997), but grazing continues in many areas.

1.6.2 Fire Management

The prevention and suppression of fires was a common policy and land management practice throughout most of the twentieth century (Sect. 1.5.2.1). The effects of fire exclusion, first by livestock grazing and then by fire management, persist today. They depend on the type of vegetation and are described in following chapters. One general impact of fire exclusion has been the buildup of fuels and tree densities that has resulted in landscape-scale crown fires (Fig. 1.33). Realization of these effects led to the development of fire management practices such as prescribed burning and use of lightning-ignited fires as management fires. Application of these practices is complicated by the increased tree densities and fuel loadings that developed with fire exclusion. Management fires are usually partially successful in countering the effects of fire exclusion (Fulé and Laughlin 2007; Vankat 2010), but sometimes result in crown fires outside the historical range of variation for the vegetation. Examples include the Outlet (2000), Poplar (2003), and Warm Fires (2006) on the Kaibab Plateau and the Cerro Grande Fire (2000) in Bandelier National Monument in north-central New Mexico. Such large, intensive fires can create habitat for invasive plants to establish.

1.6.3 Modern Climate Change

The Southwest is considered especially sensitive to climate changes (Ehleringer et al. 2000), and these changes have affected vegetation for thousands of years. Past, non-anthropogenic changes are outlined in Sect. 1.4. Recent changes (cf. Spence 2001; Lenart 2007) are treated in this book as an anthropogenic disturbance, because release of carbon dioxide and other greenhouse gases and aerosols from human activities has been identified as a driver of these changes (Houghton et al. 1996; Barnett et al. 2008; Das et al. 2009).

Modern climate change in the American Southwest is projected to involve increases in temperature, drought, and extreme events. These projections are usually based on computer modeling involving alternative scenarios for future greenhouse gas emissions and different models. For example, the Intergovernmental Panel on Climate Change (2007) used several emission scenarios and several climate models and predicted rises in average surface temperature in the Southwest of 1.5 °C (2.7 °F) by 2029 and 2.5–5 °C (4.5–9 °F) by 2099 (in comparison to temperatures in 1980–1999). A study of the Colorado River Basin synthesized results from 18 different climate models and predicted temperature rises of 1.4 °C (2.5 °F) by 2030 and 2.8 °C (5.0 °F) by 2060 (Hoerling and Eischeid 2007 in comparison to 1895–2005). Such projected increases in temperature are in line with actual increases recorded since 1976 in Arizona and New Mexico of 1.4 °C (2.5 °F) and 1 °C (1.8 °F), respectively (Lenart 2007). Also, annual and/or winter minimum temperatures have increased in the central Colorado Plateau (centered on southeastern Utah) since the 1960s (areas with longer records show increases as far back as 1925; Spence 2001).

There is weaker consensus on projected changes in precipitation, in part because precipitation is highly variable. One study using 18 global climate models predicted decreases for the twenty-first century ranging from >10 % in southern Arizona to 5–10 % in most of New Mexico and northern Arizona to 0–5 % in southwestern Colorado and most of Utah (Lenart 2007 in comparison to 1971–2000). Another study that averaged 19 global climate models predicted decreased precipitation in winter, summer, and for the year (Seager et al. 2007). In contrast, little net change in precipitation was predicted by the average of 18 climate models used by Hoerling and Eischeid (2007); they also reported much variability among simulations. Actual trends in precipitation are difficult to track because of high variability, but there was little change in the central Colorado Plateau through the late twentieth century (Spence 2001).

Regardless of recorded and predicted patterns in precipitation, most studies indicate drought has increased in degree and extent and is projected to increase in the future. A sustained, more arid climate began in the Southwest in the late twentieth and early twenty-first centuries (Seager et al. 2007; Balling and Goodrich 2010) and is predicted to continue in the twenty-first century (Hoerling and Eischeid 2007; Seager et al. 2007; Cayan et al. 2010; Seager and Vecchi 2010). Specific predictions of numerous individual models are variable, but nearly universally indicate increased aridity (Hoerling and Eischeid 2007; Seager et al. 2007; Seager and Vecchi 2010).

Another conclusion is that drought in the Southwest is no longer driven primarily by reduced precipitation and enhanced by temperature. Instead, modern drought is driven primarily by increased temperature (Hoerling and Eischeid 2007), especially in summer (Cayan et al. 2010). Although variation is predicted to continue into the future, the conditions of intense aridity experienced in the 1930s, 1950s, and early 2000s are predicted to characterize the American Southwest in the twenty-first century (Hoerling and Eischeid 2007; Seager et al. 2007). Moreover, droughts are expected to increase in length (Cayan et al. 2010). Increased extreme episodes of high temperatures and high precipitation also have been predicted (Diffenbaugh et al. 2005).

The ecological impacts of extreme events can be pronounced (e.g., Allen and Breshears 1998; Allen 2007). As stated in Sect. 1.2.5.4, forests of the American Southwest appear "particularly sensitive to drought and warmth"; therefore, changes are expected in productivity, disturbance regimes, and species ranges (Williams et al. 2010, 2012). Modern climate change has been implicated in increased large wildfire activity in the western U.S. (Westerling et al. 2006), recent large insect outbreaks (Logan et al. 2003; Breshears et al. 2005; Romme et al. 2006), and enhanced growth of invasive plants (Alward et al. 1999; Dukes and Mooney 1999; Smith et al. 2000). Effects on fire regimes include alteration of fuel accumulation, fuel combustibility, ignition rates, and fire spread (Marshall et al. 2008).

1.6.4 Air Pollution

Little research has been done on the direct effects of air pollution on the vegetation of southwestern mountains and plateaus. Nevertheless, air pollution is an important anthropogenic disturbance, at least as it has induced modern climate change, but likely also through direct impacts on vegetation. The primary sources of air pollutants in the American Southwest are likely to be large point sources (Böhm 1992), both in southern California and regional urban areas such as Phoenix, Arizona (de Paula Vasconcelos 1995; Schreiber 1996).

Although air pollution injury to trees such as ponderosa pine is well-documented in California (Arbaugh et al. 1999; Miller and McBride 1999), few reports exist for the American Southwest (but see Miller 1989; Graybill and Rose 1992). Air pollution has been postulated to reduce growth rates and affect patterns of tree mortality and regeneration, thereby altering species composition and vegetation dynamics (Binkley et al. 1992). Air pollution facilitates tree mortality by modifying environmental factors and thus stressing trees, which makes them more vulnerable to insects and pathogens (Miller 1989; Takemoto et al. 2001).

The air pollutants of greatest concern – other than greenhouse gases – are ozone, nitrogen-based compounds such as nitrate and ammonium/ammonia, and sulfate (National Park Service 2002). Ozone is one of the most phytotoxic air pollutants (National Park Service 2002; Fenn et al. 2003a). Damage has been reported for forests in southern Arizona (Miller et al. 1995) and California (Miller et al. 1997; Takemoto et al. 2001). Ozone can interact with other anthropogenic disturbances such as contemporary climate change (Wagner and Baker 2003). Ozone levels are largely stable in the American Southwest, with no statistically significant change in four of the five units of the U.S. National Park System where it was measured during the 1990s and 2000s (14–20 years, depending on site; National Park Service 2002). The exception is Mesa Verde National Park in southwestern Colorado where ozone increased during 1994–2008. Nevertheless, ozone levels have exceeded thresholds for negative effects on leaves and tree-seedling growth in some units, including Mesa Verde and Grand Canyon National Parks (National Park Service 2002).

Nitrogen deposition has many effects on terrestrial vegetation of the western United States, including altered species composition, accelerated plant growth, greater risk of wildfire (as increased plant growth leads to increased fuels), and increases in invasive species (Fenn et al. 2003a,b; Galloway et al. 2003). Deposition is highly variable, with large areas of low deposition and scattered areas of high deposition downwind from large urban and agricultural areas (Fenn et al. 2003a,b). Ammonium deposition is increasing in the Southwest, having done so in five of eight units of the U.S. National Park System where it was measured during the 1990s and 2000s; two of the three other units had trends of increased deposition (National Park Service 2010). In contrast, nitrate deposition increased in only one unit, decreased in two, and exhibited a trend of a decrease in one unit.

Sulfate is a major component of acid precipitation. Sulfate deposition appears to be in decline in the Southwest, as it significantly decreased in six of eight units of the U.S. National Park System during the 1990s and 2000s and one unit exhibited a trend of decreased deposition (National Park Service 2010).

1.6.5 Invasive Species

Invasive plants are an important anthropogenic disturbance because they can alter ecosystem structure, composition, and function (Cronk and Fuller 1995; Vitousek et al. 1996; McGlone 2010) and therefore vegetation dynamics (McGlone et al. 2012). Moreover, they are a major threat to biodiversity (Randall 1996; Chornesky and Randall 2003). Invasives are often linked to land use because they tend to colonize disturbed sites (Elton 1958; Rejmánek 1989; Vitousek et al. 1996). Establishment can be slow in relatively undisturbed areas, but land management practices that involve disturbance (e.g., prescribed burning) promote establishment (Vankat and Roy 2002). Invasive plants can affect disturbance agents such as fire (Menakis et al. 2003; Brooks et al. 2004; Brooks and Matchett 2006; Link et al. 2006).

Invasive species generally are not as abundant in the American Southwest as in other regions of the United States, primarily because extensive and intensive Euro-American land use began later in this region (Rejmánek and Randall 1994). Nevertheless, concern about growth of invasives following disturbance extends back to at least the early twentieth century (see Leiberg et al. 1904). Today, invasives are increasing in abundance and distribution and are an anthropogenic disturbance of growing importance on southwestern mountains and plateaus.

Fig. 1.34 Recreation by Boy Scouts in 1928 in or near Lincoln National Forest in south-central New Mexico (Photograph by E.S. Shipp, U.S. Forest Service)

1.6.6 Recreation

Recreation in the American Southwest (Fig. 1.34) has increased many fold over the last century. For example, annual visitation to Grand Canyon National Park increased from approximately 38,000 in 1919 to nearly 4,300,000 in 2011 (the first and last years for which data are available; National Park Service 2012). Even less well-known Zion National Park in southwestern Utah increased from less than 2,000 to more than 2,800,000 people during the same period.

Some impacts of recreation are either included within other anthropogenic disturbances (e.g., air pollution) or appear to be spatially restricted, e.g., roads and trails. Nevertheless, the total lengths and densities of roads suggest potential widespread impacts. For example, total road and trail length in San Juan National Forest in southwestern Colorado and Rio Grande National Forest in south-central Colorado is 12,211 km (7,588 miles), and road and trail density is 0.81 km/km^2 (1.30 miles/miles^2; Baker and Knight 2000; Knight 2000). The area of roads in a large, upper-elevation area of San Juan National Forest tripled from 1950 to 1993, going from 1.06 to 3.34 % of the area (McGarigal et al. 2001). Although directly affecting only 2.28 % of the area, this increase accounted for the majority of changes in landscape configuration (i.e., mean patch size, edge density, and core area) and exceeded the effects of logging. Moreover, use of roads and trails compounds their impacts through the introduction, spread, and establishment of invasive species and ignition and spread of fires (Fig. 1.33).

1.6.7 Nearby Land Use

Much of the land area on mountains and plateaus of the American Southwest is significantly disturbed by past and present logging, livestock grazing, and human development. All of these land uses can occur in the vicinity of near-natural vegetation and can be sources of invasive plants (Vankat and Roy 2002) and wildfires (e.g., Wadleigh et al. 1998). Eighty percent of Colorado's forested land (disturbed and near-natural combined) is within 3 km (2 miles) of private land, most of which has been used for agriculture (Theobald 2000). The impacts of nearby land use increase with conversion of agricultural land to residential use and with human population growth.

1.7 Conceptual Models

The above background material on drivers, processes of vegetation dynamics, and anthropogenic disturbances (Sects. 1.2, 1.3, and 1.6, respectively) can be summarized and synthesized in conceptual models. These models organize, describe, and communicate existing knowledge and hypotheses about vegetation dynamics. They explain historical changes (Sect. 1.5) and help predict future changes. Their visual format can enhance understanding. Moreover, the process of developing conceptual models highlights not only what is known, but also what is unknown, thus identifying areas of needed research. Conceptual models facilitate communication among scientists, managers, environmental planners, conservationists, environmentalists, and laypeople.

Of necessity, conceptual models are simplified representations. Therefore, "all models are wrong, but some are useful" (Box 1979). Models of intermediate complexity are most likely to be useful, as suggested by the words of French poet Paul Valéry: "All that is simple is false, all that is complex is unusable."

The models presented in the following chapters are intended to usefully and usably summarize and communicate what is known about vegetation dynamics on the mountains and plateaus of the American Southwest. The models follow the conceptual framework developed by Mark E. Miller and Lisa Thomas (cf. Miller et al. 2003; Miller 2005; O'Dell et al. 2005; Thomas et al. 2006). No single type of model could encapsulate the ranges of scale and process-specificity needed; therefore, Miller and Thomas proposed a three-tiered, nested set of conceptual models. As adapted here, they are ecosystem-characterization, vegetation-dynamics, and mechanistic models.

1.7.1 Ecosystem-Characterization Models

Ecosystem-characterization models, the most generalized of the nested set of models, identify core components and drivers of ecosystems and illustrate how they are functionally related (upper left of Fig. 1.35). These models focus on ecosystem components and interactions that affect vegetation dynamics. Conceptually, the ecosystem-characterization models are founded in Jenny's (1941) formative work on factors affecting soil characteristics, Major's (1951) extension of this work to other systems, and Chapin et al.'s (1996) interactive-control model. Jenny (1941) identified five key factors driving soil characteristics: regional climate, topography (relief), organisms (biota), soil parent material, and time. Major (1951) extended these to vegetation and stated that they also determined characteristics of ecosystems. Chapin et al. (1996) described four interactive controls (regional climate, biotic functional groups, soil resources, and disturbance regime) that affected ecosystem characteristics by intermediating relationships between the five state factors of Jenny (1941) and ecosystem processes.

The ecosystem-characterization models presented in the following chapters include three biotic components: *Soil System*, *Vegetation*, and *Animals*. They can include living and non-living elements. For example, the soil system includes such living elements as mycorrhizae and decomposers, as well as such non-living elements as rock particles, humus, water, and minerals. Vegetation includes living and non-living plants, including snags and downed logs. The biotic components correspond to divisions of the organisms (biota) factor of Jenny (1941) and Major (1951) and the biotic functional groups and soil resources interactive controls of Chapin et al. (1996). The ecosystem-characterization models also include three drivers: *Weather & Climate, Disturbance*, and *Landscape*. Weather & Climate parallels the regional climate factor of Jenny (1941) and Major (1951), which is treated as an interactive control by Chapin et al. (1996). Disturbance parallels the disturbance regime of Chapin et al. (1996). Landscape corresponds to Jenny's (1941) and Major's (1951) topography (relief) factor and also incorporates topographic position within landscapes. In addition, the ecosystem-characterization models include six drivers that are anthropogenic disturbances: *Livestock Grazing*, *Fire Management*, *Modern Climate Change*, *Invasive Species*, *Recreation*, and *Nearby Land Use*. The previously mentioned paucity of data on direct effects of air pollution currently precludes its inclusion as another anthropogenic disturbance; however, its indirect effects via climate are included in *Modern Climate Change*.

1.7.2 Vegetation-Dynamics Models

Vegetation-dynamics models, the second tier of the nested set of models, portray the characteristic dynamics of vegetation, indicating how vegetation changes and why. The vegetation dynamics-models are developed from and focus on the vegetation component (including fuel), the disturbance driver, and their interactions depicted

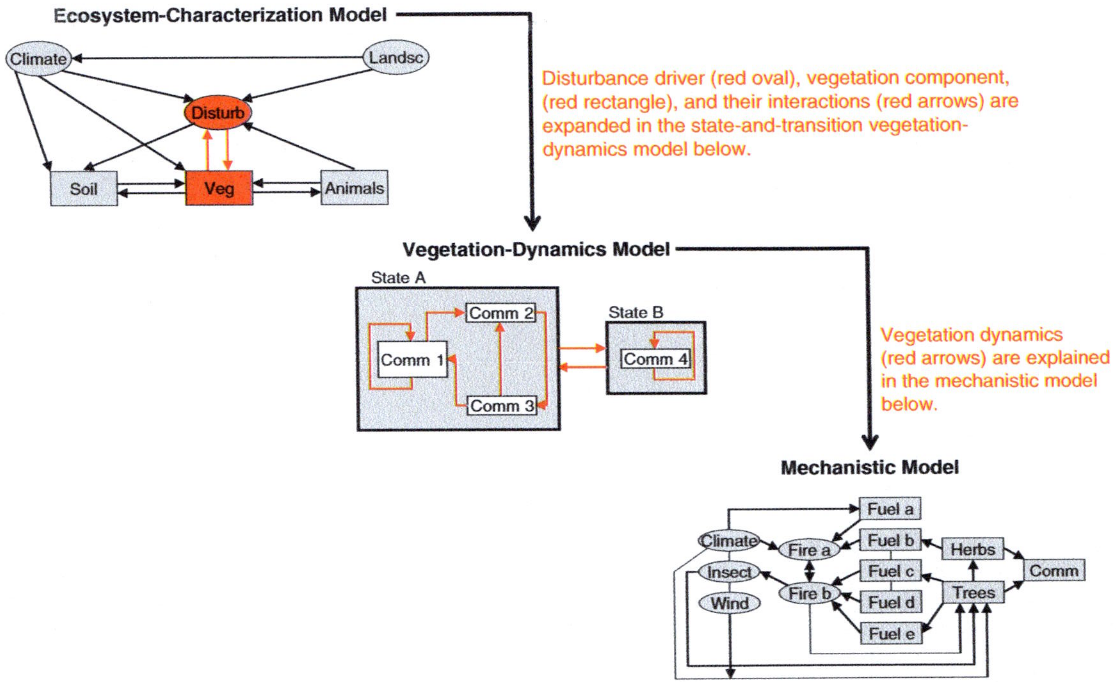

Fig. 1.35 Conceptual relationships among the three-tiers of conceptual models (individual models are reduced in complexity for clarity of relationships among models). The *ecosystem-characterization model* (*top left*) identifies the key components, including drivers, of the ecosystem (*ovals* and *rectangles*) and illustrates how the components are functionally related (*arrows*). Anthropogenic disturbance drivers are not shown in this example. Two components from the ecosystem-characterization model (disturbance and vegetation) are developed in greater detail in the *vegetation-dynamics model* (*center*), which illustrates how the vegetation changes and why. This model shows different community types (*four smaller, open rectangles*) within different ecological states (*two larger, shaded rectangles*), along with processes and transitions (*arrows*) that produce changes in community types and states. These processes and transitions are developed in greater detail in the *mechanistic model* (*bottom right*), which addresses specific underlying mechanisms of vegetation dynamics. *Landsc* landscape, *Disturb* disturbance, *Veg* vegetation, *Comm* community, *1–4* different community types, and *a–e* different attributes of fuel and fire. See following chapters for examples of full models

in the ecosystem-characterization model. The relationship between ecosystem-characterization and vegetation-dynamics models is shown schematically in Fig. 1.35. Some kinds of vegetation, such as Mixed Conifer Forest (Chap. 3), are so variable in terms of vegetation and disturbances that multiple models are needed to adequately describe the dynamics.

Miller and Thomas selected a state-and-transition format for the vegetation-dynamics models. The models identify different community types within different ecological states, along with processes resulting in shifts among community types and transitions resulting in shifts among states. The state-and-transition format has proven useful for developing and testing hypotheses about ecological thresholds for transitions among states and for the effects of land use and management activities on community and state changes (Bestelmeyer et al. 2003, 2004, 2010; Briske et al. 2005). Other modeling in the Southwest also has used this format (e.g., The Nature Conservancy 2006).

Each vegetation-dynamics model is accompanied by a bar graph to show estimated relative shifts in abundances of the community types from circa 1870 (prior to widespread Euro-American land use) to circa 1970 (end of strict fire exclusion policy) to circa the present. The estimates are based on my understanding of vegetation dynamics and should be interpreted as showing relative differences and not interpreted quantitatively (hence the absence of a scale for abundance). Although not shown on the graphs, relative abundances also shifted prior to circa 1870 as a result of variation in climate and disturbance. Localized differences also have been and are present, but are beyond the scope of these generalized bar graphs.

1.7.3 Mechanistic Models

Mechanistic models, the third tier in the nested set of models, address the specific causal mechanisms that underlie the pathways of change (i.e., processes and transitions) illustrated in the vegetation-dynamics models. The relationship between the two types of models is shown schematically in Fig. 1.35. The mechanistic models include key vegetation components such as fuel characteristics and community type (as depicted in the vegetation-dynamics model), system drivers such as fire characteristics, anthropogenic drivers such as nearby land use, and interrelationships. The number of mechanistic models can equal the number of vegetation-dynamics models or there can be fewer mechanistic models when the same mechanisms account for the processes and transitions of more than one vegetation-dynamics model.

1.8 Conclusions and Challenges

Although the vegetation on the mountains and plateaus accounts for only about a third of the area of the American Southwest, the vegetation has high ecologic, economic, and aesthetic importance. The vegetation is complex because of varied landscapes, climates, and natural disturbances. Among natural disturbances, fire is the key driver and interacts with insects and climate variation to form a disturbance complex that has major impacts. The effects of historical Native-American populations appear to have differed among landscapes from little to substantial and need additional research. Euro-American populations have had much broader impacts. Even in relatively undisturbed landscapes, anthropogenic disturbances such as historical livestock grazing and fire management are keys to understanding present vegetation structure, composition, and dynamics. Additional basic and applied research is needed to guide contemporary fire management and increase its effectiveness and efficiency at large spatial scales. The direct effects of air pollution on vegetation are largely unstudied in the American Southwest, and research should be a priority. Modern climate change likely greatly impacts vegetation and is currently receiving much research, but two precautions are necessary. Funding for climate-change research should not detract from funding in other areas of research, and funding should not become politicized. One emphasis of climate-change research should be the climate-insect-fire disturbance complex. Invasive species are increasing and appear to be influencing vegetation dynamics; research is needed on their expansion and effects. The impacts of recreation and nearby land use need study to inform land managers on their magnitude and on methods to reduce them.

References

Abruzzi WS (1995) The social and ecological consequences of early cattle ranching in the Little Colorado River Basin. Hum Ecol 23:75–98

Alcoze T (2003) First peoples in the pines: historical ecology of humans and ponderosas. In: Friederici P (ed) Ecological restoration of southwestern ponderosa pine forests. Island Press, Washington, DC

Alexander ME, Hawksworth FG (1975) Wildland fires and dwarf mistletoes: a literature review of ecology and prescribed burning. U.S. Department of Agriculture, Forest Service General Technical Report RM-14, Rocky Mountain Forest and Range Experiment Station, Fort Collins, CO

Allen CD (1989) Changes in the landscape of the Jemez Mountains, New Mexico. Dissertation, University of California, Berkeley

Allen CD (1996) Elk response to the La Mesa fire and current status in the Jemez Mountains. In: Allen CD (ed) Proceedings of the second La Mesa fire symposium: fire effects in southwestern forests. U.S. Department of Agriculture, Forest Service General Technical Report RM-286, Rocky Mountain Forest and Range Experiment Station, Fort Collins, CO

Allen CD (2001) Fire and vegetation history of the Jemez Mountains. In: Johnson PS (ed) Water, watersheds, and land use in New Mexico: impacts of population growth on natural resources. New Mexico Bureau of Mines and Mineral Resources, Socorro

Allen CD (2002) Lots of lightning and plenty of people: an ecological history of fire in the upland Southwest. In: Vale TR (ed) Fire, native peoples, and the natural landscape. Island Press, Covelo

Allen CD (2007) Interactions across spatial scales among forest dieback, fire, and erosion in northern New Mexico landscapes. Ecosystems 10:797–808

Allen CD, Breshears DD (1998) Drought-induced shift of a forest-woodland ecotone: rapid landscape response to climate variation. Proc Nat Acad Sci USA 95:14839–14842

Allen CD, Anderson RS, Jass RB, Toney JL, Baisan CH (2008) Paired charcoal and tree-ring records of high-frequency Holocene fire from two New Mexico bog sites. Int J Wildland Fire 17:115–130

Alward RD, Detling JK, Milchunas DG (1999) Grassland vegetation changes and nocturnal global warming. Science 283:229–231

Anderson RS (1989) Development of the southwestern ponderosa pine forests: what do we really know? In: Tecle A, Covington WW, Hamre RH (tech coords) Multiresource management of ponderosa pine forests. U.S. Department of Agriculture, Forest Service General Technical Report RM-185, Rocky Mountain Forest and Range Experiment Station, Fort Collins, CO

Anderson RS (1993) A 35,000 year vegetation and climate history from Potato Lake, Mogollon Rim, Arizona. Quat Res 40:351–359

Anderson RS, Shafer DS (1991) Holocene biogeography of spruce-fir forests in southeastern Arizona: implications for the endangered Mt. Graham red squirrel. Madroño 38:287–295

Anderson RS, Betancourt JL, Mead JI, Hevly RH, Adams DP (2000) Middle- and Late-Wisconsin paleobotanic and paleoclimatic records from the southern Colorado Plateau, USA. Palaeogeogr Palaeoclimatol Palaeoecol 155:31–57

Anderson RS, Jass RB, Toney JL, Allen CD, Cisneros-Dozal LM, Hess M, Heikoop J, Fessenden J (2008a) Development of the mixed conifer forest in northern New Mexico and its relationship to Holocene environmental change. Quat Res 69:263–275

Anderson RS, Allen CD, Toney JL, Jass RB, Bair AN (2008b) Holocene vegetation and fire regimes in subalpine and mixed conifer forests, southern Rocky Mountains, USA. Int J Wildland Fire 17:96–114

Arbaugh MJ, Peterson DL, Miller PR (1999) Air pollution effects on growth of ponderosa pine, Jeffrey pine, and bigcone Douglas-fir. In: Miller PR, McBride JR (eds) Oxidant air pollution impacts in the montane forests of southern California: a case study of the San Bernardino mountains. Springer, New York

Bailey LR (1980) If you take my sheep: the evolution and conflicts of Navajo pastoralism, 1630–1868. Westernlore, Pasadena

Baisan CH, Swetnam TW (1997) Interactions of fire regimes and land use in the central Rio Grande Valley. U.S. Department of Agriculture, Forest Service Research Paper RM-RP-330, Rocky Mountain Forest and Range Experiment Station, Fort Collins, CO

Baker WL (2002) Indians and fire in the Rocky Mountains: the wilderness hypothesis renewed. In: Vale TR (ed) Fire, native peoples, and the natural landscape. Island Press, Washington, DC

Baker WL, Ehle D (2001) Uncertainty in surface-fire history: the case of ponderosa pine forests in the western United States. Can J For Res 31:1205–1226

Baker WL, Knight RL (2000) Roads and forest fragmentation in the southern Rocky Mountains. In: Knight RL, Smith FW, Buskirk SW, Romme WH, Baker WL (eds) Forest fragmentation in the Southern Rocky Mountains. University Press of Colorado, Boulder

Balling RC Jr, Goodrich GB (2010) Increasing drought in the American Southwest? A continental perspective using a spatial analytical evaluation of recent trends. Phys Geogr 31:293–306

Barlein PJ, Anderson KH, Anderson PM, Edwards ME, Mock CJ, Thompson RS, Webb RS, Webb T, Whitlock C (1998) Paleoclimate simulations for North America over the past 21,000 years: features of the simulated climate and comparisons with paleoenvironmental data. Quat Sci Rev 17:549–585

Barnett TP, Pierce DW, Hidalgo HG, Bonfils C, Santer BD, Das T, Bala G, Wood AW, Nozawa T, Mirin AA, Cayan DR, Dettinger MD (2008) Human-induced changes in the hydrology of the western United States. Science 319:1080–1083

Barrett SW, Swetnam TW, Baker WL (2005) Indian fire use: deflating the legend. Fire Manag Today 65(3):31–33

Barrows JS (1978) Lightning fires in southwestern forests. Final report to USDA Forest Service, Northern Forest Fire Laboratory, Intermountain Forest and Range Experiment Station, under cooperative agreement 16–568 CA with Rocky Mountain Forest and Range Experiment Station, Fort Collins, CO

Baxter JO (1987) Las carnerada: sheep trade in New Mexico, 1700–1860. University of New Mexico Press, Albuquerque
Belsky AJ, Blumenthal DM (1997) Effects of livestock grazing on stand dynamics and soils in upland forests of the Interior West. Conserv Biol 11:315–327
Bestelmeyer BT, Brown JR, Havstad KM, Alexander R, Chavez G, Herrick JE (2003) Development and use of state-and-transition models for rangelands. J Range Manag 56:114–126
Bestelmeyer BT, Herrick JE, Brown JR, Trujillo DA, Havstad KM (2004) Land management in the American Southwest: a state-and-transition approach to ecosystem complexity. Environ Manag 34:38–51
Bestelmeyer BT, Moseley K, Shaver PL, Sanchez H, Briske DD, Fernandez-Gimenez ME (2010) Practical guidance for developing state-and-transition models. Rangelands 32:23–30
Betancourt JL (1990) Late Quaternary biogeography of the Colorado Plateau. In: Betancourt JL, Van Devender TR, Martin PS (eds) Packrat middens: the last 40,000 years of biotic change. University of Arizona Press, Tucson
Betancourt JL, Biggar N (1985) Preliminary assessment of Late Quaternary vegetation and climate of southeastern Utah based on analyses of packrat middens. Technical Report. Office of Nuclear Waste Isolation, Battelle Memorial Institute, Columbus, OH
Betancourt JL, Davis OK (1984) Packrat middens from Canyon de Chelly, northeastern Arizona: paleoecological and archaeological implications. Quat Res 21:56–64
Betancourt JL, Van Devender TR (1981) Holocene vegetation in Chaco Canyon, New Mexico. Science 214:656–658
Betancourt JL, Dean JS, Hull HM (1986) Prehistoric long-distance transport of construction beams, Chaco Canyon, New Mexico. Am Antiq 51:370–375
Betancourt JL, Van Devender TR, Martin PS (eds) (1990) Packrat middens: the last 40,000 years of biotic change. University of Arizona Press, Tucson
Binkley D, Droessler TD, Miller J (1992) Pollution impacts at the stand and ecosystem levels. In: Olson RK, Binkley D, Böhm M (eds) The responses of western forests to air pollution. Springer, New York
Binkley D, Moore MM, Romme WH, Brown PM (2006) Was Aldo Leopold right about the Kaibab deer herd? Ecosystems 9:227–241
Biondi F, Gershunov A, Cayan DR (2001) North Pacific decadal climate variability since AD 1661. J Clim 14:5–10
Böhm M (1992) Air quality and deposition. In: Olson RK, Binkley D, Böhm M (eds) The responses of western forests to air pollution. Springer, New York
Box GEP (1979) Robustness in the strategy of scientific model building. In: Launer RL, Wilkinson GN (eds) Robustness in statistics. Academic, New York
Breece CR, Kolb TE, Dickson BG, McMillin JD, Clancy KM (2008) Prescribed fire effects on bark beetle activity and tree mortality in southwestern Ponderosa Pine forests. For Ecol Manag 255:119–128
Breshears DD, Cobb NS, Rich PM, Price KP, Allen CD, Balice RG, Romme WH, Kastens JH, Floyd ML, Belnap J, Anderson JJ, Myers OB, Meyer CW (2005) Regional vegetation die-off in response to global-change-type drought. Proc Natl Acad Sci 102:15144–15148
Briske DD, Fuhlendorf SD, Smeins FE (2005) State-and-transition models, thresholds, and rangeland health: a synthesis of ecological concepts and perspectives. Rangel Ecol Manag 58:1–10
Brooks ML, Matchett JR (2006) Spatial and temporal patterns of wildfires in the Mojave Desert, 1980–2004. J Arid Environ 67:148–164
Brooks ML, D'Antonio CM, Richardson DM, Grace JB, Keeley JE, DiTomaso JM, Hobbs RJ, Pellant M, Pyke D (2004) Effects of invasive alien plants on fire regimes. Bioscience 54:677–678
Brown DE (1982) Biotic communities of the American Southwest, United States and Mexico. Desert Plants 4:1–342
Brown DE (ed) (1994) Biotic communities: southwestern United States and northwestern Mexico. University of Utah Press, Salt Lake City
Brown DE, Lowe CH, Pase CP (1980) A digitized systematic classification for ecosystems with an illustrated summary of the natural vegetation of North America. U.S. Department of

Agriculture, Forest Service General Technical Report RM-73, Rocky Mountain Forest and Range Experiment Station, Fort Collins, CO

Brown DE, Minckley WL, Collins JP (1994) Historical background to southwestern ecological studies. In: Brown DE (ed) Biotic communities: southwestern United States and northwestern Mexico. University of Utah Press, Salt Lake City

Brown DP, Comrie AC (2004) A winter precipitation 'dipole' in the western United States associated with multidecadal ENSO variability. Geophys Res Lett 31:L09203. doi:10.1029/2003GO18726

Brown M (1960) Climates of the states: Utah. Climatology of the United States, no. 60–42. U.S. Deptartment of Commerce, Weather Bureau, Washington, DC

Brown PM, Heyerdahl EK, Kitchen SG, Weber MH (2008) Climate effects on historical fires (1630–1900) in Utah. Int J Wildland Fire 17:28–39

Brugge D, Gerow PA (2000) American Indian livestock operations on tribal lands in Arizona and New Mexico. In: Jemison R, Raish C (eds) Livestock management in the American Southwest: ecology, society, and economics. Elsevier, Amsterdam

Burwell T (1999) Environmental history of the lower montane pinon (*Pinus monophylla*) treeline, eastern California. Dissertation, University of Wisconsin Madison

Cayan DR, Das T, Pierce DW, Barnett TP, Tyree M, Gershunov A (2010) Future dryness in the southwest US and the hydrology of the early 21st century drought. Proc Natl Acad Sci USA 107:21271–21276

Chapin FS III, Tom MS, Tateno M (1996) Principles of ecosystem sustainability. Am Nat 148:1016–1035

Chornesky EA, Randall JM (2003) The threat of invasive alien species to biological diversity: setting a future course. Ann Mo Bot Gard 90:67–76

Cole K (1982) Late Quaternary zonation of vegetation in the eastern Grand Canyon. Science 217:1142–1145

Cole KL (1985) Past rates of change, species richness, and a model of vegetational inertia in the Grand Canyon, Arizona. Am Nat 125:289–303

Cole KL (1990) Late Quaternary vegetation gradients through the Grand Canyon. In: Betancourt JL, Van Devender TR, Martin PS (eds) Packrat middens: the last 40,000 years of biotic change. University of Arizona Press, Tucson

Cooper CF (1960) Changes in vegetation, structure, and growth of southwestern pine forests since white settlement. Ecol Monogr 30:129–164

Crimmins MA (2006) Synoptic climatology of extreme fire-weather conditions across the southwest United States. Int J Climatol 26:1001–1016

Cronk QCB, Fuller JL (1995) Plant invaders: the threat to natural ecosystems. Chapman & Hall, London

D'Arrigo RD, Jacoby GC (1991) A 1000-yr precipitation record from northwestern New Mexico, USA: a reconstruction from tree rings and its relation to El Niño and the Southern Oscillation. Holocene 1:95–101

Dahms CW, Geils BW (tech eds) (1997) An assessment of forest ecosystem health in the Southwest. U.S. Department of Agriculture, Forest Service General Technical Report RM-GTR-295, Rocky Mountain Forest and Range Experiment Station, Fort Collins, CO

Das T, Hidalgo HG, Dettinger MD, Cayan DR, Pierce DW, Bonfils C, Barnett TP, Bala G, Mirin A (2009) Structure and detectability of trends in hydrological measures over the western United States. J Hydrometeorol 10:871–892

de Paula Vasconcelos LA (1995) Aerosol and transport climatology at the Grand Canyon. Dissertation, Washington University, St. Louis

DeBano LF, Neary DG, Ffolliott PF (2008) Soil and water resources. In: Ffolliott PF, Davis OK (eds) Natural environments of Arizona: from deserts to mountains. University of Arizona Press, Tucson

deBuys W (1985) Enchantment and exploitation: the life and hard times of a New Mexico mountain range. University of New Mexico Press, Albuquerque

Dick-Peddie WA (1993) New Mexico vegetation: past, present, and future. University of New Mexico Press, Albuquerque

Dickson BG, Prather JW, Xu Y, Hampton HM, Aumack EN, Sisk TD (2006) Mapping the probability of large fire occurrence in northern Arizona, USA. Landsc Ecol 21:747–761
Dieterich JH (1980) Chimney Spring forest fire history. U.S. Department of Agriculture, Forest Service Research Paper RM-220, Rocky Mountain Forest and Range Experiment Station, Fort Collins, CO
Diffenbaugh NS, Pal JS, Trapp RJ, Giorgi F (2005) Fine-scale processes regulate the response of extreme events to global climate change. Proc Natl Acad Sci USA 102:15774–15778
Dugan AJ (2012) A landscape fire history and a modern calibration of fire-interval methods in a ponderosa pine forest across the South Rim of Grand Canyon National Park. Thesis, University of Wyoming, Laramie
Dukes JS, Mooney HA (1999) Does global change increase the success of biological invaders? Trends Ecol Evol 14:135–139
Ehleringer JR, Schwinning S, Gebauer R (2000) Water use in arid land ecosystems. In: Press MC, Scholes JD, Baker MG (eds) Physiological plant ecology. Proceedings of the 39th symposium of the British Ecological Society, 7–9 September 1998, University of York. Blackwell Science, Boston
Elton C (1958) The ecology of invasions by plants and animals. Methuen, London
Evett RS, Mohrle CR, Hall BL, Brown TJ, Stephens SL (2008) The effect of monsoonal atmospheric moisture on lightning fire ignitions in southwestern North America. Agric For Meteorol 148:1478–1487
Farris CA, Baisan CH, Falk DA, Yool SR, Swetnam TW (2010) Spatial and temporal corroboration of a fire-scar-based fire history in a frequently burned ponderosa pine forest. Ecol Appl 20:1598–1614
Fenn ME, Baron JS, Allen EB, Rueth HM, Nydick KR, Geiser L, Bowman WD, Sickman JO, Meixner T, Johnson DW, Neitlich P (2003a) Ecological effects of nitrogen deposition in the western United States. Bioscience 53:404–420
Fenn ME, Haeuber R, Tonnesen GS, Baron JS, Grossman-Clarke S, Hope D, Jaffe DA, Copeland S, Geiser L, Rueth HM, Sickman JO (2003b) Nitrogen emissions, deposition, and monitoring in the western United States. Bioscience 53:391–403
Ffolliott PF, Davis OK (eds) (2008) Natural environments of Arizona: from deserts to mountains. University of Arizona Press, Tucson
Floyd ML, Romme WH, Hanna DD (2000) Fire history and vegetation pattern in Mesa Verde National Park, Colorado, U.S.A. Ecol Appl 10:1666–1680
Frechette JD, Meyer GA (2009) Holocene fire-related alluvial-fan deposition and climate in ponderosa pine and mixed-conifer forests, Sacramento Mountains, New Mexico, USA. Holocene 19:639–651
Fritts HC (1976) Tree rings and climate. Academic, London
Fulé PZ, Laughlin DC (2007) Wildland fire effects on forest structure over an altitudinal gradient, Grand Canyon National Park, USA. J Appl Ecol 44:136–146
Fulé PZ, Heinlein TA, Covington WW, Moore MM (2000) Continuing fire regimes in remote forests of Grand Canyon National Park. In: Cole DN, McCool SF, Borrie WT, O'Loughlin F (compilers) Proceedings: wilderness science in a time of change conference, volume 5: wilderness ecosystems, threats, and management, 23–27 May 1999, Missoula, Montana. U.S. Department of Agriculture, Forest Service Proceedings RMRS-P-15-VOL-5, Rocky Mountain Research Station, Ogden, UT
Fulé PZ, Crouse JE, Heinlein TA, Moore MM, Covington WW, Verkamp G (2003) Mixed-severity fire regime in a high-elevation forest of Grand Canyon, Arizona, USA. Landsc Ecol 18:465–486
Fulé PZ, Crouse JE, Cocke AE, Moore MM, Covington WW (2004) Changes in canopy fuels and potential fire behavior 1880–2040: Grand Canyon, Arizona. Ecol Model 175:231–248
Fulé PZ, Heinlein TA, Covington WW (2006) Fire histories in ponderosa pine forests of Grand Canyon are well supported: reply to Baker. Int J Wildland Fire 15:439–445
Galloway JN, Aber JD, Erisman JW, Seitzinger SP, Howarth RW, Cowling EB, Cosby BJ (2003) The nitrogen cascade. Bioscience 53:341–356

Gottfried GJ, Ffolliott PF, DeBano LF (1995) Forests and woodlands of the sky islands: stand characteristics and silvicultural prescriptions. In: DeBano LF, Ffolliott PF, Ortega-Rubio A, Gottfried GJ, Hamre RH, Edminster CB (tech coords) Biodiversity and management of the Madrean archipelago: the sky islands of southwestern United States and northwestern Mexico. U.S. Department of Agriculture, Forest Service, General Technical Report RM-GTR-264, Rocky Mountain Forest and Range Experiment Station, Fort Collins, CO

Graybill DA, Rose MR (1992) Coniferous forests of Arizona and New Mexico. In: Olson RK, Binkley D, Böhm M (eds) The response of western forests to air pollution. Springer, New York

Grissino-Mayer HD (1996) A 2129-year reconstruction of precipitation for northwestern New Mexico, USA. In: Dean JS, Meko DM, Swetnam TW (eds) Tree rings, environment and humanity: proceedings of the international conference, Tucson, Arizona, 17–21 May 1994. Radiocarbon, Tucson, AZ

Grissino-Mayer HD, Swetnam TW (1995) Effects of habitat diversity of fire regimes in El Malpais National Monument, New Mexico. In: Brown JK, Mutch RW, Spoon CW, Wakimoto RH (tech coords) Proceedings: symposium on fire in wilderness and park management; 1993 March 30-April 1; Missoula, Montana. U.S. Department of Agriculture, Forest Service General Technical Report INT-GTR-320, Intermountain Research Station, Ogden, UT

Grissino-Mayer HD, Baisan CH, Swetnam TW (1995) Fire history in the Pinaleño Mountains of southeastern Arizona: effects of human-related disturbances. In: DeBano LF, Ffolliott PF, Ortega-Rubio A, Gottfried GJ, Hamre RH, Edminster CB (tech coords) Biodiversity and management of the Madrean archipelago: the sky islands of southwestern United States and northwestern Mexico. U.S. Department of Agriculture, Forest Service, General Technical Report RM-GTR-264, Rocky Mountain Forest and Range Experiment Station, Fort Collins, CO

Grissino-Mayer HD, Swetnam TW, Adams RK (1997) The rare, old-aged conifers of El Malpais: their role in understanding climatic change in the American Southwest. In: Mabery K (compiler) Natural history of El Malpais National Monument. Bulletin 156, New Mexico Bureau of Mines & Mineral Resources, Socorro

Grissino-Mayer HD, Romme WH, Floyd ML, Hanna DD (2004) Climatic and human influences on fire regimes of the southern San Juan Mountains, Colorado, USA. Ecology 85:1708–1724

Haire SL, McGarigal K (2009) Changes in fire severity across gradients of climate, fire size, and topography: a landscape ecological perspective. Fire Ecol 5:86–103

Hall BL (2007) Precipitation associated with lightning-ignited wildfires in Arizona and New Mexico. Int J Wildland Fire 16:242–254

Harrington MG, Hawksworth FG (1990) Interactions of fire and dwarf mistletoe on mortality of southwestern ponderosa pine. In: Krammes JS (tech coord) Effects of fire management of southwestern natural resources. U.S. Department of Agriculture, Forest Service General Technical Report RM-191, Rocky Mountain Forest and Range Experiment Station, Fort Collins, CO

Hasbargen J (1994) A Holocene paleoclimatic and environmental record from Stoneman Lake, Arizona. Quat Res 42:188–196

Haskett B (1935) Early history of the cattle industry in Arizona. Ariz Hist Rev 6:3–42

Haskett B (1936) History of the sheep industry in Arizona. Ariz Hist Rev 7:3–49

Hebertson EG, Jenkins MJ (2003) Historic climate factors associated with major avalanche years on the Wasatch Plateau, Utah. Cold Reg Sci Technol 37:315–332

Hebertson EG, Jenkins MJ (2008) Climate factors associated with historic spruce beetle (Coleoptera: Curculionidae) outbreaks in Utah and Colorado. Environ Entomol 37:281–292

Hendricks DM (1985) Arizona soils. University of Arizona Press, Tucson

Hereford R, Webb RH (1992) Historic variation of warm-season rainfall, southern Colorado Plateau, southwestern U.S.A. Clim Chang 22:239–256

Higgins RW, Yao Y, Wang WL (1997) Influence of the North American monsoon system on the United States summer precipitation regime. J Clim 10:2600–2622

Hoerling M, Eischeid J (2007) Past peak water in the Southwest. Southwest Hydrol 6:18–19,35

Houghton JT, Meira Filho LG, Callander BA, Harris N, Kattenberg A, Maskell K (eds) (1996) Climate change 1995: the science of climate change. Cambridge University Press, Cambridge

Integrated Taxonomic Information System (2011) http://www.itis.gov. Accessed 7 July 2011

Intergovernmental Panel on Climate Change (2007) Summary for policymakers. In: Solomon S, Qin D, Manning M, Chen Z, Marquis M, Averyt KB, Tignor M, Miller HL (eds) Climate change 2007: the physical science basis. Contribution of Working Group I to the Fourth Assessment Report of the Intergovernmental Panel on Climate Change, Cambridge University Press, Cambridge, UK/New York

Jenkins MJ, Hebertson E, Page W, Jorgensen CA (2008) Bark beetles, fuels, fires and implications for forest management in the Intermountain West. For Ecol Manag 254:16–34

Jenny H (1941) Factors of soil formation: a system of quantitative pedology. McGraw-Hill, New York

Kay CE (2007) Are lightning fires unnatural? A comparison of aboriginal and lightning ignition rates in the United States. In: Masters RE, Galley KEM (eds) Proceedings of the 23rd Tall Timbers Fire Ecology Conference: fire in grassland and shrubland ecosystems. Tall Timbers Research Station, Tallahassee

Kaye MW, Swetnam TW (1999) An assessment of fire, climate, and Apache history in the Sacramento Mountains, New Mexico. Phys Geogr 20:305–330

Klemmedson JO, Smith EL (1979) Vegetation-soil relations of forests, woodlands, and grasslands of Arizona and New Mexico. In: Youngberg CT (ed) Forest soils and land use. Department of Forest and Wood Sciences, Colorado State University, Fort Collins

Knight RL (2000) Forest fragmentation and outdoor recreation in the southern Rocky Mountains. In: Knight RL, Smith FW, Buskirk SW, Romme WH, Baker WL (eds) Forest fragmentation in the Southern Rocky Mountains. University Press of Colorado, Boulder

Kou X, Baker WL (2006a) Accurate estimation of mean fire interval for managing fire. Int J Wildland Fire 15:489–495

Kou X, Baker WL (2006b) A landscape model quantifies error in reconstructing fire history from scars. Landsc Ecol 21:735–745

Leiberg JB, Rixon TF, Dodwell A (1904) Forest conditions in the San Francisco Mountains Forest Reserve, Arizona. Department of the Interior, U.S. Geological Survey Series H, Forestry, 7, Professional Paper No. 22, Washington, DC

Lenart M (2007) Global warming in the Southwest: an overview. In: Lenart M (ed) Global warming in the Southwest: projections, observations and impacts. Climate Assessment Project for the Southwest, University of Arizona, Tucson

Link SO, Keeler CW, Hill RW, Hagen E (2006) *Bromus tectorum* cover mapping and fire risk. Int J Wildland Fire 15:113–119

Logan JA, Régnière J, Powell JA (2003) Assessing the impacts of global warming on forest pest dynamics. Front Ecol Environ 1:130–137

Lowe CH Jr (1961) Biotic communities in the Sub-Mogollon Region of the inland Southwest. J Ariz Acad Sci 2:40–49

Lowe CH (1964) Arizona's natural environment: landscapes and habitats. University of Arizona Press, Tucson

Lowe CH, Brown DE (1994) Introduction. In: Brown DE (ed) Biotic communities: southwestern United States and northwestern Mexico. University of Utah Press, Salt Lake City

Major J (1951) A functional, factorial approach to plant ecology. Ecology 32:392–412

Maker HJ, Saugherty LA (1986) Soils. In: Williams JL (ed) New Mexico in maps, 2nd edn. University of New Mexico Press, Albuquerque

Mann WG, Locke SB (1931) The Kaibab deer: a brief history and recent developments. United States Department of Agriculture, Forest Service, Washington, DC

Mantua NJ, Hare SR (2002) The Pacific decadal oscillation. J Oceanogr 58:35–44

Margolis EQ, Balmat J (2009) Fire history and fire-climate relationships along a fire regime gradient in the Santa Fe Municipal Watershed, NM, USA. For Ecol Manag 258:2416–2430

Margolis EQ, Swetnam TW, Allen CD (2007) A stand-replacing fire history in upper montane forests of the southern Rocky Mountains. Can J For Res 37:2227–2241

Marshall JD, Blair JM, Peters DPC, Okin G, Rango A, Williams M (2008) Predicting and understand ecosystem responses to climate change at continental scales. Front Ecol Environ 6:273–280

McGarigal K, Romme WH, Crist M, Roworth E (2001) Cumulative effects of roads and logging on landscape structure in the San Juan Mountains, Colorado (USA). Landsc Ecol 16:327–349

McGlone, CM (2010) Cheatgrass: native plant community interactions in an invaded southwestern forest. Dissertation, Northern Arizona University, Flagstaff

McGlone CM, Stoddard MT, Springer JD, Daniels ML, Fulé PZ, Covington WW (2012) Nonnative species influence vegetative response to ecological restoration: two forests with divergent restoration outcomes. For Ecol Manag 285:195–203

Mead P (1930) An ecological description of the Kaibab Plateau, Arizona. Thesis, University of Chicago

Meko DM, Stockton CW, Boggess WR (1995) The tree-ring record of severe sustained drought. Water Resour Bull 31:789–801

Menakis JP, Osborne D, Miller M (2003) Mapping the cheatgrass-caused departure from historical natural fire regimes in the Great Basin, USA. In: Omi PN, Joyce LA (tech eds) Fire, fuel treatments, and ecological restoration: Conference proceedings. U.S. Department of Agriculture, Forest Service, Rocky Mountain Research Station, RMRS-P-29, Fort Collins, CO

Merriam CH, Stejneger L (1890) Results of a biological survey of the San Francisco Mountain region and desert of the Little Colorado, Arizona. U.S. Department of Agriculture, Division of Ornithology and Mammalogy, North American Fauna No. 3, Government Printing Office, Washington, DC

Miller ME (2005) The structure and function of dryland ecosystems: conceptual models to inform long-term ecological monitoring. U.S. Geological Survey, Biological Resources Discipline, Scientific Investigations Report 2005–5197

Miller ME, Sharrow D, Cudlip L (2003) Northern Colorado Plateau vital signs network and prototype cluster, plan for natural resources monitoring. Phase II report. National Park Service, Northern Colorado Plateau Network, Moab, UT

Miller PR (1989) Concept of forest decline in relation to western U.S. forests. In: MacKenzie JJ, El-Ashry MT (eds) Air pollution's toll of forests and crops. Yale University Press, New Haven

Miller PR, McBride JR (eds) (1999) Oxidant air pollution impacts in the montane forests of southern California: a case study of the San Bernardino Mountains. Springer, New York

Miller PR, Stolte KW, Duriscoe DM, Pronos J (tech coords) (1995) Evaluating ozone air pollution effects on pines in the western United States. U.S. Department of Agriculture, Forest Service, General Technical Report PSW-GTR-155, Pacific Southwest Research Station, Albany, CA

Miller PR, Arbaugh MJ, Temple PJ (1997) Ozone and its known and potential effects on forests in Western United States. In: Sandermann H, Wellburn AR, Heath RL (eds) Forest decline and ozone: a comparison of controlled chamber and field experiments. Springer, New York

Mitchell J, Freeman D (1993) Wildlife-livestock-fire interactions on the North Kaibab: a historical review. U.S. Department of Agriculture, Forest Service General Technical Report RM-222, Rocky Mountain Forest and Range Experiment Station, Fort Collins, CO

Murdock L (1994) Analysis of woodrat (Neotoma) middens found in Walnut Canyon, Coconino County, Arizona. Thesis, Northern Arizona University, Flagstaff

Naiman RJ, Johnston CA, Kelley JC (1988) Alteration of North American streams by beaver. Bioscience 38:753–762

National Park Service (1987) National Forest fire lookouts in the Southwestern Region, USDA Forest Service, National register of historic places: inventory and nomination form. U.S. Department of the Interior, National Park Service. http://pdfhost.focus.nps.gov/docs/NRHP/Text/64000046.pdf. Accessed 17 June 2012

National Park Service (2002) Air quality in the national parks, 2nd edn. U.S. Department of the Interior, National Park Service, Air Resources Division, Washington, DC

National Park Service (2010) Air quality in National Parks: 2009 annual performance and progress report. Natural Resource Report NPS/NRPC/ARD/NRR-2010/266. U.S. Department of the Interior, National Park Service, Natural Resource Program Center, Air Resources Division, Denver, CO. http://www.nature.nps.gov/air/pubs/pdf/gpra/AQ_Trends_In_Parks_2009_Final_Web.pdf. Accessed 23 June 2011

National Park Service (2012) NPS stats. Department of the Interior, National Park Service, Public Use Statistics Office, Denver, CO. http://www.nature.nps.gov/stats/park.cfm. Accessed 4 Sept 2012

National Weather Service (2011) Frequently asked questions about El Niño and La Niña. Climate Prediction Center http://www.cpc.ncep.noaa.gov/products/analysis_monitoring/ensostuff/ensofaq.shtml#HOWOFTEN. Accessed 24 June 2011

Nelson GC, Bennett E, Berhe AA, Cassman K, DeFries R, Dietz T, Dobermann A, Dobson A, Janetos A, Levy M, Marco D, Nakicenovic N, O'Neill B, Norgaard R, Petschel-Held G, Ojima D, Pingali P, Watson R, Zurek M (2006) Anthropogenic drivers of ecosystem change: an overview. Ecology and Society 11:29, http://www.ecologyandsociety.org/vol11/iss2/art29/. Accessed 4 Sept 2012

O'Dell TE, Garman SL, Evenden A, Beer M, Nance E, Daw S, Wight A, Powell MA, Perry D, DenBleyker R, Sharrow D, Wynn KH, Brown JB, Miller ME, Thomas L (2005) Vital signs monitoring plan. Northern Colorado Plateau Inventory and Monitoring Network, National Park Service, Moab

Orcutt JD (1999) Demography, settlement, and agriculture. In: Powers RP, Orcutt JD (eds) The Bandelier archeological survey. Vols. 1 and 2. U.S. Department of the Interior, National Park Service, Intermountain Cultural Resources Management Professional Paper No. 57, Washington, DC

Paine TD, Raffa KF, Harrington TC (1997) Interactions among Scolytid bark beetles, their associated fungi, and live host conifers. Annu Rev Entomol 42:179–206

Peet RK (2000) Forests and meadows of the Rocky Mountains. In: Barbour MG, Billings WD (eds) North American terrestrial vegetation, 2nd edn. Cambridge University Press, New York

Petersen KL (1994) Modern and Pleistocene climatic patterns in the West. In: Harper KT, StClair LL, Thorne KH, Hess WM (eds) Natural history of the Colorado plateau and Great Basin. University Press of Colorado, Niwot

Peterson CS (1975) Look to the mountains: southeastern Utah and the La Sal National Forest. Brigham Young University Press, Provo

Phillips AM III, House DA, Phillips BG (1989) Expedition to the San Francisco Peaks: C. Hart Merriam and the life zone concept. Plateau vol 60, no 2, Museum of Northern Arizona, Flagstaff

Pinchot G (1947) Breaking new ground. Harcourt, Brace & Company, New York, New York; Commemorative Edition 1998 reprint. Island Press, Washington, DC

Powell JW (1890) The non-irrigable lands of the arid region. Century Mag 39:915–922

Prior-Magee JS, Boykin KG, Bradford DF, Kepner WG, Lowry JH, Schrupp DL, Thomas KA, Thompson BC (2007) Ecoregional gap analysis of the southwestern United States: the Southwest regional gap analysis project final report. http://fws-nmcfwru.nmsu.edu/swregap/report/SWReGAP%20Final%20Report.pdf. Accessed 23 June 2012

Pyne SJ (1982) Fire in America: a cultural history of wildland and rural fire. Princeton University Press, Princeton

Pyne SJ (2001) Fire: a brief history. University of Washington Press, Seattle

Raish C, Yong W, Marzluff J (1997) Contemporary human use of southwestern ponderosa pine forests. In: Block WM, Finch DM (tech eds) Songbird ecology in southwestern ponderosa pine forests: a literature review. U.S. Department of Agriculture, Forest Service General Technical Report RM-GTR-292, Rocky Mountain Forest and Range Experiment Station, Fort Collins, CO

Randall JM (1996) Weed control for the preservation of biodiversity. Weed Technol 10:370–383

Rasmussen DI (1941) Biotic communities of Kaibab Plateau, Arizona. Ecol Monogr 11:229–275

Rejmánek M (1989) Invasibility of plant communities. In: Drake JA, Mooney HA, di Castri F, Groves RH, Kruger FJ, Rejmánek M, Williamson M (eds) Biological invasions: a global perspective. Wiley, New York

Rejmánek M, Randall JM (1994) Invasive Plants in California: 1993 summary and comparison with other areas in North America. Madroño 41:161–177

Reynolds AC, Betancourt JL, Quade J, Patchett PJ, Dean JS, Stein J (2005) $^{87}SR/^{86}SR$ sourcing of ponderosa pine used in Anasazi great house construction at Chaco Canyon, New Mexico. J Archaeol Sci 32:1061–1075

Romme WH, Floyd-Hanna L, Hana D, Grissino-Mayer H (1994) Pre-settlement range of natural variation in disturbance history and stand structure of ponderosa pine and mixed conifer forests on the San Juan National Forest: progress report for the first year's work. Unpublished report to the San Juan National Forest. Fort Lewis College, Durango

Romme WH, Clement J, Hicke JA, Kulakoswki D, MacDonald LH, Schoennagel T, Veblen TT (2006) Recent forest insect outbreaks and fire risk in Colorado forests: a brief synthesis of relevant research. Colorado Forest Restoration Institute, Colorado State University, Fort Collins. http://spot.colorado.edu/~schoenna/images/RommeEtAl2006CFRI%20.pdf. Accessed 3 July 2011

Roos CI, Sullivan AP III, McNamee C (2010) Paleoecological evidence for systematic indigenous burning in the upland Southwest. In: Dean RM (ed) The archaeology of anthropogenic environments. Center for Archaeological Investigations, Occasional Paper No. 37. Southern Illinois University, Carbondale

Salzer MW, Kipfmueller KF (2005) Reconstructed temperature and precipitation on a millennial timescale from tree-rings in the southern Colorado Plateau, U.S.A. Clim Chang 70:465–487

Savage M, Swetnam TW (1990) Early 19th-century fire decline following sheep pasturing in a Navajo ponderosa pine forest. Ecology 71:2374–2378

Sawyer GS (1976) The history of fire control on the Coconino National Forest. Thesis, Northern Arizona University, Flagstaff

Schlegel PA (1992) A history of the cattle industry in northern Arizona, 1863–1912. Thesis, Northern Arizona University, Flagstaff

Schreiber KV (1996) A synoptic climatological approach to assessment of visibility and pollutant source locations, Grand Canyon National Park area. Dissertation, University of Delaware, Newark

Scurlock D, Finch DM (1997) A historical review. In: Block WM, Finch DM (tech eds) Songbird ecology in southwestern ponderosa pine forests: a literature review. U.S. Department of Agriculture, Forest Service General Technical Report RM-GTR-292. Rocky Mountain Forest and Range Experiment Station, Fort Collins, CO

Seager R, Vecchi GA (2010) Greenhouse warming and the 21st century hydroclimate of southwestern North America. Proc Natl Acad Sci USA 107:21277–21282

Seager R, Ting M, Held I, Kushnir Y, Lu J, Vecchi G, Huang H-P, Harnik N, Leetmaa A, Lau N-C, Li C, Velez J, Naik N (2007) Model projections of an imminent transition to a more arid climate in southwestern North America. Science 316:1181–1184

Seklecki MT, Grissino-Mayer GT, Swetnam TW (1996) Fire history and the possible role of Apache-set fires in the Chiricahua Mountains of southeastern Arizona. In: Ffolliott PF, DeBano LF, Baker MB, Gottfried GJ, Solis-Garza G, Edminster CB, Neary DG, Allen LS, Hamre RH (tech coords) Effects of fire on Madrean Province ecosystems: a symposium proceedings, 11–15 Mar 1996. Tucson, Arizona. U.S. Department of Agriculture, Forest Service General Technical Report RM-GTR-289, Rocky Mountain Forest and Range Experiment Station, Fort Collins, CO

Smith SD, Huxman TE, Zitzer SF, Charlet TN, Housman DC, Coleman JS, Fenstermaker LK, Seemann JR, Nowak RS (2000) Elevated CO_2 increases productivity and invasive species success in an arid ecosystem. Nature 408:79–82

Spence JR (2001) Climate of the central Colorado Plateau, Utah and Arizona: characterization and recent trends. In: van Riper C III, Thomas KA, Stuart MA (eds) Proceedings of the fifth biennial conference of research on the Colorado Plateau. U.S. Geological Survey/FRESC Report Series USGSFRESC/COPL/2001/24

Spence JR, Romme WH, Floyd-Hanna L, Rowlands, PG (1995) A preliminary vegetation classification for the Colorado Plateau. In: van Riper C III (ed) Proceedings of the second biennial conference on research in Colorado Plateau National Parks, 25–28 Oct 1993. National Park Service Transactions and Proceedings Series NPS/NRNAU/NRTP-95/11

St. George S, Meko DM, Cook ER (2010) The seasonality of precipitation signals embedded within the North American Drought Atlas. Holocene 20:983–988

Stephens SL, Ruth LW (2005) Federal forest-fire policy in the United States. Ecol Appl 15: 532–542

Swetnam TW (1990) Fire history and climate in the southwestern United States. In: Krammes JS (tech coord) Effects of fire management of southwestern natural resources. U.S. Department of Agriculture, Forest Service General Technical Report RM-191, Rocky Mountain Forest and Range Experiment Station, Fort Collins, CO

Swetnam TW, Baisan CH (1996) Historical fire regime patterns in the southwestern United States since AD 1700. In: Allen CD (tech ed) Fire effects in southwestern forests: Proceedings of the Second La Mesa Fire Symposium, Los Alamos, New Mexico, 29–31 Mar 1994. U.S. Department of Agriculture, Forest Service General Technical Report RM-GTR-286, Rocky Mountain Forest and Range Experiment Station, Fort Collins, CO

Swetnam TW, Baisan CH (2003) Tree-ring reconstructions of fire and climate history in the Sierra Nevada and southwestern United States. In: Veblen TT, Baker WL, Montenegro G, Swetnam TW (eds) Fire and climatic change in temperate ecosystems of the western Americas. Springer, New York

Swetnam TW, Betancourt JL (1990) Fire-Southern oscillation relations in the southwestern United States. Science 249:1017–1020

Swetnam TW, Betancourt JL (1998) Mesoscale disturbance and ecological response to decadal climatic variability in the American Southwest. J Clim 11:3128–3147

Swetnam TW, Allen CD, Betancourt JL (1999) Applied historical ecology: using the past to manage for the future. Ecol Appl 9:1189–1206

Takemoto BK, Bytnerowicz A, Fenn ME (2001) Current and future effects of ozone and atmospheric nitrogen deposition on California's mixed conifer forests. For Ecol Manag 144:159–173

The Nature Conservancy (2006) Historical range of variation and state and transition modeling of historic and current landscape conditions for potential natural vegetation types of the Southwest. http://azconservation.org/downloads/category/sw_forest_assessment/. Accessed 28 June 2011

Theobald DM (2000) Fragmentation by inholdings and exurban development. In: Knight RL, Smith FW, Buskirk SW, Romme WH, Baker WL (eds) Forest fragmentation in the Southern Rocky Mountains. University Press of Colorado, Boulder

Thomas LP, Hendrie MN (ed), Lauver CL, Monroe SA, Tancreto NJ, Garman SL, Miller ME (2006) Vital signs monitoring plan for the Southern Colorado Plateau Network. Natural Resource Report NPS/SCPN/NRR-2006/002. National Park Service, Fort Collins, CO

Thompson RS, Whitlock C, Bartlein PJ, Harrison SP, Spaulding WG (1993) Climatic changes in the western United States since 18,000 yr B.P. In: Wright HE Jr, Kutzbach JE, Webb T III, Ruddiman WF, Street-Perrott FA, Bartlein PJ (eds) Global climates since the last glacial maximum. University of Minnesota Press, Minneapolis

Touchan R, Swetnam TW, Grissino-Mayer HD (1995) Effects of livestock grazing on presettlement fire regimes in New Mexico. In: Brown JK, Mutch RW, Spoon CW, Wakimoto RH (tech coords) Proceedings: symposium on fire in wilderness and park management; Mar 30–Apr 1 1993; Missoula, Montana. U.S. Department of Agriculture, Forest Service General Technical Report INT-GTR-320, Intermountain Research Station, Ogden, UT

U.S. Forest Service (2010) 2010 tornado event: large scale event recovery plan. U.S. Department of Agriculture, Forest Service, Southwest Region, Coconino National Forest, Flagstaff/Mogollon Ranger Districts. http://a123.g.akamai.net/7/123/11558/abc123/forestservic.download.akamai.com/11558/www/nepa/75678_FSPLT2_032314.pdf. Accessed 15 Feb 2012

U.S. Forest Service (2011) Forest insect and disease conditions in the Southwestern Region, 2010. U.S. Department of Agriculture, Forest Service Southwestern Region Forestry and Forest Health PR-R3-16-7. http://www.fs.usda.gov/Internet/FSE_DOCUMENTS/stelprdb5316384.pdf. Accessed 5 Mar 2012

U.S. Geological Survey National Gap Analysis Program (2005) Southwest Regional Gap Analysis Project, RS/GIS Laboratory, College of Natural Resources, Utah State University, Provo, Utah

Vale TR (1998) The myth of the humanized landscape: an example from Yosemite National Park. Nat Areas J 18:231–236

Vale TR (2002) The pre-European landscape of the United States: pristine or humanized? In: Vale TR (ed) Fire, native peoples, and the natural landscape. Island Press, Covelo

Van Devender TR (1977) Holocene woodlands in the southwestern deserts. Science 198:189–192

Van Devender TR, Spaulding WG (1979) Development of climate and vegetation in the southwestern United States. Science 204:701–710

Van Horne ML, Fulé PZ (2006) Comparing methods of reconstructing fire history using fire scars in a southwestern United States ponderosa pine forest. Can J For Res 36:855–867

Vankat JL (2010) Forest changes in Grand Canyon National Park and implications for management. Report to U.S. Department of the Interior, National Park Service, Grand Canyon National Park, Coconino

Vankat JL, Roy DG (2002) Landscape invasibility by exotic species. In: Gutzwiller K (ed) Applying landscape ecology in biological conservation. Springer, New York

Vitousek PM, D'Antonio CM, Loope LL, Westbrooks R (1996) Biological invasions as global environmental change. Am Sci 84:468–478

Wadleigh LL, Parker C, Smith B (1998) A fire frequency and comparative fuel load analysis in Gambel oak of northern Utah. In: Pruden TL, Brennan LA (eds) Fire in ecosystem management: shifting the paradigm from suppression to prescription. Tall timbers fire ecology conference proceedings, no. 20. Tall Timbers Research Station, Tallahassee, FL

Wagner DJ, Baker FA (2003) Potential effects of ozone, climate, and spruce budworm on Douglas-fir growth in the Wasatch Mountains. Can J For Res 33:910–921

Weng C, Jackson ST (1999) Late Glacial and Holocene vegetation history and paleoclimate of the Kaibab Plateau, Arizona. Palaeogeogr Palaeoclimatol Palaeoecol 153:179–201

Westerling AL, Hidalgo HG, Cayan DR, Swetnam TW (2006) Warming and earlier spring increase western U.S. forest wildfire activity. Science 313:940–943

Western Regional Climate Center (2011) Arizona climate summaries. http://www.wrcc.dri.edu/cgi-bin/cliMAIN.pl?azbrig. Accessed 24 June 2011

White MA, Vankat JL (1993) Middle and high elevation coniferous forest communities of the North Rim region of Grand Canyon National Park, Arizona, USA. Vegetatio 109:161–174

White PS, Pickett STA (1985) Natural disturbance and patch dynamics: an introduction. In: Pickett STA, White PS (eds) The ecology of natural disturbance and patch dynamics. Academic, Orlando

Wildeman G, Brock JH (2000) Grazing in the Southwest: history of land use and grazing since 1540. In: Jemison R, Raish C (eds) Livestock management in the American Southwest: ecology, society, and economics. Elsevier, Amsterdam

Williams AP, Allen CD, Millar CI, Swetnam TW, Michaelsen J, Still CJ, Leavitt SW (2010) Forest responses to increasing aridity and warmth in the southwestern United States. Proc Natl Acad Sci USA 107:21289–21294

Williams AP, Allen CD, Macalady AK, Griffin D, Woodhouse CA, Meko DM, Swetnam TW, Rauscher SA, Seager R, Grissino-Mayer HD, Dean JS, Cook ER, Gangodagamage C, Cai M, McDowell NG (2012) Temperature as a potent driver of regional forest drought stress and tree mortality. Nat Clim Chang. doi:10.1038/NCLIMATE1693. Published online: 30 September 2012

Woolsey TS Jr (1911) Western yellow pine in Arizona and New Mexico. U.S. Department of Agriculture, Forest Service Bulletin 101. Washington, DC

Wright JP, Gurney WSC, Jones CG (2004) Patch dynamics in a landscape modified by ecosystem engineers. Oikos 105:336–348

Wurtz MJ (1991) The historical geography of the Colorado Plateau 1539–1991. Thesis, Northern Arizona University, Flagstaff

Part I
Forests

Chapter 2
Spruce-Fir Forest

Abstract Spruce-Fir Forest occurs in the subalpine zone, at the highest elevation of any major forest in the American Southwest. It is dominated by Engelmann spruce, either subalpine or corkbark fir, and quaking aspen. Most research on Spruce-Fir Forest dynamics is from the central and northern Rocky Mountains, and evidence from the Southwest indicates both similarities and differences. Important drivers of vegetation dynamics are a diverse disturbance regime that is dominated by fire, wind, insects, climate variation, and anthropogenic disturbances that include livestock grazing, fire management, and nearby land use. Historical fire regimes were crown-fire in upper elevations and mixed-severity in at least some lower-elevation sites. Key processes of vegetation dynamics are succession and gap dynamics. Historical descriptions indicate Spruce-Fir Forest was dense before Euro-American influence. Fire exclusion began in the late nineteenth century. Its impacts are unknown for upper-elevation stands, but likely initially involved increases in tree density and basal area in lower-elevation stands. At least some lower-elevation stands decreased in density and basal area during the twentieth century, apparently as a result of density-dependent factors and exogenous factors such as climate change. Vegetation dynamics are illustrated in a nested, three-tiered set of conceptual models. Key conclusions and challenges for researchers and land managers are summarized.

2.1 Introduction

Spruce-Fir Forest (*Picea-Abies*) occurs in the subalpine zone, at the highest elevation of any major forest in the American Southwest (Figs. 2.1 and 2.2). On mountains that exceed its elevational range, Spruce-Fir Forest is bounded at high elevation by treeline and alpine tundra. Mixed Conifer Forest is typically adjacent at lower elevation. Spruce-Fir Forest is uncommon in the American Southwest, covering about 8,900 km^2 (3,400 miles^2), which is 1.1 % of the total area of the region (Fig. 2.3; calculations based on Prior-Magee et al. 2007). These values do not

J.L. Vankat, *Vegetation Dynamics on the Mountains and Plateaus of the American Southwest*, Plant and Vegetation 8, DOI 10.1007/978-94-007-6149-0_2,

Fig. 2.1 Spruce-Fir Forest in the San Juan Mountains of southwestern Colorado. Subalpine-Montane Grassland is in the foreground (Photograph by Laurie Thompson)

Fig. 2.2 Mosaic of Spruce-Fir Forest and Subalpine-Montane Grassland in central Utah (Photograph by author)

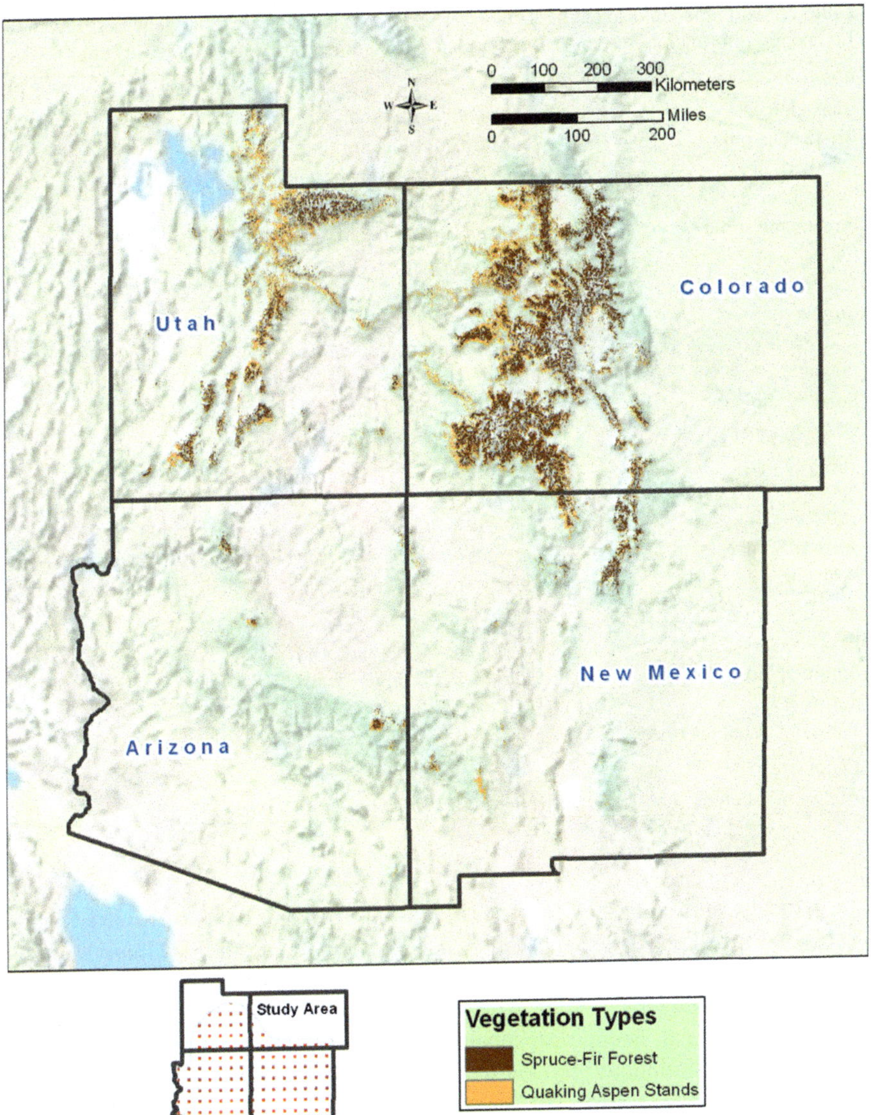

Fig. 2.3 Distribution of Spruce-Fir Forest (including subalpine and montane stands of quaking aspen) in the American Southwest. The map shows all of the states of Arizona, New Mexico, Colorado, and Utah, and the American Southwest is illustrated in *red* on the small map (Source: U.S. Geological Survey National Gap Analysis Program (2005). Map prepared by Monica Swihart)

include stands dominated by quaking aspen, which cover nearly the same area: approximately 8,100 km^2 (3,100 miles^2) and 1.0 %. Approximately 35–40 % of aspen stands are associated with Spruce-Fir Forest (most of the rest with Mixed Conifer Forest). The area covered by Spruce-Fir Forest increases northward. Stands

Table 2.1 Common and scientific names of species in this chapter. Primary sources: Integrated Taxonomic Information System (2012) for plants and animals and Bates (2006) for fungi

Plants	
Blue spruce	*Picea pungens* Engelm.
Bristlecone pine	*Pinus aristata* Engelm.
Corkbark fir	*Abies lasiocarpa* var. *arizonica* (Merriam) Lemmon
Douglas-fir	*Pseudotsuga menziesii* (Mirbel) Franco
Engelmann spruce	*Picea engelmannii* Parry ex Engelm.
Fir	*Abies* P. Mill.
Juniper	*Juniperus* L.
Limber pine	*Pinus flexilis* James
Lodgepole pine	*Pinus contorta* var. *latifolia* (Engelm. ex S. Wats.) Boivin
Pinyon	*Pinus* L.
Ponderosa pine	*Pinus ponderosa* Douglas ex P. Lawson & C. Lawson
Quaking aspen	*Populus tremuloides* Michx.
Sedge	*Carex* L.
Spruce	*Picea* A. Dietr.
Subalpine fir	*Abies lasiocarpa* var. *lasiocarpa* (Hook.) Nutt.
Western spruce dwarf mistletoe	*Arceuthobium microcarpum* (Engelm.) Hawksworth & Wiens
White fir	*Abies concolor* (Gord. & Glend.) Lindl. Ex Hildebr.
Animals	
Deer	*Odocoileus* Rafinesque, 1832
Spruce aphid	*Elatobium abietinum* (Walker, 1849)
Spruce beetle	*Dendroctonus rufipennis* (Kirby, 1837)
Western balsam bark beetle	*Dryocoetes confuses* Swaine, 1912
Fungi	
Annosum root rot	*Heterobasidion annosum* (Fr.) Bref.

are typically surrounded by or at least adjacent to larger areas of Mixed Conifer Forest. Spruce-Fir Forest landscapes also include stands of Subalpine-Montane Grassland. Subalpine landscapes of Arizona, New Mexico, and southwestern Colorado generally have Spruce-Fir Forest with scattered stands of Subalpine-Montane Grassland. In contrast, subalpine landscapes of southern and central Utah often have less continuous Spruce-Fir Forest and more extensive stands of Subalpine-Montane Grassland (Ellison 1954).

Stands are dominated by Engelmann spruce (*Picea engelmannii*), often with either subalpine fir (*Abies lasiocarpa* var. *lasiocarpa*) or corkbark fir (*Abies lasiocarpa* var. *arizonica*). The other important tree species is quaking aspen (*Populus tremuloides*), which occurs primarily in lower-elevation Spruce-Fir Forest, either as a codominant or dominant.

Stand structure and composition are influenced by elevation and other topographic factors, as well as by several natural disturbances, including fire, wind, and insects. Disturbances and vegetation dynamics such as succession result in landscape mosaics of stands with different structure and composition (Leiberg et al. 1904; Niering and Lowe 1984; Moir 1993; Patten and Stromberg 1995).

The diversity of stand composition of Spruce-Fir Forest has been detailed in vegetation classifications for different geographic areas within the American Southwest (e.g., Layser and Schubert 1979; Moir and Ludwig 1979; Youngblood and Mauk 1985; DeVelice et al. 1986; Mueggler 1988; Muldavin et al. 1996). This chapter uses a generalized vegetation classification that divides stands into three broad types: upper-elevation, lower-elevation, and quaking aspen.

Upper-elevation Spruce-Fir Forest (Fig. 2.4) is dominated by Engelmann spruce and subalpine or corkbark fir. Engelmann spruce is more common on relatively dry sites (Niering and Lowe 1984) and in the canopy of old-growth stands, in part because of larger maximum size and a longer life span (Alexander 1987). Subalpine fir is more common on mesic sites (Niering and Lowe 1984) and at lower elevations (Pfister 1972). It is shade-tolerant and generally predominates in smaller size classes.

Lower-elevation stands (Fig. 2.5) are usually dominated by the same three species, but typically have more quaking aspen and also include species such as Douglas-fir (*Pseudotsuga menziesii*), white fir (*Abies concolor*), and ponderosa pine (*Pinus ponderosa*; Moir 1993; Pase and Brown 1994). The presence of these additional species reflects a variety of factors, including surface fires before fire exclusion occurred and transition with Mixed Conifer Forest. Blue spruce (*Picea pungens*) also can be present, especially in canyons, valley bottoms, and along meadow margins (Pfister 1972; White and Vankat 1993; Pase and Brown 1994). Stands on southern aspects and shallow, rocky soils are more open and have greater abundance of quaking aspen (Pfister 1972). Lower-elevation Spruce-Fir Forest is sometimes treated as high-elevation cold-wet Mixed Conifer Forest (e.g., Romme et al. 2009a).

Quaking aspen stands of Spruce-Fir Forest (Fig. 2.6) occur primarily in sites burned by crown fire. They are favored by conditions that limit successional replacement of aspen by conifers, such as long distance to conifer seed sources (Sect. 2.3.1). They tend to be on deep, relatively fertile soils (Jones and DeByle 1985). Some stands form by aspen invasion of Montane and Subalpine Grasslands (Moir 1993; Moore and Huffman 2004). Aspen stands also occur in Mixed Conifer Forest (Sect. 3.1) and in mesic Ponderosa Pine Forest (Sect. 4.1).

Little research has been done on Spruce-Fir Forest in the American Southwest. This forest type is much more widespread in the central and northern Rocky Mountains (see Fig. 2.3), where it has been well-studied. The common – albeit usually unspoken – assumption that research from the Rocky Mountains applies across the Southwest is questionable because of differences in climate, species composition, and possibly species biology. Climatic differences between regions include timing of precipitation. The Southwest has dry springs and early summers followed by monsoonal precipitation in mid and late summers (Sect. 1.2.2). Precipitation in the central and northern Rocky Mountains is more consistent throughout spring and summer. Such differences potentially influence fire patterns, tree regeneration, etc., but have not been studied. Regional differences in species composition include lodgepole pine (*Pinus contorta* var. *latifolia*), which is absent from the Southwest but dominant on many lower- and mid-elevation sites in the central and northern

Fig. 2.4 Stand of upper-elevation Spruce-Fir Forest in central Utah (Photograph by author)

Fig. 2.5 Stand of lower-elevation Spruce-Fir Forest in central Utah (Photograph by author)

Fig. 2.6 Stand of quaking aspen on the San Francisco Peaks, north-central Arizona (Photograph by Betty J. Huffman)

Rocky Mountains (e. g., Peet 2000). Its absence at higher elevations in the central and northern Rocky Mountains suggests such upper-elevation stands have greater similarity between regions. Regional differences in species biology probably include physiological and ecological differences between corkbark fir (occurs only in the Southwest) and subalpine fir, as well as possible regional differences between populations of Engelmann spruce (Romme et al. 2009b).

With little research on Spruce-Fir Forest in the American Southwest, this chapter includes research findings from the central and northern Rocky Mountains where parallel information from the Southwest is lacking. This research is identified by its region.

2.2 Drivers

Key drivers of southwestern Spruce-Fir Forest are landscape, climate, soil, animals, natural disturbance, and anthropogenic disturbance. Each driver is important in vegetation dynamics and therefore is incorporated in the conceptual models (Sect. 2.5).

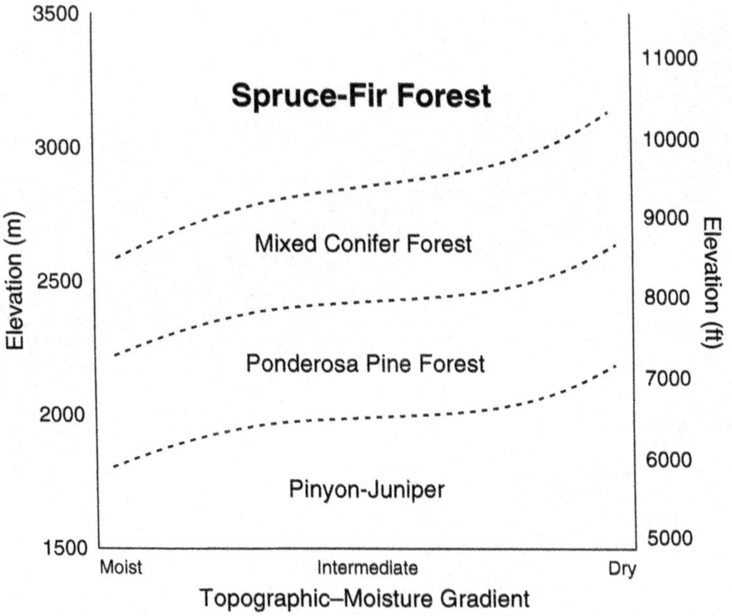

Fig. 2.7 Ecological distribution of Spruce-Fir Forest on the mountains and plateaus of the American Southwest along gradients in elevation and moisture. Elevations are approximate and generally representative of mid-latitudes of the region, i.e., landscapes in northern Arizona and northern New Mexico. Higher elevations – where present – have tree line and alpine tundra vegetation

2.2.1 *Landscape*

Spruce-Fir Forest occurs from approximately 2,500 to 3,700 m (8,200 to 12,100 ft) in the mid-latitudes of its distribution in the American Southwest (Fig. 2.7). The highest mountains capped with alpine tundra often have a treeline plant community of bristlecone pine (*Pinus aristata*) or Engelmann spruce that intergrades into Spruce-Fir Forest. Local topographic features such as slope aspect and steepness influence the position and characteristics of the treeline. The lower elevational limit of Spruce-Fir Forest is less clear because of a patchy transition with Mixed Conifer Forest in which stands of Spruce-Fir Forest become increasingly limited to moist sites such as valley bottoms and north aspects (Lang and Stewart 1910; White and Vankat 1993; Fulé et al. 2003). Historically, fire interacted with local topography to produce the transitional mosaic (Fulé et al. 2003). The effects of local topography are also evident in the grassland-dominated subalpine landscapes of southern and central Utah. Here, Spruce-Fir Forest occurs primarily on north-facing aspects and is otherwise largely limited to scattered patches in the grassland matrix (Ellison 1954). In addition, open stands dominated by limber pine (*Pinus flexilis*) occur on ridges and steep south-facing aspects (Ellison 1954; Pfister 1972).

2.2.2 Climate

Few climate data have been published for high-elevation sties in the American Southwest. The climate of the region's Spruce-Fir Forest is characterized by cool to cold temperatures. Mean annual temperature is 2–3 °C (35–36 °F), ranging from −6 to −11 °C (12 to 21 °F) in January to 10–16 °C (50–61 °F) in June–July (Pearson 1920a; Pfister 1972; Alexander 1987). The cool temperatures result in a short growing season averaging less than 80 days (Price and Evans 1937; Alexander 1987; Pase and Brown 1994); individual growing seasons can range from 54 to 117 days (Pearson 1920a; Price and Evans 1937). The diurnal range of temperatures is typically less than 11 °C (20 °F; Pearson 1920b). Temperature is influenced by topographic factors such as elevation and slope aspect.

Mean annual precipitation is 61 to >100 cm (24 to >39 in.; Pfister 1972; Alexander 1987; Pase and Brown 1994). Snowfall annually averages 3 to >5 m (10 to >16 ft; Alexander 1987; Chambers and Holthausen 2000), accumulates up to 3.7 m (12 ft) in the winter, and can persist into June in cooler microsites (Pearson 1931; Ellison 1954; Hanley et al. 1975). Extreme droughts occur occasionally (Alexander 1987). Precipitation is influenced by elevation.

Lightning is common. For example, it caused an average of 0.8 fires per year in a 4,400 ha (10,873 acres) area of mostly Spruce-Fir Forest at relatively low elevation in the North Rim region of Grand Canyon National Park, north-central Arizona (Fulé et al. 2003; during 1967–1996).

Modern climate change is described in Sect. 1.6.3 and is treated as an anthropogenic disturbance in Sect. 2.2.6.3.

2.2.3 Soil

Few data are available on soils of southwestern Spruce-Fir Forest (but see Pfister 1972 and Dye and Moir 1977). Subalpine soils in the Rocky Mountains vary according to the rock parent material, but are generally relatively deep, permeable, and capable of storing snowmelt (Leaf 1975). Southwestern subalpine soils tend to be deeper at lower elevations (Moir 1993) and have thick duff (Pase and Brown 1994). Mean depth of litter and duff combined was measured as 5.4 cm (2.1 in.) on the San Francisco Peaks in north-central Arizona, the deepest of any forest in this location (Cocke et al. 2005). The soil moisture regime tends to be udic, and the soil temperature regime is cryic (Klemmedson and Smith 1979; DeVelice et al. 1986; Moir 1993), with mean annual temperatures of approximately 2 °C (35 °F; Pearson 1920a).

Soils of Spruce-Fir Forest in Arizona are Typic Argiborolls, Argic Pachic Cryoborolls, Typic Glossoboralfs, and Lithic Cryoborolls that range from shallow to deep, fine- to medium-textured, and moderately to well-drained (Hendricks 1985). Most subalpine soils in southern New Mexico are formed from intrusive parent materials and are characterized as coarse-loamy Pachic Cryoborolls with dark A horizons (Dye and Moir 1977). Surface soils of 0–2 cm (0–1 in.) depth in old-growth stands have higher organic matter, pH, nitrogen, phosphorus, calcium, manganese, sodium, and potassium than in younger stands. Subalpine soils in Utah form from a variety of parent materials and tend to be poorly developed (Pfister 1972). Soil parent material appears of little importance to the distribution of subalpine vegetation (Ellison 1954), although relationships between soils and vegetation types are more apparent locally than regionally (Pfister 1972).

The soil system also includes annosum root rot (*Heterobasidion annosum*), which affects subalpine fir (Pfister 1972; see Sect. 2.2.5.3).

2.2.4 Animals

The animals most important to vegetation dynamics of southwestern Spruce-Fir Forest are insects, especially spruce beetle (*Dendroctonus rufipennis*) and western balsam bark beetle (*Dryocoetes confuses*). Deer (*Odocoileus*spp.) are the common large ungulates. The impacts of animals on vegetation dynamics are described in Sect. 2.2.5.3.

2.2.5 Natural Disturbance

Spruce-Fir Forest in the central and northern Rocky Mountains has infrequent, stand- to landscape-scale disturbances caused by crown fire, wind, insects, and climate variation. These disturbance agents are also present in the American Southwest, but a similar disturbance regime is less well documented.

2.2.5.1 Fire

There has been relatively little research on the historical fire regime of southwestern Spruce-Fir Forests (Moir 1993; Swetnam and Baisan 1996; Allen 2002). There are several reasons for this. First, Spruce-Fir Forest is uncommon. Second, modern fires in this type of forests are uncommon, accounting for <1 % of all lightning fires in forests and woodlands of the National Forests of Arizona and New Mexico (calculated from data in Barrows 1978 for 1960–1974). Third, fire scars, which are a source of data on fire regimes (see Sect. 1.2.5.1) are scarce (Swetnam et al. 2009), because both Engelmann spruce and subalpine fir are thin-barked and unlikely to survive fires (Bradley et al. 1992; Veblen et al. 1994). Nevertheless, fire has been and is an important driver of southwestern Spruce-Fir Forest (Leiberg et al. 1904; Ellison 1954; Pfister 1972; Niering and Lowe 1984; Stromberg and Patten 1991; Patten and Stromberg 1995; Grissino-Mayer et al. 1995; Fulé et al. 2003; Margolis et al. 2007, 2011; Margolis and Balmat 2009; Swetnam et al. 2009). The primary fire season in the Southwest is from April through June, when there is little

Fig. 2.8 Large gray area of trees killed by crown fire that reportedly occurred in about 1880 on the San Francisco Peaks in north-central Arizona (Photograph in 1910 by A.G. Varela courtesy of Coconino National Forest, U.S. Forest Service)

precipitation. The potential severity and length of the fire season in southwestern Spruce-Fir Forest depend on moisture from the snowpack of the preceding winter, as well as on longer-term climate trends (Margolis et al. 2007, 2011).

It has often been assumed – with little direct evidence – that the fire regime in the Southwest is similar to that of subalpine forests of the central and northern Rocky Mountains (cf. Merkle 1954; White and Vankat 1993; Battaglia and Shepperd 2007). The fire regime of these more northerly forests is well-documented as consisting of landscape-scale, high-intensity, high-severity, stand-replacing crown fires reoccurring every 100 to 400+ years and linked to drought (e.g., Veblen et al. 1994; Romme et al. 2006; Sibold and Veblen 2006; Sibold et al. 2006). More frequent, extensive fires are precluded by moist fuels due to snowpack melt, high precipitation that includes summer rains, and cool temperatures (Battaglia and Shepperd 2007; Romme et al. 2009b).

Some historical accounts support the assumption of a stand-replacing crown-fire regime in the Southwest. The least ambiguous accounts come from upper-elevations. Leiberg et al. (1904) wrote that a century earlier a fire had "laid waste" a dense, 245-ha (600-acres) stand of Engelmann spruce and corkbark fir on the San Francisco Peaks. Photographs document extensive high-elevation crown fire in the area reportedly dating to around 1880 (Fig. 2.8; it is likely the same fire reported by Leiberg et al. 1904).

Historical accounts from lower elevations are more open to interpretation. Lang and Stewart (1910) wrote that the Kaibab Plateau contained

> Vast denuded areas, charred stubs and fallen trunks and the general prevalence of blackened poles…

and that

> …old fires extended over large areas at higher altitudes, amounting to several square miles…

However, they possibly interpreted the Plateau's extensive meadows as "vast denuded areas" formed by fires (cf. "fire glades" of Leiberg et al. 1904). Moreover, charred stubs, blackened poles, and fallen trunks can evidence fire, but not necessarily crown fire. Most importantly, Lang and Stewart (1910) did not report extensive post-fire stands of quaking-aspen root sprouts (suckers), such as followed recent crown fire in the area (personal observation; see also Crawford 2008). Lang and Stewart (1910) were clearer on the prevalence of surface fire in forests of the Kaibab Plateau (including Mixed Conifer and Ponderosa Pine Forests):

> Evidence indicates light ground fires over practically the whole forest…

In addition, a 1911 survey mentions "openings…from small fires" in Spruce-Fir Forest in what is now the Manti-La Sal National Forest in southeast and central Utah (Ogle and DuMond 1997). This suggests small, patchy, high-severity fires of the scale associated with a mixed-severity fire regime.

Research findings also indicate both crown and mixed-severity fire regimes in southwestern Spruce-Fir Forest. Evidence from the Pinaleño Mountains of southeastern Arizona, the San Francisco Peaks of north-central Arizona, the Gila Wilderness of west-central New Mexico, and a watershed in the Sangre de Cristo Mountains of north-central New Mexico largely supports an infrequent, landscape-scale, crown-fire regime (Grissino-Mayer et al. 1995; Abolt 1997; Margolis and Balmat 2009; Swetnam et al. 2009; Margolis et al. 2011). And crown fires continue to occur (Fig. 2.9).

Past crown fire in southwestern Spruce-Fir Forest has been inferred from the presence of small to large patches of quaking aspen of uniform age (Fig. 2.10; Abolt 1997; Margolis et al. 2007, 2011; see also Romme et al. 2001). Margolis et al. (2007) targeted the largest seral aspen stands in the southern Rocky Mountains of north-central New Mexico and adjacent Colorado. Aspen stands sampled in Spruce-Fir Forest ranged from 30 to 823 ha (74–2,034 acres). Crown fires forming these stands tended to have occurred synchronously and were coincident with severe droughts and regional occurrence of surface fires. Later, Margolis et al. (2011) focused on the largest potential post-stand-replacing-fire patches in both Mixed Conifer and Spruce-Fir Forest in four mountain areas of New Mexico and Arizona; the patches included sites without aspen. The largest reconstructed patch of stand-replacing fire in Spruce-Fir Forest was 521 ha (1,287 acres).

The role of crown fire in Spruce-Fir Forest of the San Juan Mountains of southwestern Colorado was examined by a multifaceted approach involving landscape patterns, persistent evidence of fire, tree ages, and successional patterns (Romme

Fig. 2.9 Area of recent crown fire in Spruce-Fir Forest in the White Mountains of east-central Arizona. Fire lookout is at 3,313 m (10,869 ft) (Photograph by Kari Greer/U.S. Forest Service)

et al. 2009b). The average fire interval for stands was centuries, and the fire turnover time (Sect. 1.2.5.1) was approximately 300 years during the mid eighteenth to mid twentieth century, although individual stands could remain unburned for many centuries.

Another line of evidence is that charcoal sediments in lakes and wetlands of the southern Rocky Mountains indicated reoccurring crown fire on an average of once every 100–200 years throughout the Holocene (Anderson et al. 2008). The authors concluded that this was in "remarkable agreement" with modern Spruce-Fir Forest in the central Rockies.

In contrast, some research on southwestern Spruce-Fir Forest indicated a mixed-severity fire regime with both surface and crown fires (Fig. 2.11). Evidence is clearest in stands transitional with Mixed Conifer Forest. Surface fires occasionally spread from Mixed Conifer into Spruce-Fir Forest and crowned in patches. Abolt (1997) reported a 42-year historical mean surface-fire interval in such transitional stands in the Gila Wilderness (vs. 8–15 years in Mixed Conifer Forest; ≥20 % scarred). She concluded,

> The conventional wisdom of fire regimes in spruce-fir forests being primarily of the stand-replacement type is challenged by evidence suggesting that at least some moderately intense patchy surface fires occurred in [transitional] stands.

Fig. 2.10 Patches of quaking aspen that likely indicate areas of past crown fire in Spruce-Fir Forest in the San Juan Mountains of southwestern Colorado (Photograph by Betty J. Huffman)

The clearest evidence of a mixed-severity fire regime in the American Southwest is from a landscape on the North Rim of Grand Canyon National Park (Fulé et al. 2003), where the vegetation mosaic includes lower-elevation Spruce-Fir Forest transitional with Mixed Conifer Forest (Vankat 2011). Remote sensing and ground reconnaissance revealed no large, homogeneous stands of fire-originated trees, such as produced by landscape-scale crown fire (also reported by Hanley et al. 1975 for Spruce-Fir Forest stands in southern Utah). Instead, data from analyses of fire scars and forest structure were "...consistent with a mixed-severity fire regime..." (Fulé et al. 2003). The Spruce-Fir Forest part of the landscape mosaic had a mean fire interval for 1700–1879 of 8.8 years (≥10 % scarred), indicating frequent surface fires. Small-scale crown fire was implicated in that 71 % of their 0.1 ha (0.2 acre)

Fig. 2.11 Burned Spruce-Fir Forest in the Pinaleño Mountains of southeastern Arizona. Areas of living trees indicate that the fire exhibited some mixed-severity behavior (Photograph by Charles M. Truettner)

study plots in Spruce-Fir Forest were fire-initiated. Moreover, surface fires and/or small crown fires had been prevalent in the vicinity of this area for approximately the last 10,000 years, as indicated by charcoal deposition rates observed in lake sediment cores (Weng and Jackson 1999).

Fire, be it crown fire or mixed-severity fire, affects other agents of disturbance such as wind and insect outbreaks. It has been reported that crown fire in the central Rocky Mountains reduces the potential for wind damage by resulting in younger stands that are less susceptible (Kulakowski and Veblen 2002). Also, there is evidence that crown fire lessens the potential for spruce beetle outbreaks in the central Rocky Mountains (e.g., Veblen et al. 1994; Bebi et al. 2003; Kulakowski et al. 2003; Kulakowski and Veblen 2006), at least until Engelmann spruce grow into the minimum diameter impacted by beetles (Veblen et al. 1994).

In conclusion, research indicates that the historical fire regime of southwestern Spruce-Fir Forest included large and small crown fires and surface fires, with their relative importance differing by elevation. Upper-elevation stands had infrequent, large crown fires. With decreasing elevation there appears to have been a steep gradient of decreasing importance of these fires and increasing frequency and size of surface fires. As a result, lower-elevation stands (particularly stands transitional with Mixed Conifer Forest), had a mixed-severity fire regime that included both surface and crown fires.

The question of whether the historical fire regime in southwestern Spruce-Fir Forest was similar or different from that of the central and northern Rocky Mountains is not fully resolved. There are apparent similarities, with evidence indicating a crown-fire regime in upper-elevation Spruce-Fir Forest in both regions. There also are apparent differences in fire regimes, with evidence suggesting mixed-severity fire was more important in lower-elevation stands of Southwestern Spruce-Fir Forest. There are few data from the Southwest – they are scanty and spatially limited – and data on surface fire in the central and northern Rocky Mountains are nearly lacking. That lack could be due to the rarity of surface fire, but Baker and Veblen (1990) reported that the subalpine zone of Colorado has crown, mixed-severity, and surface fires. In addition, Kulakowski et al. (2003) mentioned evidence of a moderate-severity fire and a low-severity fire in a subalpine area of northwestern Colorado, and they briefly summarized information on similar fires from other areas of the central Rocky Mountains. At this time, it appears that mixed-severity fires were less common historically in the central and northern Rocky Mountains than in lower-elevation Spruce-Fir Forest in the American Southwest.

If additional research supports evidence that mixed-severity fire in lower-elevation Spruce-Fir Forest has been more common in the American Southwest, what are possible explanations for the regional difference? Here are three hypotheses: First, mixed-severity fire is more frequent in the American Southwest because of much greater abundance of lightning and lightning-ignited fires, as reported in Sects. 1.2.2 and 1.2.5.1, respectively. Second, greater mid- to late-summer precipitation in the Southwest limits the spread and severity of fires and thereby favors small surface fires over large crown fires. A third hypothesis is that latitudinal differences in key factors limiting forest fire account for the difference. In most of the Rocky Mountains, the key limiting factors change with decreasing elevation from climate-related factors at moist high-elevations, where a crown-fire regime is characteristic, to a mix of climate- and fuel-related factors at mesic mid-elevations, where a mixed-severity fire regime is characteristic (Schoennagel et al. 2004). A parallel change from northern to southern latitudes could account for what appears to be greater importance of mixed-severity fire in lower-elevation Spruce-Fir Forest in the American Southwest.

2.2.5.2 Wind

Wind is an important disturbance agent in Spruce-Fir Forests of Colorado (Alexander 1987; Veblen et al. 1989, 1991a,b, 2001; Kulakowski and Veblen 2002), where blowdowns can affect up to 92 % of old-growth stands (Veblen et al. 1991a). Most blowdowns occur in winter, when branches are snow-covered and wind speeds are greatest. Wind impacts are variable across landscapes, depending on stand structure and composition, fire history, elevation, and topographic position (Veblen et al. 2001; Baker et al. 2002; Kulakowski and Veblen 2002). Damage is positively correlated with stand density, tree height, elevation, and higher slope position. Mature Engelmann spruce are subject to windthrow because of shallow roots, and trunks of subalpine fir often break. Quaking aspen is much less frequently damaged than conifers.

Similar research has not been conducted in the American Southwest, although blowdowns occur in Spruce-Fir Forest in the region (Pfister 1972; Niering and Lowe 1984; Stromberg and Patten 1991; Moir 1993; Patten and Stromberg 1995). A blowdown of ~365 to 525 ha (900–1,300 acres) of spruce-fir-aspen forest occurred on the Kaibab Plateau in 1958 (Getsinger 1961; Russo 1964). Most trees were uprooted, but some had trunks broken off. Other blowdowns have occurred in north-central New Mexico (Figs. 2.12 and 1.24). Wind is so understudied in the Southwest that a review of forest ecosystem health (Dahms and Geils 1997) did not include it as an important disturbance factor in southwestern Spruce-Fir Forest.

The effects of blowdowns on fire are understudied. In contrast, blowdowns have been documented as important in spruce beetle outbreaks (following section).

2.2.5.3 Biotic Agents

Insects are the major biotic disturbance agent in southwestern Spruce-Fir Forest. Common species include spruce beetle on Engelmann spruce and western balsam bark beetle on subalpine fir. Another biotic disturbance agent is western spruce dwarf mistletoe (*Arceuthobium microcarpum*), which infects Engelmann spruce. Mammals also can affect vegetation dynamics, as browsing by elevated deer populations can reduce recruitment of quaking aspen saplings and trees (Merkle 1962; Pfister 1972; Fulé et al. 2002b, 2003; Moore and Huffman 2004; Binkley et al. 2006). Common pathogens are root diseases (Dahms and Geils 1997) and wood-rotting fungi, with basal decay occurring in old wounds and frost cracks (Alexander 1987). Annosum root rot, which affects subalpine fir, can impact vegetation dynamics (Pfister 1972). Biotic agents sometimes act together in a mortality complex, e.g., that for subalpine fir involves western balsam bark beetle, other beetles, and Annosum root rot, all interacting with drought stress (Blackford et al. 2010). Another mortality complex is Sudden Aspen Decline, in which repeated defoliation by insects, strong drought, warm temperatures, and late frosts facilitate increased, synchronous mortality of quaking aspen by bark beetles, other insects, and canker fungi that impact stressed trees (Fairweather et al. 2008; Worrall et al. 2008).

Multiple studies have shown that spruce beetle has had the greatest impact of any natural biotic disturbance agent affecting southwestern Spruce-Fir Forest (see Fig. 2.13), and the extent of the impact possibly surpasses that of crown fire. Outbreaks have occurred throughout recorded history (Alexander 1987; Romme et al. 2006) and perhaps account for paleoecological variations in spruce abundance (Anderson et al. 1999). Old photographs and tree-ring analysis revealed six major outbreaks in the southern Rocky Mountains since the mid nineteenth century, and an outbreak affected forests in central New Mexico between the 1850s and 1880s (Baker and Veblen 1990). Northern New Mexico was affected by large outbreaks in the 1970s (Dahms and Geils 1997). Central and southern Utah experienced outbreaks in the 1920s, 1930s, and 1990s (Dixon 1935; Pfister 1972; Dymerski et al. 2001; Battaglia and Shepperd 2007). Outbreaks around the turn of the twenty-first century in Arizona and New Mexico were detected on 163 km^2 (63 $miles^2$) in

Fig. 2.12 Blowdown in Spruce-Fir Forest in north-central New Mexico (Photograph by Arnie Friedt, New Mexico State Forestry Division)

Fig. 2.13 Spruce-Fir Forest impacted by spruce beetle in southwestern Utah (Photograph by Betty J. Huffman)

Arizona and New Mexico (U.S. Forest Service 2003; area not limited to Spruce-Fir Forest). The frequency of outbreaks is positively related to size of the area, proportion of the area unaffected by recent outbreaks, and rate of stand growth into susceptible condition, yet it is difficult to predict outbreaks (cf. Schmid and Amman 1992). An area of mostly Spruce-Fir Forest in northwestern Colorado had a mean interval of 117 years between outbreaks since 1633 and a turnover time of 259 years (Veblen et al. 1994).

Findings from the Rocky Mountains indicate that spruce beetle populations likely persist between outbreaks in trees in small windfalls (Schmid and Frye 1977; Veblen et al. 1991b; Schmid and Amman 1992) and in scattered live trees (Veblen et al. 1994). Such populations are kept at low levels by nematodes, insect parasites, and insect predators such as woodpeckers (Alexander 1987). Large population outbreaks in southern and central Utah are favored by high forest density, even-aged forests, and large Engelmann spruce (Dymerski et al. 2001). Also important are winter disturbances that provide fresh host material for spruce beetles in the spring (snow avalanches, blowdowns, and snow and ice damage; Hebertson and Jenkins 2007), as well as warm temperatures in the preceding fall and winter and drought in preceding years (Hebertson and Jenkins 2008). In the Rocky Mountains, large blowdowns can trigger outbreaks (Mielke 1950; Schmid and Frye 1977; Alexander and Shepperd 1984; Veblen et al. 1989, 1994), and factors in outbreaks are predominance of Engelmann spruce in the canopy (including in nearby stands), high stand basal area concentrated in older larger-diameter Engelmann spruce, slow tree-diameter growth, long time since fire, mild winters, and high elevation (Schmid and Hinds 1974; Schmid and Frye 1977; Veblen et al. 1994; Bebi et al. 2003).

Spruce-beetle outbreaks in the late-twentieth and early-twenty-first centuries were extensive. Those in Colorado were attributed to interaction of (a) long-term drought that stressed trees and increased their susceptibility to insects, (b) warm summers that stressed trees and accelerated insect growth, (c) warm winter temperatures that enhanced survival of insect larvae, and (d) extensive, dense forests that provided ample food for insects (Romme et al. 2006).

The most thoroughly studied spruce-beetle outbreak in the American Southwest began on the Markagunt Plateau in southwestern Utahin the late twentieth century. Scattered outbreaks occurred over an approximately 8-year period on warm sites that had low tree density, high proportion of spruce, and high potential site productivity (DeRose and Long 2012). These populations then coalesced across the landscape during a multi-year period of unusually warm summer and winter temperatures (DeRose et al. 2011; DeRose and Long 2012). Ultimately, 99 % of Engelmann spruce died, and 93 % were killed by beetles (trees ≥5 cm/2 in., diameter at breast height, dbh, i.e., at 1.4 m/4.5 ft; DeRose and Long 2007).

Insect outbreaks potentially enhance fire hazard by producing dead fuels (cf. Baker and Veblen 1990; Dahms and Geils 1997). This relationship was challenged for the Rocky Mountains by Schmid and Hinds (1974), who concluded, "the massive number of dead trees has created a large fuel buildup [but] the fire hazard seems over exaggerated." The question whether spruce beetles increase the probability of fire is difficult to address (Baker and Veblen 1990), but there appears to be little or no relationship in the Rocky Mountains (Bebi et al. 2003; Bigler et al. 2005; Romme et al. 2006) and possibly in the American Southwest. In general, the relationship of insect outbreaks and fire is complex (cf. Hicke et al. 2012; Jenkins et al. 2012). The effects of insect outbreaks on wind disturbances such as blowdowns have received little study.

2.2.5.4 Climate Variation

Drought has been identified as a disturbance factor affecting Spruce-Fir Forest, especially in stands located near the forest's southern latitudinal limit (Adams and Kolb 2005). Drought induces lagged tree mortality in Engelmann spruce and subalpine fir in northern Colorado, especially in trees with low growth rates (Bigler et al. 2007). Drought often interacts with other disturbance agents. For example, drought increases the probabilities of landscape-scale crown fires and contributes to insect outbreaks. Drought is also involved in twenty-first century increases in mortality of quaking aspen (Sect. 3.2.5.4).

2.2.6 Anthropogenic Disturbance

Land use by Native Americans in the American Southwest was concentrated in Pinyon-Juniper vegetation (*Pinus-Juniperus*; Sects. 1.5.1.1 and 5.2.6) and likely was uncommon and limited in area and impact in Spruce-Fir Forest. Information is sparse on anthropogenic drivers related primarily to Euro-American land use, but key drivers of southwestern Spruce-Fir Forest appear to be livestock grazing, fire management, modern climate change, invasive species, recreation, and nearby land use. Each of these can be important in vegetation dynamics, and therefore all are incorporated in the conceptual models (Sect. 2.5). Commercial logging was uncommon and is outside the scope of this book.

2.2.6.1 Livestock Grazing

Livestock grazing began in the second half of the nineteenth century and apparently had greater effects on lower-elevation Spruce-Fir Forest where the continuity of the herbaceous layer apparently carried surface fires. As in other forest types (Sects. 1.5.2.2, 1.6.1, 1.6.2), grazing resulted in an abrupt decrease in frequency of surface fires in these lower-elevation stands (Abolt 1997; Fulé et al. 2003). Upper-elevation Spruce-Fir Forest, which lacked surface fires, apparently did not have extensive herbaceous cover, and livestock grazing left no clear evidence of an impact. The grazing likely reduced tree regeneration in both lower- and upper-elevation stands. Today, livestock grazing in Spruce-Fir Forest is generally less extensive, although it is widespread in southern Utah (Battaglia and Shepperd 2007). Grazing is most common in stands dominated by quaking aspen (Milchunas 2006), where it affects age distributions of aspen, as well as cover and composition of the understory.

2.2.6.2 Fire Management

Fire management focused on preventing and suppressing fires throughout most of the twentieth century. While its effects on upper-elevation forest with long fire-return intervals likely were negligible, it maintained the fire exclusion initiated by livestock grazing in lower-elevation Spruce-Fir Forest where surface fires had burned (cf. Fulé et al. 2003). This changed the structure and composition of lower-elevation stands and landscapes. As older trees became more widespread, the likelihood of large-scale insect outbreaks increased. Also, where fuel loads and fuel continuity increased across lower-elevation or transitional landscapes, fires had greater potential to become landscape-scale crown fires (White and Vankat 1993; Abolt 1997; Fulé et al. 2003; Cocke et al. 2005; Swetnam et al. 2009). Fire management practices began to shift in the late-twentieth century to include management fires (both prescribed and lightning-ignited; Sect. 1.5.2.1), but such fires have been uncommon if not absent in southwestern Spruce-Fir Forest, except for lower-elevation stands where fires have burned from adjacent Mixed Conifer Forest. Additional research on the effects of fire management is needed.

2.2.6.3 Modern Climate Change

Modern climate change (Sect. 1.6.3) includes drought and warmer temperature and therefore is implicated in insect outbreaks that have affected the structure, composition, and dynamics of stands and landscapes. Air pollution is a driver of modern climate change, but pollution sources are more global than regional or local. Direct effects of air pollution on southwestern Spruce-Fir Forest are understudied, but likely include ozone phytotoxicity (Dahms and Geils 1997, see also Karnosky et al. 1999). Nevertheless, with the lack of documentation of direct effects, air pollution is not included in this chapter as a separate anthropogenic driver of southwestern Spruce-Fir Forest.

2.2.6.4 Invasive Species

Invasive plant species currently are not important in southwestern Spruce-Fir Forest. For example, non-native plants in north-central Arizona have a mean cover value of only 0.2 % (Fulé et al. 2002a) and mean species richness of approximately 5–7 species/ha (2–3 species/acre; Fisher and Fulé 2004; Laughlin et al. 2005). Nevertheless, their future introduction and colonization of high-elevation areas in general is likely with climate change and globalization (cf. Pauchard et al. 2009).

A non-native invasive insect, spruce aphid (*Elatobium abietinum*), has had outbreaks in Spruce-Fir Forest in Arizona, including defoliation of an area of 635 km^2 (245 $miles^2$; Lynch 2004). Engelmann spruce is the primary host species, followed by blue spruce. A single defoliation results in 10 % tree mortality, and an apparent synergistic relationship with western spruce dwarf mistletoe results in 70 % tree mortality. Outbreaks favor greater abundance of blue spruce where it co-occurs with Engelmann spruce; they also favor replacement of Engelmann spruce by subalpine fir or corkbark fir at higher elevations and by Douglas-fir at lower elevations. Outbreaks of spruce aphid presumably also alter the natural disturbance regime, but this is unstudied.

2.2.6.5 Recreation

Recreation has impacts on southwestern Spruce-Fir Forest that are spatially limited (e.g., construction and use of roads, trails, and ski runs), but some impacts are more extensive. Fires have been ignited by recreationists in Spruce-Fir Forest, but generally have been small in area, although the potential for larger fires exists, especially during drought. Also, recreationists appear to have introduced, spread, and helped establish invasive plants (invasives commonly occur near trails and other areas of human use). Research elsewhere in the Rocky Mountains indicates that horses used in trail-riding can introduce invasive plants by dispersing seeds through their fecal matter (Benninger-Truax et al. 1992; Wells and Lauenroth 2007), and seeds also can be introduced by off-highway vehicles (OHVs; Taylor et al. 2011). Once established on sides of trails and roads, invasive plants can spread into forest interiors (Benninger-Truax et al. 1992). Off-trail use of OHVs has additional impacts, especially on the high plateaus with Spruce-Fir Forest in southern and central Utah, where relatively flat terrain leads to widespread use.

2.2.6.6 Nearby Land Use

Nearby land use is probably less important than in other types of vegetation, because stands of Spruce-Fir Forest are often isolated at high elevation. Nevertheless, livestock grazing on adjacent Mixed Conifer Forest (and possibly on Subalpine-Montane Grassland) reduced the spread of fires into Spruce-Fir Forest. Other nearby land use can have the opposite effect. For example, a campfire started by recreationists in Mixed Conifer Forest ignited the Wallow Fire, which spread into Spruce-Fir Forest and became what at the time was the largest fire in Arizona's recorded history (Fig. 2.14).

Fig. 2.14 Recreationists camping in nearby Mixed Conifer Forest caused the crown fire that burned this area of Spruce-Fir Forest in the White Mountains of east-central Arizona (Photograph by Kari Greer/U.S. Forest Service)

2.3 Processes

The effects of fire, wind, spruce beetles, and root pathogens on vegetation dynamics have been studied across a variety of scales from small patches to landscapes in the central Rocky Mountains (e.g., Schmid and Hinds 1974; Veblen 1986; Alexander 1987; Aplet et al. 1988; Veblen et al. 1991a). Although the same disturbance agents are important in the American Southwest, there has been little parallel research on vegetation dynamics (but see DeRose and Long 2007, 2010). There are likely to be regional similarities and differences in vegetation dynamics.

Vegetation dynamics in both regions depend on the type of disturbance. Crown fires leave few trees to initiate post-fire dynamics, other than roots of quaking aspen. Blowdowns leave aspen and smaller trees that are not susceptible. Spruce beetle outbreaks leave individuals of all species, but greatly reduce the abundance of large Engelmann spruce. Surface fires and treefalls leave trees of all sizes and species. This section considers two major processes of vegetation dynamics: succession and gap dynamics. Both play key roles in the conceptual models (Sect. 2.5).

2.3.1 *Succession*

Succession following crown fire in the American Southwest involves Engelmann spruce, subalpine or corkbark fir, and sometimes quaking aspen (Pfister 1972; Niering and Lowe 1984; Stromberg and Patten 1991; Moir 1993; Patten and Stromberg 1995; Jenkins et al. 1998; Dymerski et al. 2001). Where conifers initiate succession in the San Juan Mountains, tree establishment can take several decades, depending on availability of seed sources (Romme et al. 2009b). Engelmann spruce is the primary conifer that follows stand-scale, canopy-removing disturbance such as crown fire, because it regenerates on exposed mineral soil; subalpine or corkbark fir can co-establish with it or follow several decades later (Aplet et al. 1988; Rebertus et al. 1992; Patten and Stromberg 1995). Young conifer stands generally remain free of disturbance for several decades, because low fuel loadings minimize the probability of fire and small trees are little affected by strong winds and spruce beetles. After about 70 years, spruce beetles have the potential to impact stands (Veblen et al. 1994), although the probability of an outbreak is small compared to old-growth stands. Engelmann spruce increases in canopy dominance in the absence of additional large-scale disturbance, while subalpine (or corkbark)fir increases in abundance in the understory (Aplet et al. 1988), in part because fir establishes on forest litter (Knapp and Smith 1982; Battaglia and Shepperd 2007). Post-fire succession uninterrupted by additional disturbance can take two or more centuries before change lessens.

Alternatively, succession after crown fire can begin with dominance by quaking aspen, especially in lower-elevation Spruce-Fir Forest. Aspen can remain the only dominant tree for several decades. At some point, conifers regenerate in most aspen stands in this elevational zone (Fig. 2.15; Pfister 1972; Jenkins et al. 1998), unless conifer seed sources are distant. Conifers eventually overtop aspens, and this shade-intolerant species is generally much reduced in abundance after about 150 years (Moir 1993), unless maintained by reoccurring surface fire and/or tree fall gaps (Sect. 2.3.2). Aspen can live for over 250 years in the American Southwest (Margolis et al. 2011). After conifers increase and aspen decreases, forest dynamics are similar to those described in the previous paragraph. See Sect. 3.3.2 for more on the vegetation dynamics of quaking aspen.

Succession after stand-wide blowdown is facilitated by understory trees that are relatively undisturbed. These are mostly of late-successional, shade-tolerant species. Understory trees are released with the removal of the canopy, and regeneration also occurs. Subalpine fir dominates both the understory and the regeneration in northwestern Colorado, producing a shift away from pre-blowdown dominance by Engelmann spruce (Kulakowski and Veblen 2003). Even after 65 years, the presence of fallen logs and tip-up mounds in the area of the blowdown facilitates establishment of subalpine fir and Engelmann spruce, respectively. Root-sprouting by quaking aspen also can be important following blowdowns in lower-elevation stands where aspen was present before disturbance.

Fig. 2.15 Successional stand of quaking aspen (*yellow-green leaves*) with Engelmann spruce (*dark green*) dominating the understory. Spruce-Fir Forest on the San Francisco Peaks, north-central Arizona (Photograph by Trent Larson)

Fig. 2.16 Stand of Spruce-Fir Forest in central Utah in which spruce beetles killed canopy trees, leaving understory trees that will replace them through succession (Photograph by author)

Succession after spruce beetle outbreaks is different because canopy trees other than Engelmann spruce survive the disturbance, as do understory trees of all species (Fig. 2.16). These survivors are keys to post-outbreak vegetation dynamics, and succession leads to a community that reflects the proportions of species in the canopy and understory layers. Dominance is generally shifted to subalpine (or corkbark) fir (Dymerski et al. 2001). This species is abundant among saplings and seedlings in old-growth Spruce-Fir Forest because it is shade-tolerant and regenerates on soil organic matter. If quaking aspen is present, it root sprouts following opening of the canopy and can be an important post-disturbance species unless its sprouts are reduced by ungulate browsing (cf. DeRose and Long 2010). Post-disturbance dominance of subalpine fir (and possibly quaking aspen) will persist for many decades before Engelmann spruce becomes an important species again (DeRose and Long 2007, 2010). Over time, reoccurring spruce beetle outbreaks can lead to stand-scale oscillations of abundance of Engelmann spruce paralleling but out of phase with oscillations of subalpine (or corkbark) fir (Schmid and Hinds 1974).

Regardless of the succession-initiating disturbance and the successional pattern, succession without additional disturbance leads to older stands dominated by Engelmann spruce and subalpine or corkbark fir, assuming sources of seeds for these conifers are present. Finer details of successional pathways in southwestern Spruce-Fir Forest depend on elevation and other site factors (Stromberg and Patten 1991; Moir 1993; Patten and Stromberg 1995). Site-specific patterns of succession along an elevational gradient have been outlined by Moir (1993).

2.3.2 Gap Dynamics

Even late-successional stands are always changing, because they have small-scale gap dynamics that begin with the death of one or a small cluster of canopy trees. Gaps typically form by snap-offs (trunks break) or tip-ups (trees topple, tipping the roots up). The opening in the canopy (Fig. 2.17a) alters the environment below by increasing light. Gaps formed by snap-offs leave the soil and its litter layer undisturbed, which favors regeneration of subalpine (or corkbark) fir; gaps formed by tip-ups expose mineral soil, which favors regeneration of Engelmann spruce (DeRose and Long 2010). If quaking aspen is present, gaps of either type lead to root sprouting. Subcanopy trees, tree seedlings, and aspen root sprouts in the area of the gap respond with accelerated growth (Fig. 2.17b). These individuals are thinned by various mortality agents and competition, but one or more survivors ultimately replace the dead canopy tree(s). Gap dynamics scattered throughout stands maintain the dominant species of southwestern Spruce-Fir Forest in tree understories and canopies. Gap dynamics also result in uneven-aged forests.

Fig. 2.17 (**a**) Upward view of a canopy gap formed by a small cluster of trees (including the dead tree in the photograph) and (**b**) dense undergrowth below the same canopy gap, in Spruce-Fir Forest in the North Rim region of Grand Canyon National Park (Photographs by author/National Park Service)

2.4 Historical Changes

2.4.1 Overstory

2.4.1.1 Pre-Euro-American Settlement

Historical descriptions of Spruce-Fir Forest prior to a century ago are uncommon, but each described stands as dense:

> …dense forests of spruce…almost impenetrable… (Dutton 1880, for the Markagunt Plateau)

> The tall spruces…stand so close together [that] passage would be almost impossible. (Dutton 1880, for the Aquarius Plateau of south-central Utah)

> …trees cluster so thickly together that a passage through them is extremely difficult and sometimes impossible. (Dutton 1882, for the Kaibab Plateau)

> …closely stocked stands… (Leiberg et al. 1904, for the San Francisco Peaks)

> …thick stands of spruce… (Lang and Stewart 1910, for the Kaibab Plateau)

Historical photographs support descriptions of dense stands (Fig. 2.18a, b), except near treeline and in areas of recent crown fire (Fig. 2.8).

One historical description described a vegetation mosaic:

> Part of the [forest] type is composed of densely stocked stands of small aspen, inclosing blocks of Engelmann spruce; part of it consists of Engelmann spruce set in pure stands, or mixed in varying proportions of aspen, [corkbark] fir, and bristle-cone pine. (Leiberg et al. 1904, for the San Francisco Peaks)

Unsurprisingly, quantitative data from a century or more ago are less common. Moreover, they can be misleading. Lang and Stewart (1910) sampled a "mixed type" on the Kaibab Plateau that included lower-elevation Spruce-Fir Forest and Mixed Conifer Forest. They reported an average density of 358 trees/ha (145 trees/acre) for trees ≥0.9 m (3 ft) height. Density of larger trees ≥ 15.2 cm (6 in.) dbh was 116 trees/ha (47 trees/acre).

These data sharply contrast with the historical descriptions of dense stands. The density figure that included small saplings equates to a mean area per tree of 28 m^2 (301 ft^2), and the density of larger trees equals a mean area per tree of 86 m^2 (927 ft^2) In other words, the data describe tree densities that were woodland-like, not dense forest. Lang and Stewart (1910) underestimated density by omitting quaking aspen, which can be abundant in such stands but lacked value as a timber tree. They also possibly underestimated density by including samples with areas of treeless Subalpine-Montane Grassland (Vankat 2010, 2011).

There are two other problems with using the data of Lang and Stewart (1910) to represent conditions prior to Euro-American influence (Vankat 2011). Most of the data apparently came from Mixed Conifer Forest (ponderosa pine was the dominant species). And the sampling occurred about 30 years after the beginning of livestock grazing and resultant fire exclusion. Early change with fire exclusion is suggested by what appears to be ingrowth of fire-sensitive fir by 1909: fir accounted for 27 % of trees ≥15 cm (6 in.) dbh but 59 % of seedlings and saplings <15 cm (6 in.) dbh.

Fig. 2.18 (**a, b**) Spruce-Fir Forest in 1874 in the Conejos River watershed, south-central Colorado (Photographs by Timothy H. O'Sullivan. Courtesy of the National Archives and Records Administration)

Nineteenth century structure and composition of Spruce-Fir Forest also have been estimated by a method known as forest reconstruction. This approach typically involves (a) analyzing rings of living trees and (b) using decay classes to estimate ages of snags and downed logs currently on the site. These data are used to determine/estimate which trees, snags, and logs were alive and at what dbh on a selected date in the past. Reconstruction studies have examined Spruce-Fir Forest on the North Rim (Fulé et al. 2003) and San Francisco Peaks (Cocke et al. 2005). Density estimates of trees >2.5 cm (1 in.) dbh in the nineteenth century were 150 and 263 trees/ha (61 and 106 trees/acre) for the North Rim and San Francisco Peaks, respectively. Estimates of basal area were 10 and 16 m^2/ha (44 and 70 ft^2/acre), respectively. Engelmann spruce and quaking aspen dominated on the North Rim, and bristlecone pine, Engelmann spruce, and corkbark fir dominated on the San Francisco Peaks. These estimates from forest reconstructions also sharply contrast with the historical descriptions of dense stands. Average densities reported from reconstructions equate to mean areas of 38–67 m^2/tree (409–719 ft^2/tree), despite including small saplings. Fulé et al. (2003) compared their results for the North Rim to data from Lang and Stewart (1910) and concluded their reconstruction techniques and results were verified.

Forest reconstruction underestimates historical values of density and basal area when trees from the historical date being reconstructed have been lost by combustion or decomposition (Fulé et al. 2002b, 2003; Cocke et al. 2005). Combustion has not been a factor on many Spruce-Fir Forest sites because of fire exclusion. However, losses by decomposition would have occurred (Fulé et al. 2002b) and would have been important. The decomposition constant for quaking aspen logs in a subalpine forest of northern New Mexico (Gosz 1980; see also Miller 1983; Alban and Pastor 1993) indicates 90 % loss of dry mass in only 33 years and 99 % loss in 66 years – periods far shorter than the 100+ year interval that was reconstructed in the above studies. In addition, use of decay classes to estimate year of death of downed logs appears to be imprecise. Datable Engelmann-spruce logs (dbh >30 cm/12 in.) in an old-growth subalpine forest in central Colorado had narrow ranges in years since death in decay classes 1 and 2 of a 6-class scale (~1–2 and ~7–9 years, respectively), indicating age estimations of younger logs would be precise, but the range was ~43–85 years in class 6, indicating age estimations of older logs would be imprecise (Brown et al. 1998). Loss of evidence and imprecision likely are in part why Moore et al. (2004) stated that mesic sites and higher elevations negatively affect the accuracy of forest reconstructions. These issues also account for reconstruction having substantially underestimated nineteenth century density and basal area of southwestern Spruce-Fir Forest.

In conclusion, brief qualitative descriptions and historical photographs provide the only useful information on canopy conditions in southwestern Spruce-Fir Forest approximately a century ago, and they consistently indicate dense stands. Early historical data and results of forest reconstructions indicated much more open stands, but evidence indicates that both substantially underestimated forest densities and basal areas. Similarities between early historical data and results of forest reconstructions are happenstance.

2.4.1.2 Post-Euro-American Settlement

Few studies have quantitatively examined changes in southwestern Spruce-Fir Forest since Euro-American settlement. The two reconstruction studies discussed in the previous section compared findings to contemporary data collected using standard methods of field sampling. Fulé et al. (2003) and Cocke et al. (2005) reported increases in tree density of 220 and 532 % and increases in basal area of 187 and 273 % since the nineteenth century for the North Rim and San Francisco Peaks, respectively. These increases are exaggerated by the underestimation of nineteenth century values described in the previous section. In addition, the reported increases contrast with findings obtained from two small sets of historical study plots from the North Rim that density and basal area in lower-elevation Spruce-Fir Forest decreased during 1935–2004 and/or 1984–2005 (Vankat 2011). Neither Fulé et al. (2003) nor Cocke et al. (2005) reported values for intermediate dates during the twentieth century (few reconstructions studies do), which would have indicated temporal patterns and enabled direct comparison of results with the resampling of historical plots. Another indication of problems with the Spruce-Fir Forest reconstructions is that some results are counterintuitive: both studies reported large increases in density of quaking aspen during a period of fire exclusion, yet aspen typically regenerates with fire and populations decrease without it. Indeed, resampling of historical study plots documented large decreases in aspen in lower-elevation Spruce-Fir Forest on the North Rim during 1935–2004 and 1984–2005 (Vankat 2011).

Resampling permanent plots is an inherently more accurate approach to determining historical changes, but is limited by the rarity and young age of historical plots in Spruce-Fir Forest. Findings from the above-mentioned sets of plots in lower-elevation Spruce-Fir Forest on the North Rim indicated that total density decreased 43 % from 1984 to 2005 and total basal area decreased 73 % from 1935 to 2004 (Vankat 2011 for trees ≥10 cm/4 in. dbh; one plot had burned in a surface fire). The primary tree species that decreased was quaking aspen, which is unsurprising in the absence of fire, but spruce (blue and Engelmann spruce combined) and subalpine fir also decreased. With tree mortality having opened the canopy, total sapling density increased 170 % from 1984 to 2005 and involved the same three species, which suggests cyclic change. Vankat (2011) hypothesized that lower-elevation Spruce-Fir Forest on the North Rim had rapidly increased in density and basal area following the beginning of fire exclusion in the late nineteenth century and decreased in the twentieth century, even without burning, in response to the interaction of density-dependent mortality and exogenous factors such as climate change.

Various sources have suggested that the current structure and composition of southwestern Spruce-Fir Forest are within the historical range of variation, i.e., within the range of forest conditions present before Euro-American influence (e.g., White and Vankat 1993; Dahms and Geils 1997). This was not based on data, but rather on reasoning that the ~100-year period since the beginning of fire exclusion is shorter than all but the low end of the 100 to 400+ year fire interval of a presumed crown fire regime. It also has been suggested that stands have been little affected by historical livestock grazing (Dahms and Geils 1997). This conclusion is incorrect for lower-elevation Spruce-Fir Forest stands in the Southwest, where livestock grazing reduced the frequency of fire in the mixed-severity fire regime (Abolt 1997; Fulé et al. 2003; Anderson et al. 2008). In addition, reduced fire frequencies in Ponderosa Pine Forest due to fire exclusion "...removed an important source of fires for... spruce-dominated forests" (Margolis and Balmat 2009), which suggests an indirect effect of livestock grazing on the fire regime of lower-elevation Spruce-Fir Forest. Changes in lower-elevation Spruce-Fir Forest with fire exclusion include fewer early successional stands, increases in Engelmann spruce and subalpine fir in aspen stands (Moir 1993), possibly altered successional patterns (Margolis and Balmat 2009), greater fuel loads (Fulé et al. 2004), and increased landscape homogeneity (White and Vankat 1993; Fulé et al. 2003). Stands of upper-elevation Spruce-Fir Forest that lacked surface fires are likely within the range of forest conditions present before Euro-American settlement.

Data on changes in quaking-aspen stands of southwestern Spruce-Fir Forest are unavailable. The closest approximation is in Vankat (2011), who reported data for aspen stands on the North Rim (most were at elevations below Spruce-Fir Forest). He reported large decreases in density and basal area for all tree species combined and for quaking aspen from 1935 to 2004, along with increases in ponderosa pine, i.e., changes expected with succession (Sect. 2.3.1).

In conclusion, accurate quantitative estimation of changes after Euro-American settlement is currently impossible because valid data from before or soon after that settlement are lacking. Comparison of forest reconstruction data to modern field-sampled data greatly overestimates increases in density and basal area, because forest reconstructions greatly underestimate nineteenth century values. Lower-elevation Spruce-Fir Forest possibly increased in density and basal after fire exclusion began in the nineteenth century, before peaking and decreasing in the twentieth century, but this needs verification. Upper-elevation Spruce-Fir Forest may be within the range of forest conditions present before Euro-American influence, assuming surface fires had little historical influence at such elevations. Some quaking-aspen stands of southwestern Spruce-Fir Forest appear to have shifted in composition with reduced aspen and increased conifers.

2.4.2 Understory

The understory of southwestern Spruce-Fir Forest is highly variable (Romme et al. 2009b, for north-central New Mexico and southwestern Colorado), depending on local site conditions. It can have greater variation in species composition than other forest types (Fisher and Fulé 2004). Cover values in northern Arizona average between 15 and ~25 % (White and Vankat 1993; Fulé et al. 2002a; Fisher and Fulé 2004; Laughlin et al. 2005). Cover differs greatly among stands (Dye and Moir 1977), with a range of approximately 3–65 % on the North Rim (White and Vankat 1993; Laughlin et al. 2005). Species composition has been quantitatively characterized for a few specific locations such as the San Francisco Peaks (Fisher and Fulé 2004) and the North Rim (Laughlin et al. 2005; Laughlin and Fulé 2008). Diversity values are available (Fisher and Fulé 2004; Laughlin et al. 2005; see also Dye and Moir 1977). Regional characterizations include Moir and Ludwig (1979), Youngblood and Mauk (1985), DeVelice et al. (1986), Mueggler (1988), Moir (1993), and Muldavin et al. (1996).

Early descriptions of the understory are uncommon, but include

> There is very little ground cover in the subalpine forest. (Leiberg et al. 1904, for the San Francisco Peaks)

However, this area was heavily grazed by livestock at the time.

Without more-detailed historical information, understory dynamics must be inferred from (a) observations of short-term understory changes, such as in response to fire, and (b) spatial differences related to factors that also differ temporally. As an example of using spatial differences, contemporary differences in understories among stands with different tree densities can be used to infer historical understory changes related to increases and decreases in tree density (see below). Such inferences can be problematic and at best yield only a general understanding of historical changes. Even research on contemporary understories accounts for only 20 % of the variation in cover, 18 % of the variation in species composition, and 33 % of the variation in diversity in sites on the North Rim that included Spruce-Fir Forest (Laughlin et al. 2005). Historical understory changes are likely to have been complex and differed among regions (cf. Laughlin et al. 2005). The following paragraphs assess the dynamics of understory cover, species composition, and diversity.

Understory cover would have changed with the changes in tree density described in the previous section, because it is negatively related to tree canopy cover (Merkle 1954; Hurst 1977) and is especially high in young stands after canopy-opening disturbance (Yeager and Riordan 1953; Dye and Moir 1977; Moir 1993; Chambers and Holthausen 2000). Additional insight into possible historical changes is provided by findings that understory cover is negatively related to basal area of subalpine fir (Fisher and Fulé 2004; Laughlin et al. 2005) and amount of coarse woody debris (Laughlin et al. 2005), as well as sapling density of Engelmann spruce and corkbark fir (Stromberg and Patten 1991). The relationship with basal area of Engelmann spruce has been reported as both negative and positive (Fisher and Fulé 2004; Laughlin et al. 2005). Herbs are more abundant below quaking aspen (a deciduous species) than below conifers (Chambers and Holthausen 2000). Understory cover probably decreases with succession from aspen to conifer dominance, as cover is positively associated with aspen basal area (Fisher and Fulé 2004). Cover is little affected by mixed-severity fire, at least in the short term of two years (Laughlin and Fulé 2008). These findings indicate that understory cover increased with canopy-opening disturbance and with decreases in forest density. In addition, understory cover decreased during succession and with increases in forest density.

Historical changes in understory composition (here applied mostly to growth forms) are inferable from the finding that understory species composition is related to basal areas of Engelmann spruce, subalpine fir, and quaking aspen and amount of coarse woody debris (Laughlin et al. 2005). The understory of contemporary closed-canopy stands is generally characterized by lichens, fungi, sedges (*Carex* spp.), mosses, and liverworts, and opening of the canopy leads to increases in grasses, forbs, and half-shrubs (Milchunas 2006). Shrubs are more abundant with greater light below canopy openings and in seral stands (Chambers and Holthausen 2000). Annual and biennial forbs and graminoids increased after mixed-severity fire (Laughlin and Fulé 2008). These findings indicate that understory composition changed from herbs and shrubs to non-vascular plants and sedges with succession and with increases in forest density. Moreover, decreases in forest density, including tree falls, were followed by increases in herbs – especially short-lived species – and shrubs and by decreases in non-vascular plants.

Historical changes in understory diversity are unclear. Understory species richness is positively related to basal area of Engelmann spruce and negatively related to basal area of subalpine fir and amount of coarse woody debris (Laughlin et al. 2005). Moreover, species richness of annual plants possibly decreased since the initiation of fire exclusion (Laughlin et al. 2005). However, no relationship between understory diversity and stand age was found by Dye and Moir (1977), and there are regional differences in understory richness of quaking aspen stands (Laughlin et al. 2005). Richness appears to be little affected by mixed-severity fire (Laughlin and Fulé 2008).

2.5 Conceptual Models

The following nested, three-tiered set of conceptual models is based on my interpretation of best-available information on Spruce-Fir Forest of the American Southwest. The models summarize and synthesize material on drivers (Sect. 2.2) and processes (Sect. 2.3) of vegetation dynamics, and they account for historical changes (Sect. 2.4). Their format is explained in Sect. 1.7. The models can be used to facilitate understanding of the past and the present. They also can provide insight into effects of future changes in the environment and outcomes of alternative land-management decisions. Moreover, they can guide the development of quantitative models.

2.5.1 Ecosystem-Characterization Model

The ecosystem-characterization model for southwestern Spruce-Fir Forest emphasizes *Vegetation* (a biotic component) and *Disturbance* (a driver; Fig. 2.19a, Table 2.2). Key aspects of *Vegetation* are small- to large-scale patterns, fuel, structure, and species composition. These affect various aspects of *Disturbance*. The primary agents of *Disturbance* are fire, insect outbreaks, drought, and wind. These affect tree mortality and other aspects of vegetation. The two other biotic components are *Soil*

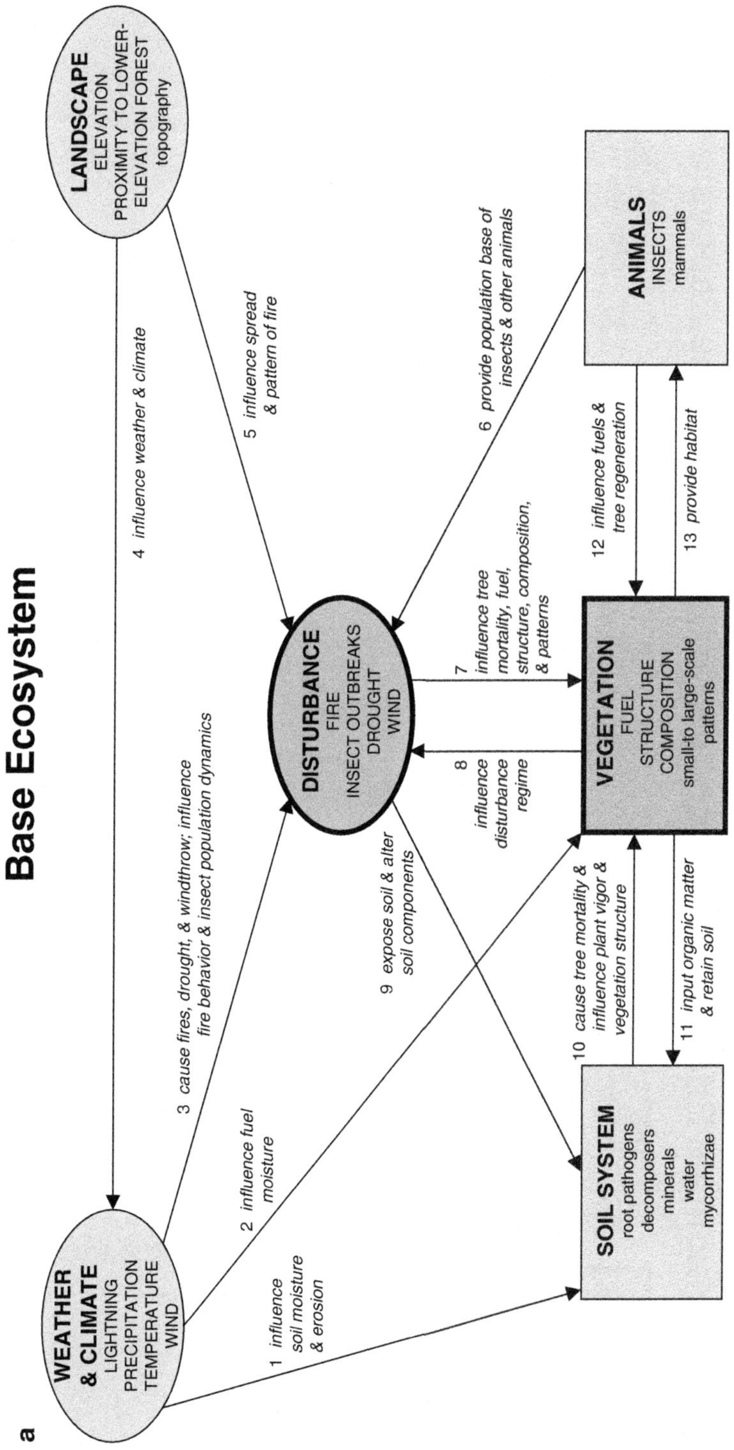
a
Base Ecosystem
WEATHER & CLIMATE
LIGHTNING
PRECIPITATION
TEMPERATURE
WIND
LANDSCAPE
ELEVATION
PROXIMITY TO LOWER-ELEVATION FOREST
topography
DISTURBANCE
FIRE
INSECT OUTBREAKS
DROUGHT
WIND
SOIL SYSTEM
root pathogens
decomposers
minerals
water
mycorrhizae
VEGETATION
FUEL
STRUCTURE
COMPOSITION
small-to large-scale patterns
ANIMALS
INSECTS
mammals
1 influence soil moisture & erosion
2 influence fuel moisture
3 cause fires, drought, & windthrow; influence fire behavior & insect population dynamics
4 influence weather & climate
5 influence spread & pattern of fire
6 provide population base of insects & other animals
7 influence tree mortality, fuel, structure, composition, & patterns
8 influence disturbance regime
9 expose soil & alter soil components
10 cause tree mortality & influence plant vigor & vegetation structure
11 input organic matter & retain soil
12 influence fuels & tree regeneration
13 provide habitat

b

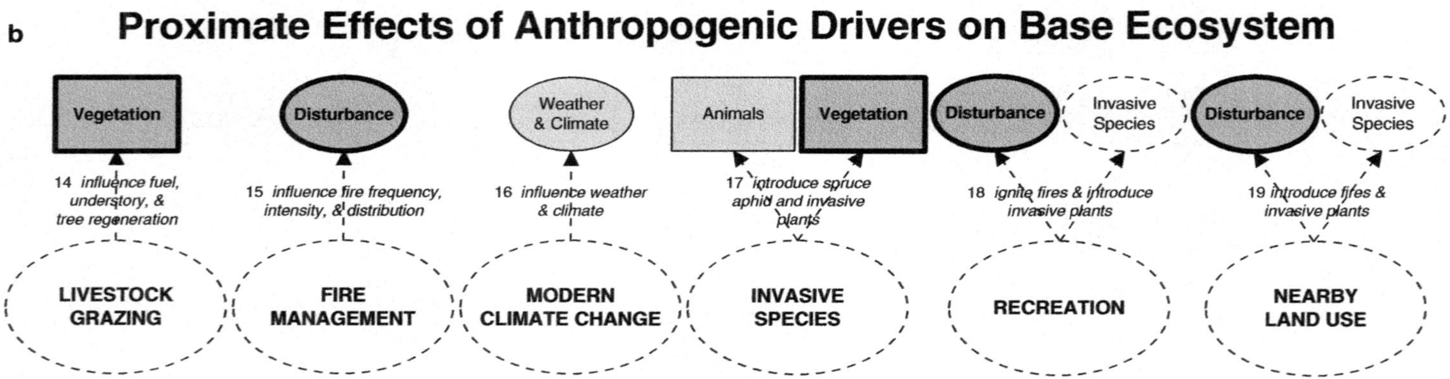

Fig. 2.19 (**a**, **b**) Ecosystem-characterization model for Spruce-Fir Forest. In the base model (**a**), biotic components are shown as *rectangles* and drivers as *solid ovals*. Text within them identifies major features in *upper case* and other important features in *lower case*. Disturbance and Vegetation are *highlighted*, because they are the basis for the vegetation-dynamics models (Figs. 2.20 and 2.21). In the anthropogenic supplement to the base model (**b**), drivers are shown as *dashed ovals*. Relationships in (**a**) and (**b**) are shown as *solid* and *dashed arrows*, respectively, and are described in Table 2.2

Table 2.2 Relationships in the ecosystem-characterization model of southwestern Spruce-Fir Forest (Fig. 2.19a, b)

Relationships	Circa 1870	Present
1	Precipitation increases soil moisture and can cause soil erosion. Winter precipitation is critical for recharging soil moisture	Same, but infiltration of precipitation, especially summer monsoonal rain, can be reduced by thicker layer of litter and duff, where unburned. Infiltration is reduced and soil erosion can be increased where this layer has been lost in high-severity fires. In addition, climate change increases the severity and duration of drought
2	Precipitation and temperature largely determine fuel moisture	Same, but climate change increases the severity and duration of drought
3	Lightning ignites fires. Consistently low precipitation and high temperature cause drought and affect insect population dynamics. Wind causes blowdowns and windthrows and influences fire behavior	Same, but climate change increases drought and insect outbreaks
4	Topography influences weather and climate	Same
5	Fires can spread from other areas of the landscape (especially from Mixed Conifer Forest), and fire patterns are influenced by elevation and topography	Same, but fire frequency is reduced
6	Increases in populations of spruce bark beetle and other insects can result in outbreaks	Same, but insects include the non-native spruce aphid and outbreaks can be more extensive
7	At lower elevations, frequent surface fires of mixed-severity fire regime thins tree regeneration, keeps fuel loadings low, and contributes to heterogeneous vegetation patterns. At all elevations, occasional crown fires kills trees in all size classes. Spruce bark beetles kill large Engelmann spruce. Drought stresses trees and increases probability of fires and outbreaks of spruce bark beetles. Infrequent strong winds cause large blowdowns in older stands	Past fire exclusion resulted in increased tree densities, fuel loadings, fuel continuity, and homogeneity of vegetation patterns at lower elevations. Surface fires are more likely to crown, and crown fires are more likely to burn larger areas. Crown fires kill trees in all size classes. Drier climate results in expansive stands of stressed trees and increased probability of fires and outbreaks of bark beetles
8	Patchy spatial heterogeneity in tree size, tree age, fuel loadings, and fuel continuity results in patchy disturbances	Greater spatial homogeneity in tree size, tree age, tree stress, fuel loadings, and fuel continuity can result in broader-scale disturbances

9	Fires at least partially and patchily expose soil surface and alter soil properties	Past fire exclusion resulted in thicker, more extensive litter and duff and likely altered other soil properties at lower elevations. Hotter, larger fires have greater effects on soil properties
10	Root pathogens cause tree mortality and influence vegetation structure. Water and minerals affect plant vigor and growth	Same, but less water is available because of drought and thicker litter and duff. Also, minerals can be less available because of greater sequestering in organic matter
11	Plants add organic matter to soil and stabilize soil, except in areas of high-severity disturbances	Same, except areas of high-severity disturbances can be larger
12	Herbivory influences fuels and tree regeneration	Same, but increased deer populations have greater effects
13	Vegetation is habitat for animals, including treefalls being habitat for spruce bark beetle	Same, but animal populations are changed where habitat is changed
14	Not applicable	Late nineteenth century livestock grazing reduced herb cover and likely influenced tree regeneration. The reduction in herb cover reduced the frequency of surface fires in lower-elevation Spruce-Fir Forest. Impacts of twenty-first century grazing are poorly known
15	Not applicable	Past fire management attempted to suppress fires throughout most of the twentieth century. More-recent fire management has made little use of prescribed or lightning-ignited management fires
16	Not applicable	Modern climate change affects weather and climate, raising temperature and increasing the frequency and extent of drought
17	Not applicable	The non-native spruce aphid has had outbreaks, but its potential impact is poorly known. Invasive plant species are uncommon
18	Not applicable	Recreationists ignite fires and introduce and spread invasive plants
19	Not applicable	Nearby land use that initiates fires and colonization by invasive plants can introduce fires and invasives into Spruce-Fir Forest

System and *Animals*, including insects that cause disturbance. A second driver is *Weather & Climate*, which ignites fires, causes drought, and influences fire behavior, insect population dynamics, and soil and fuel moisture. The third driver is *Landscape*, which includes elevation and proximity to Mixed Conifer Forest and Ponderosa Pine Forest. It influences weather, climate, and spread and pattern of fire. The model also includes six anthropogenic drivers (Fig. 2.19b, Table 2.2): *Livestock Grazing*, *Fire Management*, *Modern Climate Change*, *Invasive Species*, *Recreation*, and *Nearby Land Use*.

2.5.2 Vegetation-Dynamics Models

Separate models are necessary to illustrate the vegetation dynamics of upper-elevation and lower-elevation southwestern Spruce-Fir Forest, because of differences in fire regimes and quaking aspen. Each vegetation-dynamics model is accompanied by a bar graph to show estimated relative shifts in abundances of the community types from circa 1870 (prior to widespread Euro-American land use) to circa 1970 (end of strict fire exclusion policy) to the present. The estimates are based on my understanding of vegetation dynamics and should be interpreted as showing relative differences and not interpreted quantitatively (hence the absence of a scale for abundance). Although not shown on the graphs, relative abundances shifted prior to circa 1870 as a result of variation in climate and disturbance. Localized differences also have been and are present, but are beyond the scope of these generalized bar graphs.

2.5.2.1 Upper-Elevation Spruce-Fir Forest

The vegetation-dynamics model for upper-elevation Spruce-Fir Forest includes two states, four communities, and one transition (Fig. 2.20, Table 2.3). All states, communities, and transitions occurred historically, although *State A* has been more common than State B, and it encompasses the characteristic upper-elevation Spruce-Fir Forest communities. *State A* includes community *A1 Old Growth* dominated by Engelmann spruce and either subalpine or corkbark fir. It was the most common community historically and is maintained by gap dynamics. Crown fires, blowdowns, and insect outbreaks that kill most canopy trees change it into community *A2 Early Successional*, which is dominated by Engelmann spruce, sometimes with subalpine or corkbark fir. With succession, *A2* develops into community *A3 Mid Successional* with the same tree species; *A3* also can be formed directly from *A1* by insect outbreak that leaves many canopy trees alive. Crown fire converts *A3* into *A2*, and succession changes *A3* into *A1*.

In addition, unusually severe or repeated crown fire can change *State A* into *State B* via *Transition A↔B*. *State B* is made up of community *B1 Grassland*, which is maintained by fire. *Transition A↔B* can be reversed by invasion and establishment of Engelmann spruce and/or either subalpine or corkbark fir, converting *B1* into *A2*.

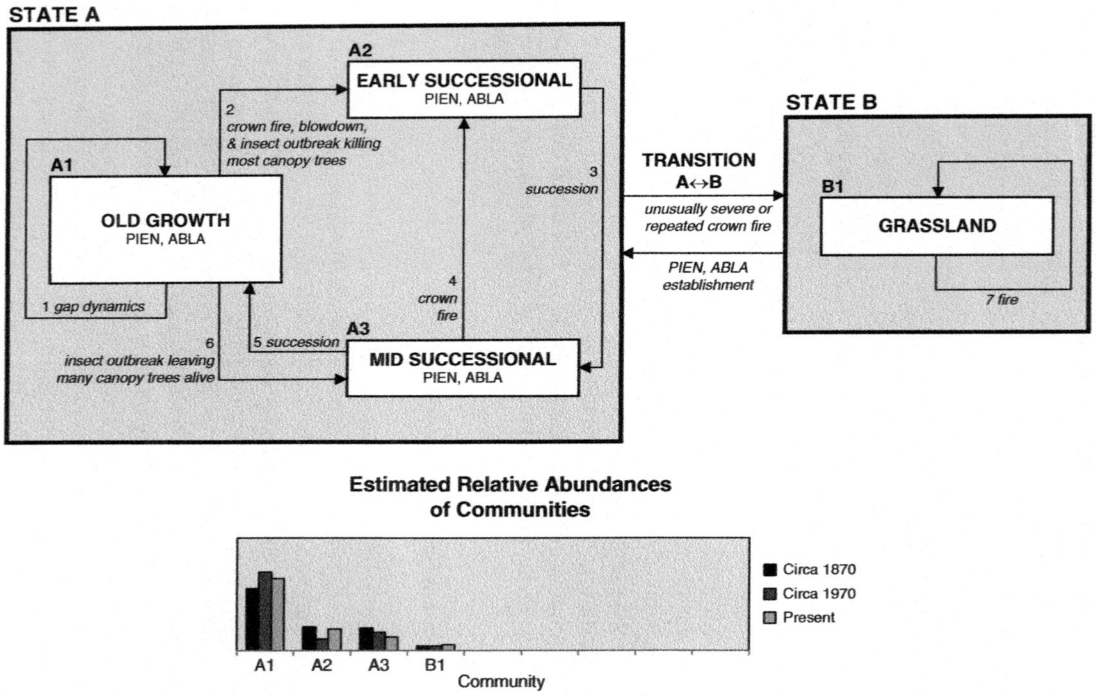

Fig. 2.20 Vegetation-dynamics model for upper-elevation Spruce-Fir Forest. Model provides details on the vegetation-disturbance portion of the ecosystem-characterization model (Fig. 2.19a, b) by showing communities (*smaller*, *open rectangles*) within states (*larger*, *shaded rectangles*), processes that form the communities, and transitions between states. Text along *arrows* identifies pathways that are described in Table 2.3. *Bar graph* shows estimated relative shifts in abundances of the community types from circa 1870 (see Sect. 2.5.2). Species: *ABLA* subalpine and corkbark fir, *PIEN* Engelmann spruce

Table 2.3 Relationships and transition shown in the vegetation-dynamics model of upper-elevation Spruce-Fir Forest (Fig. 2.20)

Relationships and transition	Circa 1870	Present
1	Gap dynamics maintain old growth	Same
2	Crown fire, blowdown, and insect outbreak that kill most canopy trees change old growth into early-successional stands	Same, but disturbance can be more common
3	Succession changes early-successional stands into mid-successional stands	Same
4	Crown fire changes mid-successional stands into early-successional stands	Same
5	Succession changes mid-successional stands into old growth	Same
6	Thinning by insect outbreak that leaves many canopy trees alive changes old growth into mid-successional stands	Same, but thinning by insect outbreaks can be more common
A ↔ B	Unusually severe or repeated crown fire converts some stands from State A to State B. Reversion to State A requires invasion and establishment of Engelmann spruce or either subalpine or corkbark fir to form early-successional stands	Same, but unusually severe crown fire is possibly more common. Larger crown fire limits invasion of Engelmann spruce and subalpine or corkbark fir
7	Fire maintains grassland stand structure and composition	Same, but fire is less frequent

2.5.2.2 Lower-Elevation Spruce-Fir Forest

The vegetation-dynamics model for lower-elevation Spruce-Fir Forest is similar, but the added presence of quaking aspen and surface fire result in an additional state, two additional communities, and two additional transitions (Fig. 2.21, Table 2.4). All states, communities, and transitions were represented historically, although *State A* has been more common than the other states, and it encompasses the characteristic lower-elevation Spruce-Fir Forest communities. Community *A1 Old Growth* is dominated by Engelmann spruce, either subalpine or corkbark fir, and quaking aspen. It is maintained by surface fire and gap dynamics. Crown fires, blowdowns, and insect outbreaks that kill most canopy trees change *A1* into one of two *Early Successional* communities, depending on the pre-disturbance abundance and vitality of quaking aspen in *A1*. Community *A2* is dominated by Engelmann spruce and sometimes has quaking aspen and either subalpine or corkbark fir. Community *A3* is dominated by quaking aspen. With succession, *A2* and *A3* develop into community *A4 Mid Successional* with the same tree species; *A4* also can be formed directly from *A1* by surface fire or insect outbreak that leaves many canopy trees alive. Crown fire converts *A4* into *A2* or *A3*, and succession changes *A4* into *A1*.

Unusually severe or repeated crown fire can change *State A* into *State B* via *Transition A*↔*B*. *State B* is made up of community *B1 Grassland*, which is maintained by fire. *Transition A*↔*B* can be reversed by invasion and establishment of Engelmann spruce, either subalpine or corkbark fir, and/or quaking aspen, converting *B1* into *A2* or *A3*, depending on the invading species.

In addition, large crown fire that isolates stands and is followed by regeneration of quaking aspen can change *State A* into *State C* via *Transition A*↔*C*. *State C* is made up of community *C1 Stable Aspen*, which is maintained by aspen regeneration in the absence of conifers. This transition can be reversed by invasion and establishment of Engelmann spruce and either subalpine or corkbark fir, converting *C1* into community *A4*.

State C also can be formed from *State B* by aspen invasion and establishment in *B1 Grassland* via *Transition B*↔*C*. This transition can be reversed by aspen mortality.

2.5.3 *Mechanistic Model*

Both vegetation-dynamics models are explained by the same mechanistic model (Fig. 2.22). It includes eight biotic components on the right side of the figure (including five aspects of fuels), five drivers on the left side, and six anthropogenic drivers at the bottom. In general, *Trees*, *Herbs & Shrubs*, and *Precipitation & Temperature* affect the five fuel characteristics. The fuel characteristics combined with *Fire Intensity*, *Fire Management*, *Nearby Land Use*, and *Recreation* influence *Fire Frequency*. *Fire Intensity*, *Precipitation & Temperature*, *Insect Outbreak*, and *Wind Blowdown* influence characteristics of *Trees*, such as species composition and tree age, size, density, and vigor. *Trees* and *Herbs & Shrubs* determine *Community Type* (of the four or six appearing in the vegetation-dynamics models).

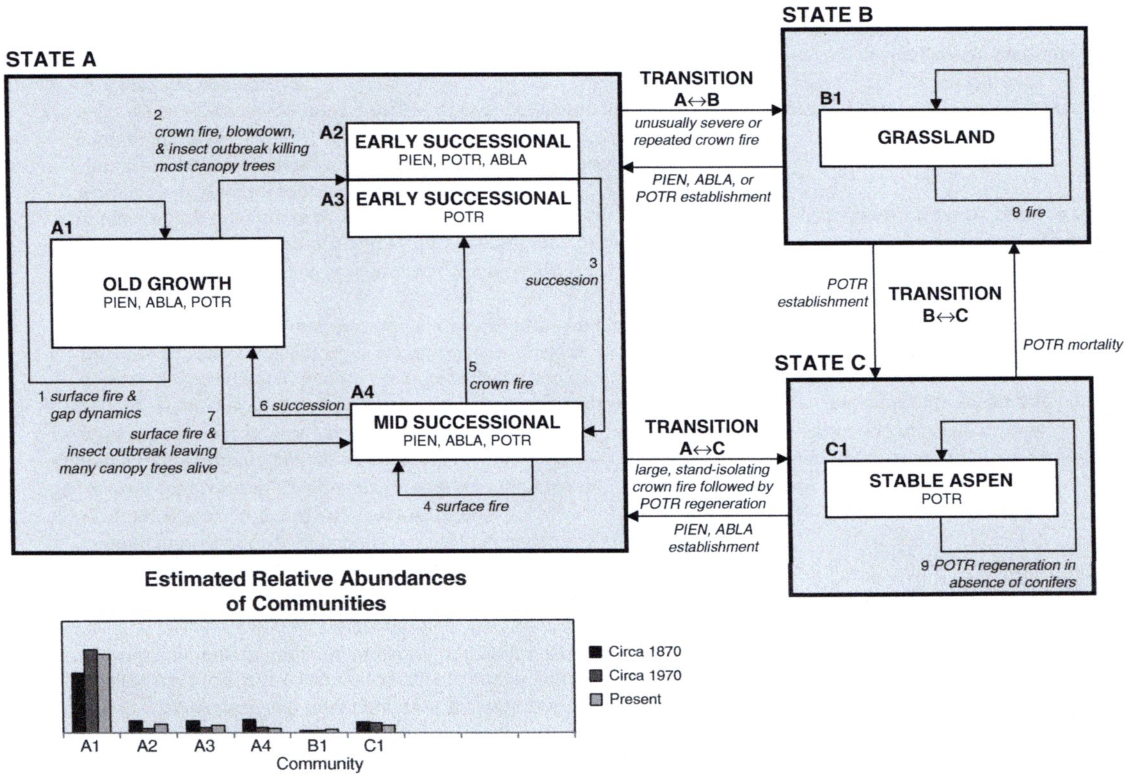

Fig. 2.21 Vegetation-dynamics model for lower-elevation Spruce-Fir Forest. Model provides details on the vegetation-disturbance portion of the ecosystem-characterization model (Fig. 2.19a, b) by showing communities (*smaller*, *open rectangles*) within states (*larger*, *shaded rectangles*), processes that form the communities, and transitions between states. Text along *arrows* identifies pathways that are described in Table 2.4. *Bar graph* shows estimated relative shifts in abundances of the community types from circa 1870 (see Sect. 2.5.2). Species: *ABLA* subalpine and corkbark fir, *PIEN* Engelmann spruce, *POTR* quaking aspen

Table 2.4 Relationships and transitions in the vegetation-dynamics model of lower-elevation Spruce-Fir Forest (Fig. 2.21)

Relationships and transitions	Circa 1870	Present
1	Surface fire and gap dynamics maintain stand structure and composition	Same, but surface fire is less frequent.
2	Crown fire, blowdown, and insect outbreak that kill most canopy trees change old growth into early-successional communities A2 or A3, depending on the abundance and vigor of quaking aspen in old growth	Same, but disturbance is possibly more common and disturbance patch size is larger
3	Succession changes early-successional stands into mid-successional stands	Same
4	Surface fire maintains stand structure	Same, but surface fire is less frequent
5	Crown fire changes mid-successional stands into early-successional communities A2 or A3, depending on the abundance and vigor of quaking aspen in the mid-successional stands	Same
6	Succession changes mid-successional stands into old growth	Same
7	Thinning by surface fire and insect outbreak that leaves many canopy trees alive changes old growth into mid-successional stands	Same, but surface fire is less frequent, and thinning by insect outbreaks is increased
A↔B	Unusually severe or repeated crown fire converts some stands from State A to State B. Reversion to State A requires invasion and establishment of Engelmann spruce, subalpine or corkbark fir, or quaking aspen to form early-successional communities A2 or A3, depending on the invading species	Same, but unusually severe and repeated crown fire is more common. Larger crown fire areas limit invasion of Engelmann spruce, subalpine or corkbark fir, and quaking aspen
8	Fire maintains grassland stand structure and composition	Same, but fire is less frequent
A↔C	Large, stand-isolating crown fire followed by regeneration of quaking aspen in the absence of conifers converts some stands from State A to State C. Reversion to State A requires invasion and establishment of Engelmann spruce or either subalpine or corkbark fir to form mid-successional community A4	Same, but crown fire is more common and likely larger in size, thereby limiting invasion of conifers
9	Quaking aspen regeneration in the absence of conifers maintains stand structure and composition	Same, but quaking aspen regeneration is reduced in some areas
B↔C	Quaking aspen invasion and establishment converts State B to State C. Reversion to State B requires quaking aspen mortality	Same, but quaking aspen mortality is more frequent in some areas

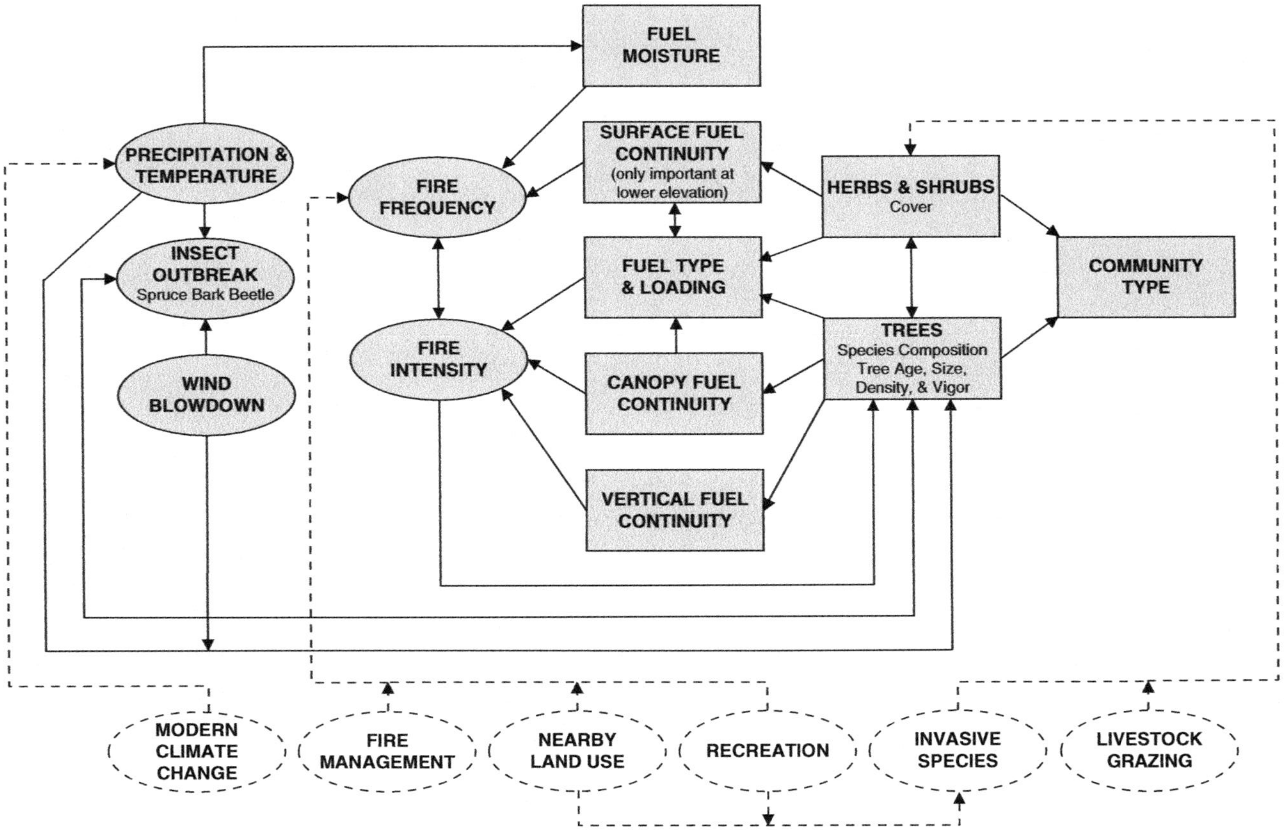

Fig. 2.22 Mechanistic model for Spruce-Fir Forest. Model provides details on the mechanisms involved in the vegetation-dynamics models (Figs. 2.20 and 2.21). *Rectangles* are biotic components (including fuels), *ovals* are drivers, and *dashed ovals* are anthropogenic drivers

2.6 Conclusions and Challenges

It is often assumed that results of research on Spruce-Fir Forest in the central and northern Rocky Mountains apply to the American Southwest, but this assumption requires testing, including studies of possible regional differences in species' biology. The disturbance regime of southwestern Spruce-Fir Forest is complex. The historical fire regime appears to have been crown-fire at upper elevations and mixed-severity fire at lower elevations, but more research is needed on both. Wind disturbance is unstudied in the American Southwest, and impacts of spruce beetle outbreaks also require additional study. Further research on the interactions of fire, wind, insects, and climate variation would enable land managers to better understand potential buffering effects among disturbances. For example, if crown fire reduces the occurrence and effects of wind and insects and vice-versa, would a mixed landscape be more resistant to disturbance events driven by modern climate change? The challenge may be educating the public on how natural disturbances are essential for healthy ecosystems. Aspects of anthropogenic disturbances needing research include the effects of fire suppression, the direct impacts of air pollution, the direct and indirect effects of modern climate change, and the biology of spruce aphid. Post-disturbance succession, including factors determining successional replacement vs. stability of stands of quaking aspen, has received little study in the American Southwest. Historical stand structure and composition have been erroneously characterized by current methods; therefore, further research is desirable – as is research on how livestock grazing, fire management, modern climate change, etc. changed historical conditions. The finding that forest density and basal area decreased in the twentieth century on the North Rim needs to be evaluated for other areas of the Southwest. Additional research on impacts of modern climate change, recreation, and nearby land use would help enhance decision-making by land managers.

References

Abolt RAP (1997) Fire histories of upper elevation forests in the Gila Wilderness, New Mexico via fire scar and stand age structure analyses. Thesis, University of Arizona, Tucson

Adams HD, Kolb TE (2005) Tree growth response to drought and temperature in a mountain landscape in northern Arizona, USA. J Biogeor 32:1629–1640

Alban DH, Pastor J (1993) Decomposition of aspen, spruce, and pine boles on two sites in Minnesota. Can J For Res 23:1744–1749

Alexander RR (1987) Ecology, silviculture, and management of the Engelmann spruce-subalpine fir type in the central and southern Rocky Mountains. U.S. Department of Agriculture, Forest Service, Agricultural Handbook 659, Washington, DC

Alexander RR, Shepperd WD (1984) Silvical characteristics of Engelmann spruce. U.S. Department of Agriculture, Forest Service General Technical Report RM-114, Rocky Mountain Forest and Range Experiment Station, Fort Collins, CO

Allen CD (2002) Lots of lightning and plenty of people: an ecological history of fire in the upland Southwest. In: Vale TR (ed) Fire, native peoples, and the natural landscape. Island Press, Covelo

Anderson RS, Hasbargen J, Koehler PA, Feiler EJ (1999) Late Wisconsin and Holocene subalpine forests of the Markagunt Plateau of Utah, southwestern Colorado Plateau, U.S.A. Arct Antarct Alp Res 31:366–378

Anderson RS, Allen CD, Toney JL, Jass RB, Bair AN (2008) Holocene vegetation and fire regimes in subalpine and mixed conifer forests, southern Rocky Mountains, USA. Int J Wildland Fire 17:96–114

Aplet GH, Laven RD, Smith FW (1988) Patterns of community dynamics in Colorado Engelmann spruce-subalpine fir forests. Ecology 62:312–319

Baker WL, Veblen TT (1990) Spruce beetles and fire in the nineteenth-century subalpine forests of western Colorado, U.S.A. Arct Alp Res 22:65–80

Baker WL, Flaherty PH, Lindemann JD, Veblen TT, Eisenhart KS, Kulakowski D (2002) Effect of vegetation on the impact of a severe blowdown in the Southern Rocky Mountains, USA. For Ecol Manag 168:63–75

Barrows JS (1978) Lightning fires in southwestern forests. Final report to USDA Forest Service, Northern Forest Fire Laboratory, Intermountain Forest and Range Experiment Station, under cooperative agreement 16–568 CA with Rocky Mountain Forest and Range Experiment Station, Fort Collins, CO

Bates ST (2006) A preliminary checklist of Arizona macrofungi. Canotia 2:47–78

Battaglia MA, Shepperd WD (2007) Ponderosa pine, mixed conifer, and spruce-fir forests. In: Hood SM, Miller M (eds) Fire ecology and management of the major ecosystems of southern Utah. U.S. Department of Agriculture, Forest Service General Technical Report RMRS-GTR-202, Rocky Mountain Research Station, Fort Collins, CO

Bebi P, Kulakowski D, Veblen TT (2003) Interactions between fire and spruce beetles in a subalpine Rocky Mountain forest landscape. Ecology 84:362–371

Benninger-Truax M, Vankat JL, Schaefer RL (1992) Trail corridors as habitat and conduits for movement of plant species in Rocky Mountain National Park, Colorado, USA. Landsc Ecol 6:269–278

Bigler C, Kulakowski D, Veblen TT (2005) Multiple disturbance interactions and drought influence fire severity in Rocky Mountain subalpine forests. Ecology 86:3018–3029

Bigler C, Gavin DG, Gunning C, Veblen TT (2007) Drought induces lagged tree mortality in a subalpine forest in the Rocky Mountains. Oikos 116:1983–1994

Binkley D, Moore MM, Romme WH, Brown PM (2006) Was Aldo Leopold right about the Kaibab deer herd? Ecosystems 9:227–241

Blackford D, Skov K, DeBlander V, Keyes C (2010) Utah forest insect and disease conditions report 2005–2009. U.S. Department of Agriculture, Forest Service, Forest Health Protection, Ogden, UT. http://www.fs.usda.gov/Internet/FSE_DOCUMENTS/stelprdb5196536.pdf. Accessed 5 Mar 2012

Bradley AF, Noste NV, Fischer WC (1992) Fire ecology of forests and woodlands in Utah. U.S. Department of Agriculture, Forest Service General Technical Report INT-287, Intermountain Research Station, Ogden, UT

Brown PM, Shepperd WD, Mata SA, McClain DL (1998) Longevity of windthrown logs in a subalpine forest of central Colorado. Can J For Res 28:932–936

Chambers CL, Holthausen RS (2000) Montane ecosystems used as rangelands. In: Jemison R, Raish C (eds) Livestock management in the American Southwest: ecology, society, and economics. Elsevier, New York

Cocke AE, Fulé PZ, Crouse JE (2005) Forest change on a steep mountain gradient after extended fire exclusion: San Francisco Peaks, Arizona, USA. J Appl Ecol 42:814–823

Crawford JA (2008) Five years of vegetation change following high-severity fire and fire-fighting activities in Grand Canyon National Park. In: van Riper C III, Sogge MK (eds) The Colorado Plateau III: Integrating research and resources management for effective conservation. Proceedings of the 8th biennial conference of research on the Colorado Plateau. University of Arizona Press, Tucson

Dahms CW, Geils BW (tech eds) (1997) An assessment of forest ecosystem health in the Southwest. U.S. Department of Agriculture, Forest Service General Technical Report RM-GTR-295, Rocky Mountain Forest and Range Experiment Station, Fort Collins, CO

DeRose RJ, Long JN (2007) Disturbance, structure, and composition: spruce beetle and Engelmann spruce forests on the Markagunt Plateau, Utah. For Ecol Manag 244:16–23

DeRose RJ, Long JN (2010) Regeneration response and seedling bank dynamics on a *Dendroctonus rufipennis*-killed *Picea engelmannii* landscape. J Veg Sci 21:377–387

DeRose RJ, Long JN (2012) Factors influencing the spatial and temporal dynamics of Engelmann spruce mortality during a spruce beetle outbreak on the Markagunt Plateau, Utah. For Sci 58:1–14

DeRose RJ, Long JN, Ramsey RD (2011) Combining dendrochronological data and the disturbance index to assess Engelmann spruce mortality caused by a spruce beetle outbreak in southern Utah, USA. Remote Sens Environ 115:2342–2349

DeVelice RL, Ludwig JA, Moir WH, Ronco F Jr (1986) A classification of forest habitat types of northern New Mexico and southern Colorado. U.S. Department of Agriculture, Forest Service General Technical Report RM-131, Rocky Mountain Forest and Range Experiment Station, Fort Collins, CO

Dixon H (1935) Ecological studies on the high plateaus of Utah. Bot Gaz 97:272–320

Dutton CE (1880) Report on the geology of the High Plateaus of Utah. Department of the Interior, U.S. Geographical and Geological Survey of the Rocky Mountain Region, Government Printing Office, Washington, DC

Dutton CE (1882) The physical geology of the Grand Cañon district. In: Powell JW (ed) Second annual report of the United States Geological Survey to the Secretary of the Interior 1880–'81. Government Printing Office, Washington, DC

Dye AJ, Moir WH (1977) Spruce-fir forest at its southern distribution in the Rocky Mountains, New Mexico. Am Midl Nat 97:133–146

Dymerski AD, Anhold JA, Munson AS (2001) Spruce beetle (*Dendroctonus rufipennis*) outbreak in Engelmann spruce (*Picea engelmannii*) in central Utah, 1986–1998. West N Am Nat 61:19–24

Ellison L (1954) Subalpine vegetation of the Wasatch Plateau, Utah. Ecol Monogr 24:89–184

Fairweather ML, Geils BW, Manthei M (2008) Aspen decline on the Coconino National Forest. In: McWilliams M, Palacios P (compilers) Proceedings of the 55th annual western international forest disease work conference. Oregon Department of Forestry, Salem

Fisher MA, Fulé PZ (2004) Changes in forest vegetation and arbuscular mycorrhizae along a steep elevation gradient in Arizona. For Ecol Manag 200:293–311

Fulé PZ, Crouse JE, Daniels M, Springer J, Heinlein TA, Moore MM, Covington WW, Verkamp G (2002a) Mixed-severity fire regime in a high-elevation forest: Grand Canyon, Arizona. Final Report to the Joint Fire Science Program, CA-1200-99-009-NAU 04 (Part 1)

Fulé PZ, Covington WW, Moore MM, Heinlein TA, Waltz AEM (2002b) Natural variability in forests of the Grand Canyon, USA. J Biogeogr 29:31–47

Fulé PZ, Crouse JE, Heinlein TA, Moore MM, Covington WW, Verkamp G (2003) Mixed-severity fire regime in a high-elevation forest of Grand Canyon, Arizona, USA. Landsc Ecol 18:465–486

Fulé PZ, Crouse JE, Cocke AE, Moore MM, Covington WW (2004) Changes in canopy fuels and potential fire behavior 1880–2040: Grand Canyon, Arizona. Ecol Modelling 175:231–248

Getsinger FR (1961) Blow down in the Kaibab forest. Westways 53(December):26–27

Gosz JR (1980) Biomass distribution and production budget for a nonaggrading forest ecosystem. Ecology 61:507–514

Grissino-Mayer HD, Baisan CH, Swetnam TW (1995) Fire history in the Pinaleño Mountains of southeastern Arizona: effects of human-related disturbances. In: DeBano LF, Ffolliott PF, Ortega-Rubio A, Gottfried GJ, Hamre RH, Edminster CB (tech coords) Biodiversity and management of the Madrean archipelago: the sky islands of southwestern United States and northwestern Mexico. U.S. Department of Agriculture, Forest Service, General Technical Report RM-GTR-264, Rocky Mountain Forest and Range Experiment Station, Fort Collins, CO

Hanley DP, Schmidt WC, Blake GM (1975) Stand structure and successional status of two spruce-fir forest in southern Utah. U.S. Department of Agriculture, Forest Service Research Paper INT-176, Intermountain Forest and Range Experiment Station, Ogden, UT

Hebertson EG, Jenkins MJ (2007) The influence of fallen tree timing on spruce beetle brood production. West N Am Nat 67:452–460

Hebertson EG, Jenkins MJ (2008) Climate factors associated with historic spruce beetle (Coleoptera: Curculionidae) outbreaks in Utah and Colorado. Environ Entomol 37:281–292

Hendricks DM (1985) Arizona soils. University of Arizona Press, Tucson

Hicke JA, Johnson MC, Hayes JL, Preisler HK (2012) Effects of bark beetle-caused tree mortality on wildfire. For Ecol Manag 271:81–90

Hurst M (1977) An ecological description of forest communities on the North Rim of Grand Canyon National Park. U.S. Department of the Interior, National Park Service manuscript on file, Grand Canyon National Park, AZ

Integrated Taxonomic Information System (2012) http://www.itis.gov. Accessed 27 July 2012

Jenkins MJ, Dicus CA, Hebertson EG (1998) Post-fire succession and disturbance interactions on an Intermountain subalpine spruce-fir forest. In: Pruden TL, Brennan LA (eds) Fire in ecosystem management: shifting the paradigm from suppression to prescription. Proceedings of the 20th Tall Timbers Fire Ecology Conference. Tall Timbers Research Station, Tallahassee, FL

Jenkins MJ, Page WG, Hebertson EG, Alexander ME (2012) Fuels and fire behavior dynamics in bark beetle-attacked forests in western North America and implications for fire management. For Ecol Manag 275:23–34

Jones JR, DeByle NV (1985) Soils. In: DeByle NV, Winokur PR (eds) Aspen: ecology and management in the western United States. U.S. Department of Agriculture, Forest Service General Technical Report RM-119, Rocky Mountain Forest and Range Experiment Station, Fort Collins, CO

Karnosky DF, Mankovska B, Percy K, Dickson RE, Podila GK, Sober J, Noormets A, Hendrey G, Coleman MD, Kubiske M, Pregitzer KS, Isebrands JG (1999) Effects of tropospheric O_3 on trembling aspen and interaction with CO_2: results from an O_3-gradient and a FACE experiment. Water Air Soil Pollut 116:311–322

Klemmedson JO, Smith EL (1979) Vegetation-soil relations of forests, woodlands, and grasslands of Arizona and New Mexico. In: Youngberg CT (ed) Forest soils and land use. Department of Forest and Wood Sciences, Colorado State University, Fort Collins

Knapp AK, Smith WK (1982) Factors influencing understory seedling establishment of Engelmann spruce (*Picea engelmannii*) and subalpine fir (*Abies lasiocarpa*) in southeast Wyoming. Can J Bot 60:2753–2761

Kulakowski D, Veblen TT (2002) Influences of fire history and topography on the pattern of a severe blowdown in a Colorado subalpine forest. J Ecol 90:806–819

Kulakowski D, Veblen TT (2003) Subalpine forest development following a blowdown in the Mount Zirkel Wilderness, Colorado. J Veg Sci 14:653–660

Kulakowski D, Veblen TT (2006) The effect of fires on susceptibility of subalpine forests to a 19th century spruce beetle outbreak in western Colorado. Can J For Res 36:2974–2982

Kulakowski D, Veblen TT, Bebi P (2003) Effects of fire and spruce beetle outbreak legacies on the disturbance regime of a subalpine forest in Colorado. J Biogeogr 30:1445–1456

Lang DK, Stewart SS (1910) Reconnaissance of the Kaibab National Forest. U.S. Department of Agriculture, Forest Service, Unpublished report. http://archive.library.nau.edu/cdm4/document.php?CISOROOT=/cpa&CISOPTR=57641&CISOSHOW=57583. Accessed 7 July 2011

Laughlin DC, Fulé PZ (2008) Wildland fire effects on understory plant communities in two fire-prone forests. Can J For Res 38:133–142

Laughlin DC, Bakker JD, Fulé PZ (2005) Understory plant community structure in lower montane and subalpine forests, Grand Canyon National Park, USA. J Biogeogr 32:2083–2102

Layser EF, Schubert GH (1979) Preliminary classification for the coniferous forest and woodland series of Arizona and New Mexico. U.S. Department of Agriculture, Forest Service Research Paper RM-208, Rocky Mountain Forest and Range Experiment Station, Fort Collins, CO

Leaf CF (1975) Watershed management in the Rocky Mountain subalpine zone: the status of our knowledge. U.S. Department of Agriculture, Forest Service General Technical Report RM-137, Rocky Mountain Forest and Range Experiment Station, Fort Collins, CO

Leiberg JB, Rixon TF, Dodwell A (1904) Forest conditions in the San Francisco Mountains Forest Reserve, Arizona. Department of the Interior, U.S. Geological Survey Series H, Forestry, 7, Professional Paper No. 22, Washington, DC

Lynch AM (2004) Fate and characteristics of *Picea* damaged by *Elatobium abietinum* (Walker) (Homoptera: Aphididae) in the White Mountains of Arizona. West N Am Nat 64:7–17

Margolis EQ, Balmat J (2009) Fire history and fire-climate relationships along a fire regime gradient in the Santa Fe Municipal Watershed, NM, USA. For Ecol Manag 258:2416–2430

Margolis EQ, Swetnam TW, Allen CD (2007) A stand-replacing fire history in upper montane forests of the southern Rocky Mountains. Can J For Res 37:2227–2241

Margolis EQ, Swetnam TW, Allen CD (2011) Historical stand-replacing fire in upper montane forests of the Madrean sky islands and Mogollon Plateau, southwestern USA. Fire Ecol 7:88–107

Merkle J (1954) An analysis of the spruce-fir community on the Kaibab Plateau, Arizona. Ecology 35:316–322

Merkle J (1962) Plant communities of the Grand Canyon area, Arizona. Ecology 43:698–710

Mielke JL (1950) Rate of deterioration of beetle-killed Engelmann spruce. J For 48:882–888

Milchunas DG (2006) Responses of plant communities to grazing in the southwestern United States. U.S. Department of Agriculture, Forest Service General Technical Report RMRS-GTR-169, Rocky Mountain Research Station, Fort Collins, CO

Miller WE (1983) Decomposition rates of aspen bole and branch litter. For Sci 29:351–356

Moir WH (1993) Alpine tundra and coniferous forest. In: Dick-Peddie WA New Mexico vegetation: past, present, and future. University of New Mexico Press, Albuquerque

Moir WH, Ludwig JA (1979) A classification of spruce-fir and mixed conifer habitat types of Arizona and New Mexico. U.S. Department of Agriculture, Forest Service Research Paper RM-207, Rocky Mountain Forest and Range Experiment Station, Fort Collins, CO

Moore MM, Huffman DW (2004) Tree encroachment on meadows of the North Rim, Grand Canyon National Park, Arizona, U.S.A. Arct Antarct Alp Res 36:474–483

Moore MM, Huffman DW, Fulé PZ, Covington WW, Crouse JE (2004) Comparison of historical and contemporary forest structure and composition on permanent plots in southwestern ponderosa pine forests. For Sci 50:162–176

Mueggler WF (1988) Aspen community types of the Intermountain Region. U.S. Department of Agriculture, Forest Service General Technical Report INT-250, Intermountain Research Station, Ogden, UT

Muldavin EH, DeVelice RL, Ronco F Jr (1996) A classification of forest habitat types: southern Arizona and portions of the Colorado Plateau. U.S. Department of Agriculture, Forest Service General Technical Report RM-GTR-287, Rocky Mountain Forest and Range Experiment Station, Fort Collins, CO

Niering WA, Lowe CH (1984) Vegetation in the Santa Catalina Mountains: community types and dynamics. Vegetatio 58:3–28

Ogle K, DuMond V (1997) Historical vegetation on National Forest lands in the Intermountain Region. U.S. Department of Agriculture Forest Service, Rocky Mountain Research Station, Ogden

Pase CP, Brown DE (1994) Rocky Mountain (Petran) subalpine conifer forest. In: Brown DE (ed) Biotic communities: southwestern United States and northwestern Mexico. University of Utah Press, Salt Lake City

Patten DT, Stromberg JC (1995) Dynamics of the landscape patches in the old growth forest on Mt. Graham (Pinaleno Mountains), Arizona. In: DeBano LF, Ffolliott PF, Ortega-Rubio A, Gottfried GJ, Hamre RH, Edminster CB (tech coords) Biodiversity and management of the Madrean Archipelago: the sky islands of southwestern United States and northwestern Mexico, 19–23 Sept 1994, Tucson, Arizona. U.S. Department of Agriculture, Forest Service General Technical Report RM-GTR-264, Rocky Mountain Forest and Range Experiment Station, Fort Collins, CO

Pauchard A, Kueffer C, Dietz H, Daehler CC, Alexander J, Edwards PJ, Arévalo JR, Cavieres LA, Guisan A, Haider S, Jakobs G, McDougall K, Millar CI, Naylor BJ, Parks CG, Rew LJ, Seipel T (2009) Ain't no mountain high enough: plant invasions reaching new elevations. Front Ecol Environ 7:479–486

Pearson GA (1920a) Factors controlling the distribution of forest trees: part I. Ecology 1:139–159

Pearson GA (1920b) Factors controlling the distribution of forest trees: part II. Ecology 1:289–308

Pearson GA (1931) Forest types in the Southwest as determined by climate and soil. Technical Bulletin 247. U.S. Department of Agriculture, Forest Service, Washington, DC

Peet RK (2000) Forests and meadows of the Rocky Mountains. In: Barbour MG, Billings WD (eds) North American terrestrial vegetation, 2nd edn. Cambridge University Press, New York
Pfister RD (1972) Vegetation and soils in the subalpine forests of Utah. Dissertation, Washington State University, Pullman
Price R, Evans RB (1937) Climate of the west front of the Wasatch Plateau in central Utah. Mon Weather Rev 65:291–301
Prior-Magee JS, Boykin KG, Bradford DF, Kepner WG, Lowry JH, Schrupp DL, Thomas KA, Thompson BC (2007) Ecoregional gap analysis of the southwestern United States: the Southwest regional gap analysis project final report. http://fws-nmcfwru.nmsu.edu/swregap/report/SWReGAP%20Final%20Report.pdf. Accessed 11 July 2012
Rebertus AJ, Veblen TT, Roovers LM, Mast JN (1992) Structure and dynamics of old-growth Engelmann spruce-subalpine fir in Colorado. In: Kaufmann MR, Moir WH, Bassett RL (techcoords) Old-growth forests in the Southwest and Rocky Mountain regions. Proceedings of a workshop, 9–13 Mar 1992, Portal, Arizona. U.S. Department of Agriculture, Forest Service General Technical Report RM-213, Rocky Mountain Forest and Range Experiment Station, Fort Collins, CO
Romme WH, Floyd-Hanna L, Hana DD, Bartlett E (2001) Aspen's ecological Role in the West. In: Sheppard WD, Binkley D, Bartos DL, Stohlgren TJ, Eskew LG (compilers) U.S. Department of Agriculture, Forest Service Proceedings RMRS-P-18, Rocky Mountain Research Station, Fort Collins, CO
Romme WH, Clement J, Hicke JA, Kulakowski D, MacDonald LH, Schoennagel T, Veblen TT (2006) Recent forest insect outbreaks and fire risk in Colorado forests: a brief synthesis of relevant research. Colorado Forest Restoration Institute, Colorado State University, Fort Collins. http://spot.colorado.edu/~schoenna/images/RommeEtAl2006CFRI%20.pdf. Accessed 3 July 2011
Romme WH, Redders JS, Floyd ML, Hanna D (2009a) Mixed conifer forests. In: Romme WH, Floyd ML, Hanna D (eds) Historical range of variability and current landscape condition analysis: south central highlands section, southwestern Colorado & northwestern New Mexico. Colorado Forest Restoration Institute, Colorado State University and Region 2, U.S. Forest Service, Fort Collins
Romme WH, Floyd ML, Hanna D, Redders JS, Lindsey JP (2009b) Spruce-fir forests. In: Romme WH, Floyd ML, Hanna D (eds) Historical range of variability and current landscape condition analysis: south central highlands section, southwestern Colorado & northwestern New Mexico. Colorado Forest Restoration Institute, Colorado State University and Region 2, U.S Forest Service, Fort Collins
Russo JP (1964) The Kaibab North deer herd: its history, problems and management. State of Arizona, Game and Fish Department Wildlife Bulletin No. 7, Phoenix
Schmid JM, Amman GD (1992) *Dendroctonus* beetles and old-growth forests in the Rockies. In: Kaufmann MR, Moir WH, Bassett RL (tech coords) Old-growth forests in the Southwest and Rocky Mountain regions. Proceedings of a workshop, 9–13 Mar 1992, Portal, Arizona. U.S. Department of Agriculture, Forest Service General Technical Report RM-213, Rocky Mountain Forest and Range Experiment Station, Fort Collins, CO
Schmid JM, Frye RH (1977) Spruce beetle in the Rockies. U.S. Department of Agriculture, Forest Service General Technical Report RM-49, Rocky Mountain Forest and Range Experiment Station, Fort Collins, CO
Schmid JM, Hinds TE (1974) Development of spruce-fir stands following spruce beetle outbreaks. U.S. Department of Agriculture, Forest Service Research Paper RM-131, Rocky Mountain Forest and Range Experiment Station, Fort Collins, CO
Schoennagel T, Veblen TT, Romme WH (2004) The interaction of fire, fuels, and climate across Rocky Mountain forests. Bioscience 54:661–676
Sibold JS, Veblen TT (2006) Relationships of subalpine forest fires in the Colorado Front Range with interannual and multidecadal scale climatic variation. J Biogeogr 33:833–842
Sibold JS, Veblen TT, Gonzalez ME (2006) Spatial and temporal variation in historic fire regimes in subalpine forests across the Colorado Front Range in Rocky Mountain National Park, Colorado, USA. J Biogeogr 33:631–647

Stromberg JC, Patten DT (1991) Dynamics of the spruce-fir forests on the Pinaleno Mountains, Graham Co., Arizona. Southwest Nat 36:37–48

Swetnam TW, Baisan CH (1996) Historical fire regime patterns in the southwestern United States since AD 1700. In: Allen CD (tech ed) Fire effects in southwestern forests: proceedings of the Second La Mesa Fire Symposium, Los Alamos, New Mexico, 29–31 Mar 1994. U.S. Department of Agriculture, Forest Service General Technical Report RM-GTR-286, Rocky Mountain Forest and Range Experiment Station, Fort Collins, CO

Swetnam TW, Baisan CH, Grissino-Mayer HD (2009) Tree-ring perspectives on fire regimes and forest dynamics in mixed conifer and spruce-fir forests on Mt. Graham. In: Sanderson HR, Koprowski JL (tech eds) The last refuge of the Mt. Graham red squirrel: ecology of endangerment, University of Arizona Press, Tucson

Taylor K, Mangold J, Rew LJ (2011) Weed dispersal by vehicles. MontGuide MT201105AG. http://msuextension.org/publications/AgandNaturalResources/MT201105AG.pdf. Accessed 30 Jan 2012

U.S. Forest Service (2003) Forest insect and disease conditions in the Southwestern Region, 2002. U.S. Department of Agriculture, Forest Service Southwestern Region Forestry and Forest Health R3-03-01. http://www.fs.usda.gov/Internet/FSE_DOCUMENTS/stelprdb5238438.pdf. Accessed 5 Mar 2012

U.S. Geological Survey National Gap Analysis Program (2005) Southwest Regional Gap Analysis Project, RS/GIS Laboratory, College of Natural Resources, Utah State University, Logan, Utah

Vankat JL (2010) Forest changes in Grand Canyon National Park and implications for management. Report to U.S. Department of the Interior, National Park Service, Grand Canyon National Park, AZ

Vankat JL (2011) Post-1935 changes in forest vegetation of Grand Canyon National Park, Arizona, USA: Part 2: Mixed conifer, spruce-fir, and quaking aspen forests. For Ecol Manag 261:326–341

Veblen TT (1986) Treefalls and the coexistence of conifers in subalpine forests of the central Rockies. Ecology 67:644–649

Veblen TT, Hadley KS, Reid MS, Rebertus AJ (1989) Blowdown and stand development in a Colorado subalpine forest. Can J For Res 19:1218–1225

Veblen TT, Hadley KS, Reid MS (1991a) Disturbance and stand development of a Colorado subalpine forest. J Biogeogr 18:707–716

Veblen TT, Hadley KS, Reid MS, Rebertus AJ (1991b) The response of subalpine forests to spruce beetle outbreak in Colorado. Ecology 72:213–231

Veblen TT, Hadley KS, Nel EM, Kitzberger T, Reid M, Villalba R (1994) Disturbance regime and disturbance interactions in a Rocky Mountain subalpine forest. J Ecol 82:125–135

Veblen TT, Kulakowski D, Eisenhart KS, Baker WL (2001) Subalpine forest damage from a severe windstorm in northern Colorado. Can J For Res 31:2089–2097

Wells FH, Lauenroth WK (2007) The potential for horses to disperse alien plants along recreational trails. Rangel Ecol Manag 60:574–577

Weng C, Jackson ST (1999) Late Glacial and Holocene vegetation history and paleoclimate of the Kaibab Plateau, Arizona. Palaeogeogr, Palaeoclimatol, Palaeoecol 153:179–201

White MA, Vankat JL (1993) Middle and high elevation coniferous forest communities of the North Rim region of Grand Canyon National Park, Arizona, USA. Vegetatio 109:161–174

Worrall JJ, Egeland L, Eager T, Mask RA, Johnson EW, Kemp PA, Shepperd WD (2008) Rapid mortality of *Populus tremuloides* in southwestern Colorado, USA. For Ecol Manag 255:686–696

Yeager LE, Riordan LE (1953) Effects of beetle-killed timber on range and wildlife in Colorado. Trans N Am Nat Resour Wildl Conf 18:596–616

Youngblood AP, Mauk RL (1985) Coniferous forest habitat types of central and southern Utah. U.S. Department of Agriculture, Forest Service General Technical Report INT-187, Intermountain Research Station, Ogden, UT

Chapter 3
Mixed Conifer Forest

Abstract Mixed Conifer Forest occurs in an elevational band below Spruce-Fir Forest and above Ponderosa Pine Forest. It has diverse stands reflecting elevation, topography, moisture availability, disturbance history, and successional state. Trees include ponderosa pine, Douglas-fir, white fir, quaking aspen, southwestern white pine, limber pine, blue spruce, and others. A primary driver of Mixed Conifer Forest has been fire. The historical, mixed-severity fire regime combined frequent, widespread, low-severity surface fires and infrequent, patchy, high-severity crown fires. Important anthropogenic disturbances are livestock grazing, fire management, modern climate change, invasive species, recreation, and nearby land use. Gap dynamics and succession are key vegetation dynamics. Historical conditions are poorly known. Forest density and basal area increased as a result of exclusion of fire, which began in the second half of the nineteenth century. At least some stands decreased in density and basal area during the twentieth century, apparently as a result of density-dependent factors and exogenous factors such as climate change. Fire exclusion also increased landscape homogeneity. Greater fuel loads and more continuous fuels have resulted in landscape-scale crown fires. Vegetation dynamics are illustrated in a nested, three-tiered set of conceptual models. Key conclusions and challenges for researchers and land managers are summarized.

3.1 Introduction

The term "mixed conifer" in the American Southwest applies to forests that have various combinations of ponderosa pine (*Pinus ponderosa*), Douglas-fir (*Pseudotsuga menziesii*), white fir (*Abies concolor*), quaking aspen (*Populus tremuloides*), southwestern white pine (*Pinus strobiformis*), limber pine (*Pinus flexilis*), blue spruce (*Picea pungens*), Engelmann spruce (*Picea engelmannii*), and either subalpine fir (*Abies lasiocarpa* var. *lasiocarpa*) or corkbark fir (*Abies lasiocarpa* var. *arizonica*; Figs. 3.1 and 3.2). With so many tree species, stands have different combinations of

J.L. Vankat, *Vegetation Dynamics on the Mountains and Plateaus of the American Southwest*, Plant and Vegetation 8, DOI 10.1007/978-94-007-6149-0_3,

Fig. 3.1 Mixed Conifer Forest in the San Juan Mountains of southwestern Colorado (Photograph by Betty J. Huffman)

Fig. 3.2 Mixed Conifer Forest along the border of New Mexico and Colorado (Photograph by author)

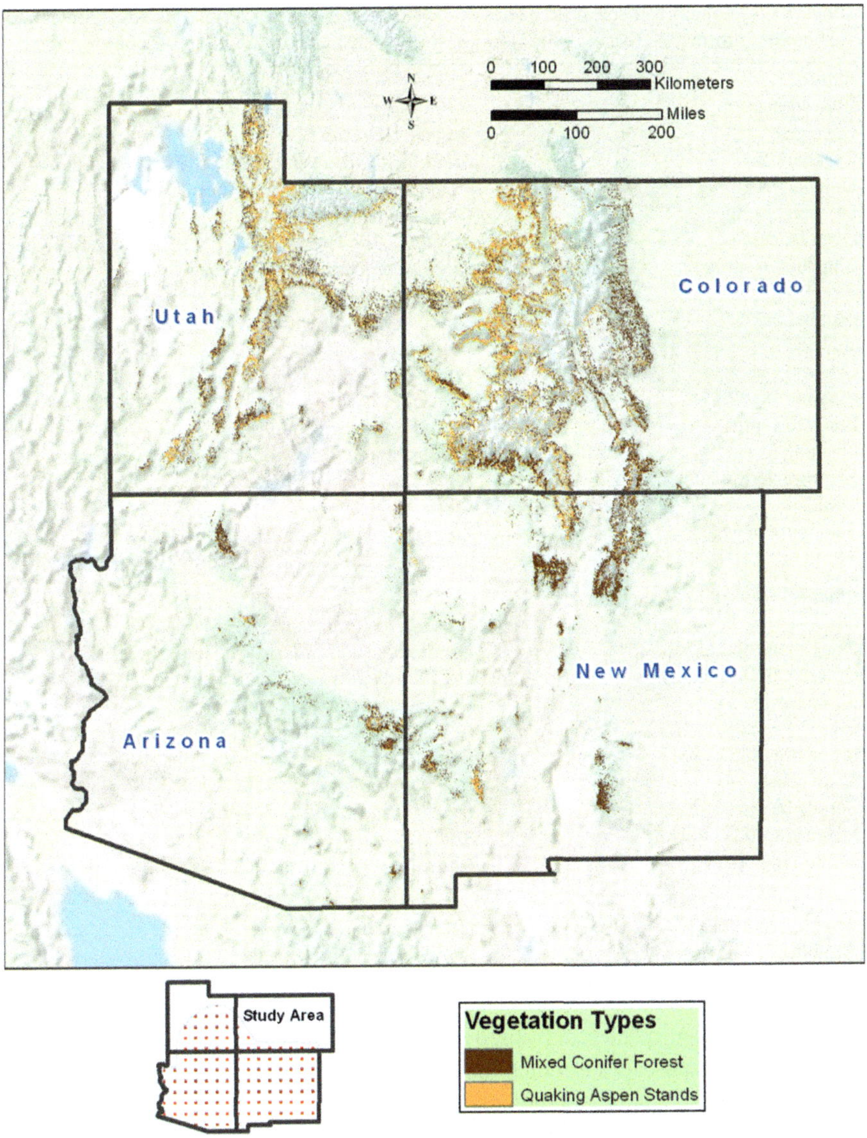

Fig. 3.3 Distribution of Mixed Conifer Forest (including subalpine and montane stands of quaking aspen) in the American Southwest. The map shows all of the states of Arizona, New Mexico, Colorado, and Utah, and the American Southwest is illustrated in *red* on the small map (Source: U.S. Geological Survey National Gap Analysis Program 2005 Southwest Regional Gap Analysis Project, RS/GIS Laboratory, College of Natural Resources, Utah State University, Logan, Utah. Map prepared by Monica Swihart)

trees that reflect elevation, topography, moisture availability, disturbance history, successional state, and other factors (Moir and Ludwig 1979; Youngblood and Mauk 1985; Bradley et al. 1992; Romme et al. 1992, 2009b; Moir 1993; White and Vankat 1993; Muldavin et al. 1996; Wu 1999; Fulé et al. 2003a; Aoki 2010). With its

Table 3.1 Common and scientific names of species in this chapter. Primary sources: Integrated Taxonomic Information System (2012) for plants and most animals and Bates (2006) for fungi

Plants	
Blue spruce	*Picea pungens* Engelm.
Cheatgrass	*Bromus tectorum* L.
Corkbark fir	*Abies lasiocarpa* var. *arizonica* (Merriam) Lemmon
Douglas-fir	*Pseudotsuga menziesii* (Mirbel) Franco
Dry spike sedge	*Carex foenea* Willd.
Dwarf mistletoe	*Arceuthobium* Bieb.
Engelmann spruce	*Picea engelmannii* Parry ex Engelm.
Fir	*Abies* P. Mill.
Gambel oak	*Quercus gambelii* Nutt.
Juniper	*Juniperus* L.
Limber pine	*Pinus flexilis* James
Pinyon	*Pinus* L.
Ponderosa pine	*Pinus ponderosa* Douglas ex P. Lawson & C. Lawson
Quaking aspen	*Populus tremuloides* Michx.
Southwestern white pine	*Pinus strobiformis* Engelm.
Spruce	*Picea* A. Dietr.
Subalpine fir	*Abies lasiocarpa* var. *arizonica* (Hook.) Nutt.
White fir	*Abies concolor* (Gord. & Glend.) Lindl. Ex Hildebr.
Animals	
Deer	*Odocoileus* Rafinesque, 1832
Douglas-fir beetle	*Dendroctonus pseudotsugae* Hopkins, 1905
Douglas-fir tussock moth	*Orgyia pseudotsugata* McDunnough, 1921
Fir engraver	*Scolytus ventralis* LeConte, 1868
Elk	*Cervus elaphus* Linnaeus, 1758
Mountain pine beetle	*Dendroctonus ponderosae* Hopkins, 1902
Mule deer	*Odocoileus hemionus* Rafinesque, 1817
Spruce beetle	*Dendroctonus rufipennis* Kirby, 1837
Western spruce budworm	*Archips fumiferana* Clemens
White-tailed deer	*Odocoileus virginianus* Zimmermann, 1780
Fungi	
Annosus root rot	*Heterobasidion annosum* (Fr.) Bref.
Armillaria root rot	*Armillaria* (Fr.:Fr.) Staude

multiple dominant species and varied site conditions, the mosaic of stands in Mixed Conifer Forest landscapes is more pronounced and diverse than in other southwestern coniferous forests. Stands dominated by Douglas-fir or aspen are often treated as separate forest types (e.g., Merriam 1890; Alexander et al. 1984; MacMahon 1988; Bradley et al. 1992; Moir 1993; Gottfried et al. 1995; Villanueva-Díaz and McPherson 1995; Hood and Miller 2007; Romme et al. 2009a).

Mixed Conifer Forest occurs in an elevational band below Spruce-Fir Forest (*Picea-Abies*), where present, and above Ponderosa Pine Forest. This position makes it transitional in terms of environment, species composition, and disturbance regimes, and this partially accounts for its complexity (Romme et al. 2009b). Mixed Conifer Forest is the third most extensive vegetation on the mountains and plateaus of the American Southwest after Pinyon-Juniper vegetation (*Pinus-Juniperus*) and Ponderosa Pine Forest. Mixed Conifer Forest covers about 13,000 km^2 (5,000 miles^2), which is 1.7 % of the total area of the region (Fig. 3.3; calculations based on

Prior-Magee et al. 2007). These values do not include stands dominated by quaking aspen, which cover 8,100 km^2 (3,100 $miles^2$) and 1.0 %. Approximately 50–60 % of aspen stands are associated with Mixed Conifer Forest (most of the rest with Spruce-Fir Forest). Mixed Conifer Forest landscapes also have stands of Subalpine-Montane Grassland. The area covered by Mixed Conifer Forest increases northward into southwestern Colorado and southern and central Utah. In Arizona and New Mexico, it occurs in relatively small areas usually surrounded by larger areas of Ponderosa Pine Forest; however, Mixed Conifer Forest is much more extensive than Ponderosa Pine Forest in Utah (O'Brien 1999). The forest has increased in area historically, as fire exclusion beginning in the nineteenth century resulted in increases in abundance of shade-tolerant conifers such as white fir in higher-elevation stands of Ponderosa Pine Forest, converting them into Mixed Conifer Forest (Dahms and Geils 1997; Swetnam et al. 2001; Mast and Wolf 2004; Evans et al. 2011).

The diverse stands of Mixed Conifer Forest have been the subject of detailed vegetation classifications for different geographic areas within the American Southwest (e.g., Layser and Schubert 1979; Moir and Ludwig 1979; Alexander et al. 1984; Youngblood and Mauk 1985; DeVelice et al. 1986; Mueggler and Campbell 1986; Mueggler 1988; Muldavin et al. 1996). In simpler form, stands are divided into three broad types: moist-mesic and dry-mesic stands dominated by conifers (Vankat 2011) and quaking aspen. All three types can occur in the same landscape, particularly at mid-elevations. Moist-mesic and dry-mesic are also referred to as cool-moist and warm-dry, respectively, and sometimes low-elevation Spruce-Fir Forest is treated as high-elevation cold-wet Mixed Conifer Forest (e.g., Romme et al. 2009b).

Moist-mesic stands of Mixed Conifer Forest (Fig. 3.4) occur on north and east aspects, lower slopes, and forested valley bottoms and are more common at high elevations. Stands generally have dense structure and are dominated by Engelmann spruce, blue spruce, ponderosa pine, quaking aspen, and subalpine or corkbark fir (White and Vankat 1993; Vankat 2011). Blue spruce often dominates low slope positions near drainages and margins of meadows (Moir 1993; White and Vankat 1993). Stands have greater fuel loadings and more vertically continuous fuels than other stand types.

Dry-mesic stands (Fig. 3.5) occur on south and west aspects, ridgetops, and mid slopes and are more common at low elevations. Stands generally have a more open structure than moist-mesic stands. They are dominated by ponderosa pine, white fir, quaking aspen, and Douglas-fir (Dieterich 1983; White and Vankat 1993; Vankat 2011). Stands have lower fuel loadings and less vertically continuous fuels than moist-mesic stands.

Quaking aspen stands (Fig. 3.6) occur primarily in areas burned by crown fire and other sites where conditions limit successional replacement of aspen by conifers, such as long distance to conifer seed sources (Sect. 3.3.2). They tend to be on deep, relatively fertile soils (Jones and DeByle 1985b). Some stands form by aspen invasion of Subalpine and Montane Grasslands (Moir 1993; Moore and Huffman

Fig. 3.4 Moist-mesic Mixed Conifer Forest in the North Rim region of Grand Canyon National Park, north-central Arizona (Photograph by author)

Fig. 3.5 Dry-mesic Mixed Conifer Forest in the North Rim region of Grand Canyon National Park, north-central Arizona (Photograph by author)

Fig. 3.6 Quaking-aspen Mixed Conifer Forest in the North Rim region of Grand Canyon National Park, north-central Arizona (Photograph by Betty J. Huffman)

2004). Aspen stands also occur in lower-elevation Spruce-Fir Forest and in mesic Ponderosa Pine Forest (Sects. 2.1 and 4.1, respectively).

Diversity of tree species, stands, and landscapes is a key aspect of southwestern Mixed Conifer Forest. The diversity of trees is important in succession following stand-initiating disturbances such as crown fire, because nearby undisturbed stands provide varied seed sources (Wu 1999). In addition, the tree species diversity – which includes the dominants of Spruce-Fir and Ponderosa Pine Forests – facilitates responses to climate change. With climate warming, an upward shift in elevation of Mixed Conifer Forest can occur as species of moist-mesic Mixed Conifer Forest increase in relative abundance in lower-elevation Spruce-Fir Forest (Sect. 2.1), converting it into moist-mesic Mixed Conifer Forest. At low elevations, most species of dry-mesic Mixed Conifer Forest decline in relative abundance but ponderosa pine increases and forms new stands of moist Ponderosa Pine Forest (Sect. 4.1). Conversely, with climate cooling a downward shift in elevation of Mixed Conifer Forest can occur as some species of moist-mesic Mixed Conifer Forest decline in relative abundance at high elevation but Engelmann spruce and subalpine or corkbark fir increase and form new stands of lower-elevation Spruce-Fir Forest. At low elevations, dry-mesic Mixed Conifer

Forest species increase in relative abundance in moist Ponderosa Pine Forest, converting it into dry-mesic Mixed Conifer Forest. These climate-driven shifts are facilitated by broad transition zones between adjacent forest types, as described in Sect. 3.2.1.

In the American Southwest, Mixed Conifer Forest has been researched more than Spruce-Fir Forest, but substantially less than Ponderosa Pine Forest. Research on mixed conifer forests elsewhere in North America, such as in the Sierra Nevada, Cascades, and the central and northern Rocky Mountains, is not necessarily applicable to the Southwest because of substantial differences in species composition and climate. Therefore, this chapter is based nearly entirely on research from the American Southwest.

3.2 Drivers

Key drivers of southwestern Mixed Conifer Forest are landscape, climate, soil, animals, natural disturbance, and anthropogenic disturbance. Each driver is important in vegetation dynamics, and therefore all are incorporated in the conceptual models (Sect. 3.5).

3.2.1 Landscape

Mixed Conifer Forest occurs mainly from approximately 2,200 to 3,000 m (7,200 to 9,800 ft) in the mid-latitudes of its distribution in the American Southwest (Fig. 3.7). The upper and lower elevational boundaries are indistinct because of overlap in tree species between adjacent forest types.

The upper-elevational limit of Mixed Conifer Forest is often a patchy transition with Spruce-Fir Forest. For example, Mixed Conifer Forest on higher elevations of the Kaibab Plateau of north-central Arizona becomes increasingly limited to relatively dry sites such as ridgetops and south and west aspects (Lang and Stewart 1910; White and Vankat 1993; Fulé et al. 2003a). Before fire exclusion, local topography interacted with fire to produce the mosaic within this transition (Fulé et al. 2003a).

At its lower-elevational limit, Mixed Conifer Forest transitions into Ponderosa Pine Forest. This ecotone can be broad, and in areas of diverse topography can be a mosaic (Plummer 1904; Greenamyre 1913). Transitional stands with mixed species dominance are included in this book as Mixed Conifer Forest.

Within the core of its elevational band, Mixed Conifer Forest consists of stands of different species composition that are related to topography, specifically gradients in elevation and topographically determined moisture, as well as disturbance

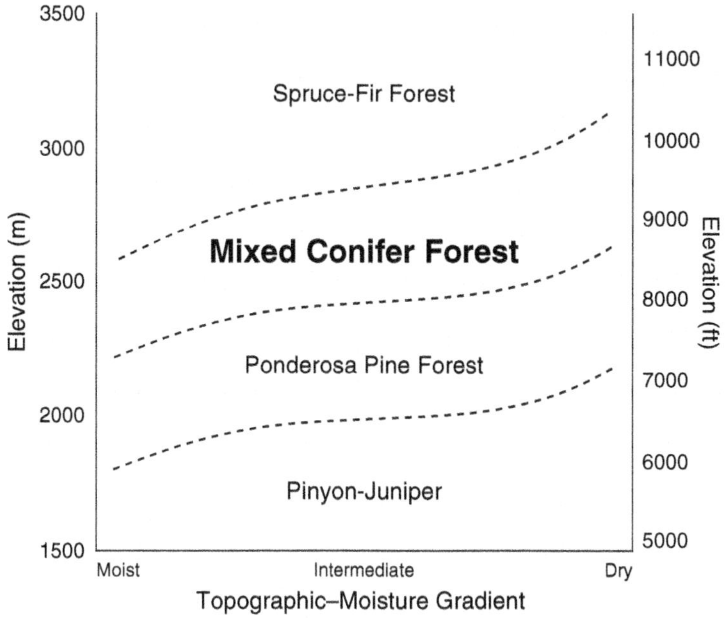

Fig. 3.7 Ecological distribution of Mixed Conifer Forest on the mountains and plateaus of the American Southwest along gradients in elevation and moisture. Elevations are approximate and generally representative of mid-latitudes of the region, i.e., landscapes in northern Arizona and northern New Mexico

(Figs. 3.8 and 3.9a, b; Lowe 1964; Lowe and Brown 1973; Klemmedson and Smith 1979; White and Vankat 1993; Wu 1999; Brown et al. 2001; Fulé et al. 2003a; Vankat et al. 2005). Most lower-elevation stands are dominated by ponderosa pine, with other species subdominant or dominant on more mesic topographic sites such as north and east aspects and lower slope positions. At mid-elevations, the mosaic includes stands dominated by ponderosa pine on dry topographic sites such as ridgetops, by mixed species on mesic sites, and by spruces, firs, and quaking aspen on moist sites such as valley bottoms. At higher elevations, mixed stands occupy drier topographic sites and spruces, either subalpine or corkbark fir, and quaking aspen dominate on more mesic sites.

Although stands dominated by ponderosa pine could be considered Ponderosa Pine Forest and stands dominated by spruces and subalpine or corkbark fir as Spruce-Fir Forest, they are included as part of the Mixed Conifer Forest in this book because (a) these stands occur throughout much of the elevational range of the Mixed Conifer Forest, not just in transition areas and (b) their ecosystem processes, such as fire, are closely integrated with adjacent stands. Superimposed, largely congruent, and reinforcing of this topography-driven mosaic is a second mosaic determined by disturbance and succession (Sects. 3.2.5 and 3.3.2).

Fig. 3.8 Topographic effects on stand structure and composition across a Mixed Conifer Forest landscape. Note the open forest structure of the south-facing slope in the mid-ground, compared to denser stands on the opposite north-facing slope. Also note the distribution of quaking aspen (*light green color*), which indicates past crown fire, on portions of the slopes and uplands in the background (Photograph along the border between New Mexico and Colorado by author)

3.2.2 Climate

Few long-term climate data have been published for Mixed Conifer Forest in the American Southwest. The climate is generally characterized by cool to cold temperatures producing short to moderately long growing seasons averaging 87 days over two decades in central Utah (range: 61–114 days; Price and Evans 1937). Temperature is influenced by topographic factors such as elevation and slope aspect. Mean annual temperature is <2–6 °C (<36–43 °F), and mean annual precipitation typically 600–1,120 mm (24–44 in.; Pearson 1920; Moir 1993; Chambers and Holthausen 2000). July and August are the months with highest average

Fig. 3.9 (**a**, **b**) Topographic effects illustrated by differences in structure and composition of nearby stands of Mixed Conifer Forest. (**a**) The relatively open stand dominated by ponderosa pine is on a ridgetop. (**b**) The dense stand of quaking aspen, white fir, and spruce is on the east-facing slope of the same landform approximately 75 m (250 ft) away and 30 m (100 ft) lower in elevation (Photographs in North Rim region of Grand Canyon National Park, north-central Arizona, by author (**a**) and by Isabella Colombari, American Conservation Experience (**b**) (Courtesy of Grand Canyon National Park))

precipitation (Gottfried and Ffolliott 1992). Mean annual snowfall in Arizona stands is 2.5–4.0 m (8–13 ft), as estimated from Beschta (n.d.). Lightning is common; a large area that includes Mixed Conifer Forest in the Jemez Mountains of north-central New Mexico annually averaged 2.1 cloud-to-ground lightning strikes/ha (0.8 strikes/acre; Allen 2002 for 1985–1994).

Unlike many other vegetation types in the American Southwest, Mixed Conifer Forest is typically not limited by moisture because there is ample precipitation (Moir 1993). The combination of moisture availability and warm daytime temperatures during the growing season results in Mixed Conifer Forest being the Southwest's most productive forest (Moir 1993) and most productive terrestrial vegetation (Whittaker and Niering 1975).

Modern climate change is described in Sect. 1.6.3 and is treated as an anthropogenic driver of Mixed Conifer Forest in Sect. 3.2.6.3.

3.2.3 Soil

Southwestern Mixed Conifer Forest occurs on a variety of soils that have formed from several types of parent material. The soils are generally deep, permeable, and capable of storing snowmelt. Litter and duff depth combined depends on the species dominating the canopy. On the San Francisco Peaks of north-central Arizona, stands dominated by a mix of conifers have greater depths similar to Spruce-Fir Forest, and stands dominated by quaking aspen have shallower depths similar to Ponderosa Pine Forest (Cocke et al. 2005). Soil moisture is generally much greater than at lower elevations (Pearson 1931). The soil moisture regime is udic, as water is available all or most of the growing season, and the soil temperature regime ranges from frigid to cryic (Klemmedson and Smith 1979; Moir 1993). Soils with Mixed Conifer Forest in Arizona are Alfisols (Typic Eutroboralfs, Typic Glossoboralfs, and Mollic Eutroboralfs) and Entisols (Typic Cryorthents and Typic Ustorthents), with most being well-drained, shallow to deep, and fine to moderately coarse textured (Hendricks 1985). Soils in New Mexico are mostly Alfisols with some Mollisols (Maker and Saugherty 1986).

Soils also include fungal diseases that can affect vegetation dynamics, including Armillaria root rot (*Armillaria* spp.) and annosus root rot (*Heterobasidion annosum*).

3.2.4 Animals

Animal species important in the dynamics of southwestern Mixed Conifer Forest include insect species such as western spruce budworm (*Archips fumiferana*), mountain pine beetle (*Dendroctonus ponderosae*), Douglas-fir beetle (*D. pseudotsugae*), spruce beetle (*D. rufipennis*), fir engraver (*Scolytus ventralis*), and

Fig. 3.10 Complex mosaic of stands on the San Francisco Peaks, Coconino National Forest, north-central Arizona. Differences in proportions of quaking aspen (*yellow* and *light green*) and conifers (*dark green*) in the canopy reflect differences among stands relative to past disturbance and succession (Photograph by Betty J. Huffman)

Douglas-fir tussock moth (*Orgyia pseudotsugata*). Large ungulates sometimes important in vegetation dynamics are mule deer (*Odocoileus hemionus*), white-tailed deer (*O. virginianus*), and elk (*Cervus elaphus*). The impacts of animals are described in Sects. 3.2.5.3 and 3.2.5.4.

3.2.5 Natural Disturbance

Most natural disturbance occurs in Southwestern Mixed Conifer Forest as a result of fire and biotic agents such as insects. Disturbance was more frequent prior to fire exclusion and occurred in a mosaic of different intensities across landscapes. This disturbance-driven mosaic was superimposed on and largely congruent with the topography-driven mosaic described in Sect. 3.2.1 (Swetnam and Lynch 1989; Swetnam 1990; Moir 1993; White and Vankat 1993). Moreover, the disturbance- and topography-driven mosaics were mutually reinforcing. Differences in tree species' dominance in the disturbance-driven mosaic were characterized largely by differences in the abundance of quaking aspen (Fig. 3.10). Aspen is a shade-intolerant, early successional species that rapidly responds to burning and other disturbance of its stems by sprouting (suckering) from its shallow, spreading root system (e.g., Frey et al. 2003); aspen rarely regenerates sexually (Bartos 2007). Aspen is the sole dominant in early-successional and stable (persistent) stands, co-dominant with conifers in mid-successional stands, and absent or sub-dominant in late-successional coniferous stands (Sect. 3.3.2). Patches of aspen and mixed aspen-conifers are used to determine past disturbance (e.g., Margolis et al. 2007, 2011).

3.2.5.1 Fire

Fire is a key driver of Mixed Conifer Forest in the American Southwest. This forest type had 12 % of all lightning fires in forests and woodlands of the National Forests of Arizona and New Mexico, compared to <1 % in Spruce-Fir Forest (calculated from data in Barrows 1978, for 1960–1974). The historical fire regime is characterized as mixed-severity, having combined frequent, widespread, relatively low-severity surface fires and infrequent, patchy, high-severity crown fires (Figs. 3.11 and 3.12; Allen 1989; Allen et al. 1995; Touchan et al. 1996; Abolt 1997; Wu 1999; Swetnam et al. 2001, 2009; Fulé et al. 2003a; Grissino-Mayer et al. 2004; Margolis et al. 2007, 2011; Frechette and Meyer 2009; Margolis and Balmat 2009; Aoki 2010; Bigio et al. 2010; Jenkins et al. 2011). Not all fires were mixed-severity, but the regime was mixed-severity both temporally and spatially (Sect. 1.2.5.1). Moist-mesic stands had both surface and crown fires, but dry-mesic stands had predominantly surface fires. The complexity of the fire regime necessitates lengthy description.

Fire intensities were related to variables such as stand structure and composition, fuels, elevation, topography (especially aspect), weather-climate, and fire history (Swetnam and Brown 1992; Touchan et al. 1996; Abolt 1997; Brown et al. 2001; Jenkins et al. 2011). Fires generally burned as surface fires across landscapes, especially at lower elevations and in relatively dry, open areas such as ridgetops and south and west aspects. In dry years, fires occasionally crowned in areas of more fuels and more vertically continuous fuels, such as at higher elevations and in mesic, dense sites on north and east aspects. The limiting factor for surface fire was generally moisture, not fuel (Allen et al. 1995; Swetnam and Baisan 1996; Touchan et al. 1996; Wu 1999; Fulé et al. 2009; Margolis and Balmat 2009). Crown fire was limited by both moisture and fuel (cf. Frechette and Meyer 2009; Bigio et al. 2010; Jenkins et al. 2011). The primary fire season is from April through June, when there is little precipitation (Sect. 1.2.2). The potential severity and length of the fire season depend on moisture from the snowpack of the preceding winter and long-term climate (Margolis et al. 2007, 2011), as well as weather of the current monsoon season.

The surface-fire component of the historical mixed-severity fire regime is better known than the crown-fire component. Analyses of fire scars show seasonal variation, with most historical surface fires having occurred in late spring to early summer (May–June) when fuels were drier (Dieterich 1983; Grissino-Mayer et al. 1995; Wolf and Mast 1998; Heinlein et al. 2005; Margolis and Balmat 2009). Seasonal patterns differed regionally (cf. Grissino-Mayer et al. 2004) and topographically (Heinlein et al. 2005).

Fig. 3.11 Surface fire in Mixed Conifer Forest in north-central New Mexico (Photograph by Kari Greer/U.S. Forest Service)

Fig. 3.12 Crown fire in Mixed Conifer Forest in east-central Arizona (Photograph by Jayson Coil)

Historical surface fires tended to occur in drier years (Touchan et al. 1996; Grissino-Mayer et al. 2004; Brown et al. 2008; Fulé et al. 2009). Studies have differed on whether fires were associated with wet conditions in the preceding 1–2 years. This association is generally true for Ponderosa Pine Forest, where the precipitation increases the cover and continuity of herbaceous fuels (Sect. 4.2.5.1). Some studies found no such relationship in Mixed Conifer Forest (Swetnam and Baisan 1996; Touchan et al. 1996; Margolis et al. 2007; Fulé et al. 2009), as would be expected for a fire regime ostensibly not limited by fuel. In contrast, other studies found this relationship (Baisan and Swetnam 1990; Grissino-Mayer et al. 2004; Allen et al. 2008; Margolis and Balmat 2009), a counterintuitive finding for a forest where fuel is generally not limiting. A possible explanation is that increased fuels from prior wet years are required to carry fire from lower-elevation, drier Ponderosa Pine Forest into Mixed Conifer Forest, as well as to carry fire from drier sites such as south-facing slopes to more moist sites such as north-facing slopes within topographically heterogeneous Mixed Conifer Forest landscapes (Margolis and Balmat 2009).

Mean intervals for surface fire determined from fire scars ranged from 9 to 26 years (median of 16 years; ≥10 % scarred) from 1700 to 1900 in nine low-elevation Mixed Conifer Forest sites in the Southwest (Swetnam and Baisan 1996; the wide range of reported mean intervals in this section is partly accounted for by differences in area and intensity of sampling). Other studies across the Southwest have reported intervals that fall into or near this range (e.g., Dieterich 1983; Stein 1988; Touchan et al. 1996; Jenkins et al. 1998; Wu 1999; Grissino-Mayer et al. 2004; Fulé et al. 2003a, b, 2009; Heinlein et al. 2005; Allen et al. 2008; Margolis and Balmat 2009; Bigio et al. 2010). Mean fire intervals varied spatially (and temporally). They generally decreased with elevation. For example, relatively high-, mid-, and low-elevation sites had historical mean intervals (≥10 % scarred) of 12, 9, and 5 years (Wolf and Mast 1998) and 16, 7, and 6 years (Fulé et al. 2003a) in the North Rim region of Grand Canyon National Park in north-central Arizona. Fire intervals also differed by slope aspect. North aspects on the North Rim had the longest historical mean intervals and south aspects the shortest (13 vs. 6 years; ≥10 % scarred; Fulé et al. 2003a). In the San Juan Mountains of southwestern Colorado, mesic aspects had much longer historical mean intervals than xeric aspects (42 vs. 12 years, respectively; ≥10 % scarred; Wu 1999). In contrast, different aspects had similar mean fire intervals in an area of the Santa Catalina Mountains of southeastern Arizona (Iniguez et al. 2008).

The range of historical mean fire intervals for surfaces fires in Mixed Conifer Forest overlapped that of Ponderosa Pine Forest. In general, intervals were longer in Mixed Conifer Forest (e.g., Touchan et al. 1996; Swetnam et al. 2001; Margolis and Balmat 2009). Similarity in historical mean fire intervals can be related to fire synchrony because of proximity of the two forest types (Allen et al. 1995) and likely spread of fires from low to higher elevation. Also, results could be biased toward similarity because most studies utilized fire scars from the same species, ponderosa pine. However, synchrony was not always present: only 24 % of fires in Ponderosa Pine Forest were recorded in Mixed Conifer Forest in a watershed of the Sangre de

Cristo Mountains in north-central New Mexico (Margolis and Balmat 2009; see also Heinlein et al. 2005).

Differences in historical mean fire intervals between moist-mesic and dry-mesic Mixed Conifer Forest ranged from small to large. On the North Rim, mean fire intervals were 9 and 7 years for mesic mixed stands and drier ponderosa-pine-dominated stands, respectively (Fulé et al. 2003a; ≥10 % scarred). Data from 32 sites in the Southwest also showed that mean fire intervals were only 2 years different between mesic and drier sites (Evans et al. 2011; 15 years for "mixed conifer" and 13 years for "ponderosa pine-mixed conifer"). In contrast, moist-mesic stands in the San Juan Mountains had very infrequent fires (an assumption based on few trees having more than one fire scar; Aoki 2010). And mean fire intervals decreased in increasingly drier stands from 52 to 22 to 19 years (compiled from Wu 1999). The large regional differences among moist-mesic stands appear related to several factors. First, differences in mean fire intervals can be an artifact of sampling at a landscape scale (North Rim) vs. smaller site-scale studies (San Juan Mountains), because more fires determined with larger-scale sampling shorten historical mean fire intervals. Second, the relatively high precipitation of the San Juan Mountains reduces the number of fires. Third, the higher landscape heterogeneity of the North Rim results in more frequent fires in moist-mesic stands because of their close proximity to dry-mesic stands with relatively high fire frequencies.

Fire intervals in quaking aspen stands are more difficult to determine and are understudied. Fire scars in typical aspen stands in Ephraim Canyon in central Utah suggested relatively frequent "small, light fires" occurred before Euro-American settlement (Baker 1925). Fire-scarred conifers in or adjacent to an aspen patch in Mixed Conifer Forest in north-central New Mexico indicated a mean fire interval of 7 years during 1847–1873, but other aspen stands studied had no more than one fire in addition to the stand-initiating fire (Margolis et al. 2007).Further insight is that a 76-km^2 (29-$miles^2$) aspen-dominated landscape in the western San Juan Mountains had a 140-year fire rotation period at the time of Euro-American settlement (Romme et al. 2001).

Another important parameter of fire regimes is lengths of individual fire intervals, i.e., gaps in fire occurrence. These fire-free periods are especially significant for Mixed Conifer Forest, because its many tree species have different requirements for establishment (see below). Variation in fire-free periods is essential to the diversity of Mixed Conifer Forest (Wu 1999). Individual fire-free periods differ over time and space (Wu 1999), including among microsites within stands and among regions across the Southwest. For example, historical fire-free periods were 2–15 years and 1–28 years in two areas of the North Rim, compared to 3–50 years in an area of southwestern Colorado (Fulé et al. 2003a, b, 2009; ≥10 % scarred; see also Grissino-Mayer et al. 2004; Margolis and Balmat 2009; Swetnam et al. 2009; again, such differences are can be partially accounted for by differences in sampling area and intensity). Spatial and temporal variability in fire-free periods is related to continuity of fuels and topography (Swetnam and Brown 1992; Brown et al. 2001), as well as climate and the vagaries of weather. For example, south and north slopes on the North Rim had maximum fire-free periods of 13 and 34 years, respectively (Fulé et al. 2003a). This was greater variation than reported for elevation and forest type (differences in sampling area and intensity likely influenced results).

Historically, fire-free periods tended to be longer in Mixed Conifer Forest than in Ponderosa Pine Forest (Baisan and Swetnam 1990; Swetnam et al. 2001; Grissino-Mayer et al. 2004; Margolis and Balmat 2009). For example, historical fire-free periods ranged from 4 to 48 years for Mixed Conifer Forest vs. 4– 27 years for Ponderosa Pine Forest within a small drainage in the Jemez Mountains (Allen et al. 2008). Longer individual and mean fire-free periods are the result of more mesic conditions in Mixed Conifer Forest, where fuel moisture is usually higher as a result of greater precipitation from orographic uplift, lower evapotranspiration rates from cooler temperature, and greater spring melt from snowpack (Grissino-Mayer et al. 2004). When Mixed Conifer Forest fuels were dry, historical surface fires were generally more widespread in Mixed Conifer Forest than in Ponderosa Pine Forest (Fulé et al. 2003b; Grissino-Mayer et al. 2004), presumably because of greater fuel loadings and fuel continuity. Fire years in Mixed Conifer Forest were significantly drier than in Ponderosa Pine Forest in a watershed in the Sangre de Cristo Mountains (Margolis and Balmat 2009).

Differences in fire-free periods between moist-mesic and dry-mesic Mixed Conifer Forest range from relatively small to large. On the heterogeneous landscape of the North Rim, ranges were similar for mesic mixed stands (1–28 years) and drier ponderosa-pine-dominated stands (2–25 years; Fulé et al. 2003a). Data from 27 sites in the Southwest showed overall ranges of 1–50 for mesic "mixed conifer" and 1–33 years for drier "ponderosa pine-mixed conifer" (compiled from Evans et al. 2011, but averages of the shortest periods and longest periods resulted in similar ranges: 2–23 and 1–21 years, respectively). Fire-free periods in increasingly drier Mixed Conifer Forest stands in the San Juan Mountains were 24–125, 1–65, and 2–97 years (compiled from Wu 1999; averages of the shortest and longest periods of the ranges were 24–95, 9–47, and 5–59 years, respectively).

Little information is available on size of historical surface fires, but it is likely that size was highly variable. Sizes generally have been characterized as small in moist-mesic sites in the San Juan Mountains (Romme et al. 2009b), possibly based on the finding that many fires scarred only one tree in moist-mesic stands (Aoki 2010).

Although the crown-fire component of the historical mixed-severity fire regime is not as well understood as the surface-fire component (Margolis et al. 2011), there is clear evidence of infrequent, patchy crown fires. The evidence includes fire-originated stands: even-aged aspen, even-aged conifers, and some stands of Subalpine-Montane Grasslands. Crown-fire occurrence was affected by fuel conditions, including steep slopes with vertical stacking of tree crowns (Jenkins et al. 2011). Fuel loadings increased during moist periods and where topographic fuel breaks reduced the spread of surface fires (the relationship to insect outbreaks is complex; cf. Hicke et al. 2012; Jenkins et al. 2012). Research in the Gila Wilderness of west-central New Mexico indicated that severe fires tended to occur in Mixed Conifer and Spruce-Fir Forests, on steep slopes, on north-facing slopes, and in cool, wet sites (Holden et al. 2009). The amount of area severely burned was correlated with variability in precipitation in the previous several months (Holden et al. 2007). Slope aspect also was a critical factor in surface vs. crown fire in Mixed Conifer Forest in a watershed in the Sangre de Cristo Mountains (Margolis and Balmat 2009) and on the San Francisco Peaks, where even stands on adjacent slopes below the same ridge top differed in surface vs. crown fire (Margolis et al. 2011).

Crown-fire patches were highly variable in size. Some fire-originated stands of aspen and grasslands in the Jemez Mountains are very large, extending across relatively homogeneous slopes (Allen 1984, 1989; Touchan et al. 1996). A study of aspen stands originated by historical crown fires in Mixed Conifer Forest of north-central New Mexico and south-central Colorado included patches ranging from 66 to 1,173 ha (163 to 2,899 acres; Margolis et al. 2007; stand selection favored large patches). Patches produced by a nineteenth century crown fire in a watershed of the Sangre de Cristo Mountains were determined to have been smaller, ranging from 34 to 110 ha (84 to 272 acres; Margolis and Balmat 2009). Patches from a single, extensive fire in the eastern San Juan Mountains were heterogeneous in size and mostly <25 ha (<62 acres; Aoki 2010). On the highly heterogeneous topography of the North Rim, patches of trees reflecting past crown fires were limited to 2 ha (5 acres; Fulé et al. 2003b). This wide range of patch sizes in the American Southwest appears strongly positively related to the degree of topographic homogeneity, including the absence of fuel breaks. Maximum patch size was smaller in Mixed Conifer Forest than in Spruce-Fir Forest on three mountains in Arizona and New Mexico: 286 vs. 521 ha (706 vs. 1,287 acres; Margolis et al. 2011).

Dating crown fires is challenging. If trees that originated following a crown fire are still present, especially in an even-aged stand, tree-ring counts provide the approximate date (cf. Margolis et al. 2007, 2011). Intervals within a landscape can be as short as 29 years or much longer (Aoki 2010). Maximum tree ages in patches of old-growth conifers have been used to estimate minimum time since crown fire. Trees 300-years old in the White Mountains of east-central Arizona (Dieterich 1983) and 500–800 years in the southern Rocky Mountains (Lynch and Swetnam 1992) have been interpreted as indicating rarity of crown fire. Crown fires also have been dated from Holocene charcoal deposits in alluvial sediments (Frechette and Meyer 2009; Bigio et al. 2010; Jenkins et al. 2011). Unfortunately, these studies are unable to determine the frequency of pre-historical crown fires because alluvial sediments probably do not record all crown fires, and fire size is not documented because sediments reflect the entire drainage, not just the area burned. Crown fires recorded in alluvium on Kendrick Mountain in north-central Arizona suggest intervals averaged 200–400 years during the last 2,000 years, with variation depending on long-term patterns in climate (Jenkins et al. 2011). Crown fires appear to have occurred during severe droughts that followed multi-decadal wet periods during which fuels accumulated (Margolis et al. 2007; Fulé et al. 2009; Frechette and Meyer 2009; Bigio et al. 2010; Jenkins et al. 2011).

Differences in crown fires between moist-mesic and dry-mesic Mixed Conifer Forest need clarification. Age structures in moist-mesic stands in the eastern San

Juan Mountains indicated crown fires had been widespread (evidence was found in 70 % of randomly selected plots), but patches were small (<25 ha/62 acres; Aoki 2010). These findings and others that indicated few surface fires led to the conclusion that the historical fire regime of moist-mesic stands in the study area in the eastern San Juan Mountains had been similar to that of Spruce-Fir Forest (Aoki 2010; Sect. 2.2.5.1). Studies have not focused explicitly on historical crown fires in dry-mesic Mixed Conifer Forest, but they have been hypothesized as rare (e.g., Romme et al. 2009b). This and the high frequency of surface fires indicate the historical fire regime of dry-mesic stands was similar to that of Ponderosa Pine Forest (e.g., Wu 1999; Sect. 4.2.5.1).

Effects of the surface-fire portion of the mixed-severity fire regime likely were similar to those of surface fire in Ponderosa Pine Forest, which have been studied more thoroughly (Sect. 4.2.5.1). In general, frequent surface fires kept forest structure more open, favored tree regeneration by reducing herb cover, thinned cohorts of tree seedlings and saplings but increased growth of survivors, and kept fuel loads relatively constant spatially and temporally. Different lengths of fire-free intervals affected tree regeneration. Much of the variation in tree regeneration is related to bark thickness, which is a key factor statistically related to species composition across the range of woodlands and forests on uplands of the American Southwest (Laughlin et al. 2011). Short fire-free intervals favor ponderosa pine and Douglas-fir, species that develop fire-resistant bark at a relatively young age. Moderately long fire-free intervals are necessary for regeneration of white fir and limber pine, species that take longer to develop such bark. Long intervals are required for species that do not develop fire-resistant bark, including Engelmann spruce, blue spruce, subalpine fir, and corkbark fir.

Effects of the crown-fire portion of the mixed-severity fire regime include loss of canopy and subcanopy trees, reduction of fuels, and regeneration of quaking aspen and other early successional species. Patchy crown fires initiated stands that accentuated the disturbance mosaic of southwestern Mixed Conifer Forest landscapes (White and Vankat 1993; Abolt 1997; Wu 1999).

As in other forest types in the American Southwest, the historical fire regime of Mixed Conifer Forest changed in the second half of the nineteenth century (see Sects. 3.2.6.1 and 3.2.6.2).

3.2.5.2 Wind

Wind disturbance is less evident in Mixed Conifer Forest than in Spruce-Fir Forest (see Sect. 2.2.5.2). Stand-scale blowdowns have not been reported. Windthrows of single to a few trees occur (Fig. 3.13; Pearson 1950), but infrequently for quaking aspen (Jones and DeByle 1985a). Windthrows initiate gap dynamics (Sect. 3.3.1) and are more common on coarsely textured soils (Fulé et al. 2002). Wind damage (sometimes augmented by root pathogens) is common in Douglas-fir, corkbark fir, and Engelmann spruce (Gottfried 1978).

Fig. 3.13 Small blowdown of Douglas-fir in 1928 in Lincoln National Forest, south-central New Mexico (Photograph by E.S. Shipp/U.S. Forest Service)

3.2.5.3 Biotic Agents

Many species impact trees in southwestern Mixed Conifer Forest. Examples of host-specific insect species include mountain pine beetle on ponderosa pine, Douglas-fir beetle on Douglas-fir, and spruce beetle on Engelmann spruce. Examples of generalist insects include western spruce budworm on Douglas-fir, firs, and spruces, fir engraver on firs, and Douglas-fir tussock moth on Douglas-fir, white fir, and spruces. Important fungal diseases are Armillaria root rot on ponderosa pine, Douglas-fir, spruces, and subalpine fir and annosus root rot on ponderosa pine, white fir, and subalpine fir. Dwarf mistletoes (*Arceuthobium* spp.) affect most tree species. Large ungulates impact quaking-aspen regeneration.

The primary insect affecting southwestern Mixed Conifer Forest is the western spruce budworm, a defoliator (Fig. 3.14). It is considered the most destructive insect in coniferous forests of the western United States (Fellin and Dewey 1986 in Ryerson et al. 2003). In the American Southwest, western spruce budworm feeds mostly on Douglas-fir and white fir (Linnane 1986). Western spruce budworm can weaken trees, making them more susceptible to infestation by Douglas-fir beetle and fir engraver (Lynch and Swetnam 1992). Forests most susceptible to outbreaks

(a) are old, dense, closed-canopied, and multi-layered, (b) have Douglas-fir and white fir as canopy dominants, (c) have shade-tolerant species in the understory, and (d) are stressed by drought, high density, dwarf mistletoe, root disease, or marginal site conditions (Linnane 1986; Fellin et al. 1990 in Moir 1993; Lynch and Swetnam 1992). Outbreaks can be extensive, e.g., defoliation was detected in 2,266 km^2 (875 $miles^2$) in Arizona and New Mexico in 2009 (U.S. Forest Service 2010; area not limited to Mixed Conifer Forest). Regional outbreaks have occurred in the Southwest at intervals of 20–50 years and up to 83 years for larger outbreaks (Swetnam 1987; Swetnam and Lynch 1989, 1993; Ryerson et al. 2003). Outbreaks lasted about 11 years in northern New Mexico (Swetnam and Lynch 1993). Budworm activity generally increased in wetter periods and decreased in drier periods (Swetnam and Lynch 1993; Ryerson et al. 2003).

It has been stated that fire exclusion made southwestern Mixed Conifer Forest more susceptible to severe, extensive, spatially synchronous outbreaks of western spruce budworm (Moir 1993; Battaglia and Shepperd 2007). Similar statements have been made about mountain pine beetle, Douglas-fir beetle, Douglas-fir tussock moth, Armillaria and annosus root rots, and dwarf mistletoe (Battaglia and Shepperd 2007). Studies of western spruce budworm found: severity either increased or remained unchanged with fire exclusion, extent increased, spatial synchrony increased, frequency remained unchanged, and duration either remained unchanged or somewhat decreased (Swetnam 1987; Lynch and Swetnam 1992; Swetnam and Lynch 1993; Ryerson et al. 2003).

Outbreaks of western spruce budworm have a variety of impacts on southwestern Mixed Conifer Forest. This defoliator feeds primarily on understory trees (Brookes et al. 1987), especially individuals in a weakened condition, and therefore acts as a thinning agent (Moir 1993). Stands can survive multiple outbreaks (Ryerson et al. 2003), but overstory mortality can occur following repeated defoliations or interaction with other insects and pathogens (Linnane 1986). Outbreaks can alter forest structure, composition, and dynamics (Lynch and Swetnam 1992; Moir 1993). Forest structure is changed by the loss of understory trees, and several defoliations can produce single-storied stands of large Douglas-fir and white fir (Moir 1993). Forest composition is changed by mortality resulting from selective feeding on Douglas-fir and white fir, as well as by regeneration of ponderosa pine, quaking aspen, and southwestern white pine in subsequent canopy gaps (Moir 1993). Such changes in forest composition can affect forest dynamics, e.g., the loss of shade-tolerant trees can slow succession on drier sites (Moir 1993). In addition, outbreaks of western spruce budworm and other insects can increase the probability, severity, and extent of fires; however, research has not shown that these effects are strong (Romme et al. 2006).

Dwarf mistletoe infects approximately half of the area of southwestern Mixed Conifer Forest, where it can affect fire behavior and is affected by fire (Conklin and Fairweather 2010; Evans et al. 2011). Dwarf mistletoe affects fire by increasing tree mortality, which increases fuel loadings (Mathiasen et al. 1990). Dwarf mistletoe also causes dense clusters of branches ("witches brooms") and accumulation of flammable resins on live trees, and these can facilitate the laddering of fire into

Fig. 3.14 Tree mortality caused mostly by western spruce budworm, in Carson National Forest, north-central New Mexico (Photograph by Daniel Ryerson, Forest Health, New Mexico Zone, U.S. Forest Service)

Fig. 3.15 Deer and/or elk produce a browse line below which all aspen branches and leaves are consumed, as shown in a stand of Mixed Conifer Forest burned 12 years earlier near the San Francisco Peaks in north-central Arizona. Browsing can cause mortality of shorter aspens (Photograph by Clarissa Thorne)

forest canopies (Alexander and Hawksworth 1975; Evans et al. 2011; Fig. 1.26). Fire's effects on dwarf mistletoe include reducing its populations by both scorch-pruning and killing infected trees and reducing the spread of dwarf mistletoe through maintaining more open forests (Conklin and Fairweather 2010).

Mule deer, white–tailed deer, and elk are common large ungulates in much of the Mixed Conifer Forest of the American Southwest. Their population dynamics before Euro-American influence are unknown, but likely affected tree regeneration patterns, especially in quaking aspen, whose sprouts are heavily browsed (Fig. 3.15; cf. McHenry 1935; Rasmussen 1941; Merkle 1954; Fulé et al. 2002, 2003a; Mast and Wolf 2006). In addition, deer (*Odocoileus* spp.) and elk population dynamics possibly have influenced the frequency of surface fires by affecting the abundance and spatial continuity of grasses and forbs, although this has not been documented. The mixed-severity fire regime likely favored deer and elk by stimulating aspen root-sprouting and maintaining open stands with significant grass cover (Dieterich 1983).

3.2.5.4 Climate Variation

Drought acts as a disturbance agent in southwestern Mixed Conifer Forest primarily in interaction with other disturbance agents. For example, drought is associated with surface and crown fires, as well as the population dynamics of western spruce budworm (Sects. 3.2.5.1 and 3.2.5.3, respectively). Drought in the early twenty-first century in north-central Arizona was associated with a >200 % increase in tree mortality in 2002–2007 (Ganey and Vojta 2011). Mortality was disproportionately greater for quaking aspen and white fir and lower for all other species, including ponderosa pine and Douglas-fir. Mortality occurred in all study plots, but was spatially variable – although uncorrelated with elevation and stand density. Absolute mortality was greater in smaller diameter classes, but relative mortality was generally greater in larger diameter classes. Relative mortality was higher than in Ponderosa Pine Forest (Sect. 4.2.5.4).

Drought is involved with declines in aspen, and these have been reported for southwestern Mixed Conifer Forest and other forests of North America (Shepperd et al. 2001; Fairweather et al. 2008; Worrall et al. 2008; Rehfeldt et al. 2009). For example, large decreases in aspen density and basal area occurred from 1935 to 2004 in conifer-aspen stands in Mixed Conifer Forest on the North Rim (Vankat 2011). The declines were attributed to the interaction of several factors, many of which have broad geographic applicability: (a) Sudden Aspen Decline (Fig. 3.16) in which strong drought, warm temperatures, late frosts, and repeated defoliation by insects facilitate increased, synchronous aspen mortality by bark beetles, other insects, and canker fungi that impact stressed trees (Fairweather et al. 2008; Worrall et al. 2008; Zegler et al. 2012), (b) reduced regeneration of aspen because of fire exclusion (Moir 1993; White and Vankat 1993), (c) reduced recruitment of saplings and trees because of herbivory by elevated deer populations (e.g., Merkle 1962; Fulé et al. 2002, 2003a; Mast and Wolf 2006; Moore and Huffman 2004; Binkley et al. 2006), (d) increased competition from shade-tolerant conifers such as white fir

Fig. 3.16 Patch of dying and dead quaking aspen (*center*) killed by Sudden Aspen Decline (SAD) in San Juan National Forest, southwestern Colorado (Photograph by Phil Kemp, U.S. Forest Service)

during fire exclusion (Sect. 3.4.1.2), (e) ozone concentrations that have the potential to negatively impact many aspects of aspen biology (e.g., Karnosky et al. 1999), and (f) succession from aspen to conifer dominance in stands and across landscapes (Sect. 3.3.2), as facilitated by fire exclusion.

3.2.6 Anthropogenic Drivers

Land use by Native Americans in the American Southwest was concentrated in Pinyon-Juniper vegetation (Sects. 1.5.1.1 and 5.2.6) and likely was uncommon and limited in area and impact in Mixed Conifer Forest. Key anthropogenic drivers related primarily to Euro-American land use are livestock grazing, fire management, modern climate change, invasive species, recreation, and nearby land use. Each of these can be important in vegetation dynamics, and therefore all are incorporated in the conceptual models (Sect. 3.5). Logging occurred, but is outside the scope of this book.

3.2.6.1 Livestock Grazing

Livestock grazing became widespread in the American Southwest in the late nineteenth century. It reduced the biomass and continuity of the herbaceous layer that had carried surface fires, and the frequency of surface fires in southwestern Mixed Conifer Forest abruptly decreased. Today, livestock grazing is generally less widespread and less intensive, although it is still impactful (Battaglia and Shepperd 2007). Grazing has occurred most commonly in stands with quaking aspen (Reynolds 1969; Milchunas 2006), and it has been the primary economic use of aspen stands in the western United States (DeByle 1985). Grazing reduces aspen regeneration and understory cover, affecting tree age distributions and understory composition (DeByle 1985; Mueggler 1985a; Bartos 2007).

3.2.6.2 Fire Management

Fire management throughout most of the twentieth century focused on preventing and suppressing fires. This continued the exclusion of fire that began with livestock grazing (previous section). Without frequent fires, southwestern Mixed Conifer Forest changed in structure and composition at both stand and landscape scales. Younger trees, especially of shade-tolerant species such as white fir, increased, and shade-intolerant species such as quaking aspen decreased. Overall, there were large increases in tree density, fuel loading, and horizontal and vertical fuel continuity in individual stands and across landscapes. Therefore, conditions for the crown-fire component of the mixed-severity fire regime increased across landscapes, making them at risk of large crown fires (White and Vankat 1993; Abolt 1997; Wu 1999). The fire regime shifted from patchy, mixed-severity fires toward landscape-scale crown fires, i.e., fires that burned across landscapes with heterogeneous topography and historically heterogeneous vegetation. The potential spread of fire was enhanced by the successional loss of stands of quaking aspen, which formerly had been natural fire breaks in landscapes because of their low flammability (Battaglia and Shepperd 2007).

Fire management practices began to shift focus in the late twentieth century to include management fires (both prescribed and lightning-ignited). High fuel loadings and fuel continuity make such fire management practices risky in Mixed Conifer Forest, and some management fires have grown into landscape-scale crown fires requiring suppression. Examples include the Outlet, Poplar, and Warm Fires on the Kaibab Plateau during 2000–2006. Similar wildfires have occurred elsewhere in the American Southwest (see Battaglia and Shepperd 2007). It appears that these landscape-scale crown fires are highly anomalous when compared to the history of fire in southwestern Mixed Conifer Forest (Swetnam et al. 2001; see also discussion of crown-fire size in Sect. 3.2.5.1).

3.2.6.3 Modern Climate Change

Modern climate change is implicated in many changes, including increased drought, widespread decreases in aspen, and decreases in Mixed Conifer Forest density and basal area on the North Rim (Sects. 3.2.5.4 and 3.4.1.2). Therefore, it has affected the structure, composition, and dynamics of southwestern Mixed Conifer Forest stands and landscapes. Air pollution is a driver of modern climate change, but pollution sources are more global than regional or local, and direct effects of air pollution on southwestern Mixed Conifer Forest are understudied. However, there is circumstantial evidence of likely impacts. For example, ozone levels in Grand Canyon National Park (Bowman 2003) have exceeded thresholds for foliar injury on ponderosa pines observed in California (National Park Service 2002). Ozone also can affect quaking aspen (Sect. 3.2.5.4). Nevertheless, with the lack of documentation of direct effects, air pollution is not included in this chapter as a separate anthropogenic driver of southwestern Mixed Conifer Forest.

3.2.6.4 Invasive Species

Invasive species of current concern are mostly plants, but few data have been published for southwestern Mixed Conifer Forest. Study of a stand in southwestern Colorado determined that exotic plants accounted for 3.6 % of the plant species (Korb et al. 2007). In unburned areas on the North Rim, exotic plants had 0–0.1 % cover, made up 0–3 % of the flora, and had low richness (Huisinga et al. 2005; Laughlin et al. 2005). Exotics increase with disturbance. In an area of intense prescribed fire on the North Rim, exotics had greater cover (1 %) and made up a higher percentage of the flora (3–5 %) compared to the unburned area (Huisinga et al. 2005). Also, the cover and richness of exotics increase with decreases in canopy cover (Fisher and Fulé 2004). Therefore, with increases in mixed-severity and landscape-scale crown fires in the twenty-first century, invasive plants have potential for growing importance as an anthropogenic disturbance. For example, cheatgrass (*Bromus tectorum*) increased following the Outlet Fire on the North Rim (Crawford 2008). Moreover, future introduction and colonization of high-elevation areas in general is likely with climate change and globalization (Pauchard et al. 2009).

3.2.6.5 Recreation

Recreation has many impacts on southwestern Mixed Conifer Forest that are spatially limited (e.g., construction and use of roads, trails, campsites, and ski runs), but some impacts are more extensive. Fires ignited by recreationists in Mixed Conifer Forest have ranged from small to large in area and include the Wallow Fire of 2011 in the White Mountains, which at the time was the largest fire in recorded Arizona history at 2,177 km^2 (841 $miles^2$). Recreationists also appear to have introduced, spread, and helped establish invasive plants (invasives are most commonly found along trails and other areas of human use). Research elsewhere in the Rocky Mountains indicates that horses used in trail-riding can introduce invasives by dispersing seeds through their fecal matter (Benninger-Truax et al. 1992; Wells and

Lauenroth 2007), and seeds also can be introduced by off-highway vehicles (OHVs; Taylor et al. 2011). Once established on sides of trails and roads, invasive plants can spread into forest interiors (Benninger-Truax et al. 1992). Off-trail use of OHVs has additional impacts, especially on the high plateaus of Utah where moderately sloped terrain is conducive to their widespread use.

3.2.6.6 Nearby Land Use

Nearby land use is an important anthropogenic driver because it has introduced fires that spread into Mixed Conifer Forest. Examples include the Warm Fire on the Kaibab Plateau. Nearby land use also facilitates establishment of invasive species that can spread into Mixed Conifer Forest.

3.3 Processes

The mixed-severity fire regime of frequent, widespread surface fires and infrequent, patchy crown fires requires consideration of vegetation dynamics at two scales: canopy gap and stand. Canopy-gap dynamics follow surface fires and other low-severity disturbances. Stand-scale dynamics (succession) follow crown fires and other high-severity disturbances. Both processes play key roles in the conceptual models (Sect. 3.5).

3.3.1 Gap Dynamics

Canopy gaps form where the death of one or a few canopy trees creates an opening in the otherwise intact forest canopy. This alters the environment below it by increasing light. Subcanopy trees, tree seedlings, and aspen root sprouts in the area of the gap respond with accelerated growth. They are thinned by surface fires, other mortality agents such as defoliating insects, and competition, but one or more survivors ultimately grow and fill the gap in the canopy. Gap dynamics occurring throughout stands result in uneven-aged forests.

Numerous factors affect which tree species fill which canopy gaps. One important factor under the historical fire regime was the length of fire-free intervals (Sect. 3.2.5.1). At lower elevations and on relatively dry sites with short fire-free intervals, ponderosa pine and Douglas-fir were common replacement trees because they form fire-resistant bark at relatively young ages. At mid elevations and on mesic sites, intermediate-length fire-free intervals allowed white fir into canopies, as this species has thin bark when young, but develops thicker, more fire-resistant bark with age. At higher elevations and on moist sites, long fire-free intervals resulted in gap closure by thin-barked Engelmann spruce and subalpine fir (Bradley et al. 1992). Quaking aspen is unique. It was favored by fire nearly regardless of the historical range of fire frequencies because it root sprouts. Gaps also facilitate the persistence of understory aspen and possibly play a role in the stability of some aspen stands (next section).

3.3.2 Succession

Vegetation dynamics at the stand scale are dominated by succession. Pathways of succession are diverse in southwestern Mixed Conifer Forest (Jones 1974; Bradley et al. 1992), because they are influenced by many factors, including species present before fire, seed- and bud-banks, distances from seed sources, plant life-history strategies, local site conditions, severity of disturbance, and post-fire animal use. As mentioned in Sect. 3.1, the diversity of trees in individual stands of Mixed Conifer Forest can mean that undisturbed stands are sources of varied seeds and thereby influence succession (Wu 1999). The absence or presence of quaking aspen before crown fire makes a critical difference in the post-fire successional pathway.

An example of succession in the absence of aspen occurs in the Sacramento Mountains of south-central New Mexico (Hanks and Dick-Peddie 1974). The initial post-crown-fire community is dominated by herbs for a year or two. Then sprouts of Gambel oak (*Quercus gambelii*) and other shrubs gradually become dominant. Later, conifers such as ponderosa pine, southwestern white pine, Douglas-fir, and white fir invade and gradually form a closed canopy 100–200 years following crown fire. A Gambel oak stage has also been described for succession in dry-mesic stands in the San Juan Mountains (Romme et al. 2009b)

Where quaking aspen is present before fire, its roots are stimulated to sprout by fire. Aspen sprouts usually appear in the first year (Fig. 3.17a), but can be delayed (Bradley et al. 1992) or reduced in number by high-severity fires that cause heat damage to roots. Root sprouting results in rapid development of a post-fire community dominated by small aspens and herbs (Fig. 3.17b). Even if only scattered aspens had been present in the pre-fire forest, their clusters of root sprouts can coalesce within a few years (Pearson 1914; Jones and DeByle 1985a). Seed-regeneration by aspen is rare, but can occur during unusually cool, moist years in the Southwest (Elliott and Baker 2004). Conifers reproduce only by seed.

Recruitment of aspen stems can continue for several decades after fire (Romme et al. 2001), but their high density leads to stem mortality and patches thin. Herbivory by deer, elk, and livestock also can reduce the density of aspen root sprouts (e.g., Smith et al. 2011). Sprouts can be dense. For example, they averaged 37,000 stems/ha (~15,000 stems/acre) 7 years after fire in a mixed aspen-conifer landscape in south-central Utah (Smith et al. 2011). Density was most strongly correlated with pre-fire stand composition: an average of <5,000 stems/ha (~2,000 stems/acre) in stands formerly dominated by conifers and ~60,000 stems/ha (~24,000 stems/acre) in former aspen stands.

After establishment, aspen can maintain itself on some sites with little or no ingrowth of conifers (Fig. 3.6). Indicators of long-term stand stability are

Fig. 3.17 (**a, b**) Sprouts from the roots of quaking aspen can dominate succession after crown fire, Kaibab Plateau, north-central Arizona. (**a**) Sprouts can appear within weeks after crown fire (Photograph by author). (**b**) Sprouts grow rapidly and can form dense stands, as shown here during the fall season, 8 years after crown fire (Photograph courtesy of Marc E. Gottlieb)

uneven-age distribution of canopy aspen and of course lack or rarity of conifers (Mueggler 1985a). The percentage of stable aspen stands appears to vary widely among southwestern landscapes. On Cedar Mountain, a plateau in southwestern Utah, 84 % of aspen stands had characteristics of stable stands (Rogers et al. 2010; Oukrop et al. 2011). Only ~150 km (~95 miles) away, no aspen stands appear stable on the North Rim (personal observation).

Environmental conditions associated with aspen stability vs. successional replacement by conifers are unclear (Mueggler 1985a). One factor is elevation, as persistent aspen stands tend to be more common at lower elevations (Pfister 1972; Romme et al. 2001). Historical fire intervals can account for this observation, as they are likely to be shorter in low-elevation aspen stands near Ponderosa Pine Forest and limit the establishment of conifers (Romme et al. 2009a). In contrast, historical fire intervals presumably are longer in higher-elevation aspen stands, and shade-tolerant conifers can establish, overtop, and reduce or replace shade-intolerant aspen in succession (Romme et al. 2009a). A second factor is soil (Layser and Schubert 1979). It interacts with climate to produce dense understory vegetation that limits conifer invasion in Utah (Pfister 1972). A third possible factor is fire. Some stable stands have continuous aspen regeneration, which indicates a connection to reoccurring fire (Margolis et al. 2007). A fourth factor is episodic aspen regeneration apparently related to senescence of cohorts of canopy trees (Kurzel et al. 2007).

It appears that some of today's stable aspen stands have persisted for perhaps thousands of years (Bartos 2007). Moreover, stands have the appearance of stability where replacement of aspen by invading conifers takes centuries or even millennia (cf. Mueggler 1985a, 1989; Romme et al. 2001). It is debatable whether such stands should be considered stable or successional (and whether this question is meaningful at that time scale). Regardless, aspen stands eventually can become decadent and convert to other vegetation (a) in the absence of fire, (b) in the presence of excessive grazing by deer, elk, and livestock, and (c) with mortality of trees (Bartos 2007). Forty-one percent of the stable aspen stands mentioned above for Cedar Mountain were considered unhealthy (Oukrop et al. 2011).

Alternatively, many aspen stands are invaded and can be replaced by conifers in succession. Conifers generally invade aspen stands 15–20 years after fire (Kleinman 1973; in Mueggler 1985b). Yet aspen typically persists as a dominant tree species for about 75–100 years (Moir 1993). Successional aspen stands >150 years of age are uncommon (Moir 1993), although aspen can live for over 250 years in the American Southwest (Margolis et al. 2011). The specific species of colonizing conifers depend largely on elevation and site conditions. Conifer invasion can be facilitated by nearby, undisturbed stands having multiple species as possible seed sources (Wu 1999). Succession to conifer dominance is facilitated by deer browsing that eliminates aspen regeneration for lengthy periods, such as has occurred on the North Rim (Fulé et al. 2002). In addition, high mortality of aspen, as reported for north-central Arizona (cf. Vankat 2011; Zegler et al. 2012), likely enhances the rate of succession to conifers.

As additional time passes, either conifer dominance continues to increase as quaking aspen is replaced or conifers and aspen remain as co-dominants. Both of these patterns were observed by repeat photography of historical photographs in the

San Juan Mountains, even in areas lacking evidence of fire (Zier and Baker 2006). Gap dynamics are possibly important in maintaining aspen in the absence of fire (see previous section), but periodic surface fire followed by root sprouting appears to be a more important factor. For example, without surface fire, aspen lost ~70 % of its density and basal area from 1935 to 2004 in Mixed Conifer Forest on the North Rim (Vankat 2010). A third possible pattern is that conifers and quaking aspen have alternating cycles of abundance, even in the absence of fire. This is suggested by the finding that the loss of aspen trees in unburned stands on the North Rim was accompanied by an increase in mean density of aspen saplings from 3 to 160 individuals/ha (1–65 individuals/acre) during 1984–2005 (Vankat 2010).

3.4 Historical Changes

3.4.1 Overstory

3.4.1.1 Pre-Euro-American Settlement

There are no known historical descriptions of nineteenth century southwestern Mixed Conifer Forest. Powell (1879), in describing elevational ranges of tree species in Utah, listed ponderosa pine, Douglas-fir, white fir, and blue spruce as overlapping at 2,130–2,740 m (~7,000–9,000 ft), but he did not describe forest conditions. The absence of nineteenth century descriptions contrasts with multiple accounts of Spruce-Fir Forest at higher elevation and Ponderosa Pine Forest at lower elevation. This suggests either early observers found Mixed Conifer Forest to be unremarkable – neither as dense as Spruce-Fir Forest nor as open as Ponderosa Pine Forest – or they considered it only as a transition between these two major forest types.

There are descriptions from the early twentieth century, but they are brief:

> The forest…consists of closely stocked stands, whose density is due to great quantities of aspen of all ages in which the coniferous trees are set in small groups, thin lines, or as isolated trees. (Leiberg et al. 1904, for aspen-dominated stands on the San Francisco Peaks)

> The aspen stands in every case represent primary restocking after exceedingly destructive fires which wiped out most of the original coniferous growth. (Leiberg et al. 1904, for the San Francisco Peaks, including Spruce-Fir Forest)

> …in the mixed forests, the litter caused by fallen timber is very heavy. This is particularly true on the Blue Mountain plateau, where it is not uncommon to find a stand of 200 trees to the acre [~500/ha]. (Plummer 1904, for the White Mountains)

> The Douglas-fir subtype is composed of varying proportions of Douglas-fir and [white fir], a little spruce and yellow pine. Usually the stands are moderately dense with small amount of ground cover and fair reproduction, particularly of…white fir. (Lang and Stewart 1910, for the Kaibab Plateau)

No photographs showing undisturbed nineteenth century stands are known to me. Photographs from the first decade of the twentieth century indicate a range of stand densities, but generally more open stand structure than at present (Figs. 3.18 and 3.19), particularly in dry-mesic Mixed Conifer Forest.

Fig. 3.18 Moist-mesic Mixed Conifer Forest in 1910 in Apache National Forest, east-central Arizona (Photograph by G.A. Pearson, courtesy of Fort Valley Experimental Forest Historic Images, U.S. Forest Service)

The earliest known quantitative data on southwestern Mixed Conifer Forest (other than the density estimate by Plummer 1904 quoted above) were provided by Lang and Stewart (1910). In 1909, they sampled a "mixed type" on the Kaibab Plateau that combined Mixed Conifer Forest and low-elevation Spruce-Fir Forest (they provided no data on the "Douglas-fir subtype" mentioned in the preceding quotation). They reported an average density of 360 trees/ha (146 trees/acre) for trees ≥0.9 m (3 ft) height and 117 trees/ha (47 trees/acre) for trees ≥15 cm (6 in.) diameter at breast height (dbh, i.e., at 1.4 m/4.5 ft), but the accuracy of these values is questionable (Sect. 2.4.1.1). Lang and Stewart (1910) underestimated density because they did not sample quaking aspen, which is abundant in Mixed Conifer Forest but lacked value for lumber. They also possibly underestimated density by including samples with areas of treeless Subalpine-Montane Grassland (Vankat 2010, 2011). In addition, the data do not reflect conditions prior to Euro-American influence, which began with livestock grazing about 30 years earlier (Vankat 2011). Early change with fire exclusion is suggested by what appears to be ingrowth of fire-sensitive fir by 1909: fir accounted for 27 % of trees ≥15 cm (6 in.) dbh but 59 % of seedlings and saplings <15 cm (6 in.) dbh.

Fig. 3.19 Dry-mesic Mixed Conifer Forest in 1910 in Apache National Forest, east-central Arizona (Photograph by G.A. Pearson courtesy of Fort Valley Experimental Forest Historic Images, U.S. Forest Service)

Nineteenth century structure and composition of Mixed Conifer Forest have been estimated by forest reconstruction. This approach usually involves (a) analyzing rings of living trees and (b) using decay classes to estimate ages of snags and downed logs currently on the site. These data are used to determine/estimate which trees, snags, and logs were alive and at what dbh on a selected date in the past. Reconstruction studies of Mixed Conifer Forest have been done on the North Rim (Fulé et al. 2002, 2003a), the San Francisco Peaks (Cocke et al. 2005; Heinlein et al. 2005), southwestern Colorado (Fulé et al. 2009), and Utah (Heyerdahl et al. 2011). The data for Utah are not directly comparable and therefore are excluded from the following discussion.

Findings from forest reconstructions for the nineteenth century differ widely, suggesting great variation in Mixed Conifer Forest. These differences are in part attributable to elevational differences, but even studies at overlapping elevations produced divergent results. For example, Fulé et al. (2002) reported density of 246 trees/ha (100 trees/acre) and basal area of 29 m^2/ha (126 ft^2/acre) for a site on the North Rim, and Heinlein et al. (2005) reported densities of 52 trees/ha (21 trees/acre) and basal areas of 9 and 12 m^2/ha (39 and 52 ft^2/acre) for two sites on the San Francisco Peaks. All sites were between 2,370 and 2,700 m (7,776–8,858 ft).

Reconstruction studies often differ in terms of minimum diameters reported, which prevents direct comparison of densities. In studies mentioned in the previous paragraph, Fulé et al. (2002) included trees ≥2.5 cm (1 in.) dbh, and Heinlein et al. (2005) appear to have included trees >0 cm dbh. Also, reconstruction studies often do not include diameter-class data that would facilitate comparison of results of different studies (as well as enable inference of nineteenth century stand dynamics).

Forest reconstructions likely underestimated – possibly substantially underestimated – nineteenth century density and basal area of southwestern Mixed Conifer Forest. Forest reconstructions underestimate when evidence of trees from the historical date being reconstructed has been lost by combustion or decomposition (Fulé et al. 2002, 2003a; Cocke et al. 2005). Combustion has not been a factor on many Mixed Conifer Forest sites because of fire exclusion. In contrast, losses by decomposition would have occurred (Fulé et al. 2002) and would have been important. The decomposition constant for quaking aspen logs in a subalpine forest of northern New Mexico (Gosz 1980; cf. Miller 1983; Alban and Pastor 1993) indicates 90 % loss of dry mass in only 33 years and 99 % loss in 66 years – periods that are far shorter than the 100+ year interval reconstructed in the above studies. In addition, decomposition is likely a factor with other species as well, because mean wood density of trees across the range of woodlands and forests on uplands of the American Southwest is lowest in moist-mesic Mixed Conifer Forest (Laughlin et al. 2011). One-third of 834 conifers ≥20 cm (8 in.) dbh sampled in upper-elevation Mixed Conifer Forest in the eastern San Juan Mountains could not be dated to a specific age due to "advanced deterioration of the wood" (Aoki 2010). Loss of evidence likely accounts in part for why Moore et al. (2004) stated that mesic sites and higher elevations negatively affect the accuracy of forest reconstructions.

Another approach that has been used to estimate nineteenth century structure and composition of southwestern vegetation is sampling of relict stands little disturbed by Euro-Americans (see Sect. 4.4.1.1). However, relict areas have not been described for southwestern Mixed Conifer Forest, presumably because fire exclusion and its impacts on forest structure and composition appear to have been universal.

In conclusion, with no detailed historical descriptions, no useful nineteenth century photographs, flawed early data, inaccurate forest reconstructions, and lack of relict sites, the historical conditions of southwestern Mixed Conifer Forest are unclear. The best alternative is estimation of nineteenth century conditions from knowledge of current forest conditions, the historical disturbance regime, and patterns of vegetation dynamics. In general, moist-mesic stands were typically dominated by Engelmann or blue spruce, ponderosa pine, quaking aspen, Douglas-fir, and either subalpine or corkbark fir. These stands had open understories, as maintained by frequent surface fires, but occasional, long fire-free periods resulted in increased densities and allowed regeneration of spruce and fir, particularly on more moist sites. Surface fires also maintained aspen in stands. Crown fires resulted in early successional stands, especially of aspen. In contrast, dry-mesic stands were dominated by ponderosa pine, white fir, Douglas-fir, and quaking aspen. With their higher frequency of surface fires, dry-mesic stands were more open in both the understory and canopy than moist-mesic stands. Surface fires maintained more consistent densities, as well as aspen as a component of stands. White fir entered the canopy on cooler, moister sites where longer fire-free intervals occasionally occurred.

Understanding the historical conditions of southwestern Mixed Conifer Forest requires more than compilation of information on stands. It also requires landscape and time-scape perspectives. Historically, it appears that Mixed Conifer Forest landscapes – especially at mid-elevations – were complex mosaics of patches of vegetation, disturbance, and succession (Fig. 3.20a, b). Vegetation was diverse with stands having different sets of dominant tree species. Disturbance was diverse with a mixed-severity fire regime and various other disturbances. Succession was diverse with multiple pathways. Landscape patch sizes were directly proportional to topographic homogeneity: large in relatively homogeneous areas, small in topographically heterogeneous areas, and variable in landscapes that combined both. Moreover, the diverse mosaics of vegetation, disturbance, and succession – including the proportions of moist-mesic, dry-mesic, early-successional, mid-successional, and late-successional/old-growth patch types – shifted through time in response to environmental changes.

3.4.1.2 Post-Euro-American Settlement

Fire exclusion initially resulting from livestock grazing and later from fire suppression activities resulted in changes in structure and composition of southwestern Mixed Conifer Forest stands and landscapes (Dieterich 1983; Allen 1989; Moir 1993; White and Vankat 1993; Fulé et al. 2002, 2003a, 2009; Mast and Wolf 2004; Cocke et al. 2005; Heinlein et al. 2005; Vankat et al. 2005; Vankat 2010, 2011).

As discussed in the preceding section, reconstructions of nineteenth century Mixed Conifer Forest are likely inaccurate. Nevertheless, they have been used as a base for comparison to contemporary field data collected from the same areas. These studies indicated total tree densities increased 283–3,026 % and total basal areas increased 45–458 %, with the largest change in stands dominated by quaking aspen (Fulé et al. 2002, 2003a, 2009; Cocke et al. 2005; Heinlein et al. 2005). Comparing conifer-dominated stands on the San Francisco Peaks, total density and percentage increases in density and basal area tended to be larger at lower elevation (Cocke et al. 2005; Heinlein et al. 2005).

These increases in density and basal area reported in reconstruction studies are exaggerated by the underestimation of nineteenth century values of density and

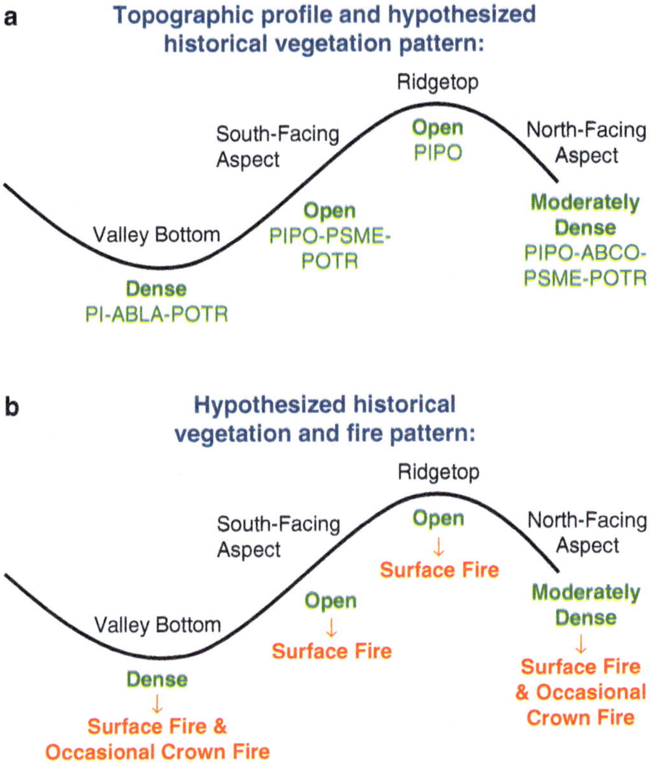

Fig. 3.20 (**a**, **b**) Heterogeneous conditions across a topographic profile in Mixed Conifer Forest prior to Euro-American settlement. (**a**) *Top* diagram shows hypothesized pattern in stand structure and composition. (**b**) *Bottom* diagram shows hypothesized fire pattern relative to vegetation structure. The elevational range from ridgetop to valley bottom can be as little as 20 m (66 ft). Species: *ABCO* white fir, *ABLA* subalpine or corkbark fir, *PI* spruce, *PIPO* ponderosa pine, *POTR* quaking aspen, *PSME* Douglas-fir

basal area described in the previous section. In addition, some changes shown using data from reconstruction studies are counterintuitive. For example, all the above reconstruction studies reported large increases in density of quaking aspen during a long period of fire exclusion, yet aspen regeneration occurs primarily with fire and populations decrease without it. Furthermore, the above forest reconstructions did not report data for intermediate dates during the twentieth century (few reconstruction studies do), which would have indicated temporal patterns. Without such data, reconstructions of Mixed Conifer Forest give the impression of unidirectional increases in density and basal area from the nineteenth to the twenty-first century.

Findings of increases in densities and basal areas on the North Rim using reconstruction data (Fulé et al. 2002, 2003a) contrast with findings of decreases during 1935–2004 and/or 1984–2005 obtained by resampling two sets of historical study

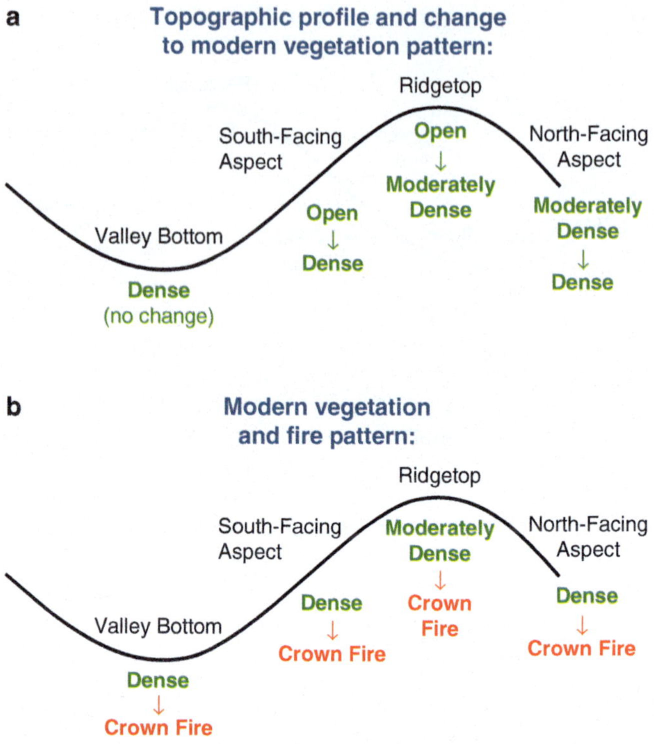

Fig. 3.21 (**a, b**) Homogeneous conditions across the same topographic profile shown in Fig. 3.20a, b occurring after Euro-American settlement and fire exclusion. (**a**) *Top* diagram shows changes in stand structure with fire exclusion. (**b**) *Bottom* diagram shows observed fire pattern relative to current vegetation pattern

plots in the same region. Total density in the study plots decreased 34 % from 1984 to 2005, and total basal area decreased 45 % from 1935 to 2004 (Vankat 2011). Some plots had recently burned, but even unburned stands decreased 24 and 48 %, respectively (Vankat 2010). Decreases in density and basal area were more evident in dry-mesic than in moist-mesic stands (Vankat 2011). The primary tree species that decreased were quaking aspen and white fir. With the canopy opening, presumably as a result of these decreases, total sapling density increased 132 % from 1984 to 2005 and involved aspen, white fir, and subalpine fir.

In addition, Vankat (2011) used data sets from various dates from Mixed Conifer Forest on the North Rim to analyze changes since the late nineteenth century. Findings suggested that Mixed Conifer Forest had rapidly increased in total density and basal area following the beginning of fire exclusion in the late nineteenth century and later decreased, likely in response to the interaction of density-dependent mortality and exogenous factors such as climate change.

Overall, density and basal area increased, but not in the unidirectional pattern implied in reconstruction studies.

Regardless of post-1935 dynamics, it is evident that southwestern Mixed Conifer Forest increased in density with recruitment of smaller trees beginning with fire exclusion. Many of these were white fir, Douglas-fir, Engelmann spruce, and corkbark or subalpine fir, i.e., species that are fire-sensitive when young and formerly had been thinned by surface fires (Merkle 1962; White and Vankat 1993; Abolt 1997; Bastian 2001; Mast and Wolf 2004). For example, 75 % of canopy white firs in a watershed of the Sangre de Cristo Mountains were recruited since the last widespread fire, and those trees produced seedlings and saplings that are now ladder fuels (Margolis and Balmat 2009). Increases of shade-tolerant species in adjacent stands of Ponderosa Pine Forest have resulted in their reclassification as Mixed Conifer Forest (Evans et al. 2011).

These changes in forest structure and tree composition have had several impacts. For example, they increased the likelihood of larger-scale outbreaks of insects and pathogens (Lynch and Swetnam 1992; Moir 1993; Swetnam and Lynch 1993; Heinlein et al. 2005; Fulé et al. 2003a; Sect. 3.2.5.3). Increased forest densities also likely reduced tree vigor, as reported for Ponderosa Pine Forest (Sect. 4.4.1.2), although greater moisture availability in Mixed Conifer Forest probably ameliorated this impact. Increased forest densities also increased horizontal and vertical fuel continuity (White and Vankat 1993; Heinlein et al. 2005), as well as canopy fuel (Fulé et al. 2004). Another major change due to fire exclusion has been reduced regeneration of quaking aspen (Sect. 3.2.5.4). Fire exclusion prevented initiation of new stands, and on-going succession in historical aspen stands led to overtopping and replacement of aspens by conifers (e.g., Bartos 2001; Battaglia and Shepperd 2007).

The changes in forest structure, composition, and fuel loadings homogenized formerly heterogeneous southwestern Mixed Conifer Forest landscapes (Fig. 3.21a, b). The increases in fuel loadings as well as horizontal and vertical fuel continuity occurred across landscapes in both moist-mesic and dry-mesic stands. Therefore, the former mixed-severity fire regime with occasional patchy crown fires changed to a crown fire regime with landscape-scale fires. And as predicted by White and Vankat (1993), large crown fires extending across slope positions and aspects further homogenized southwestern Mixed Conifer Forest landscapes by replacing formerly complex mosaics with extensive early-successional stands.

The shift to increased landscape homogeneity has ramifications for post-fire succession. Large patches of crown fire are likely to have areas where heat sterilization of soils reduced or removed seed- and bud-banks important in succession. Also, succession can be slowed or prevented by greater distances to conifer seed sources. However, changes in succession are hypothetical, because landscape-scale crown fires are a recent phenomenon in the Southwest and therefore are understudied. Information on the first few years of post-fire succession on the North Rim indicate an initial pulse of ruderal species, followed by increasing abundance and dominance of dry spike sedge (*Carex foenea*) and quaking aspen, as well as increasing abundance of cheatgrass, an invasive exotic (Crawford 2008).

3.4.2 Understory

The understory of southwestern Mixed Conifer Forest is highly variable (Moir 1993) and depends on local site conditions influenced by topography (Laughlin et al. 2005; Korb et al. 2007). Cover values in various sites in northern Arizona and southwestern Colorado average between 11 and 60 % (White and Vankat 1993; Fisher and Fulé 2004; Huisinga et al. 2005; Laughlin et al. 2005; Korb et al. 2007). Cover differs greatly among stands, with a range of approximately 3-85 % on the North Rim (White and Vankat 1993; Laughlin et al. 2005). Species composition has been quantitatively characterized for a few specific locations such as the San Francisco Peaks (Fisher and Fulé 2004), the North Rim (Laughlin et al. 2005; Huisinga et al. 2005), and southwestern Colorado (Korb et al. 2007). Diversity values are available (Fisher and Fulé 2004; Laughlin et al. 2005; Huisinga et al. 2005). Regional characterizations include Moir and Ludwig (1979), Alexander et al. (1984), Youngblood and Mauk (1985), DeVelice et al. (1986), Mueggler and Campbell (1986), Mueggler (1988), Moir (1993), and Muldavin et al. (1996).

Early descriptions of the understory are uncommon, but include:

> The [Mixed Conifer Forest] has a cover which in the more open stands does not greatly differ from that of the [Ponderosa Pine Forest], while in the close-set stands, where the grass growth is nearly choked out, a thin layer of decaying leaves forms the ground cover. (Leiberg et al. 1904, for the San Francisco Peaks, an area that was grazed by livestock at the time.)

Other descriptions include Douglas-fir dominated stands on the Kaibab Plateau having a "small amount of ground cover" (Lang and Stewart 1910). In contrast, another commented about thick "growth of grass over all of [a mixed conifer] type" in the White Mountains (Greenamyre 1913).

With such limited information, historical understory dynamics must be inferred from (a) observations of short-term understory changes, such as in response to fire, and (b) spatial differences related to factors that also differ temporally. As an example of using spatial differences, contemporary differences in understories among stands dominated by different tree species can be used to infer historical understory changes related to fire exclusion (see below). Such inferences can be problematic and at best yield only a general understanding of historical changes. Even research on contemporary understories accounts for only 20 % of the variation in cover, 18 % of the variation in species composition, and 33 % of the variation in diversity in upper-elevation sites on the North Rim that included Mixed-Conifer Forest (Laughlin et al. 2005). Moreover, historical understory changes are likely to have been complex and to have differed among regions (cf. Laughlin et al. 2005). The following paragraphs consider the dynamics of understory cover, species composition, and diversity.

Understory cover would have changed with overstory successional dynamics. Understory cover is projected to have been especially high in young stands after crown fire (Laughlin et al. 2005) and presumably other canopy-opening disturbances. Decreases in understory cover with succession from aspen to conifer dominance are suggested by findings that understory cover is greater in stands dominated by quaking aspen than in stands dominated by conifers (Fisher and Fulé 2004; Laughlin et al. 2005; Korb et al. 2007). Also, understory cover is positively related with aspen basal area (Fisher and Fulé 2004) and negatively related to basal area of ponderosa pine, Douglas-fir, and subalpine fir (Fisher and Fulé 2004; Laughlin et al. 2005). These findings indicate understory cover increased with canopy-opening disturbance and then decreased during succession (except in stable stands of quaking aspen).

Understory cover also would have decreased with overstory changes in response to fire exclusion. Evidence indicates that understory cover is negatively related to tree canopy cover (Hurst 1977; Huisinga et al. 2005), amount of coarse woody debris (Laughlin et al. 2005), duff depth (Huisinga et al. 2005), sapling density of Engelmann spruce and corkbark fir (Stromberg and Patten 1991; in Spruce-Fir Forest), and basal area of ponderosa pine, Douglas-fir, and subalpine fir (Fisher and Fulé 2004; Laughlin et al. 2005). Understory cover was likely reduced in the absence of fire (Huisinga et al. 2005), especially where white fir increased (Merkle 1962). Therefore, understory cover likely decreased with increasing forest density during fire exclusion and later increased in those areas where forest density decreased.

Change in understory composition is inferable from (a) findings that understory species composition is related to amount of coarse woody debris and basal areas of Engelmann spruce, subalpine fir, and quaking aspen (Laughlin et al. 2005) and (b) differences between stands dominated by aspen vs. stands dominated by conifers (Korb et al. 2007). An example of change in understory species composition is the above-mentioned increase in dry spike sedge and cheatgrass in an area of extensive crown fire on the North Rim (Crawford 2008).

Historical changes in understory diversity are inferable from studies of modern understories. Understory species richness is positively related to basal area of Engelmann spruce and negatively related to basal area of subalpine fir and amount of coarse woody debris (Laughlin et al. 2005). It is also negatively related to overstory canopy cover and duff depth (Huisinga et al. 2005). Understory diversity (including species richness) was likely highest following crown fire (Laughlin et al. 2005). The absence of surface fires probably reduced species richness (Huisinga et al. 2005). Therefore, understory diversity increased with canopy-opening disturbance and then decreased with succession. Understory diversity also likely decreased with increases in forest density during fire exclusion and later increased in those areas where forest density decreased.

3.5 Conceptual Models

The following nested, three-tiered set of conceptual models is based on my interpretation of best-available information on Mixed Conifer Forest of the American Southwest. The models summarize and synthesize material on drivers (Sect. 3.2) and processes (Sect. 3.3) of vegetation dynamics, and they account for historical changes (Sect. 3.4). Their format is explained in Sect. 1.7. The models can be used to facilitate understanding of the past and the present. They also can provide insight into effects of future changes in the environment and outcomes of alternative land-management decisions. Moreover, they can guide the development of quantitative models.

3.5.1 Ecosystem-Characterization Model

The ecosystem-characterization model for southwestern Mixed Conifer Forest emphasizes *Vegetation* (a biotic component) and *Disturbance* (a driver; Fig. 3.22a, Table 3.2). Key aspects of *Vegetation* are small- to large-scale patterns, fuel, structure, and species composition. These affect various aspects of *Disturbance*. The primary agents of *Disturbance* are fire and drought, and these affect tree mortality, fuel, and vegetation patterns, structure, and composition. The two other biotic components are *Soil System* and *Animals*, including insects that cause disturbance. A second driver is *Weather & Climate*, which ignites fires, causes drought, and influences fire behavior, insect population dynamics, and soil and fuel moisture. The third driver is *Landscape*, which includes topography, elevation, and proximity to Ponderosa Pine Forest. It influences weather, climate, and spread and pattern of fire. The model also includes six anthropogenic drivers that affect various ecosystem components and drivers, especially vegetation and disturbance: *Livestock Grazing*, *Fire Management*, *Modern Climate Change*, *Invasive Species*, *Recreation*, and *Nearby Land Use* (Fig. 3.22b, Table 3.2).

3.5.2 Vegetation-Dynamics Models

Separate models are needed to illustrate the vegetation dynamics of moist-mesic and dry-mesic southwestern Mixed Conifer Forest, because of the greater importance of quaking aspen and the greater occurrence of crown fire in moist-mesic stands. Each vegetation-dynamics model is accompanied by a bar graph to show estimated relative shifts in abundances of the community types from circa 1870 (prior to widespread Euro-American land use) to circa 1970 (end of strict fire exclusion policy) to the present. The estimates are based on my understanding of vegetation dynamics and should be interpreted as showing relative differences and not interpreted quantitatively (hence the absence of a scale for abundance). Although not shown on the graphs, relative abundances shifted prior to circa 1870 as a result of variation in climate and disturbance. Localized differences also have been and are present, but are beyond the scope of these generalized bar graphs.

3.5.2.1 Moist-Mesic Mixed Conifer Forest

The vegetation-dynamics model for moist-mesic Mixed Conifer Forest has three states, eight communities, and three transitions (Fig. 3.23, Table 3.3). All occurred historically. *State A* has been much more common than the other states, and it

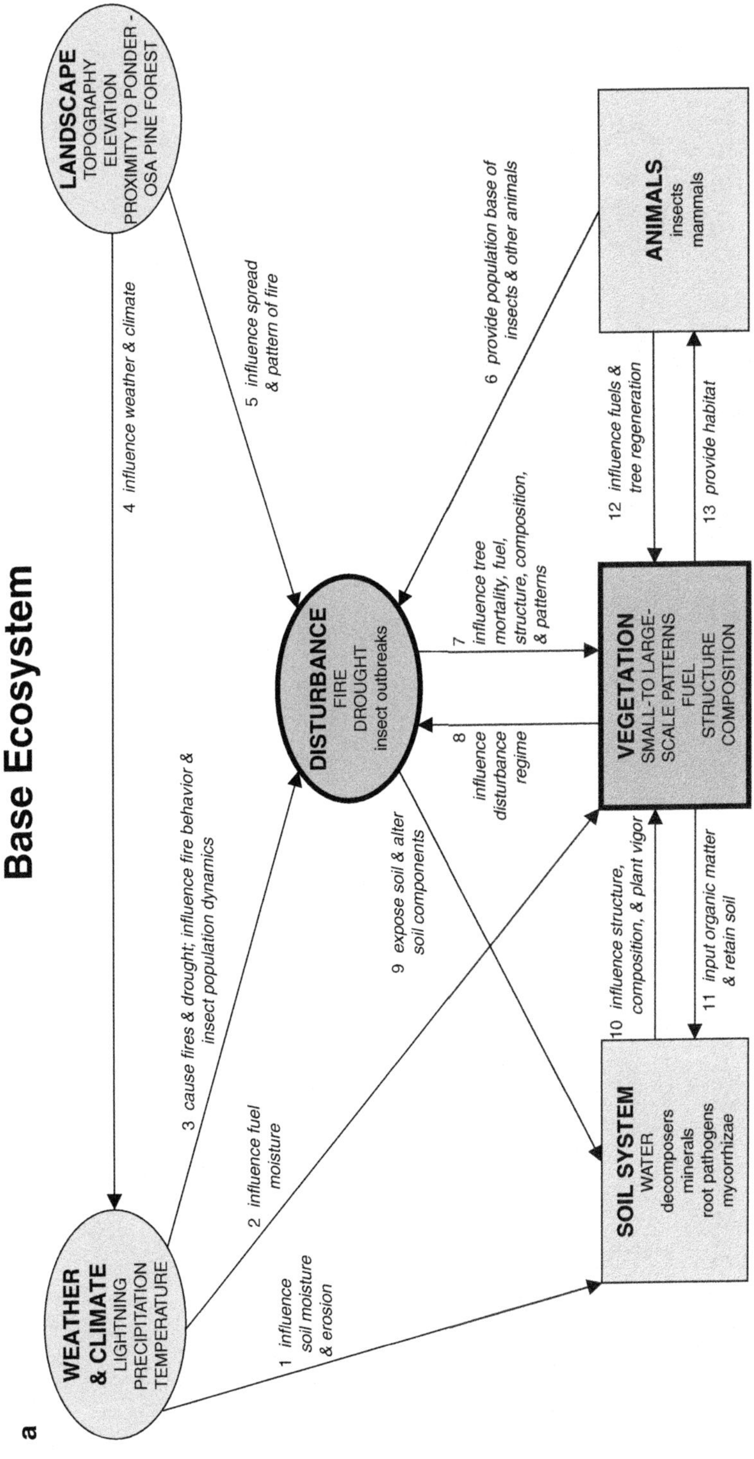
a
Base Ecosystem
WEATHER & CLIMATE
LIGHTNING
PRECIPITATION
TEMPERATURE
LANDSCAPE
TOPOGRAPHY
ELEVATION
PROXIMITY TO PONDEROSA PINE FOREST
DISTURBANCE
FIRE
DROUGHT
insect outbreaks
VEGETATION
SMALL-TO LARGE-SCALE PATTERNS
FUEL
STRUCTURE
COMPOSITION
SOIL SYSTEM
WATER
decomposers
minerals
root pathogens
mycorrhizae
ANIMALS
insects
mammals
1 influence soil moisture & erosion
2 influence fuel moisture
3 cause fires & drought; influence fire behavior & insect population dynamics
4 influence weather & climate
5 influence spread & pattern of fire
6 provide population base of insects & other animals
7 influence tree mortality, fuel, structure, composition, & patterns
8 influence disturbance regime
9 expose soil & alter soil components
10 influence structure, composition, & plant vigor
11 input organic matter & retain soil
12 influence fuels & tree regeneration
13 provide habitat

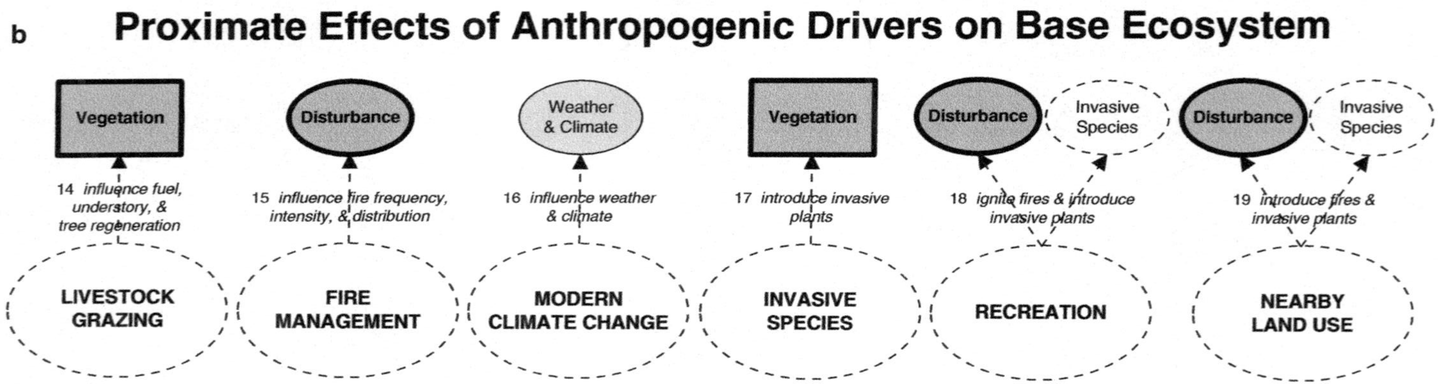

Fig. 3.22 (**a**, **b**) Ecosystem-characterization model for Mixed Conifer Forest. In the base model (**a**), biotic components are shown as *rectangles* and drivers as *solid ovals*. Text within them identifies major features in *upper case* and other important features in *lower case*. Disturbance and Vegetation are *highlighted*, because they are the basis for the vegetation-dynamics models (Figs. 3.23 and 3.24). In the anthropogenic supplement to the base model (**b**), drivers are shown as *dashed ovals*. Relationships in (**a**) and (**b**) are shown as *solid* and *dashed arrows*, respectively, and are described in Table 3.2

Table 3.2 Relationships in the ecosystem-characterization model of southwestern Mixed Conifer Forest (Fig. 3.22a, b)

Relationships	Circa 1870	Present
1	Precipitation increases soil moisture and can cause soil erosion. Winter precipitation is critical for recharging soil moisture	Same, but infiltration of precipitation, especially summer monsoonal rain, can be reduced by thicker layer of litter and duff, where unburned. Infiltration is reduced and soil erosion can be increased where this layer has been lost in high-severity fires. In addition, climate change increases the severity and duration of drought
2	Precipitation and temperature largely determine fuel moisture	Same, but climate change increases the severity and duration of drought
3	Lightning ignites fires. Consistently low precipitation and high temperature cause drought and affect insect population dynamics. Wind influences fire behavior	Same, but climate change increases drought and can result in more extensive insect outbreaks
4	Topography influences weather and climate	Same, but the effects on forest microclimate can be reduced with dense tree cover
5	Fires can spread from other areas of the landscape (especially from Ponderosa Pine Forest), and fire patterns are influenced by elevation and topography	Same, but the effects of topography are reduced in areas with higher fuel loadings and more continuous fuels
6	Increases in populations of western spruce budworm, other insects, elk, and deer can lead to disturbance	Same
7	Mixed-severity fire regime contributes to heterogeneous vegetation patterns. Frequent surface fires thin tree regeneration and keep fuel loadings low. Occasional patchy crown fires kill trees in all size classes. Insects kill understory and overstory trees, producing fine and coarse fuels. Drought increases stress in trees, surface fire frequency, probability of patchy crown fires, and outbreaks of some insects	Past fire exclusion resulted in increased tree densities, fuel loadings, fuel continuity, and homogeneity of vegetation patterns. Surface fires are more likely to crown, and crown fires are more likely to burn across landscapes. Crown fires kill trees in all size classes and reinforce homogeneous vegetation patterns. Drier climate results in expansive stands of stressed trees and increased probability of fires and insect outbreaks
8	Spatial heterogeneity in fuel loadings and fuel continuity results in mixed-severity fire regime of frequent surface fires and occasional patchy crown fires	Greater spatial homogeneity of fuel loadings and fuel continuity results in greater probability of landscape-scale crown fires

9	Fires at least partially and patchily expose soil surface and alter soil properties	Past fire exclusion resulted in thicker, more extensive litter and duff and likely altered other soil properties. Hotter, larger fires likely have greater effects on soil properties
10	Water and minerals in soil affect plant vigor and growth	Same, but less water is available because of drought, higher temperatures, and thicker litter and duff. Minerals can be less available because of greater sequestering in organic matter
11	Plants add organic matter to soil and stabilize soil, except in areas of crown fires	Same, except areas of crown fires are larger
12	Herbivory influences fuels and tree regeneration	Same, but increased deer and elk populations have greater effects
13	Vegetation is habitat for animals	Same, but animal populations are changed where habitat is changed
14	Not applicable	Late nineteenth century livestock grazing reduced herb cover and likely influenced tree regeneration. The reduction in herb cover reduced the frequency of surface fires. Impacts of twenty-first century grazing are poorly known
15	Not applicable	Past fire management attempted to suppress fires throughout most of the twentieth century. This resulted in large changes in forest structure and composition. Beginning in the late twentieth century, fire management included prescribed and lightning-ignited management fires. These fires do not always remain surface fires
16	Not applicable	Modern climate change affects weather and climate, raising temperature and increasing the frequency and extent of drought
17	Not applicable	Invasive plants have potential to modify the understory
18	Not applicable	Recreationists ignite fires and introduce and spread invasive plants
19	Not applicable	Nearby land use that initiates fires and colonization by invasive plants can introduce fires and invasives into Mixed Conifer Forest

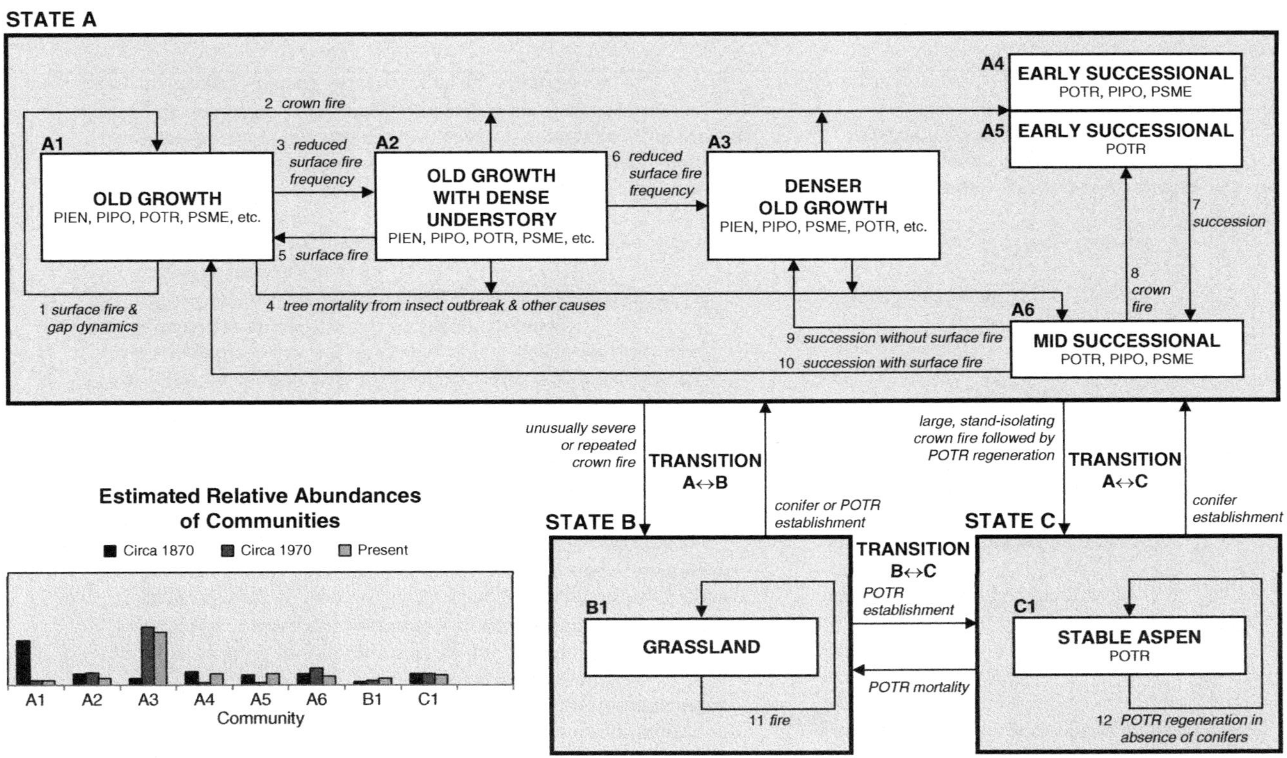

Fig. 3.23 Vegetation-dynamics model for moist-mesic Mixed Conifer Forest. Model provides details on the vegetation-disturbance portion of the ecosystem-characterization model (Fig. 3.22a, b) by showing communities (*smaller*, *open rectangles*) within states (*larger*, *shaded rectangles*), processes that form the communities, and transitions between states. Text along *arrows* identifies pathways that are described in Table 3.3. *Bar graph* shows estimated relative shifts in abundances of the community types from circa 1870 (see Sect. 3.5.2). Species: *PIEN* Engelmann spruce, *PIPO* ponderosa pine, *POTR* quaking aspen, *PSME* Douglas-fir

Table 3.3 Relationships and transitions in the vegetation-dynamics model of moist-mesic Mixed Conifer Forest (Fig. 3.23)

Relationships and transitions	Circa 1870	Present
1	Surface fire and gap dynamics maintain old growth	Same, but surface fire is less frequent
2	Crown fire changes all three types of old growth stands into early successional communities A4 or A5, depending on the abundance and vigor of quaking aspen in the old growth stands	Same, but crown fire is more common and more widespread
3	Periods of reduced frequency of surface fire leave tree understory unthinned, producing old growth with dense understory	Same and surface fire is less frequent
4	Extensive thinning by insect outbreak and other causes of tree mortality changes all types of old growth stands into mid successional stands	Same, but possibly is more common
5	Surface fire thins tree understory, changing stands into old growth without a dense understory	Same, but surface fire is less frequent
6	Periods of reduced frequency of surface fire allow understory trees to grow into the canopy, changing stands into denser old growth	Same and surface fire is less frequent
7	Succession changes early successional stands into mid successional community	Same
8	Crown fire changes mid successional stands into early successional communities A4 or A5, depending on the abundance and vigor of quaking aspen in the mid successional stands	Same
9	Succession without surface fire changes mid successional stands into denser old growth stands	Same and surface fire is less frequent
10	Succession with surface fire changes mid successional stands into old growth community A1	Same, but surface fire is less frequent
A↔B	Unusually severe or repeated crown fire converts some stands from State A to State B. Reversion to State A requires invasion and establishment of conifers or quaking aspen to form early successional communities A4 or A5, depending on the invading species	Same, but unusually severe and repeated crown fire is more common. Larger crown fire limits invasion of conifers and quaking aspen
11	Fire maintains grassland stand structure and composition	Same, but fire is less frequent

(continued)

Table 3.3 (continued)

Relationships and transitions	Circa 1870	Present
A ↔ C	Large, stand-isolating crown fire followed by regeneration of quaking aspen in the absence of conifers converts some stands from State A to State C. Reversion to State A requires invasion and establishment of conifers to form mid successional community A4	Same, but stand-isolating crown fire is more common and larger in size, thereby limiting invasion of conifers
12	Quaking aspen regeneration in the absence of conifers maintains stand structure and composition	Same, but quaking aspen regeneration is reduced in some areas
B ↔ C	Quaking aspen invasion and establishment converts State B to State C. Reversion to State B requires quaking aspen mortality	Same, but quaking aspen mortality is more frequent in some areas

encompasses the characteristic moist-mesic Mixed Conifer Forest communities, which are dominated by species such as Engelmann spruce, ponderosa pine, quaking aspen, and Douglas-fir. *State A* includes community *A1 Old Growth*, which was the most abundant community circa 1870, but is uncommon today. It is maintained by surface fire and gap dynamics. With reduced frequency of surface fire, community *A1* changes to community *A2 Old Growth with Dense Understory*. Surface fire changes community *A2* into community *A1*. Community *A2* changes with ongoing reduced frequency of surface fire to community *A3 Denser Old Growth*. In addition, all three old-growth communities can be converted into community *A6 Mid Successional* by insect outbreak and other causes of tree mortality.

Because crown fire was a significant part of the historical fire regime of moist-mesic Mixed Conifer Forest, it and the successional communities formed by it are included within *State A*. Crown fire changes all three old-growth communities into an *Early Successional* community, either *A4* dominated by species such as quaking aspen, ponderosa pine, and Douglas-fir or *A5* dominated by quaking aspen. The early successional community formed depends on the pre-fire abundance and vitality of aspen. With succession, *A4* and *A5* develop into community *A6 Mid Successional* with species such as quaking aspen, ponderosa pine, and Douglas-fir. The dynamics of *A6* depend on fire. With crown fire, *A6* changes to one of the two *Early Successional* communities. Without fire, succession converts *A6* to *A3*. But with surface fire, succession converts *A6* to *A1*.

Unusually severe or repeated crown fire can change *State A* into *State B* by *Transition A↔B*. *State B* is made up of community *B1 Grassland*, which is maintained by fire. *Transition A↔B* can be reversed by invasion and establishment of conifers and/or quaking aspen, converting *B1* into either *A4* or *A5*, depending on the invading species.

In addition, large crown fire that isolates stands and is followed by regeneration of quaking aspen can change *State A* into *State C* via *Transition A↔C*. *State C* is made up of community *C1 Stable Aspen*, which is maintained by aspen regeneration in the absence of conifers. *Transition A↔C* can be reversed by invasion and establishment of conifers, converting *C1* into community *A6*.

State C also can be formed from *State B* by invasion and establishment of aspen in *B1 Grassland*. This *Transition B↔C* can be reversed by aspen mortality.

3.5.2.2 Dry-Mesic Mixed Conifer Forest

The vegetation-dynamics model for dry-mesic Mixed Conifer Forest (Fig. 3.24, Table 3.4) is similar, but importantly divides *State A* of moist-mesic sites into two states: *A* and *B*. This reflects less frequent crown fire in dry-mesic Mixed Conifer Forest and a transition from one state to another when it occurs. The dry-mesic model is similar to the vegetation-dynamics models of Ponderosa Pine Forest (Sect. 4.5.2).

The model for dry-mesic Mixed Conifer Forest has five states, ten communities, and five transitions. All occurred historically. *State A* has been much more common than the other states, and it encompasses the three most characteristic dry-mesic Mixed Conifer Forest communities, all of which are dominated by species such as

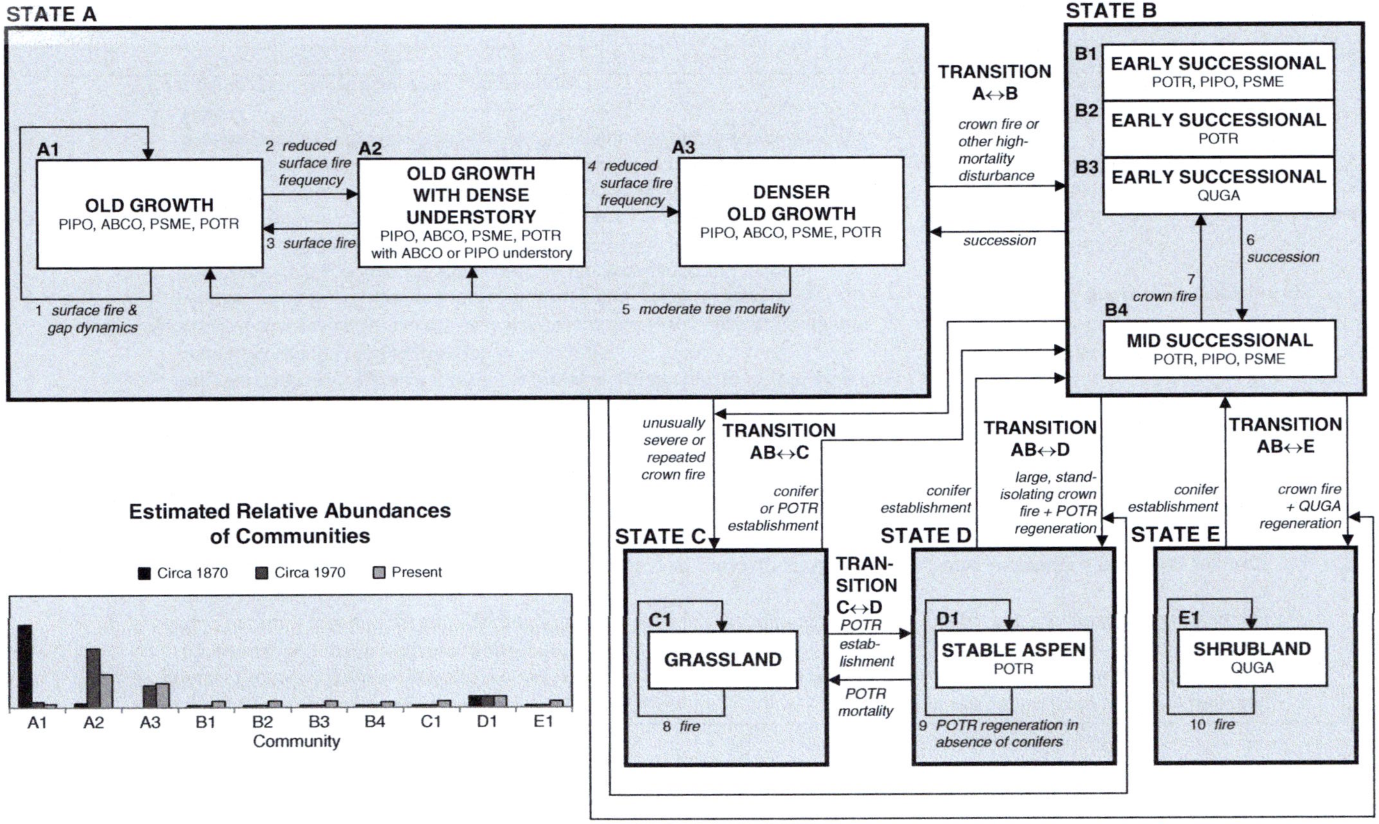

Fig. 3.24 Vegetation-dynamics model for dry-mesic Mixed Conifer Forest. Model provides details on the vegetation-disturbance portion of the ecosystem-characterization model (Fig. 3.22a, b) by showing communities (*smaller*, *open rectangles*) within a states (*larger*, *shaded rectangles*), processes that form the communities, and transitions between states. Text along *arrows* identifies pathways that are described in Table 3.4. *Bar graph* shows estimated relative shifts in abundances of the community types from circa 1870 (see Sect. 3.5.2). Species: *ABCO* white fir, *PIPO* ponderosa pine, *POTR* quaking aspen, *PSME* Douglas-fir, *QUGA* Gambel oak

Table 3.4 Relationships and transitions in the vegetation-dynamics model of dry-mesic Mixed Conifer Forest (Fig. 3.24)

Relationships and transitions	Circa 1870	Present
1	Surface fire and gap dynamics maintain old growth	Same but surface fire is less frequent
2	Periods of reduced frequency of surface fire leave tree understory unthinned, producing old growth with dense understory	Same and surface fire is less frequent
3	Surface fire thins tree understory, changing stands into old growth without a dense understory	Same, but surface fire is less frequent
4	Periods of reduced frequency of surface fire allow understory trees to grow into the canopy, changing stands into denser old growth	Same and surface fire is less frequent
5	Moderate tree mortality, such as with some insect outbreaks, reduces density of canopy trees, changing stands into old growth or old growth with dense understory	Same and insect outbreaks are more frequent
A↔B	Crown fire or other high-mortality disturbance changes old growth communities of State A into early successional communities B1, B2, or B3, depending on the abundance and vigor of quaking aspen and Gambel oak in the old growth stands. Reversion to State A requires succession	Same and crown fire is more extensive
6	Succession changes early successional stands into mid successional community	Same
7	Crown fire changes mid successional stands into early successional communities B1, B2, or B3, depending on the abundance and vigor of quaking aspen and Gambel oak in the mid successional stands	Same and crown fire is more extensive
AB↔C	Unusually severe or repeated crown fire converts some stands from States A and B into State C. Reversion to State B requires invasion and establishment of conifers or quaking aspen to form early successional communities B1 or B2, depending on the invading species. Reversion to State A has to be via State B	Same, but unusually severe and repeated crown fire is more extensive. Larger size of crown fire limits invasion of conifers and quaking aspen
8	Fire maintains stand structure and composition	Same, but fire is less frequent

(continued)

Table 3.4 (continued)

Relationships and transitions	Circa 1870	Present
AB ↔ D	Large, stand-isolating crown fire followed by regeneration of quaking aspen in the absence of conifers converts some stands from States A and B to State D. Reversion to State B requires invasion and establishment of conifers to form mid successional community B4. Reversion to State A has to be via State B	Same, but large, stand-isolating crown fire is more common. Larger crown fire limits invasion of conifers
9	Quaking aspen regeneration in the absence of conifers maintains stand structure and composition	Same, but quaking aspen regeneration is reduced in some areas
AB ↔ E	Crown fire followed by regeneration of Gambel oak converts some stands from States A and B into State E. Reversion to State B requires invasion and establishment of conifers to form early successional community B3 transitioning into B4. Reversion to State A has to be via State B	Same, but crown fire is more extensive. Larger crown fire limits invasion of conifers
10	Fire maintains stand structure and composition	Same, but fire is less frequent
C ↔ D	Quaking aspen invasion and establishment converts State C to State D. Reversion to State C requires aspen mortality	Same, but quaking aspen mortality is more frequent in some areas

ponderosa pine, white fir, Douglas-fir, and quaking aspen. Community *A1 Old Growth*, which is maintained by surface fire and gap dynamics, was the most common community circa 1870. With reduced frequency of surface fires, community *A1* forms *A2 Old Growth with Dense Understory*. Therefore, *A2* increased with fire exclusion as ponderosa pine and/or white fir became more abundant in the understory without thinning by frequent surface fire. Surface fire can return community *A2* to *A1*, and consequently *A2* decreased since circa 1970 because of management fires. Continued reduced frequency of surface fire results in the maturation of understory trees and changes community *A2* into community *A3 Denser Old Growth*. Moderate tree mortality, as for example with some insect outbreaks, changes community *A3* into *A1* or *A2*.

Transition A↔B changes *State A* to *State B* with crown fire or other disturbance that results in high mortality of trees. One of three *Early Successional* communities is formed: *B1* with quaking aspen, ponderosa pine, and Douglas-fir, *B2* with quaking aspen, or *B3* with shrubs of Gambel oak. The community formed depends on the pre-fire abundance and vitality of aspen and Gambel oak. Succession changes *B1*, *B2*, and *B3* into community *B4 Mid Successional* dominated by quaking aspen, ponderosa pine, and Douglas-fir. Crown fire changes *B4* into *B1*, *B2*, or *B3*, also depending on the pre-fire abundance and vitality of aspen and Gambel oak. *Transition A↔B* can be reversed by succession.

Unusually severe or repeated crown fire can change *States A* and *B* into *State C* via *Transition AB↔C*. *State C* is made up of community *C1 Grassland*, which is maintained by fire. *Transition AB↔C* can be reversed by invasion and establishment of conifers and/or aspen, converting *C1* into either *B1* or *B2* depending on the invading species. Reversion to State A occurs only via State B.

Large crown fire that isolates stands and is followed by regeneration of quaking aspen can change *States A* and *B* into *State D* via *Transition AB↔D*. *State D* is made up of community *D1 Stable Aspen*, which is maintained by aspen regeneration in the absence of conifers. *Transition AB↔D* can be reversed by invasion and establishment of conifers, converting *D1* into community *B4*. Reversion to State A occurs only via State B.

State D also can be formed from *State C* by aspen invasion and establishment in *C1 Grassland*. This *Transition C↔D* can be reversed by aspen mortality.

Crown fire followed by regeneration of Gambel oak can change *States A* and *B* into *State E* via *Transition AB↔E*. *State E* consists of community *E1 Shrubland*, which is dominated by Gambel oak and maintained by fire (Gambel oak resprouts after fire). *Transition AB↔E* can be reversed by invasion and establishment of conifers, converting *E1* into *B3* transitioning into *B4*. Reversion to State A occurs only via State B.

3.5.3 *Mechanistic Model*

Both vegetation-dynamics models are explained by the same mechanistic model (Fig. 3.25). It has eight biotic components on the right side of the figure (including five aspects of fuels), four drivers on the left side, and six anthropogenic factors

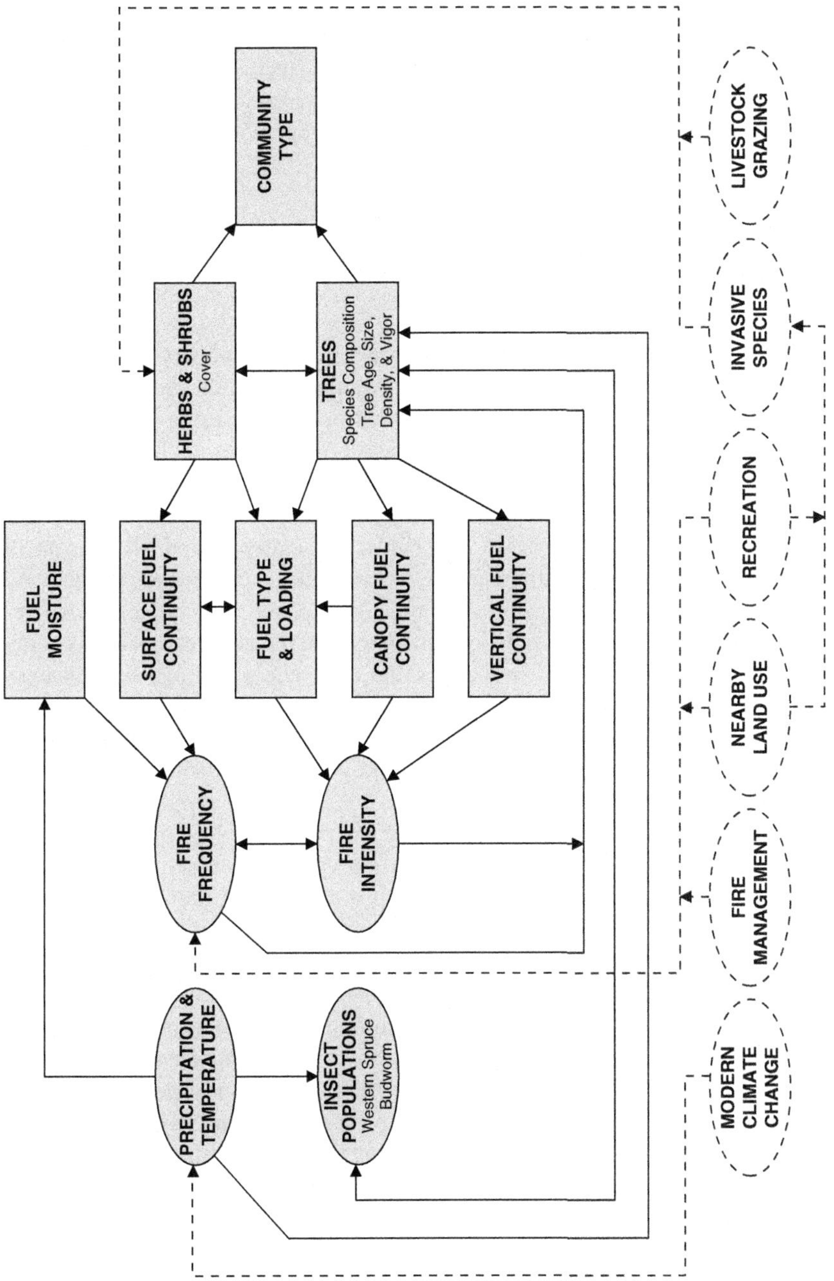

Fig. 3.25 Mechanistic model for Mixed Conifer Forest. Model provides details on the mechanisms involved in the vegetation-dynamics models (Figs. 3.23 and 3.24). *Rectangles* are biotic components (including fuels), *ovals* are drivers, and *dashed ovals* are anthropogenic drivers

at the bottom. In general, *Trees*, *Herbs & Shrubs*, and *Precipitation & Temperature* affect the five fuel characteristics. The fuel characteristics combined with *Fire Intensity*, *Fire Management*, *Nearby Land Use*, and *Recreation* influence *Fire Frequency*. *Fire Frequency*, *Fire Intensity*, *Precipitation & Temperature*, and *Insect Populations* influence characteristics of *Trees*, such as species composition and tree age, size, density, and vigor. *Trees* and *Herbs & Shrubs* determine *Community Type* (of the eight/ten appearing in the vegetation-dynamics models).

3.6 Conclusions and Challenges

The key characteristic of Mixed Conifer Forest is its complexity of vegetation, topography, and disturbance. Additional research is needed on interrelationships among these different aspects of diversity. Of particular interest is how vegetation and disturbance have changed through time and how they are likely to respond to future changes in climate and land management. More specifically, the mixed-severity fire regime needs further research, particularly on regional variations, relationships between fire severity and landscape features, and all aspects of crown fire. Differences in the fire regimes of moist-mesic, dry-mesic, and quaking aspen stands of Mixed Conifer Forest require clarification. The influence of sampling area and intensity on Mixed Conifer Forest fire return intervals needs study to better understand apparent similarities and differences among research findings. From an applied standpoint, it is critically important to develop fire management techniques to reintroduce fire into Mixed Conifer Forest landscapes in ways that ultimately mimic the historical fire regime. This is essential to avoid landscape-scale crown fires that remove vestiges of historical conditions and alter landscapes for centuries. Wind disturbance, impacts of insects, and interactions of disturbance agents across landscapes require more study. Human impacts needing research include the spread of invasive species – especially in relationship to fire management and recreation – and the direct impacts of air pollution on vegetation. Post-disturbance succession, specifically at the landscape scale, requires additional study. For example, it is important to better understand the role of conifer seed dispersal, as well as factors determining successional replacement vs. stability of stands of quaking aspen. Historical stand structure and composition have been poorly characterized and therefore need further study, perhaps utilizing modeling. Research on how Mixed Conifer Forest has been and is being impacted by climate change must be a priority. The finding that forest density and basal area decreased on the North Rim in the twentieth century should be evaluated for other areas of the Southwest and the causes clearly identified. Additional research on impacts of recreation and nearby land use would help enhance decision-making by land managers.

References

Abolt RAP (1997) Fire histories of upper elevation forests in the Gila Wilderness, New Mexico via fire scar and stand age structure analyses. Thesis, University of Arizona, Tucson

Alban DH, Pastor J (1993) Decomposition of aspen, spruce, and pine boles on two sites in Minnesota. Can J For Res 23:1744–1749

Alexander BG Jr, Ronco F Jr, White AS, Ludwig JA (1984) Douglas-fir habitat types of northern Arizona. U.S. Department of Agriculture, Forest Service General Technical Report RM-108, Rocky Mountain Forest and Range Experiment Station, Fort Collins, CO

Alexander ME, Hawksworth FG (1975) Wildland fires and dwarf mistletoes: a literature review of ecology and prescribed burning. U.S. Department of Agriculture, Forest Service General Technical Report RM-14, Rocky Mountain Forest and Range Experiment Station, Fort Collins, CO

Allen CD (1984) Montane grassland in the landscape of the Jemez Mountains, New Mexico. Thesis, University of Wisconsin, Madison

Allen CD (1989) Changes in the landscape of the Jemez Mountains, New Mexico. Dissertation, University of California, Berkeley

Allen CD (2002) Lots of lightning and plenty of people: an ecological history of fire in the upland Southwest. In: Vale TR (ed) Fire, native peoples, and the natural landscape. Island Press, Covelo

Allen CD, Touchan R, Swetnam TW (1995) Landscape-scale fire history studies support fire management action at Bandelier. Park Sci 15:18–19

Allen CD, Anderson RS, Jass RB, Toney JL, Baisan CH (2008) Paired charcoal and tree-ring records of high-frequency Holocene fire from two New Mexico bog sites. Int J Wildland Fire 17:115–130

Aoki CF (2010) Fire history and serotiny in the Rocky Mountains of Colorado. Thesis, Colorado State University, Fort Collins

Baisan CH, Swetnam TW (1990) Fire history on a desert mountain range: Rincon Mountain Wilderness, Arizona, USA. Can J For Res 20:1559–1569

Baker FS (1925) Aspen in the central Rocky Mountain region. U.S. Department of Agriculture Bulletin No. 1291, Washington, DC

Barrows JS (1978) Lightning fires in southwestern forests. Final report to USDA Forest Service, Northern Forest Fire Laboratory, Intermountain Forest and Range Experiment Station, under cooperative agreement 16–568 CA with Rocky Mountain Forest and Range Experiment Station, Fort Collins, CO

Bartos DL (2001) Landscape dynamics of aspen and conifer forests. In: Shepperd WD, Binkley D, Bartos DL, Stohlgren TJ, Eskew LG (compilers) Sustaining aspen in western landscapes: symposium proceedings. U.S. Department of Agriculture, Forest Service Proceedings RMRS-P-18, Rocky Mountain Research Station, Fort Collins, CO

Bartos DL (2007) Aspen. In: Hood SM, Miller M (eds) Fire ecology and management of the major ecosystems of southern Utah. U.S. Department of Agriculture, Forest Service General Technical Report RMRS-GTR-202, Rocky Mountain Research Station, Fort Collins, CO

Bastian HV (2001) The effects of a low intensity fire on a mixed conifer forest in Bryce Canyon National Park, Utah. In: Vance RK, Edminster CB, Covington WW, Blake JA (compilers) Ponderosa pine ecosystems restoration and conservation: steps toward stewardship; 25–27 Apr 2000; Flagstaff, Arizona. U.S. Department of Agriculture, Forest Service Proceedings RMRS-P-22, Rocky Mountain Research Station, Ogden, UT

Bates ST (2006) A preliminary checklist of Arizona macrofungi. Canotia 2:47–78

Battaglia MA, Shepperd WD (2007) Ponderosa pine, mixed conifer, and spruce-fir forests. In: Hood SM, Miller M (eds) Fire ecology and management of the major ecosystems of southern Utah. U.S. Department of Agriculture, Forest Service General Technical Report RMRS-GTR-202, Rocky Mountain Research Station, Fort Collins, CO

Benninger-Truax M, Vankat JL, Schaefer RL (1992) Trail corridors as habitat and conduits for movement of plant species in Rocky Mountain National Park, Colorado, USA. Landsc Ecol 6:269–278

Beschta RL (n.d.) Climatology of the ponderosa pine type in central Arizona. Agricultural Experiment Station Technical Bulletin 228, College of Agriculture, University of Arizona, Tucson

Bigio E, Swetnam TW, Baisan CH (2010) A comparison and integration of tree-ring and alluvial records of fire history at the Missionary Ridge Fire, Durango, Colorado, USA. Holocene 20:1047–1061

Binkley D, Moore MM, Romme WH, Brown PM (2006) Was Aldo Leopold right about the Kaibab deer herd? Ecosystems 9:227–241

Bowman CB (2003) Potential for ozone stress on vegetation at Grand Canyon National Park. Unpublished report, Science Center, Grand Canyon National Park, AZ

Bradley AF, Noste NV, Fischer WC (1992) Fire ecology of forests and woodlands in Utah. U.S. Department of Agriculture, Forest Service General Technical Report INT-287, Intermountain Research Station, Ogden, UT

Brookes MH, Campbell RW, Colbert JJ, Mitchell RG, Stark RW (tech coords) (1987) Western spruce budworm. U.S. Department of Agriculture, Forest Service, Technical Bulletin 1694. Washington, DC

Brown PM, Kaye MW, Huckaby LS, Baisan CH (2001) Fire history along environmental gradients in the Sacramento Mountains, New Mexico: influences of local patterns and regional processes. Ecoscience 8:115–126

Brown PM, Heyerdahl EK, Kitchen SG, Weber MH (2008) Climate effects on historical fires (1630–1900) in Utah. Int J Wildland Fire 17:28–39

Chambers CL, Holthausen RS (2000) Montane ecosystems used as rangelands. In: Jemison R, Raish C (eds) Livestock management in the American Southwest: ecology, society, and economics. Elsevier, New York

Cocke AE, Fulé PZ, Crouse JE (2005) Forest change on a steep mountain gradient after extended fire exclusion: San Francisco Peaks, Arizona, USA. J Appl Ecol 42:814–823

Conklin DA, Fairweather ML (2010) Dwarf mistletoes and their management in the southwest. U.S. Department of Agriculture, Forest Service, Southwestern Region, R3-FH-10-01. http://www.fs.fed.us/r3/publications/documents/dm_management.pdf. Accessed 12 Jan 2012

Crawford JA (2008) Five years of vegetation change following high-severity fire and fire-fighting activities in Grand Canyon National Park. In: van Riper C III, Sogge MK (eds) The Colorado Plateau III: Integrating research and resources management for effective conservation. Proceedings of the 8th biennial conference of research on the Colorado Plateau. University of Arizona Press, Tucson

Dahms CW, Geils BW (tech eds) (1997) An assessment of forest ecosystem health in the Southwest. U.S. Department of Agriculture, Forest Service General Technical Report RM-GTR-295, Rocky Mountain Forest and Range Experiment Station, Fort Collins, CO

DeByle NV (1985) Animal impacts. In: DeByle NV, Winokur PR (eds) Aspen: ecology and management in the western United States. U.S. Department of Agriculture, Forest Service General Technical Report RM-119, Rocky Mountain Forest and Range Experiment Station, Fort Collins, CO

DeVelice RL, Ludwig JA, Moir WH, Ronco F Jr (1986) A classification of forest habitat types of northern New Mexico and southern Colorado. U.S. Department of Agriculture, Forest Service General Technical Report RM-131, Rocky Mountain Forest and Range Experiment Station, Fort Collins, CO

Dieterich JH (1983) Fire history of southwestern mixed conifer: a case study. For Ecol Manag 6:13–31

Elliott GP, Baker WL (2004) Quaking aspen (*Populus tremuloides* Michx.) at treeline: a century of change in the San Juan Mountains, Colorado, USA. J Biogeogr 31:733–745

Evans AM, Everett RG, Stephens SL, Youtz JA (2011) Comprehensive fuels treat practices guide for mixed conifer forests: California, central and southern Rockies, and the Southwest. Forest Guild, Santa Fe

Fairweather ML, Geils BW, Manthei M (2008) Aspen decline on the Coconino National Forest. In: McWilliams M, Palacios P (compilers) Proceedings of the 55th annual western international forest disease work conference. Oregon Department of Forestry, Salem

Fellin DG, Dewey JE (1986) Western spruce budworm. U.S. Department of Agriculture, Forest Service Forest Insect Disease Leaflet 53. Washington, DC

Fellin DG, Moir WH, Linnane JF, Lynch AM (1990) Western spruce budworm in Arizona and New Mexico: outbreak history and effects of direct suppression programs. U.S. Department of Agriculture, Forest Service manuscript on file, Rocky Mountain Forest and Range Experiment Station, Fort Collins, CO

Fisher MA, Fulé PZ (2004) Changes in forest vegetation and arbuscular mycorrhizae along a steep elevation gradient in Arizona. For Ecol Manag 200:293–311

Frechette JD, Meyer GA (2009) Holocene fire-related alluvial-fan deposition and climate in ponderosa pine and mixed-conifer forests, Sacramento Mountains, New Mexico, USA. Holocene 19:639–651

Frey BR, Lieffers VJ, Landhäusser SM, Comeau PG, Greenway KJ (2003) An analysis of sucker regeneration of trembling aspen. Can J For Res 33:1169–1179

Fulé PZ, Covington WW, Moore MM, Heinlein TA, Waltz AEM (2002) Natural variability in forests of the Grand Canyon, USA. J Biogeogr 29:31–47

Fulé PZ, Crouse JE, Heinlein TA, Moore MM, Covington WW, Verkamp G (2003a) Mixed-severity fire regime in a high-elevation forest of Grand Canyon, Arizona, USA. Landsc Ecol 18:465–486

Fulé PZ, Heinlein TA, Covington WW, Moore MM (2003b) Assessing fire regimes on Grand Canyon landscapes with fire-scar and fire-record data. Int J Wildland Fire 12:129–145

Fulé PZ, Crouse JE, Cocke AE, Moore MM, Covington WW (2004) Changes in canopy fuels and potential fire behavior 1880–2040: Grand Canyon, Arizona. Ecol Model 175:231–248

Fulé PZ, Korb JE, Wu R (2009) Changes in forest structure of a mixed conifer forest, southwestern Colorado, USA. For Ecol Manag 258:1200–1210

Ganey JL, Vojta SC (2011) Tree mortality in drought-stressed mixed-conifer and ponderosa pine forests, Arizona, USA. For Ecol Manag 261:162–168

Gosz JR (1980) Biomass distribution and production budget for a nonaggrading forest ecosystem. Ecology 61:507–514

Gottfried GJ (1978) Five-year growth and development in a virgin Arizona mixed conifer stand. U.S. Department of Agriculture, Forest Service Research Paper RM-203, Rocky Mountain Forest and Range Experiment Station, Fort Collins, CO

Gottfried GJ, Ffolliott PF (1992) Effects of moderate timber harvesting in an old-growth Arizona mixed conifer watershed. In: Kauffman MR, Moir WH, Bassett RL (tech coords) Old growth forests in the Southwest and Rocky Mountain regions, proceedings of a workshop, 9–13 Mar 1992, Portal, Arizona. U.S. Department of Agriculture, Forest Service General Technical Report RM-213, Rocky Mountain Forest and Range Experiment Station, Fort Collins, CO

Gottfried GJ, Ffolliott PF, DeBano LF (1995) Forests and woodlands of the sky islands: stand characteristics and silvicultural prescriptions. In: DeBano LF, Ffolliott PF, Ortega-Rubio A, Gottfried GJ, Hamre RH, Edminster CB (tech coords) Biodiversity and management of the Madrean archipelago: the sky islands of southwestern United States and northwestern Mexico. U.S. Department of Agriculture, Forest Service, General Technical Report RM-GTR-264, Rocky Mountain Forest and Range Experiment Station, Fort Collins, CO

Greenamyre HH (1913) The composite type on the Apache National Forest. U.S. Department of Agriculture, Forest Service, Bulletin 125. Fort Valley Experiment Station, Washington, DC

Grissino-Mayer HD, Baisan CH, Swetnam TW (1995) Fire history in the Pinaleño Mountains of southeastern Arizona: effects of human-related disturbances. In: DeBano LF, Ffolliott PF, Ortega-Rubio A, Gottfried GJ, Hamre RH, Edminster CB (tech coords) Biodiversity and management of the Madrean archipelago: the sky islands of southwestern United States and northwestern Mexico. U.S. Department of Agriculture, Forest Service, General Technical Report RM-GTR-264, Rocky Mountain Forest and Range Experiment Station, Fort Collins, CO

Grissino-Mayer HD, Romme WH, Floyd ML, Hanna DD (2004) Climatic and human influences on fire regimes of the southern San Juan Mountains, Colorado, USA. Ecology 85:1708–1724

Hanks JP, Dick-Peddie WA (1974) Vegetation patterns on the White Mountains, New Mexico. Southwest Nat 18:372–382

Heinlein TA, Moore MM, Fulé PZ, Covington WW (2005) Fire history and stand structure of two ponderosa pine-mixed conifer sites: San Francisco Peaks, Arizona, USA. Int J Wildland Fire 14:307–320

Hendricks DM (1985) Arizona Soils. University of Arizona Press, Tucson

Heyerdahl EK, Brown PM, Kitchen S, Weber MH (2011) Multicentury fire and forest histories at 19 sites in Utah and eastern Nevada. U.S. Department of Agriculture, Forest Service General Technical Report RMRS-GTR-261WWW, Rocky Mountain Research Station, Fort Collins, CO. http://www.fs.fed.us/rm/pubs/rmrs_gtr261.pdf. Accessed 12 Jan 2011

Hicke JA, Johnson MC, Hayes JL, Preisler HK (2012) Effects of bark beetle-caused tree mortality on wildfire. For Ecol Manag 271:81–90

Holden ZA, Morgan P, Crimmins MA, Steinhorst RK, Smith AMS (2007) Fire season precipitation variability influences fire extent and severity in a large southwestern wilderness area, United States. Geophys Res Lett 34:1–5

Holden ZA, Morgan P, Evans JS (2009) A predictive model of burn severity based on 20-year satellite-inferred burn severity data in a large southwestern US wilderness area. For Ecol Manag 258:2399–2406

Hood SM, Miller M (eds) (2007) Fire ecology and management of the major ecosystems of southern Utah. U.S. Department of Agriculture, Forest Service General Technical Report RMRS-GTR-202, Rocky Mountain Research Station, Fort Collins, CO

Huisinga KD, Laughlin DC, Fulé PZ, Springer JD, McGlone CM (2005) Effects of an intense prescribed fire on understory vegetation in a mixed conifer forest. J Torrey Bot Soc 132:590–601

Hurst M (1977) An ecological description of forest communities on the North Rim of Grand Canyon National Park. U.S. Department of the Interior, National Park Service manuscript on file, Grand Canyon National Park, AZ

Iniguez JM, Swetnam TW, Yool SR (2008) Topography affected landscape fire history patterns in southern Arizona, USA. For Ecol Manag 256:295–303

Integrated Taxonomic Information System (2012) http://www.itis.gov. Accessed 29 July 2012

Jenkins MJ, Dicus CA, Godfrey JE (1998) Restoration of mixed conifer communities using prescribed fire in Bryce Canyon National Park. In: Pruden TL, Brennan LA (eds) Fire in ecosystem management: shifting the paradigm from suppression to prescription. Proceedings of the 20th Tall Timbers Fire Ecology Conference. Tall Timbers Research Station, Tallahassee, FL

Jenkins MJ, Page WG, Hebertson EG, Alexander ME (2012) Fuels and fire behavior dynamics in bark beetle-attacked forests in western North America and implications for fire management. For Ecol Manag 275:23–34

Jenkins SE, Sieg CH, Anderson DE, Kaufman DS, Pearthree PA (2011) Late Holocene geomorphic record of fire in ponderosa pine and mixed-conifer forests, Kendrick Mountain, northern Arizona, USA. Int J Wildland Fire 20:125–141

Jones JR (1974) Silviculture of southwestern mixed conifer and aspen: the status of our knowledge. U.S. Department of Agriculture, Forest Service Research Paper RM-122, Rocky Mountain Forest and Range Experiment Station, Fort Collins, CO

Jones JR, DeByle NV (1985a) Fire. In: DeByle NV, Winokur PR (eds) Aspen: ecology and management in the western United States. U.S. Department of Agriculture, Forest Service General Technical Report RM-119, Rocky Mountain Forest and Range Experiment Station, Fort Collins, CO

Jones JR, DeByle NV (1985b) Soils. In: DeByle NV, Winokur PR (eds) Aspen: ecology and management in the western United States. U.S. Department of Agriculture, Forest Service General Technical Report RM-119, Rocky Mountain Forest and Range Experiment Station, Fort Collins, CO

Karnosky DF, Mankovska B, Percy K, Dickson RE, Podila GK, Sober J, Noormets A, Hendrey G, Coleman MD, Kubiske M, Pregitzer KS, Isebrands JG (1999) Effects of tropospheric O_3 on trembling aspen and interaction with CO_2: results from an O_3-gradient and a FACE experiment. Water Air Soil Pollut 116:311–322

Kleinman LH (1973) Community characteristics of six burned aspen-conifer sites and their related animal use. Thesis, Brigham Young University, Provo

Klemmedson JO, Smith EL (1979) Vegetation-soil relations of forests, woodlands, and grasslands of Arizona and New Mexico. In: Youngberg CT (ed) Forest soils and land use. Department of Forest and Wood Sciences, Colorado State University, Fort Collins
Korb JE, Daniels ML, Laughlin DC, Fulé PZ (2007) Understory communities of warm-dry, mixed-conifer forests in southwestern Colorado. Southwest Nat 52:493–503
Kurzel BP, Veblen TT, Kulakowski D (2007) A typology of stand structure and dynamics of quaking aspen in northwest Colorado. For Ecol Manag 252:176–190
Lang DK, Stewart SS (1910) Reconnaissance of the Kaibab National Forest. U.S. Department of Agriculture, Forest Service, unpublished report. http://archive.library.nau.edu/cdm4/document.php?CISOROOT=/cpa&CISOPTR=57641&CISOSHOW=57583. Accessed 7 July 2011
Laughlin DC, Bakker JD, Fulé PZ (2005) Understory plant community structure in lower montane and subalpine forests, Grand Canyon National Park, USA. J Biogeogr 32:2083–2102
Laughlin DC, Fulé PZ, Huffman DW, Crouse J, Laliberté E (2011) Climatic constraints on trait-based forest assembly. J Ecol 99:1489–1499
Layser EF, Schubert GH (1979) Preliminary classification for the coniferous forest and woodland series of Arizona and New Mexico. U.S. Department of Agriculture, Forest Service Research Paper RM-208, Rocky Mountain Forest and Range Experiment Station, Fort Collins, CO
Leiberg JB, Rixon TF, Dodwell A (1904) Forest conditions in the San Francisco Mountains Forest Reserve, Arizona. Department of the Interior, U.S. Geological Survey Series H, Forestry, 7, Professional Paper No. 22, Washington, DC
Linnane JP (1986) Integrated forest protection guide, western spruce budworm in the Southwest. U.S. Department of Agriculture, Forest Service Southwestern Region Forest Pest Management Report R3-86-7, Albuquerque
Lowe CH (1964) Arizona's natural environment: landscapes and habitats. University of Arizona Press, Tucson
Lowe CH, Brown DE (1973) The natural vegetation of Arizona. Arizona Resources Information System, Cooperative Publication No. 2, Phoenix
Lynch AM, Swetnam TW (1992) Old-growth mixed-conifer and western spruce budworm in the southern Rocky Mountains. In: Kauffman MR, Moir WH, Bassett RL (tech coords) Old growth forests in the Southwest and Rocky Mountain regions, proceedings of a workshop, 9–13 Mar 1992, Portal, Arizona. U.S. Department of Agriculture, Forest Service General Technical Report RM-213, Rocky Mountain Forest and Range Experiment Station, Fort Collins, CO
MacMahon JA (1988) Vegetation of Utah. In: Albee BJ, Schultz LM, Goodrich S (eds) Atlas of the vascular plants of Utah. Utah Museum of Natural History Occasional Publication No. 7, Salt Lake City, UT
Maker HJ, Saugherty LA (1986) Soils. In: Williams JL (ed) New Mexico in maps, 2nd edn. University of New Mexico Press, Albuquerque
Margolis EQ, Balmat J (2009) Fire history and fire-climate relationships along a fire regime gradient in the Santa Fe Municipal Watershed, NM, USA. For Ecol Manag 258:2416–2430
Margolis EQ, Swetnam TW, Allen CD (2007) A stand-replacing fire history in upper montane forests of the southern Rocky Mountains. Can J For Res 37:2227–2241
Margolis EQ, Swetnam TW, Allen CD (2011) Historical stand-replacing fire in upper montane forests of the Madrean sky islands and Mogollon Plateau, southwestern USA. Fire Ecol 7:88–107
Mast JN, Wolf JJ (2004) Ecotonal changes and altered tree spatial patterns in lower mixed-conifer forests, Grand Canyon National Park, Arizona, U.S.A. Landsc Ecol 19:167–180
Mast JN, Wolf JJ (2006) Spatial patch patterns and altered forest structure in middle elevation versus upper ecotonal mixed-conifer forests, Grand Canyon National Park, Arizona, USA. For Ecol Manag 236:241–250
Mathiasen RL, Hawksworth FG, Edminster CB (1990) Effects of dwarf mistletoe on growth and mortality of Douglas-fir in the Southwest. Gt Basin Nat 50:173–179
McHenry DE (1935) Quaking aspen: its future in the park. Gd Canyon Nat Notes 9:361–365
Merkle J (1954) An analysis of the spruce-fir community on the Kaibab Plateau, Arizona. Ecology 35:316–322

Merkle J (1962) Plant communities of the Grand Canyon area, Arizona. Ecology 43:698–710

Merriam CH (1890) Part I: General results of a biological survey of the San Francisco Mountain region in Arizona, with special reference to the distribution of species. In: Merriam CH, Stejneger L (eds) Results of a biological survey of the San Francisco Mountain region and desert of the Little Colorado, Arizona. U.S. Department of Agriculture, Division of Ornithology and Mammalogy, North American Fauna No. 3, Government Printing Office, Washington, DC

Milchunas DG (2006) Responses of plant communities to grazing in the southwestern United States. U.S. Department of Agriculture, Forest Service General Technical Report RMRS-GTR-169, Rocky Mountain Research Station, Fort Collins, CO

Miller WE (1983) Decomposition rates of aspen bole and branch litter. For Sci 29:351–356

Moir WH (1993) Alpine tundra and coniferous forest. In: Dick-Peddie WA New Mexico vegetation: past, present, and future. University of New Mexico Press, Albuquerque

Moir WH, Ludwig JA (1979) A classification of spruce-fir and mixed conifer habitat types of Arizona and New Mexico. U.S. Department of Agriculture, Forest Service Research Paper RM-207, Rocky Mountain Forest and Range Experiment Station, Fort Collins, CO

Moore MM, Huffman DW (2004) Tree encroachment on meadows of the North Rim, Grand Canyon National Park, Arizona, U.S.A. Arct Antarct Alp Res 36:474–483

Moore MM, Huffman DW, Fulé PZ, Covington WW, Crouse JE (2004) Comparison of historical and contemporary forest structure and composition on permanent plots in southwestern ponderosa pine forests. For Sci 50:162–176

Mueggler WF (1985a) Forage. In: DeByle NV, Winokur PR (eds) Aspen: ecology and management in the western United States. U.S. Department of Agriculture, Forest Service General Technical Report RM-119, Rocky Mountain Forest and Range Experiment Station, Fort Collins, CO

Mueggler WF (1985b) Vegetation associations. In: DeByle NV, Winokur PR (eds) Aspen: ecology and management in the western United States. U.S. Department of Agriculture, Forest Service General Technical Report RM-119, Rocky Mountain Forest and Range Experiment Station, Fort Collins, CO

Mueggler WF (1988) Aspen community types of the Intermountain Region. U.S. Department of Agriculture, Forest Service General Technical Report INT-250, Intermountain Research Station, Ogden, UT

Mueggler WF (1989) Age distribution and reproduction of Intermountain aspen stands. West J Appl For 4:41–45

Mueggler WF, Campbell RB Jr (1986) Aspen community types of Utah. U.S. Department of Agriculture, Forest Service Research Paper INT-362, Intermountain Research Station, Ogden, UT

Muldavin EH, DeVelice RL, Ronco F Jr (1996) A classification of forest habitat types: southern Arizona and portions of the Colorado Plateau. U.S. Department of Agriculture, Forest Service General Technical Report RM-GTR-287, Rocky Mountain Forest and Range Experiment Station, Fort Collins, CO

National Park Service (2002) Air quality in the National Parks, 2nd edn. Air Resources Division, U.S. Department of the Interior, National Park Service, Washington, DC

O'Brien RA (1999) Comprehensive inventory of Utah's forest resources, 1993. U.S. Department of Agriculture, Forest Service Resource Bulletin RMRS-RB-1, Rocky Mountain Research Station, Ogden, UT

Oukrop CM, Evans DM, Bartos DL, Ramsey RD, Ryel RJ (2011) Moderate-scale mapping methods of aspen stand types: a case study for Cedar Mountain in southern Utah. U.S. Department of Agriculture, Forest Service General Technical Report RMRS-GTR-259, Rocky Mountain Research Station, Fort Collins, CO

Pauchard A, Kueffer C, Dietz H, Daehler CC, Alexander J, Edwards PJ, Arévalo JR, Cavieres LA, Guisan A, Haider S, Jakobs G, McDougall K, Millar CI, Naylor BJ, Parks CG, Rew LJ, Seipel T (2009) Ain't no mountain high enough: plant invasions reaching new elevations. Front Ecol Environ 7:479–486

Pearson GA (1914) The role of aspen in the reforestation of mountain burns in Arizona and New Mexico. Plant World 17:249–260

Pearson GA (1920) Factors controlling the distribution of forest trees, part I. Ecology 1:139–159

Pearson GA (1931) Forest types in the Southwest as determined by climate and soil. Technical Bulletin 247. U.S. Department of Agriculture, Forest Service, Washington, DC

Pearson GA (1950) Management of ponderosa pine in the Southwest, as developed by research and experimental practice. U.S. Department of Agriculture Agricultural Monograph 6, Washington, DC

Pfister RD (1972) Vegetation and soils in the subalpine forests of Utah. Dissertation, Washington State University, Pullman

Plummer FG (1904) Forest conditions in the Black Mesa Forest Reserve, Arizona. U.S. Geological Survey Series H, Forestry, 8, Professional Paper No. 23, Washington, DC

Powell JW (1879) Report on the lands of the arid region of the United States with a more detailed account of the lands of Utah, 2nd edn. Government Printing Office, Washington, DC

Price R, Evans RB (1937) Climate of the west front of the Wasatch Plateau in central Utah. Mon Weather Rev 65:291–301

Prior-Magee JS, Boykin KG, Bradford DF, Kepner WG, Lowry JH, Schrupp DL, Thomas KA, Thompson BC (2007) Ecoregional gap analysis of the southwestern United States: the Southwest regional gap analysis project final report. http://fws-nmcfwru.nmsu.edu/swregap/report/SWReGAP%20Final%20Report.pdf. Accessed 11 July 2012

Rasmussen DI (1941) Biotic communities of Kaibab Plateau, Arizona. Ecol Monogr 11:229–275

Rehfeldt GE, Ferguson DE, Crookston NL (2009) Aspen, climate, and sudden decline in western USA. For Ecol Manag 258:2353–2364

Reynolds HG (1969) Aspen grove use by deer, elk, and cattle in southwestern coniferous forests. U.S. Department of Agriculture, Forest Service Research Note RM-138, Rocky Mountain Forest and Range Experiment Station, Fort Collins, CO

Rogers PC, Leffler AJ, Ryel RJ (2010) Landscape assessment of a stable aspen community in southern Utah, USA. For Ecol Manag 259:487–495

Romme WH, Jamieson DW, Redders JS, Bigsby G, Lindsey JP, Kendall D, Cowen R, Kreykes T, Spencer AW, Ortega JC (1992) Old-growth forests of the San Juan National Forest in southwestern Colorado. In: Kauffman MR, Moir WH, Bassett RL (tech coords) Old growth forests in the Southwest and Rocky Mountain regions, proceedings of a workshop, 9–13 Mar 1992, Portal, Arizona. U.S. Department of Agriculture, Forest Service General Technical Report RM-213, Rocky Mountain Forest and Range Experiment Station, Fort Collins, CO

Romme WH, Floyd-Hanna L, Hana DD, Bartlett E (2001) Aspen's ecological role in the West. In: Shepperd WD, Binkley D, Bartos DL, Stohlgren TJ, Eskew LG (compilers) Sustaining aspen in western landscapes: symposium proceedings. U.S. Department of Agriculture, Forest Service Proceedings RMRS-P-18, Rocky Mountain Research Station, Fort Collins, CO

Romme WH, Clement J, Hicke JA, Kulakowski D, MacDonald LH, Schoennagel T, Veblen TT (2006) Recent forest insect outbreaks and fire risk in Colorado forests: a brief synthesis of relevant research. Colorado Forest Restoration Institute, Colorado State University, Fort Collins. http://spot.colorado.edu/~schoenna/images/RommeEtAl2006CFRI%20.pdf. Accessed 3 July 2011

Romme WH, Floyd ML, Hanna D, Bartlett EJ (2009a) Aspen forests. In: Romme WH, Floyd ML, Hanna D (eds) Historical range of variability and current landscape condition analysis: south central highlands section, southwestern Colorado & northwestern New Mexico. Colorado Forest Restoration Institute, Colorado State University and Region 2, U.S. Forest Service, Fort Collins

Romme WH, Redders JS, Floyd ML, Hanna D (2009b) Mixed conifer forests. In: Romme WH, Floyd ML, Hanna D (eds) Historical range of variability and current landscape condition analysis: south central highlands section, southwestern Colorado & northwestern New Mexico. Colorado Forest Restoration Institute, Colorado State University and Region 2, U.S. Forest Service, Fort Collins

Ryerson DE, Swetnam TW, Lynch AM (2003) A tree-ring reconstruction of western spruce budworm outbreaks in the San Juan Mountains, Colorado, U.S.A. Can J For Res 33:1010–1028

Shepperd WD, Binkley D, Bartos DL, Stohlgren TJ, Eskew LG (compilers) (2001) Sustaining aspen in western landscapes: symposium proceedings. U.S. Department of Agriculture, Forest Service Proceedings RMRS-P-18, Rocky Mountain Research Station, Fort Collins, CO

Smith EA, O'Loughlin D, Buck JR, St. Clair SB (2011) The influences of conifer succession, physiographic conditions and herbivory on quaking aspen regeneration after fire. For Ecol Manag 262:325–330

Stein SJ (1988) Fire history of the Paunsaugunt Plateau in southern Utah. Gt Basin Nat 48:58–63

Stromberg JC, Patten DT (1991) Dynamics of the spruce-fir forests on the Pinaleno Mountains, Graham Co., Arizona. Southwest Nat 36:37–48

Swetnam TW (1987) Western spruce budworm outbreaks in northern New Mexico: tree-ring evidence of occurrence and radial growth impacts from 1700–1983. In: Jacoby GC, Hornbeck JW (compilers) Proceedings of the international symposium on ecological aspects of tree-ring analysis, August 17–21, 1986, Marymount College, Tarrytown, New York. US Department of Energy, CONF-8608144, Washington, DC

Swetnam TW (1990) Fire history and climate in the southwestern United States. In: Krammes JS (tech coord) Effects of fire management of southwestern natural resources. U.S. Department of Agriculture, Forest Service General Technical Report RM-191, Rocky Mountain Forest and Range Experiment Station, Fort Collins, CO

Swetnam TW, Baisan CH (1996) Historical fire regime patterns in the southwestern United States since AD 1700. In: Allen CD (tech ed) Fire effects in southwestern forests: Proceedings of the Second La Mesa Fire Symposium, Los Alamos, New Mexico, 29–31 Mar 1994. U.S. Department of Agriculture, Forest Service General Technical Report RM-GTR-286, Rocky Mountain Forest and Range Experiment Station, Fort Collins, CO

Swetnam TW, Brown PM (1992) Oldest known conifers in the southwestern United States: temporal and spatial patterns of maximum age. In: Kauffman MR, Moir WH, Bassett RL (tech coords) Old growth forests in the Southwest and Rocky Mountain regions, proceedings of a workshop, 9–13 Mar 1992, Portal, Arizona. U.S. Department of Agriculture, Forest Service General Technical Report RM-213, Rocky Mountain Forest and Range Experiment Station, Fort Collins, CO

Swetnam TW, Lynch AM (1989) A tree-ring reconstruction of western spruce budworm history in the southern Rocky Mountains. For Sci 35:962–986

Swetnam TW, Lynch AM (1993) Multicentury, regional-scale patterns of western spruce budworm outbreaks. Ecol Monogr 63:399–424

Swetnam TW, Baisan CH, Kaib JM (2001) Forest fire histories of the sky islands of La Frontera. In: Webster GL, Bahre CJ (eds) Changing plant life of La Frontera: observations on vegetation in the US/Mexico borderlands. University of New Mexico Press, Albuquerque

Swetnam TW, Baisan CH, Grissino-Mayer HD (2009) Tree-ring perspectives on fire regimes and forest dynamics in mixed conifer and spruce-fir forests on Mt. Graham. In: Sanderson HR, Koprowski JL (tech eds) The last refuge of the Mt. Graham red squirrel: ecology of endangerment, University of Arizona Press, Tucson

Taylor K, Mangold J, Rew LJ (2011) Weed dispersal by vehicles. MontGuide MT201105AG. http://msuextension.org/publications/AgandNaturalResources/MT201105AG.pdf. Accessed 30 Jan 2012

Touchan R, Allen CD, Swetnam TW (1996) Fire history and climatic patterns in ponderosa pine and mixed-conifer forests of the Jemez Mountains, northern New Mexico. In: Krammes JS (tech coord) Effects of fire management of southwestern natural resources. U.S. Department of Agriculture, Forest Service General Technical Report RM-191, Rocky Mountain Forest and Range Experiment Station, Fort Collins, CO

U.S. Forest Service (2010) Forest insect and disease conditions report in the Southwestern Region, 2009. U.S. Department of Agriculture, Forest Service Southwestern Region Forestry and Forest Health PR-R3-16-6. http://www.fs.usda.gov/Internet/FSE_DOCUMENTS/stelprdb5238446.pdf. Accessed 5 Mar 2012

Vankat JL (2010) Forest changes in Grand Canyon National Park and implications for management. Report to U.S. Department of the Interior, National Park Service, Grand Canyon National Park, AZ

Vankat JL (2011) Post-1935 changes in forest vegetation of Grand Canyon National Park, Arizona, USA: Part 2—Mixed conifer, spruce-fir, and quaking aspen forests. For Ecol Manag 261:326–341

Vankat JL, Crocker-Bedford DC, Bertolette DR, Leatherbury P, McKinnon T, Sipe CL (2005) Indications of large changes in mixed conifer forests of Grand Canyon National Park. In: van Riper C III, Mattson DJ (eds) The Colorado Plateau II: biophysical, socioeconomic, and cultural research. Proceedings of the 7th biennial conference of research on the Colorado Plateau. University of Arizona Press, Tucson

Villanueva-Díaz J, McPherson GR (1995) Forest stand structure in mountains of Sonora, Mexico and New Mexico, USA. In: DeBano LF, Ffolliott PF, Ortega-Rubio A, Gottfried GJ, Hamre RH, Edminster CB (tech coords) Biodiversity and management of the Madrean archipelago: the sky islands of southwestern United States and northwestern Mexico. U.S. Department of Agriculture, Forest Service, General Technical Report RM-GTR-264, Rocky Mountain Forest and Range Experiment Station, Fort Collins, CO

Wells FH, Lauenroth WK (2007) The potential for horses to disperse alien plants along recreational trails. Rangel Ecol Manag 60:574–577

White MA, Vankat JL (1993) Middle and high elevation coniferous forest communities of the North Rim region of Grand Canyon National Park, Arizona, USA. Vegetatio 109:161–174

Whittaker RH, Niering WA (1975) Vegetation of the Santa Catalina Mountains, Arizona. V. Biomass, production, and diversity along the elevation gradient. Ecology 56:771–790

Wolf JJ, Mast JN (1998) Fire history of mixed-conifer forests on the North Rim, Grand Canyon National Park, Arizona. Phys Geogr 19:1–14

Worrall JJ, Egeland L, Eager T, Mask RA, Johnson EW, Kemp PA, Shepperd WD (2008) Rapid mortality of *Populus tremuloides* in southwestern Colorado, USA. For Ecol Manag 255:686–696

Wu R (1999) Fire history and forest structure in mixed conifer forests of Southwest Colorado. Thesis, Colorado State University, Fort Collins

Youngblood AP, Mauk RL (1985) Coniferous forest habitat types of central and southern Utah. U.S. Department of Agriculture, Forest Service General Technical Report INT-187, Intermountain Research Station, Ogden, UT

Zegler TJ, Moore MM, Fairweather ML, Ireland KB, Fulé PZ (2012) *Populus tremuloides* mortality near the southwestern edge of its range. For Ecol Manag 282:196–207

Zier JL, Baker WL (2006) A century of vegetation change in the San Juan Mountains, Colorado: an analysis using repeat photography. For Ecol Manage 228:251–262

Chapter 4
Ponderosa Pine Forest

Abstract Ponderosa Pine Forest is the lowest-elevation, most extensive forest in the American Southwest. It occurs in an elevational band below Mixed Conifer Forest and above Pinyon-Juniper vegetation, Gambel Oak Shrubland, and Interior Chaparral Shrubland. Stands are dominated by ponderosa pine, and are divided into moist, mesic, and dry stand types with decreasing elevation and moisture availability. A historical fire regime of frequent, low-severity surface fires is widely documented, but there is growing evidence of historical mixed-severity and high-severity fires, especially for steep slopes in areas of heterogeneous topography. Other important natural disturbances include insect outbreaks and drought. Late nineteenth century livestock grazing initiated fire exclusion, which was continued by fire management through most of the twentieth century. Other anthropogenic drivers are modern climate change, invasive species, recreation, and nearby land use. Vegetation dynamics are dominated by tree regeneration, thinning, and succession. Historical conditions ranged from open-canopied stands with a well-developed, often grass-dominated understory – more woodland than forest – to denser stands. Stand densities increased during the twentieth century because of the exclusion of surface fires. Vegetation dynamics are illustrated in a nested, three-tiered set of conceptual models. Key conclusions and challenges for researchers and land managers are summarized.

4.1 Introduction

Ponderosa Pine Forest (*Pinus ponderosa*) is the lowest-elevation coniferous forest of the American Southwest (Figs. 4.1 and 4.2). It occurs in an elevational band below Mixed Conifer Forest and above Pinyon-Juniper vegetation (*Pinus-Juniperus*), Gambel Oak Shrubland (*Quercus gambelii*), and Interior Chaparral Shrubland. With this elevational position, Ponderosa Pine Forest has the most extensive, contiguous distribution of any forest in the Southwest. It covers about 49,000 km^2 (18,900 $miles^2$), which is 6.3 % of the total area of the region (Fig. 4.3; calculations

J.L. Vankat, *Vegetation Dynamics on the Mountains and Plateaus of the American Southwest*, Plant and Vegetation 8, DOI 10.1007/978-94-007-6149-0_4,

Fig. 4.1 Landscape of Ponderosa Pine Forest in Santa Fe National Forest, north-central New Mexico (Photograph by Betty J. Huffman)

Fig. 4.2 Landscape of Ponderosa Pine Forest with surface fire in Santa Fe National Forest, north-central New Mexico. Note variation in tree densities (Photograph by Kari Greer/U.S. National Forest)

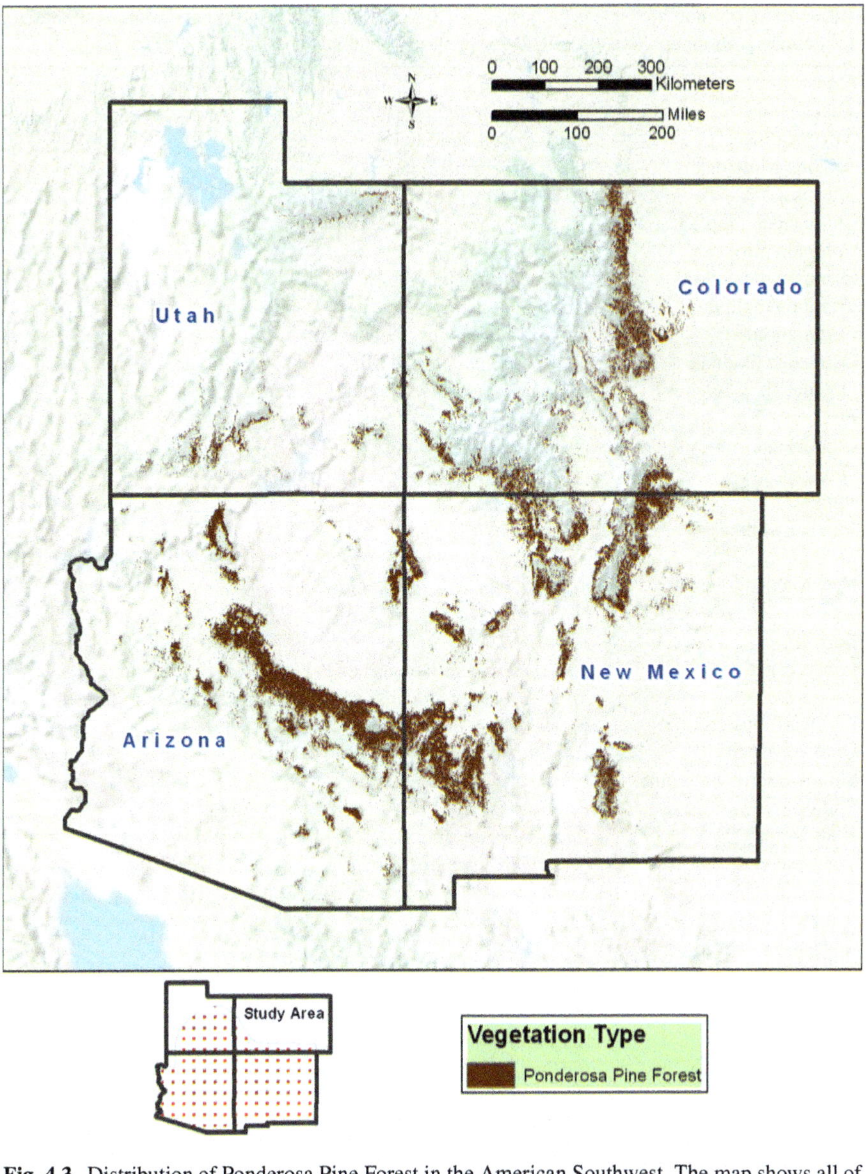

Fig. 4.3 Distribution of Ponderosa Pine Forest in the American Southwest. The map shows all of the states of Arizona, New Mexico, Colorado, and Utah, and the American Southwest is illustrated in *red* on the small map (Source: U.S. Geological Survey National Gap Analysis Program 2005 Southwest Regional Gap Analysis Project, RS/GIS Laboratory, College of Natural Resources, Utah State University, Logan, Utah. Map prepared by Monica Swihart)

based on Prior-Magee et al. 2007). These values do not include stands dominated by quaking aspen (*Populus tremuloides*), which cover 8,100 km^2 (3,100 miles^2) and 1.0 % of the region. Only approximately 5 % of aspen stands are associated with Ponderosa Pine Forest; the rest occur with Spruce-Fir (*Picea-Abies*) and Mixed Conifer Forests

Table 4.1 Common and scientific names of species in this chapter. Primary sources: Integrated Taxonomic Information System (2012) for plants and animals and Bates (2006) for fungi

Plants	
Apache pine	*Pinus engelmannii* Carrière
Arizona pine	*Pinus arizonica* Engelm.
Arizona white oak	*Quercus arizonica* Sarg.
Cheatgrass	*Bromus tectorum* L.
Chihuahuan pine	*Pinus leiophylla* Schiede & Deppe
Colorado pinyon	*Pinus edulis* Engelm.
Common dandelion	*Taraxacum officinale* F.H. Wigg.
Common mullein	*Verbascum thapsis* L.
Common salsify	*Tragopogon dubius* Scop.
Dalmatian toadflax	*Linaria dalmatica* (L.) Mill.
Douglas-fir	*Pseudotsuga menziesii* (Mirbel) Franco
Fir	*Abies* P. Mill.
Gambel oak	*Quercus gambelii* Nutt.
Gray oak	*Quercus grisea* Liebm.
Juniper	*Juniperus* L.
Kentucky bluegrass	*Poa pratensis* L.
Manzanita	*Arctostaphylos* Adans.
New Mexico locust	*Robinia neomexicana* A. Gray
Oak	*Quercus* L.
Pinyon	*Pinus* L.
Ponderosa pine	*Pinus ponderosa* Douglas ex P. Lawson & C. Lawson
Ponderosa pine dwarf mistletoe	*Arceuthobium vaginatum* spp. *cryptopodum* (Engelm.) Hawksworth & Wiens
Quaking aspen	*Populus tremuloides* Michx.
Southwestern white pine	*Pinus strobiformis* Engelm.
Spruce	*Picea* A. Dietr.
Utah juniper	*Juniperus osteosperma* (Torr.) Little
Wavyleaf oak	*Quercus* X *undulata* Torr.
White fir	*Abies concolor* (Gord. & Glend.) Lindl. Ex Hildebr.
Wild lettuce	*Lactuca serriola* L.
Animals	
Bark beetle	*Dendroctonus* Erichson, 1846 and *Ips* De Geer, 1775
Deer	*Odocoileus* Rafinesque, 1832
Elk	*Cervus elaphus* Linnaeus, 1758
Ips engraver beetle	*Ips* De Geer, 1775
Pandora moth	*Coloradia pandora* Blake, 1863
Pine sawfly	*Neodiprion* Rohwer, 1918
Sheep	*Ovis aries* Linnaeus, 1758
Western pine beetle	*Dendroctonus brevicomis* LeConte, 1876
Fungi	
Armillaria root rot	*Armillaria* (Fr.:Fr.) Staude
Annosus root rot	*Heterobasidion annosum* (Fr.) Bref.

(see Sects. 2.1 and 3.1, respectively). The only vegetation type that covers more area than Ponderosa Pine Forest on southwestern mountains and plateaus is Pinyon-Juniper. Southwestern Ponderosa Pine Forest decreases in area northward and becomes uncommon in central Utah (Fig. 4.3).

Fig. 4.4 Example of moist Ponderosa Pine Forest in the North Rim region of Grand Canyon National Park, north-central Arizona (Photograph by Betty J. Huffman)

The structure of undisturbed stands of southwestern Ponderosa Pine Forest is characterized by large ponderosa pines. Overstory cover is often ~15–33 % (e.g., Pearson 1923, 1950; White 1985; Covington and Sackett 1986). Although such an open overstory suggests woodland physiognomy, the large size of the trees and the presence of stands with greater overstory cover result in the vegetation being described as forest. Stand variation results from differences in density of ponderosa pine, as well as species of other canopy trees (at upper elevations), subcanopy trees, and understory plants. Detailed vegetation classifications of Ponderosa Pine Forest are available for different geographic areas within the American Southwest (Layser and Schubert 1979; Hanks et al. 1983; Alexander et al. 1984, 1987; Youngblood and Mauk 1985; DeVelice et al. 1986; Fitzhugh et al. 1987; Muldavin et al. 1996). More generally, stands are divisible into three broad types: moist, mesic, and dry (Vankat 2011).

Moist stands of Ponderosa Pine Forest (Fig. 4.4) are common at relatively high elevation and extend to lower elevation in drainages. They are transitional with Mixed Conifer Forest. Historically, these stands had scattered individuals of white fir (*Abies concolor*), Douglas-fir (*Pseudotsuga menziesii*), and southwestern white pine (*Pinus strobiformis*) in the canopy, quaking aspen in the subcanopy, and Gambel oak in both the subcanopy and shrub layers. Quaking aspen is most abundant in recently disturbed stands.

Mesic stands of Ponderosa Pine Forest (Fig. 4.5) dominate mid elevations. Ponderosa pine is usually the only canopy tree. Gambel oak occurs in the subcanopy and shrub layers.

Dry stands of Ponderosa Pine Forest (Fig. 4.6) occur primarily at low elevations. They also have Gambel oak in the subcanopy and shrub layers. Other subcanopy species are pinyons such as Colorado pinyon (*Pinus edulis*) and junipers such as Utah juniper (*Juniperus osteosperma*). Stands are transitional with Pinyon-Juniper and sometimes other vegetation.

In addition to this elevational, moisture-driven gradation within Ponderosa Pine Forest, there is substantial latitudinal variation. In Utah and southwestern Colorado, Gambel oak and sometimes other shrubs are often more abundant. In southern Arizona and New Mexico, ponderosa pine is replaced in dry forests of similar overstory physiognomy by Apache pine (*Pinus engelmannii*), Arizona pine (*P. arizonica*), or Chihuahuan pine (*P. leiophylla*), and the understory can be dominated by oaks (*Quercus* spp.) such as Arizona white oak (*Q. arizonica*), Gray oak (*Q. grisea*), or wavyleaf oak (*Q.* X *undulata*).

Ponderosa Pine Forest is also extensive elsewhere in the western United States, as for example in north-central Colorado (Fig. 4.3). However, research findings from other regions are not always applicable to the Southwest, because there are differences in climate, associated species, and other factors. This chapter is based nearly entirely on research done on the mountains and plateaus of the American Southwest, where Ponderosa Pine Forest is the most thoroughly studied vegetation.

4.2 Drivers

Key drivers of southwestern Ponderosa Pine Forest are landscape, climate, soil, animals, natural disturbance, and anthropogenic disturbance. Each driver is important in vegetation dynamics, and therefore all are incorporated in the conceptual models (Sect. 4.5).

4.2.1 Landscape

Ponderosa Pine Forest occurs at approximately 1,800–2,500 m (5,900–8,200 ft) elevation in the mid-latitudes of its distribution in the American Southwest (Fig. 4.7). The topography ranges from relatively level plateaus to steep mountain slopes. At its upper elevational limit, Ponderosa Pine Forest intergrades with Mixed Conifer Forest. In areas of diverse topography, this transition is a mosaic (Plummer 1904; Greenamyre 1913) in which stands of Ponderosa Pine Forest occur on drier sites

Fig. 4.5 Example of mesic Ponderosa Pine Forest in the South Rim region of Grand Canyon National Park, north-central Arizona (Photograph by author)

Fig. 4.6 Example of dry Ponderosa Pine Forest in the South Rim region of Grand Canyon National Park, north-central Arizona (Photograph by Betty J. Huffman)

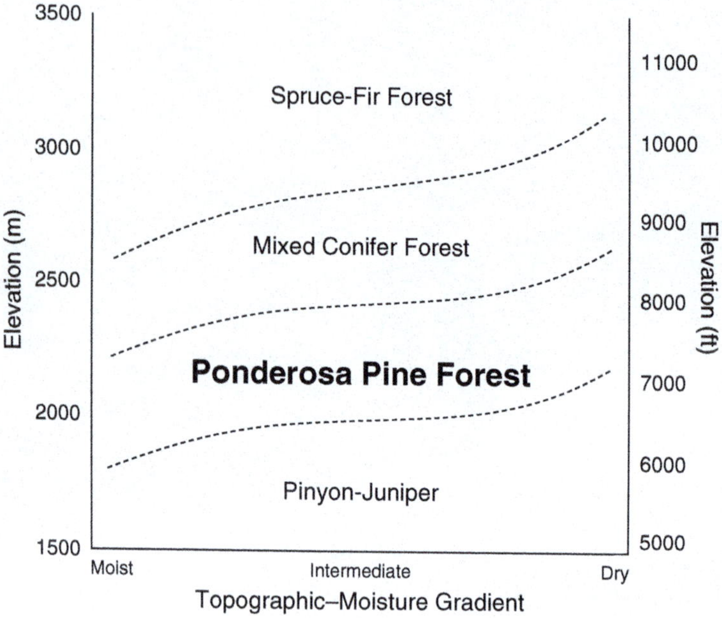

Fig. 4.7 Ecological distribution of Ponderosa Pine Forest on the mountains and plateaus of the American Southwest along gradients in elevation and moisture. Elevations are approximate and generally representative of mid-latitudes of the region, i.e., landscapes in northern Arizona and northern New Mexico

such as south aspects and ridgetops, and stands of Mixed Conifer Forest occur on more mesic sites such as north aspects and drainages. In areas where the topography is less diverse, this transition is gradual. The clarity of the mosaic and position of the transition changed as white fir, which is shade-tolerant and fire-intolerant, increased with fire exclusion during the twentieth century. This essentially converted stands into Mixed Conifer Forest (Dahms and Geils 1997; Swetnam et al. 2001; Mast and Wolf 2004; Evans et al. 2011).

At its lower elevational limit, Ponderosa Pine Forest intergrades primarily with Pinyon-Juniper vegetation but also with Gambel Oak Shrubland and Interior Chaparral Shrubland. These transitions also can be gradual or as a mosaic, but the mosaic is not always clearly related to topography. Fire appears to be the dominant factor determining the mosaic, with patterns both related and unrelated to topography.

4.2.2 *Climate*

Climate influences southwestern Ponderosa Pine Forest directly and indirectly. Direct effects include influences of moisture availability on tree recruitment. Indirect effects occur through influences on disturbances such as fire and insects. The climate is characterized by cool temperatures and relatively long growing seasons of 120–180 days (Schubert 1974; Moir 1993; Pase and Brown 1994), during which moisture likely limits plant growth (Moir 1993). The climate can be considered borderline for forest, because it is among the driest of any forest area in North America (Moir 1993; Pase and Brown 1994). Annual precipitation averages 400 to >760 mm (16–30 in.) and is divided between winter and the summer monsoon season separated by dry springs and moderately dry falls (Pase and Brown 1994; Chambers and Holthausen 2000). Summer precipitation is possibly necessary for the existence of this forest in areas that average <640 mm (25 in.) annual precipitation (Pase and Brown 1994). Winters are relatively mild compared to upper-elevation forest types. Winter precipitation falls mostly as snow, which can prevent deep soil freezing and saturates soils when melting (Schubert 1974). Low-elevation stands have an average annual temperature of 6 °C (43 °F), while mid-elevation stands average 5 °C (41 °F; Moir 1993). Lightning is common; for example, a large area of primarily Ponderosa Pine Forest in the Jemez Mountains of north-central New Mexico received an average of 2.1 cloud-to-ground lightning strikes/ha (0.8 strikes/acre) annually from 1985 to 1994 (Allen 2002). More details on climate are available in Beschta (n.d.).

Modern climate change is described in Sect. 1.6.3 and is treated as an anthropogenic driver of Ponderosa Pine Forest in Sect. 4.2.6.3.

4.2.3 Soil

Ponderosa Pine Forest occurs on a wide range of soils formed from a variety of parent materials (Pase and Brown 1994), but soils are generally deep, permeable, and capable of storing snowmelt. Soils in the southern Rocky Mountains of Colorado and New Mexico range from fine to moderately coarse-textured sandy loams with gravel (Paulsen 1975). Litter and duff depth is lower than in conifer-dominated stands of Mixed Conifer Forest and Spruce-Fir Forest (cf. Cocke et al. 2005, for the San Francisco Peaks of north-central Arizona). The soil moisture regime is ustic, and the soil temperature regime is frigid (Klemmedson and Smith 1979; Moir 1993). Ponderosa Pine Forest soils in Arizona are Alfisols (Typic Eutroboralfs, Typic Glossoboralfs, and Mollic Eutroboralfs) and Entisols (Typic Cryorthents and Typic Ustorthents), with most being well-drained, shallow to deep, and fine- to moderately coarse-textured (Hendricks 1985). Soils in New Mexico are mostly Alfisols with some Mollisols (Maker and Saugherty 1986). Soil characteristics that affect moisture availability are critical to Ponderosa Pine Forest, with porous soils being most productive (Clary 1975; Paulsen 1975).

The soil system also includes root-decay fungi. Species important in vegetation dynamics are annosus root rot (*Heterobasidion annosum*) and Armillaria root rot (*Armillaria* spp.; Dahms and Geils 1997; Moir et al. 1997).

4.2.4 Animals

The animals most important to vegetation dynamics of southwestern Ponderosa Pine Forest are insects, especially bark beetles (*Dendroctonus* spp. and *Ips* spp.). Elk (*Cervus elaphus*) and deer (*Odocoileus* spp.) are the common large ungulates. The impacts of animals on vegetation dynamics are described in Sect. 4.2.5.3.

4.2.5 Natural Disturbance

The natural disturbance regime of southwestern Ponderosa Pine Forest is dominated by fire. Wind disturbance occurs but is less widespread than at higher elevations (e.g., Sect. 2.2.5.2). Insect outbreaks can have major impacts. Climate variations such as drought are critically important, primarily through interactions with fire and insect outbreaks.

4.2.5.1 Fire

Fire is likely to have been a key driver of Ponderosa Pine Forest in the American Southwest for as long as ponderosa pine has dominated landscapes (cf. Weng and Jackson 1999). Ponderosa pine is well-adapted to fire with deep roots, fire-resistant bark, self-pruned lower branches, branches and cones distant from the ground, open arrangement of branches and needles unfavorable to spread of fire, needles with high moisture content, thick bud scales, and longevity of seed production (Moir et al. 1997; Keeley and Zedler 1998; Covington 2003; Kaufmann et al. 2005). These enable trees to survive and regenerate on landscapes with frequent surface fires. Moreover, the open structure of historical stands resulted in a generally warm, dry microenvironment on the forest floor that kept fuel moisture very low, facilitating the ignition and spread of surface fires (Harrington 1982). Fire is more common than in higher-elevation forests. Nearly 80 % of all lightning-ignited fires in forests and woodlands of National Forests of Arizona and New Mexico occurred in Ponderosa Pine Forest (calculated from data in Barrows 1978, for 1960–1974).

Identification of the importance of fire in Ponderosa Pine Forest dates back to the mid nineteenth century, when American expeditions surveyed resources of the region. Economic interest in timber resources led to a negative view of fire, because fire killed trees, reduced lumber quality by scarring trees, and limited tree regeneration. In addition, cultural biases of the time led observers to believe that Native Americans were the primary source of ignitions and that fire was unnatural. This precluded early understanding of the essential role of fire in this forest type (Allen 2002; Swetnam and Baisan 2003). Scientific understanding of fire in southwestern Ponderosa Pine Forest dates to Weaver (1951a), who provided data documenting frequent surface fires and suggested that such fires had been critical to maintaining healthy, open forest.

Today, most authorities concur that the historical fire regime was characterized by frequent, low-severity surface fires (Fig. 4.8; e.g., Weaver 1951a; Swetnam and Baisan 1996) that occasionally crowned in relatively small areas via fuel ladders (Swetnam et al. 2001; Iniguez et al. 2009). Surface fires were carried by fine fuels on the forest floor, especially herbaceous plants but also abundant annual needle fall (3,900 kg/ha or 3,500 lb/acre; Bradley et al. 1992). Given the relatively dry climate, the limiting factor for surface fire was generally fuel, not moisture. Factors that control fires and fire intervals can change over time (e.g., Iniguez et al. 2009). Today, dense stands with thick litter are not fuel-limited; instead fire is limited by moisture, as in higher elevation forests (Holden et al. 2007a).

The importance of fine fuels links fire and weather, because fine fuels dry and pick up moisture quickly (Bradley et al. 1992). The primary fire season is from April or May through June, when there is little precipitation (e.g., Margolis and Balmat 2009), although the fire season can extend into summer (e.g., Fulé et al. 1997). July can be a key month, because fires are more prevalent if the monsoon season is delayed or initially has below-average rainfall (Grissino-Mayer et al. 2004). Years with abundant surface fire are correlated with drought, especially when preceded by 1–3 years of high precipitation during which herbaceous fine fuels increased (Swetnam and Baisan 1996; Touchan et al. 1996; Grissino-Mayer et al. 2004; Allen 2007; Allen et al. 2008; Brown et al. 2008; Margolis and Balmat 2009). During such years, fires are not always synchronous among sites at fine spatial scales within a landscape, where local conditions of site productivity and fuel continuity can be more important, but fire synchrony is apparent at broad spatial scales (Ireland et al. 2012). In contrast, years with little fire are correlated with high precipitation. Given the importance of weather, episodic climatic events such as the El Niño-Southern Oscillation (ENSO; Sect. 1.2.2) have large impacts on fire regimes (Swetnam and Betancourt 1990, 1998; Grissino-Mayer and Swetnam 2000).

A key parameter of fire regimes of southwestern Ponderosa Pine Forest is mean interval for surface fire. Historical intervals determined from fire scars for 31 sites ranged from 4 to 36 years (median of 13 years; ≥10 % scarred) from 1700 to 1900 (Swetnam and Baisan 1996). This wide range in intervals is partly accounted for by differences in area and intensity of sampling (Sect. 1.2.5.1). More recent studies in the Southwest have reported intervals that fall into or near this range (e.g., Fulé et al. 1997; Brown et al. 2001; Swetnam et al. 2001; Grissino-Mayer et al. 2004; Allen et al. 2008; Iniguez et al. 2009; Margolis and Balmat 2009). Frequent fires reflect the dryness of the American Southwest in that the annual inputs of organic matter (herbaceous shoots and needles) accumulate because of slow decomposition rates, and these fuels are often sufficiently dry to carry fire.

Mean intervals for historical fires varied temporally, depending on climate (e.g., Swetnam and Baisan 1996; Touchan et al. 1996; Fulé et al. 2000). Intervals also varied spatially, as longer intervals are associated with sites of (a) topographic isolation, which reduces the spread of fires, (b) low elevation, apparently because of lower production of fine fuels to carry fires, (c) high elevation, apparently because of higher moisture levels, and (d) more moisture (Allen et al. 1995; Swetnam and Baisan 1996). Mean fire intervals in Ponderosa Pine Forest of the San Juan

Mountains of southwestern Colorado decreased with elevation from 8–13 years at high elevation to 6–11 years at middle elevation to 6 years at low elevation (Grissino-Mayer et al. 2004). Intervals tend to be longer on less productive sites, because of reduced, less-continuous surface fuels (Rollins et al. 2002).

A second important parameter of fire regimes is the length of individual fire intervals, i.e., gaps in fire occurrence. These fire-free periods also differ temporally and spatially, including among microsites within stands, landscapes within regions, and regions within the American Southwest. For example, historical fire-free periods were up to 11 and 22 years in two areas of the North Rim of Grand Canyon National Park in north-central Arizona (Fulé et al. 2002; ≥10 % scarred). Again, such differences are likely partially accounted for by differences in sampling area and intensity. Spatial and temporal variability in fire-free periods is related to the factors that influence mean fire intervals, as described in the preceding paragraph, as well as the vagaries of weather. Another possible factor is differences in the flammability of leaf litter among tree species (e.g., see Abella and Fulé 2008b for Gambel oak having less flammable litter than conifers). The lengths of fire-free periods are important because they affect tree regeneration and persistence. Short periods favor species that develop fire-resistant bark at a relatively young age such as ponderosa pine and Douglas-fir, as well as sprouts of species such as Gambel oak. Longer fire-free periods are necessary for regeneration of species that require more time to develop fire-resistant bark, such as white fir, and for the development and persistence of larger individuals of Gambel oak (Abella and Fulé 2008b). Although the length of fire-free periods was critical to the survival of seedlings and small saplings lacking fire-resistant bark, entire stands did not need to be fire-free, only some sites within stands. This suggests that fire-free periods can be more important at the scale of single trees or clusters of trees than at the scale of stands. Nevertheless, such data are rarely reported in fire-history studies.

The role of crown fire (Fig. 4.9) in the historical fire regime has received much less study. Of course, surface fires resulted in the torching of single trees or small clusters of trees, but the question is whether crowning historically occurred at stand or landscape scales. The consensus has been that large crown fires were absent or rare both temporally and spatially (Woolsey 1911; Cooper 1960; Moir et al. 1997; Fulé et al. 2003); however, evidence of historical crown fire can be difficult to document in southwestern Ponderosa Pine Forest. In forests such as Mixed Conifer Forest, past crown fires are usually identified by post-fire cohorts of early-successional trees, e.g., even-aged stands of quaking aspen (Sect. 3.2.5.1). Even-aged cohorts can be difficult to identify in southwestern Ponderosa Pine Forest, where they have been thinned by frequent surface fires and intermixed with cohorts of regeneration related to climate (Baker 2006; Margolis and Balmat 2009).

Nevertheless, there is evidence of historical crown fire in southwestern Ponderosa Pine Forest. For example, a 60-ha (~150-acres) patch of crown fire was documented for a mixed-severity fire on Rincon Peak in southeastern Arizona in the mid twentieth century (Iniguez et al. 2009). Also, a mixed-severity fire in the late twentieth century that produced crown-fire patches of ~200–500 ha (~500–1,250 acres) was considered similar to fires that occurred before Euro-American settlement in the Animas Mountains

Fig. 4.8 Surface fire in mesic Ponderosa Pine Forest in Grand Canyon National Park, north-central Arizona (Photograph by Fire and Aviation Management, Grand Canyon National Park)

Fig. 4.9 Crown fire in mesic Ponderosa Pine Forest in Gila National Forest, west-central New Mexico (Photograph by Steven Meister and U.S. Forest Service, Gila National Forest)

of southwestern New Mexico (Swetnam et al. 2001). In addition, crown fires have been documented from Holocene charcoal deposits in alluvial sediments at Kendrick Mountain in north-central Arizona (Jenkins et al. 2011). Crown-fire intervals at this site averaged 200–400 years during the last 2,000 years, but this likely underestimates frequency because alluvial sediments do not record all crown fires (nor do they document fire size).

The above evidence suggests crown fire was spatially and temporally limited in southwestern Ponderosa Pine Forest. It also indicates that historical crown fire occurred primarily on steep slopes in areas of heterogeneous topography. Steep slopes have vertically stacked tree crowns that facilitate upslope burning of crown fire (Jenkins et al. 2011) and have been characterized as "breeders of very large fires" (Barrows 1978). Heterogeneous topography results in fuel breaks that limit the spread of surface fires, resulting in isolated stands with infrequent fires and heavy fuels (Swetnam et al. 2001; Iniguez et al. 2009; Jenkins et al. 2011). Crown-fire intervals varied depending on long-term patterns in climate (Jenkins et al. 2011). Crown fires likely occurred after wet periods during which surface fires were reduced, fuel accumulated, and tree densities increased (Roos and Swetnam 2012). Historical crown fires at Kendrick Mountain are thought to have been both fuel- and moisture-limited (Jenkins et al. 2011). Additional research could reveal other sites of historical crown fires, such as drainages where moisture can reduce the frequency of surface fires, resulting in dense stands, high fuel loadings, and vertical fuel continuity.

Determining the prevalence of crown fire (including the crown-fire portion of mixed-severity fire) is challenging for large areas. Evidence of past fire regimes has been extrapolated from vegetation data recorded in historical land surveys dating to circa 1880–1904 in three large areas of north-central and east-central Arizona (Williams and Baker 2011, 2012, 2013). The proportions of small and large trees recorded in the surveys were interpreted as indicating that mixed and high-severity fire structured about 38, 41, 88 % of the three areas and low-severity fire structured 62, 59, and 12 %. The differences among study areas reflect differences in vegetation. For example, Williams and Baker (2013) examined the Coconino Plateau of north-central Arizona, an area of Ponderosa Pine Forest intermixed with Pinyon-Juniper vegetation (only 34 % of the historical landscape was continuous Ponderosa Pine Forest). This intermixing affected the fire regime because differences between Ponderosa Pine Forest and Pinyon-Juniper vegetation limit movement of fire from one to the other (Huffman et al. 2008; Sect. 5.2.5.1).

The fire regime of Ponderosa Pine Forest on the portion of the Coconino Plateau included in the South Rim region of Grand Canyon National Park was examined in greater detail by Dugan (2012). The lack of logging in this area – in contrast to the mostly disturbed areas studied by Williams and Baker (2012, 2013) – enabled (a) censusing fire scars to date past fires and (b) utilizing modern age structures to reconstruct severity levels of past fires. Findings for pre-1880 fires (i.e., prior to fire exclusion) indicated that mixed-severity fire accounted for 23 % of the total area burned and low-severity fire for 77 %. Historical fire turnover times (fire rotations) were 24.9 years in the South Rim area dominated by Ponderosa Pine Forest,

50.4 years in the area where Ponderosa Pine Forest was more intermixed with Pinyon-Juniper vegetation, and 33.2 years overall.

Findings that the fire regimes of the areas studied by Williams and Baker (2012, 2013) and Dugan (2012) included mixed-severity and/or high-severity fires raise a critical question. Why have most other fire-history studies, including some done in portions of the same areas, reported only low-severity fires? One explanation is that other researchers (a) selected small study areas with open stands and numerous fire scars, i.e., areas likely to have been structured by surface fires, and (b) did not link fire-scar and age-structure data (Baker 2009; Dugan 2012).

With surface fires being the focus of most researchers, the effects of surface fires have been extensively researched. High-frequency, low-severity surface fires kept forest structure relatively open. Microsites for tree regeneration were produced as reductions in the herbaceous layer decreased competition for germinating seedlings (Sackett et al. 1996). Cohorts of seedlings and saplings were thinned, but the growth of survivors increased (Dahms and Geils 1997; see also Holden et al. 2007b). Fuel loads were kept relatively low and constant both spatially and temporally. In addition, studies of prescribed burns in modern forests indicate that soil properties were altered. The layer of duff on the soil surface was kept thin and patchy, enhancing moisture availability (Covington et al. 1997; Feeney et al. 1998) and exposing mineral soil. In addition, mineralization was increased (White 1986, 1996), which increased nutrient mobilization (Covington and Sackett 1984) and nutrient concentrations on the soil surface (Covington and Sackett 1990), including nitrogen (e.g., Harris and Covington 1983; Covington and Sackett 1990, 1992). Populations of arbuscular mycorrhizae in the soil possibly increased (Korb et al. 2003).

As with other forest types in the American Southwest, the portion of the historical fire regime characterized by frequent, low-severity surface fires changed late in the nineteenth century (Sects. 4.2.6.1, 4.2.6.2, and 4.4.1.2).

4.2.5.2 Wind

Windthrows of scattered trees occur (Pearson 1950; Kolb et al. 2001), but there is little in the scientific literature about stand-scale effects of wind on southwestern Ponderosa Pine Forest. Woolsey (1911) described "considerable windfall" when soils were saturated, but possibly was referring to areas that had been partially logged. There also is mention of a major blowdown in the Jemez Mountains (Allen 1989).

A well-documented example of wind damage occurred in north-central Arizona in 2010, when several tornados affected 2,375 ha (5,868 acres) of mostly Ponderosa Pine Forest (Fig. 4.10; U.S. Forest Service 2010). Damage to stands ranged from none to complete blowdown. Within the core area of damage, average stands decreased in density from 1,006 to 12 trees/ha (407 to 5 trees/acre), including 57 to 7 trees/ha (23 to 3 trees/acre) for trees ≥51 cm (16 in.) diameter at breast height (dbh), i.e., at 1.4 m (4.5 ft). Basal area decreased from 31 to 2 m^2/ha (137– 10 ft^2/acre), and canopy cover decreased from 65 to 10 % (U.S. Forest Service 2011a).

Impacts on the forest were such that the paths of individual tornados were evident from aircraft (see Fig. 4.7 in U.S. Forest Service 2010).

4.2.5.3 Biotic Agents

Insects, especially bark beetles, are important disturbance agents in southwestern Ponderosa Pine Forest (Fig. 4.11). Bark beetles affect stand structure and possibly were important historically in maintaining low tree densities, especially following surface fire (cf. Breece et al. 2008) and drought (Allen 1989; Negrón et al. 2009). Bark beetles also have affected vegetation distribution, as they caused mortality of ponderosa pine in the Jemez Mountains that moved the ecotone between Ponderosa Pine Forest and Pinyon-Juniper vegetation upslope (Allen and Breshears 1998).

Stands of Ponderosa Pine Forest in northern Arizona maintain several species of bark beetles, possibly persisting in lightning-scarred trees (Sánchez-Martínez and Wagner 2002). Bark beetles typically attack scattered, small clusters of trees, but larger outbreaks also occur (cf. Douglas and Stevens 1979). Extensive outbreaks have been reported for the Kaibab Plateau of northern Arizona (Lang and Stewart 1910; Wilson and Tkacz 1996), Bandelier National Monument in north-central New Mexico in 1955–1958 (Allen 1989), and much of the Southwest in the first few years of the twenty-first century (U.S. Forest Service 2011b). Insect-caused tree mortality in the twenty-first century outbreak peaked in Ponderosa Pine Forest in 2003 with about 3,087 km^2 (1,192 $miles^2$) affected in Arizona and New Mexico (U.S. Forest Service 2011b). Ponderosa pine mortality was caused primarily by ips engraver beetles (*Ips* spp.) in Arizona and western pine beetle (*Dendroctonus brevicomis*) in New Mexico (Negrón et al. 2009; U.S. Forest Service 2011b). *Ips* species have greatest impacts on dense stands, at low elevations, and on trees with diameters of 10–35 cm (4–14 in.) dbh, i.e., trees that had established during the period of fire exclusion (Negrón et al. 2009). Western pine beetle appears to impact primarily larger, more mature trees (Miller and Keen 1960).

Inference from stands with different tree densities in north-central Arizona suggested that populations of *Dendroctonus* (but not *Ips*) increased with tree densities during fire exclusion (Zausen et al. 2005). However, the relationship between tree density and outbreaks is unclear. Battaglia and Shepperd (2007) suggested that increased tree density and reduced tree vigor made Ponderosa Pine Forest more susceptible to outbreaks in southern Utah, but Sánchez-Martínez and Wagner (2002) reported no effect of increased tree density on outbreaks in north-central Arizona.

Other important insects include pine sawflies (*Neodiprion* spp.; McMillin and Wagner 1993), but little information is available for the American Southwest. Pandora moth (*Coloradia pandora*) defoliates ponderosa pine at a landscape scale, but does not appear to significantly impact tree growth and vigor (Bennett et al. 1987; Schmid and Bennett 1988; Miller and Wagner 1989).

Another biotic disturbance agent is ponderosa pine dwarf mistletoe (*Arceuthobium vaginatum* spp. *cryptopodum*), a parasite plant that infects approximately one-third

Fig. 4.10 Wind damage caused by a tornado in mesic Ponderosa Pine Forest in Coconino National Forest, north-central Arizona (Photograph by U.S. Forest Service, Coconino National Forest)

Fig. 4.11 Mortality of ponderosa pine (note *reddish-brown color*) caused by engraver beetles in Prescott National Forest, central Arizona (Photograph by Joel McMillin, Forest Health, Arizona Zone, U.S. Forest Service)

Fig. 4.12 Mistletoe-caused growth of a "witches broom" on a ponderosa pine in Coconino National Forest, north-central Arizona (Photograph by Betty J. Huffman)

of the area of Ponderosa Pine Forest in Arizona and New Mexico (Andrews and Daniels 1960). This species causes host trees to accumulate resins and form dense clusters of branches ("witches brooms"; Fig. 4.12). When burned, these increase crown scorching, which is a primary factor in tree mortality following surface fire (Harrington and Hawksworth 1990; Breece et al. 2008). The resins and witches brooms also facilitate the laddering of fire into forest canopies (Alexander and Hawksworth 1975). Research in north-central Arizona determined that severely infected stands had higher total fuel loadings but no differences in canopy fuels (Hoffman et al. 2007). These stands also required lower wind speeds for surface fires to crown than did uninfected stands.

Other biotic disturbance agents are root-decay fungi and large ungulates. Root decay fungi such as annosus root rot and Armillaria root rot reduce growth and weaken trees, which increases the likelihood of bark beetle infestation and windthrow. Interactions of these fungi with insects and pathogens generally accounted for about one-third of tree mortality in several National Forests of Arizona and New Mexico (Wood 1983, in Moir et al. 1997).

Elk and deer, the common large ungulates, likely have affected tree regeneration patterns, especially of quaking aspen, a heavily browsed species (see Sects. 3.2.5 and 3.3.2).

4.2.5.4 Climate Variation

Periodic drought is a critically important disturbance factor affecting Ponderosa Pine Forest. It acts primarily through interactions with other disturbance agents such as fire and insects. The importance of drought in the American Southwest was recognized early in the twentieth century. Woolsey (1911) reported 10 % of the standing trees (all or mostly ponderosa pine) in a large timber sale in Coconino National Forest in north-central Arizona were dead, as "the result largely of unfavorable moisture conditions." He also noted that drought interacted with insects and pathogens.

Drought in the early twenty-first century in north-central Arizona was associated with a 74 % increase in tree mortality in 2002–2007 (Ganey and Vojta 2011). Mortality was disproportionately greater for Gambel oak and quaking aspen and lower for ponderosa pine. Mortality occurred in 98 % of study plots but the degree of mortality was spatially variable. Mortality was uncorrelated with stand density and elevation (although negative impacts of drought on tree growth are greater in low-elevation forests; Adams and Kolb 2005). Absolute mortality was greater in smaller diameter classes, but relative mortality (i.e., mortality relative to species abundance) was generally greater in larger diameter classes. Relative mortality was lower than in Mixed Conifer Forest (Sect. 3.2.5.4). Drought also increased the elevation of the ecotone between Ponderosa Pine Forest and Pinyon-Juniper vegetation by causing mortality of ponderosa pine by bark beetles (Allen and Breshears 1998; Koepke et al. 2010), as mentioned in the previous section. See Sect. 3.2.5.3 for the role of drought (and other factors) in aspen decline.

4.2.6 Anthropogenic Disturbance

Land use by Native Americans in the American Southwest was concentrated in Pinyon-Juniper vegetation (Sects. 1.5.1.1 and 5.2.6), but Ponderosa Pine Forest was commonly used for hunting and sometimes for settlements (Roos et al. 2010). Major anthropogenic disturbances related primarily to Euro-American land use and affecting relatively undisturbed stands of southwestern Ponderosa Pine Forest are livestock grazing, fire management, modern climate change, invasive species, recreation, and nearby land use. Each of these can be important in vegetation dynamics,

Fig. 4.13 Domestic sheep (*Ovis aries*) grazing in 1925 in mesic Ponderosa Pine Forest, north-central Arizona (Photograph by E.W. Loveridge, U.S. Forest Service, Coconino National Forest)

and therefore all are incorporated in the conceptual models (Sect. 4.5). For a review of several contemporary human uses, see Raish et al. (1997). Logging has been widespread, but is outside the scope of this book.

4.2.6.1 Livestock Grazing

Livestock grazing became widespread in the late nineteenth century and had both direct and indirect effects on southwestern Ponderosa Pine Forest (Fig. 4.13). Direct effects included consumption of palatable plants, which reduced understory cover and altered the competitive balance among plant species, thereby affecting the composition of the forest understory. However, documentation of such effects can be challenging, because changes in forest canopy cover can dominate understory changes (cf. Bakker and Moore 2007). The major indirect effect of livestock grazing was that it reduced the biomass and continuity of the herbaceous layer, which formerly carried surface fires, and thereby caused an abrupt decrease in the frequency of surface fires in the late nineteenth century.

Livestock grazing continues today in many areas, but is generally less intensive (Battaglia and Shepperd 2007) and is not allowed in most protected areas. Long-term light to moderate livestock grazing appears to have little impact (Milchunas

2006). Grazing is most common in stands with quaking aspen (Reynolds 1969; Milchunas 2006), where it reduces aspen regeneration and understory cover, affecting tree-age distributions and understory composition (DeByle 1985; Mueggler 1985; Bartos 2007). Effects of heavy grazing can persist for more than 40 years, even after the withdrawal of livestock (Hanks et al. 1983). Overgrazing also can result in increases in invasive species (e.g., Korb et al. 2005; Battaglia and Shepperd 2007). Study of grazing exclosures over a 25-year period documented that removal of livestock grazing from Ponderosa Pine Forest greatly increased grass cover and regeneration of ponderosa pine (Potter and Krenetsky 1967).

4.2.6.2 Fire Management

Fire management throughout most of the twentieth century focused on preventing and suppressing fires. This continued the exclusion of surface fires that was initiated by livestock grazing (previous section). Just as southwestern Ponderosa Pine Forest was "profoundly shaped by fire" (Romme et al. 2009), it was also profoundly altered by fire exclusion (Fig. 4.14a–c; Sect. 4.4.1.2). Without frequent fires, southwestern Ponderosa Pine Forest increased in tree density, fuel loadings, and horizontal and vertical fuel continuities across landscapes, which led to increased frequency and size of crown fires (Fig. 4.15, e.g., Fulé et al. 2004; Romme et al. 2009). Examples

Fig. 4.14 (**a–c**) Repeat photography (1909, 1949, and 2012) in old-growth Ponderosa Pine Forest of the Fort Valley Experimental Forest, Coconino National Forest, north-central Arizona. (**a**) In the 1909 photograph, note absence of seedlings and saplings of ponderosa pine.

Fig. 4.14 (continued) (**b**) In the 1949 photograph, note ponderosa pine regeneration had occurred, probably in 1919. (**c**) In the 2012 photograph, note growth but little thinning of ponderosa pine regeneration during the period of fire exclusion (Photographs by W.R. Matton (**a**) and F.R. Herman (**b**) courtesy of Fort Valley Experimental Forest Historic Images, U.S. Forest Service, and by author (**c**))

of large crown fires affecting southwestern Ponderosa Pine Forest in the twenty-first century include the Rodeo-Chediski Fire in east-central Arizona (1,897 km^2 or 732 miles^2), the Whitewater Baldy Fire Complex in the Gila Mountains of west-central New Mexico (1,205 km^2 or 465 miles^2), and the Missionary Ridge Fire in the San Juan Mountains (286 km^2 or 110 miles^2). Shifts in climate could be related to this change in fire behavior, but increased fuel is the principal cause, based on observations of lower fire severity in sites less changed by fire exclusion (Stephens and Fulé 2005) and modeling of fire behavior (Fulé et al. 2004). Fire management practices began to shift focus in the late twentieth century to include management fires (both prescribed and lightning-ignited). This has successfully changed fire behavior in some areas (Fig. 4.16). However, the area affected by management fires has been small, and tree densities, fuel loadings, and fuel continuity continue to result in landscape-scale crown fires in many areas. Fire management during 1979–2011 in the South Rim region of Grand Canyon National Park has been so active that the modern fire rotation is 17.4 years vs. the historical rotation of 33.2 years (moreover, the rotation period was only 10.3 years in 1998–2011; Dugan 2012; see Sect. 4.2.5.1).

4.2.6.3 Modern Climate Change

Modern climate change (Sect. 1.6.3) is implicated in fires, insect outbreaks, and drought. Therefore, it has affected the structure, composition, and dynamics of southwestern Ponderosa Pine Forest. Air pollution is a driver of modern climate change, but pollution sources are more global than regional or local, and direct effects of air pollution on southwestern Ponderosa Pine Forest are understudied. However, there is circumstantial evidence of likely impacts. For example, ozone levels in Grand Canyon National Park (Bowman 2003) have exceeded thresholds for foliar injury on ponderosa pines observed in California (National Park Service 2002). Ozone also can affect quaking aspen (e.g., Karnosky et al. 1999). Nevertheless, with the lack of documentation of direct effects, air pollution is not included in this chapter as a separate anthropogenic driver of southwestern Ponderosa Pine Forest.

4.2.6.4 Invasive Species

Invasive plant species are more abundant and much more thoroughly studied in Ponderosa Pine Forest than in higher-elevation forests. A review of exotic species in Ponderosa Pine Forest in Arizona reported that cover values are typically <10 % (McGlone and Egan 2009). However, the abundance of invasives is highly variable. For example, mean cover of exotic plants in a regional study of ten Ponderosa Pine Forest ecosystem types in central and north-central Arizona ranged from <0.1 to 7 %, and their relative cover was <1 to 16 % (Abella et al. 2012). Mean exotic species richness ranged from ~0 to 1.7 species/m^2 (species/10.8 ft^2). Multiple regression

Fig. 4.15 A century or more of fire exclusion has led to landscape-scale crown fires. Note evidence of past fire in the foreground and a current fire in the background in Gila National Forest, west-central New Mexico (Photograph by Mormon Lake Interagency Hotshot Crew and U.S. Forest Service, Gila National Forest)

Fig. 4.16 Smoke from surface fire in Gila National Forest, west-central New Mexico (Photograph courtesy of U.S. Forest Service, Gila National Forest)

determined that about half of the variation in exotic species cover and richness was related to native species and environmental variables. Moist ecosystem types were most heavily invaded. Disturbance, which usually leads to increases of invasives, only entered the regional-scale regression models when native species were dropped from the analysis. This surprising result is likely accounted for by widespread anthropogenic disturbance, including livestock grazing.

At the stand scale, invasives generally increase following disturbances such as fire, tree thinning for fire management, and livestock grazing (e.g., Sackett et al. 1996; Korb et al. 2005; Battaglia and Shepperd 2007; McGlone and Egan 2009; McGlone et al. 2009a, 2012b; Stoddard et al. 2011). Invasives are generally uncommon in areas of little disturbance (Laughlin et al. 2004; Korb et al. 2005) and abundant where disturbance is severe (Sackett et al. 1996; Crawford et al. 2001; Griffis et al. 2001; Stoddard and McGlone 2008), but some studies have shown no correlation with severity (e.g., Kuenzi et al. 2008). Also, invasives generally decrease with time since disturbance (Stoddard and McGlone 2008). For example, relative cover of exotics dropped from 6 % immediately after disturbance (i.e., understory thinning and prescribed burning) to ~3 % after 6 years (Stoddard et al. 2011), but this can be negated by the introduction of other invasive species (cf. McGlone et al. 2009b).

The most widespread species of invasives differ among studies (cf. Sackett et al. 1996; Scurlock and Finch 1997; Sieg et al. 2003; McGlone and Egan 2009; McGlone et al. 2009b). The regional study of Ponderosa Pine Forest ecosystem types by Abella et al. (2012) mentioned above listed the following invasives as having ≥20 % frequency values averaged across the types: common dandelion (*Taraxacum officinale*; 42 %), common salsify (*Tragopogon dubius*; 42 %), cheatgrass (*Bromus tectorum*; 36 %), common mullein (*Verbascum thapsus*; 35 %), Kentucky bluegrass (*Poa pratensis*; 33 %), wild lettuce (*Lactuca serriola*; 27 %), and Dalmatian toadflax (*Linaria dalmatica*; 20 %).

The invasive of perhaps greatest concern is cheatgrass (cf. Sieg et al. 2003), an annual with no native ecological equivalent. Its broad ecological amplitude allows it to grow and reproduce in a variety of environments. This and dispersal attached to animals, including humans, give it a wide geographic distribution and the ability to expand it. Cheatgrass is a cool-season species that can capture resources before most native species (but see McGlone et al. 2011, 2012a). Its growth can produce extensive cover of fine fuels every year, giving it the potential to shorten fire intervals. Cheatgrass has had a larger, more-widespread impact on lower-elevation vegetation, including Pinyon-Juniper vegetation (Sect. 5.2.6.4). Nevertheless, it can dominate successional patterns in Ponderosa Pine Forest following restoration treatments (McGlone et al. 2012b). Moreover, its role in promoting fire in Ponderosa Pine Forest in the western United States goes back at least to 1938:

> A lush, early spring growth of cheatgrass (*Bromus tectorum*), followed by dry, hot, late spring and early summer weather set the stage for this fire. (Weaver 1959, for north-central Oregon)

Dry, hot weather in late spring and early summer is atypical for north-central Oregon, but is characteristic of the American Southwest (Sects. 1.2.2 and 4.2.2).

Fig. 4.17 Campground illustrating recreation in 1928 near Ruidoso in south-central New Mexico (Photograph by E. S. Shipp/U.S. Forest Service)

4.2.6.5 Recreation

Impacts of recreation are also greater in Ponderosa Pine Forest than higher-elevation forests in the American Southwest. This is because of greater accessibility and use by recreationists (Fig. 4.17). Some impacts are spatially limited (e.g., construction and use of roads, trails, and campsites, but others are more extensive). Fires ignited by recreationists in Ponderosa Pine Forest have included the Chediski portion of the Rodeo-Chediski Fire (Sect. 4.2.6.2) and the Schultz Fire, which burned 61 km^2 (24 $miles^2$) in the area of the San Francisco Peaks. Recreationists also appear to have introduced and spread invasive plants (invasives are often more common near roads and campsites). In addition, the use of off-highway vehicles by recreationists negatively impacts forest understories and soils.

4.2.6.6 Nearby Land Use

Nearby land use is an important anthropogenic driver. It too has greater effects on Ponderosa Pine Forest than higher-elevation forests. Ponderosa Pine Forest borders more adjacent lands, especially disturbed lands, and has more cities, towns, and housing developments embedded within it. Such proximity facilitates the spread of fires and invasives species into southwestern Ponderosa Pine Forest.

4.3 Processes

Ponderosa Pine Forest of the American Southwest has two major types of vegetation dynamics. One involves the paired processes of tree regeneration and thinning, and the other is succession. Research on both has concentrated on woody plants. These processes play key roles in the conceptual models (Sect. 4.5).

4.3.1 Regeneration and Thinning

The process of tree regeneration and thinning in southwestern Ponderosa Pine Forest has been referred to as gap dynamics. This can be misleading, at least for old-growth stands with open canopies, because the process is dissimilar to the gap dynamics of forests such as southwestern Spruce-Fir Forest (Sect. 2.3.2) and broad-leaved deciduous forests of the eastern United States. In such closed-canopied forests, gap dynamics begin with the death of one or a small cluster of canopy trees. The resultant opening in the tree canopy alters the microenvironment below it by increasing light. Smaller trees that were present before or established after gap formation respond with accelerated growth. Within a few decades, the canopy gap is filled by one or more of these trees or by ingrowth of crowns of adjacent canopy trees.

Open-canopied stands of southwestern Ponderosa Pine Forest do not have such gap dynamics. Instead, their woodland-like structure is characterized by persistent spaces between tree crowns. Long-term stability of both open and tree patches is evidenced by the finding that the soils of about 70 % of tree patches within stands in north-central Arizona are Alfisols, and soils in about 70 % of open patches just 5 m (16 ft) away are Mollisols (Scott R. Abella, personal communication). Persistence of open patches contrasts with canopy gap dynamics in which gaps continuously form and close in different locations over decadal time spans. Therefore, the old-growth structure of stands of southwestern Ponderosa Pine Forest depends not on gap dynamics but instead on the on-going processes of tree regeneration and thinning, as described below. Closed-canopied stands can have gap dynamics (Moir et al. 1997), but such stands also depend on regeneration-thinning dynamics.

Discussion of regeneration of ponderosa pine began with forest surveyors and lumbermen in the early twentieth century, who were concerned about the presence of replacement trees after logging:

> …mature trees occur among good patches of saplings and poles…isolated or in groups… Underneath the old trees and in open areas between the groups are often numerous small seedlings… (Lang and Stewart 1910, for the Kaibab Plateau)

See also Woolsey (1911) for Arizona and New Mexico.

Historical patterns of ponderosa-pine regeneration have been studied by examining dates of recruitment of current trees (Cooper 1960; White 1985; Savage et al. 1996; Mast et al. 1999; Brown and Wu 2005; Sánchez Meador et al. 2009; Dugan 2012).

Findings show broad pulses of recruitment separated by periods of less regeneration (Mast et al. 1999; Brown and Wu 2005). The broad pulses occurred during multi-decadal periods when surface-fire frequency was reduced by wet conditions or by lack of wet/dry oscillations important for the buildup and drying of fuels associated with surface fires (Brown and Wu 2005; Sect. 4.2.5.1). However, regeneration pulses also can be associated with fire and drought (Dugan 2012), which can be associated with overstory mortality and release of resources. The maximum period without regeneration in a stand in north-central Arizona was two decades (Mast et al. 1999; see also Villanueva-Díaz and McPherson 1995).

The last two regional regeneration cohorts (1910s–1930 and mid 1970s–1980s) occurred with wetter conditions and also with fire exclusion and reduced livestock grazing (Fig. 4.18; Kaufmann et al. 2007; see also Dugan 2012). The lack of surface fires increased ponderosa pine establishment by orders of magnitude in a north-central Arizona forest in 1919 (Mast et al. 1999). More-recent seedlings in the same area have died under a now-dense overstory (P.Z. Fulé, personal observation, cited in Mast et al. 1999). Years of abundant regeneration occur with heavy seed crops, moist spring and summer weather, and absence of fire (Pearson 1950; Bradley et al. 1992; Savage et al. 1996). Such years can be infrequent outside of the multi-decadal periods mentioned above, and can be site specific (Cooper 1960). A stand in north-central Arizona went 73 years (1919–1992) between confluences of conditions necessary for abundant regeneration (Savage et al. 1996).

Study of regeneration in a north-central Arizona forest that had been partially logged in the late nineteenth century indicated establishment was highest in canopy gaps and other forest interspaces (Sánchez Meador et al. 2009). Initial recruitment occurred near the center of natural grass openings (livestock grazing had eliminated competition from herbs, according to Pearson 1942). Regeneration later filled in other spaces. Ponderosa pine regeneration in Utah occurs as scattered individuals, scattered clumps, or dense stands (Bradley et al. 1992). Clumps can vary greatly in size, but in the Southwest are generally 0.02–0.3 ha (0.05–0.7 acres), with some as large as 0.8 ha (2 acres; White 1985; Kaufmann et al. 2007).

The thinning portion of the regeneration-thinning vegetation dynamic of southwestern Ponderosa Pine Forest historically occurred as surface fire reduced cohorts of seedling and sapling ponderosa pines (Fig. 4.19). Substantial self-thinning from intra-specific competition is not common today, at least within unburned, twentieth century cohorts (personal observation). Surface fires were a density-independent driver of ponderosa pine density, because they occurred largely independently of the density of overstory and understory trees (Brown and Wu 2005). Prescribed fire in east-central Arizona that resulted in the loss of only 0–5 % of trees with a dbh of ≥30 cm (12 in.) caused much greater mortality of smaller trees: 98–99 % of seedlings <30 cm (12 in.) height and 63–76 % of saplings from >30 cm (12 in.) height up to 8 cm (3 in.) dbh (Gaines et al. 1958).

With the presence of different cohorts, stands were uneven-aged. Regeneration between major pulses of regeneration also contributes to the uneven-age distribution of stands. The age-distribution of individual patches has been reported both as even (Cooper 1960, 1961) and uneven (e.g., White 1985).

Fig. 4.18 Ponderosa pine seedlings and saplings that likely date to a regeneration pulse in 1919. Trees have not been thinned by fire because of fire exclusion. In Coconino National Forest, north-central Arizona (Photograph by Betty J. Huffman)

Fig. 4.19 Fire thinning ponderosa pine regeneration in Grand Canyon National Park, north-central Arizona (Photograph by Fire and Aviation Management, Grand Canyon National Park)

4.3.2 Succession

Succession in Ponderosa Pine Forest of the American Southwest occurs after high-severity, stand-scale disturbance. Despite increases in crown fires following decades of fire exclusion, there is little research on post-fire succession in southwestern Ponderosa Pine Forest. Successional pathways that have been described indicate succession is greatly affected by pre-fire stand composition (e.g., Savage and Mast 2005), especially where sprouting species were present (e.g., Haire and McGarigal 2008). Successional patterns are also influenced by soil erosion, site moisture conditions, elevation, proximity to seed sources, and disturbance during succession.

Succession after crown fire in moist Ponderosa Pine Forest, i.e., at higher elevations, often involves quaking aspen (Bradley et al. 1992; Savage and Mast 2005), a species that root sprouts following fire (Sect. 3.3.2). Herbs and aspen sprouts dominate early in succession (Fig. 4.20), but ponderosa pine seedlings soon appear if seed sources are nearby. Subsequent surface fire favors aspen, until ponderosa pine saplings develop fire-resistant bark. If aspen and ponderosa pine continue to codominate, ponderosa pine eventually overtops aspen and aspen either senesces or remains as a subcanopy species in ponderosa pine-dominated stands. High mortality of aspen in moist Ponderosa Pine Forest, as reported for north-central Arizona (cf. Vankat 2011; Zegler et al. 2012), likely enhances the rate of succession to ponderosa pine (see Sect. 3.2.5.4).

Crown fire in moist and mesic stands of Ponderosa Pine Forest in which Gambel oak is abundant can have successional stands dominated by that sprouting species (Fig. 4.21; Harper et al. 1985; Romme et al. 1992; Bradley et al. 1992; Moir 1993; Savage and Mast 2005; Strom and Fulé 2007; Haire and McGarigal 2008). Gambel oak, sometimes with other shrubs such as New Mexico locust (*Robinia neomexicana*) and manzanita (*Arctostaphylos* spp.), can persist for long periods. Where nearby seeds sources are present, ponderosa pine invades and establishes. Before its saplings develop fire-resistant bark, subsequent fire favors shrubs. After ponderosa pine saplings develop fire-resistant bark, light fires favor it and kill back stems of the shrubs. Succession continues to forest, and the density of trees in late-successional stands negatively affects the density of the understory shrubs that persist (see Sect. 7.3.2).

Stands of dry Ponderosa Pine Forest and related forests with Apache, Arizona, and Chihuahuan pines often have pinyons, junipers, and evergreen oaks such as wavyleaf oak in the understory. After crown fire, the understory species dominate from early- through mid-succession (Bradley et al. 1992; Moir 1993; Barton 2002). Mid-successional stands can persist for long periods, but where seed sources available, pines can invade, establish, and eventually overtop the mid-succession dominants. (see Sect. 8.3.2).

Alternatively, succession in moist, mesic, and dry Ponderosa Pine Forest can be dominated by ponderosa pine, beginning early in succession. This is likely to occur when sprouting trees and shrubs are absent or uncommon before disturbance and when seed sources of ponderosa pine are nearby. Long-distance seed dispersal supplements reseeding of ponderosa pine from edges of burned patches (Haire and McGarigal 2010).

Succession following crown fire in stands in Arizona and New Mexico can lead to a variety of possible outcomes: forests, shrublands, and grasslands (Fig. 4.22a, b; Savage and Mast 2005; Strom and Fulé 2007; Iniguez et al. 2009; Roccaforte

Fig. 4.20 Succession in moist Ponderosa Pine Forest often begins with rapid sprouting of quaking aspen. This six-year-old patch of quaking aspen originated after crown fire in the North Rim region of Grand Canyon National Park, north-central Arizona (Photograph by Betty J. Huffman)

Fig. 4.21 Succession in mesic and dry Ponderosa Pine Forest can involve rapid sprouting of Gambel oak. These approximately two-year-old sprouts originated after forest fire in the North Rim region of Grand Canyon National Park, north-central Arizona (Photograph courtesy of Marc E. Gottlieb)

Fig. 4.22 (**a**, **b**) Modern landscape-scale crown fires can cause conversion of southwestern Ponderosa Pine Forest into shrublands and grasslands. (**a**) The photograph shows shrubland in the foreground and a patch of unburned Ponderosa Pine Forest in the mid-ground 19 years after fire in Tonto National Forest, central Arizona. (**b**) The photograph shows grassland 15 years after fire in Coconino National Forest in north-central Arizona. There is little or no evidence of succession to forest in either photograph (Photographs by Jackson Leonard/U.S. Forest Service (**a**) and Betty J. Huffman (**b**))

et al. 2012). Sites that have changed from forest to shrubland or grassland possibly will persist as non-forest if subsequent fire maintains them and/or seed sources of ponderosa pine are distant (Savage and Mast 2005). Also, stands that regrew as forest have little evidence of self-thinning decades later, making them susceptible to future crown fire and possible long-term conversion to grassland or shrubland. Impediments to recovery of sustainable Ponderosa Pine Forest include competition with sprouting species such as quaking aspen, oaks, New Mexico locust, and grasses, long distance to seed sources of ponderosa pine, and modern climate change toward higher temperatures and drier conditions (Roccaforte et al. 2012; see Sect. 1.6.3). Modeling has indicated that it will take centuries before Ponderosa Pine Forest recovers across a landscape burned by crown fire in east-central Arizona (Strom and Fulé 2007).

4.4 Historical Changes

4.4.1 Overstory

4.4.1.1 Pre-Euro-American Settlement

Most descriptions of nineteenth century southwestern Ponderosa Pine Forest characterize it as open and park-like, with large, widely spaced trees and a dense, grass-dominated herbaceous understory. Such descriptions are typically justified by reference to historical accounts such as:

> We came to a glorious forest of lofty pines…every foot being covered with the finest grass, and beautiful broad grassy vales extended in every direction. The forest was perfectly open… (Beale 1858, for east of the San Francisco Peaks)

> All the morning passing through a fine open forest of tall pines, with extensive open glades and meadows at short distances. (Beale 1858, for the Zuni Mountains of northwestern/west-central New Mexico)

> The trees are large and noble in aspect and stand widely apart… Instead of dense thickets where we are shut in by impenetrable foliage, we can look far beyond and see the tree trunks vanishing away like an infinite colonnade. (Dutton 1882, for the Kaibab Plateau)

> The lofty pine forest…is a noteworthy forest, not alone on account of the size and beauty of the single species of tree of which it is composed (*Pinus ponderosa*), but also because of its openness, freedom from undergrowth, and its grassy carpet… (Merriam 1890a, for the region of the San Francisco Peaks)

> …the history of the forests of Arizona…gather[ed] from many of the oldest reliable pioneers, is that when first invaded by the white man the forests were open… (Holsinger 1902)

See similar comments by Ives (1861) for near Bill Williams Mountain in north-central Arizona, Rothrock (1875) for the Zuni Mountains, Powell (1890) for the San Francisco Peaks and near the South Rim of the Grand Canyon in north-central Arizona, Bailey (1904; in Allen 2002) for the Jemez Mountains, and Pinchot (1947) for north-central Arizona in 1900.

It has been argued that ecologists and land managers have overemphasized historical descriptions of open forests and underemphasized descriptions of dense forests (Pollock and Suckling 1997). While this is possibly true, Beale's (1858) comments such as "a heavy forest of pine" can be misinterpreted, because "heavy" can refer to an abundance of large trees, not to overall stand density (cf. Woolsey 1911). In addition, failure to differentiate between Beale's (1858) observations of Ponderosa Pine Forest and Pinyon-Juniper vegetation can lead to confusion. Perhaps it is more likely that early observers, many of whom were more familiar with dense forests of the eastern United States, focused on what was unfamiliar to them, i.e., open stands.

Regardless of that debate, some historical accounts indicated variation in density:

> ...thick woods, which for the sake of our wagons we were obliged to avoid as much as possible... (Möllhausen 1858, for the area of the San Francisco Peaks)

> On small areas in various places in these mountains the stand is much more dense [sic]. (Kellogg 1902, for the Chiricahua Mountains of southeastern Arizona)

although open stands were indicated to be predominant in some areas:

> The pine forest is...rarely crowded. (Merriam 1890a, for the area of the San Francisco Peaks)

> The pine occurs mostly in open stand[s] park-like or even isolated in character. Practically all of the [Ponderosa Pine Forest] is open... (Lang and Stewart 1910, for the Kaibab Plateau)

Historical descriptions also included an overview of stand structure:

> The stand embodies all age classes in varying density and proportions [with] the old trees fairly uniformly distributed among the young growth over which they tower with large spreading crowns... (Lang and Stewart 1910, for the Kaibab Plateau)

as well as a quantitative description of stand structure:

> ...pines standing at intervals varying from 50 to 100 feet [15 to 30 m]... (Dutton 1882, for the Kaibab Plateau)

Whether this estimate was based on measurement or visually estimated, there is reason to accept it as the first quantitative description of southwestern Ponderosa Pine Forest structure and the only quantitative description dating to near the beginning of Euro-American influence. The source, Clarence Edward Dutton, was experienced with observation and measurement. He previously had served as an army ordnance officer, published multiple scientific papers, and participated in lengthy field expeditions to describe and map geological features in the western United States (Longwell 1958). Using Dutton's (1882) estimate of the mean distance between trees, density of canopy pines in Ponderosa Pine Forest on the Kaibab Plateau in the late nineteenth century was 11–43 trees/ha (4–17 trees/acre). Therefore, it too indicated variation in stand density.

There are few nineteenth century photographs of undisturbed southwestern Ponderosa Pine Forest (see Figs. 4.23a, b and 4.24). They and later photographs through at least the first decade of the twentieth century indicate open stand structure, but with variation among and within stands (Fig. 4.25). Of course, any extrapolation to Ponderosa Pine Forest throughout the American Southwest is risky, because

a

b

Fig. 4.23 (**a**) Photograph taken in 1873 showing southwestern Ponderosa Pine Forest on the plateau and slopes across the canyon. (**b**) The open structure of this forest is more apparent in the enlargement of a portion of the image. Location is given as North Fork Cañon, Sierra Blanca Creek, Arizona, but this place name is not currently used. Likely in east-central Arizona (Photograph by Timothy H. Sullivan; enlargement by author. Source: U.S. National Archives and Records Administration)

Fig. 4.24 Photograph taken in 1885 showing variation in density of southwestern Ponderosa Pine Forest near the San Francisco Peaks in north-central Arizona. Open areas are Subalpine-Montane Grassland (Photograph by J. K. Hillers. Source: U.S. Geological Survey)

Fig. 4.25 Stand of mesic Ponderosa Pine Forest in 1909 in the Fort Valley Experimental Forest, Coconino National Forest, north-central Arizona. Although surface fires ended with fire exclusion approximately 30 years earlier, the stand has remained open because of the lack of a regeneration pulse of ponderosa pine and possibly because of livestock grazing (Photograph by G. A. Pearson, courtesy of Fort Valley Experimental Forest Historic Images, U.S. Forest Service)

the sample is small and non-random (open stands likely were considered more photogenic than dense, closed stands).

Detailed quantitative descriptions are in forest surveys dating to the early twentieth century. Lang and Stewart (1910) sampled Ponderosa Pine Forest on the Kaibab Plateau in 1909 and reported the average density of trees ≥15 cm (6 in.) dbh was 111 trees/ha (45 trees/acre). However, Lang and Stewart (1910) underestimated density because they did not sample quaking aspen, pinyon, juniper, or Gambel oak, apparently because they lacked value for lumber. They also possibly included samples that extended into treeless meadows (Vankat 2010, 2011). In addition, their data do not reflect conditions prior to Euro-American influence, which began with livestock grazing and resultant fire exclusion about 30 years earlier (Vankat 2011). In fact, their data indicate ingrowth of fire-sensitive fir (probably white fir), which accounted for only 6 % of trees but 37 % of seedlings and saplings <15 cm (6 in.) dbh. Therefore, Lang and Stewart's (1910) data for larger trees better reflect conditions prior to Euro-American influence: the density of ponderosa pines >46 cm (18 in.) dbh was 30 trees/ha (12 trees/acre), which falls near the middle of the range extrapolated from Dutton (1882).

Another early report included densities of Ponderosa Pine Forest in Arizona and New Mexico. Woolsey (1911) described "average" stands in Tusayan (now part of Kaibab) and Coconino National Forests in north-central Arizona as having 27 and 40 pines/ha (11 and 16 pines/acre), respectively, for individuals ≥15 cm (6 in.) dbh. Variation in stand structure was indicated by his characterization of the 764-ha (1,888-acres) area averaged for Coconino National Forest as "frequently very dense". Considering pines >46 cm (18 in.) dbh, Tusayan and Coconino National Forests averaged 12 and 22 pines/ha (5 and 9 pines/acre), respectively, values within the range extrapolated from Dutton (1882).

Nineteenth century structure and composition of Ponderosa Pine Forest also have been estimated by forest reconstruction. This approach usually involves analyzing rings of living trees and using decay classes to estimate ages of snags and downed logs currently on the site to determine/estimate which were alive and at what dbh on a selected date in the past. This method has been used for southwestern Ponderosa Pine Forest primarily in north-central Arizona (e.g., Covington et al. 1997; Fulé et al. 1997, 2002; Menzel and Covington 1997; Mast et al. 1999; Abella and Denton 2009; Sánchez Meador et al. 2010), but also see Romme et al. (2009) for southwestern Colorado and Heyerdahl et al. (2011) for Utah. Reconstruction studies often differ in terms of minimum diameters reported, preventing direct comparison of reconstructed densities. In addition, most reconstruction studies do not include diameter-class data that would facilitate comparison of results of different studies and enable inference of nineteenth century stand dynamics.

The accuracy of tree-ring-based forest reconstruction depends on evidence of all trees from the historical date, e.g., 1880, persisting to the modern sampling date in the form of living trees, snags, downed logs, etc. Forest reconstructions underestimate historical density and basal area when evidence of trees has been lost by combustion or decomposition (Fulé et al. 2002, 2003) or is otherwise missing. Combustion is a factor on some sites of reconstruction studies of southwestern

Ponderosa Pine Forest (Moore et al. 2004), but of course not where fire has been totally excluded. Decomposition is a factor for young trees and small trees, as well as species with decay-susceptible wood, including pinyons (Kearns et al. 2005), quaking aspen (Gosz 1980; see Sect. 2.4.1.1), and white fir. Losses by tree cutting are important if evidence such as stumps is no longer present.

The accuracy of reconstruction of southwestern Ponderosa Pine Forest has been examined by resampling study plots where tree stems >9.1 cm (3.6 in.) dbh and other structures such as stumps had been mapped in the early twentieth century. Moore et al. (2004) relocated of 91 % of the mapped tree structures and therefore suggested that reconstruction in southwestern Ponderosa Pine Forest is reliable within 10 %. The authors acknowledged that trees too small to have been included on the original stem maps and no longer evidenced in the plot were a source of additional error (therefore, the accuracy of the many reconstructions that include trees with a dbh of ≤9.1 cm/3.6 in. is unexamined). Also, Moore et al. (2004) included stump holes as evidence of mapped tree structures, even though structures that cannot be dated would have limited use in reconstructions where historical stem maps are unavailable. In addition, thinning of mature trees in the study plots near the time of the original mapping likely increased the vigor of the remaining trees and thereby increased persistence of structural evidence. Nevertheless, reconstructions of nineteenth century southwestern Ponderosa Pine Forest apparently are more accurate than reconstructions of Mixed Conifer and Spruce-Fire Forests (see Sects. 2.4.1.1 and 3.4.1.1, respectively).

Reconstruction studies of densities and basal areas of nineteenth century southwestern Ponderosa Pine Forest have produced widely different results. This is partly because of differences in minimum diameters included (at least for densities), but it also reflects differences in sites. The effects of site conditions were highlighted by a study in north-central Arizona in which sites were randomly selected within nine Ponderosa Pine Forest ecosystem types that reflected a broad range of soils (Abella and Denton 2009). Each site was sampled for reconstruction using a 1 ha (2.5 acres) plot. Among the ecosystem types, mean tree densities reconstructed for 1880 ranged from 5 to 99 trees/ha (2 to 40 trees/acre) for individuals ≥9 cm (3.5 in.) dbh (diameter limit: Scott R. Abella, personal communication). Moreover, there was substantial variation among sites within the same ecosystem type: 54–85 trees/ha (22–34 trees/acre) in the least-variable type and 39–143 trees/ha (16–58 trees/acre) in the most-variable. Statistical analysis of factors influencing density indicated that soils and climate variables were more important than topography (including elevation).

Two other studies also reported high variation in forest density circa 1880–1900 across two large areas in north-central and one in east-central Arizona (Williams and Baker 2012, 2013). Mean tree densities reconstructed from original land survey records were remarkably similar among the areas: 142–144 trees/ha (57–58 trees/acre) for trees >10 cm (4 in.) dbh. Densities varied within each the areas: 19–33 % of the areas had open forests (<100 trees/ha; 40 trees/acre) and 15–17 % had dense forests (>200 trees/ha; 81 trees/acre), suggesting spatially complex forests. Another study based on land survey records dating to 1878–1879 also indicated substantial variation in forest density across a large area of mostly Ponderosa Pine Forest near Flagstaff, Arizona (Arundel 2000).

Tree-ring-based reconstructions done on the Kaibab Plateau are of particular interest, because other historical data are available from the area (cf. Dutton 1882; Lang and Stewart 1910; Vankat 2011), and some areas have been considered reference sites for late nineteenth century conditions (as discussed below). Reconstructions of stand structure to 1879 were done for three protected, relatively remote sites on the North Rim of Grand Canyon National Park, where surface fires had burned during the twentieth century (Fulé et al. 2002). Total densities were 153–160 trees/ha (62–65 trees/acre) and total basal areas were 17–21 m^2/ha (74–91 ft^2/acre) for trees ≥2.5 cm (1 in.) dbh. Ponderosa pine accounted for ≥96 % of the density and basal area. Ranges for individual plots indicated great stand heterogeneity. Fulé et al. (2002) acknowledged that their reconstructed 1879 values were likely underestimates because of loss of evidence in surface fires. They reported reconstructed densities for ponderosa pines >15.2 cm dbh (6 in.) as 125–141 pines/ha (51–57 pines/acre), but erroneously stated that this range matched historical data. Lang and Stewart (1910) actually reported 99 pines/ha (40 pines/acre).

Another protected forest approximately 130 km (80 miles) to the south is the Gus Pearson Natural Area, which has been the site of extensive research on Ponderosa Pine Forest (e.g., Biondi 1996; Mast et al. 1999; Wallin et al. 2004). Ponderosa pine density reconstructed for 1876 was 57 pines/ha (23 pines/acre; for trees >0 dbh; Covington et al. 1997). Basal area was 26 m^2/ha (112 ft^2/acre), as calculated from diameter-class data provided by Covington et al. (1997). Therefore, reconstructed density was much lower, and basal area was higher than the North Rim sites. The density of pines ≥51 cm (20 in.) dbh was reported as 44 pines/ha (18 pines/acre), slightly above the range extrapolated from Dutton (1882). The stand had not burned since before the reconstructed date of 1876.

A second approach to estimating nineteenth century structure and composition of southwestern Ponderosa Pine Forest is through sampling relict stands little disturbed by Euro-Americans. The three North Rim sites described above were said to provide "…a contemporary example of the forest characteristics that might have been extant…" without Euro-American influence (Fulé et al. 2002). However, this conclusion was challenged because mean fire intervals at the sites increased by ~4 to 9 times during 1880–1997 (Vankat 2011). In addition, sampling of contemporary forests at the sites indicated substantial increases in tree densities and basal areas over reconstructed values (Fulé et al. 2002; see next section). Such increases would not be expected for true relict sites.

Two isolated mesa tops without livestock grazing have been examined as relict areas with Ponderosa Pine Forest in Zion National Park in southwestern Utah (Madany and West 1980, 1983, 1984). Forest structure was described as a savanna, with an open canopy and herb-dominated groundlayer (Madany and West 1984). Tree density was 163 trees/ha (66 trees/acre) for individuals >5 cm (2 in.) dbh (Madany and West 1984). It is questionable whether the data are broadly representative of southwestern Ponderosa Pine Forest. The mesa tops are only 85 and 150 ha

(210 and 370 acres) in area (and included other vegetation types). The total sample was only 0.21 ha (0.5 acres). In addition, the mean fire interval of 69 years on the one mesa researched for fire history was much longer than the typical interval of approximately 4–36 years reported in Sect. 4.2.5.1. The long interval likely reflects limited ignitions and spread of lightning fires on the small mesa top.

Stands of Ponderosa Pine Forest in El Malpais National Monument in west-central New Mexico have been mentioned as possible relict sites (e.g., Covington 2003). These stands have been studied primarily for climate and fire history (e.g., Grissino-Mayer and Swetnam 1997; Grissino-Mayer et al. 1997). Use as relict sites is inappropriate because partial fire exclusion began in ~1880 and altered forest structure (Lewis 2003).

In conclusion, southwestern Ponderosa Pine Forest before Euro-American influence was more varied than many historical accounts suggest. Open, park-like conditions were common, but apparently so were sites where trees were denser. Stands were uneven-aged and consisted of clusters of ponderosa pines that established during widely separated periods of exceptional regeneration and were thinned by surface fires (Sect. 4.3.1). Historical forest structure and composition are best approximated by forest reconstructions, but reconstructions based on tree rings likely underestimate late nineteenth century densities and basal areas.

4.4.1.2 Post-Euro-American Settlement

Stands of southwestern Ponderosa Pine Forest – even stands in protected areas such as National Parks – have greatly changed since the late nineteenth century (e.g., Weaver 1951a; Harrington and Sackett 1990; Covington and Moore 1994a, b; Fulé et al. 2002; Vankat 2011). There are many facets to the changes, including changes in forest density, diameter distribution, structural diversity, tree vigor, and species composition.

Increases in tree density have been attributed to fire exclusion, which began when livestock grazing consumed herbaceous fuels that had carried surface fires (Sect. 4.2.6.1). It continued throughout most of the twentieth century because of fire-management-suppression activities (Sect. 4.2.6.2). Possible roles of other factors such as direct impacts of livestock grazing, wildlife population dynamics, tree cutting, and climate fluctuations were considered by Fulé et al. (2002) for Grand Canyon National Park, but they concluded that forest structure had changed primarily because of fire exclusion. Climate appears to have been an additional, essential factor in some areas. For example, although the fire regime was altered when livestock grazing began in the Chuska Mountains of northeastern Arizona and adjacent New Mexico in about 1830, forest structure didn't change until the early twentieth century. At that time, warm, wet weather coupled with the lack of surface fires enabled tree regeneration (Savage 1991). Many areas of the Southwest experienced a major pulse of regeneration in 1919 (Moir 1993) that was correlated with uncommon seasonal and interannual climatic factors, along with the decline of livestock grazing and ongoing fire exclusion (Savage et al. 1996).

Research on changes in southwestern Ponderosa Pine forest has focused on tree density – it is visually obvious and easily quantified. A common approach has been to compare data from reconstructions of nineteenth century forest structure to recent samples from the same sites. Nearly all such studies have reported large increases in density (e.g. Covington and Moore 1994a; Menzel and Covington 1997; Sánchez Meador et al. 2009), but see Abella (2008) for a dry stand that was stable. The results of broad-scale reconstructions by Abella and Denton (2009) and Williams and Baker (2012) do not provide insight into changes in relatively undisturbed forest, because their research areas were extensively logged in the late nineteenth and twentieth centuries.

Reconstruction studies done in protected areas include Fulé et al. (2002) for the three relatively remote North Rim sites mentioned in the previous section. These sites had surface fires (but at reduced frequencies) during the twentieth century. Densities at these sites in 1997–1998 were 2.5–5.9 times reconstructed values, with modern values of 389–936 trees/ha (157–379 trees/acre) for trees ≥2.5 cm (1 in.) dbh. These values include high densities of Gambel oak and New Mexico locust, species that are not amenable to accurate reconstruction because individuals are generally small and prone to decay (Fulé et al. 2002). Considering ponderosa pine only, densities at the three sites were 1.3–1.6 times reconstructed values, with modern values of 193–249 pines/ha (78–100 pines/acre). Basal areas were similarly 1.4–1.6 times reconstructed values, with modern values of 22–31 m^2/ha (96–135 ft^2/acre). The accuracy of these estimates of increases were affected by twentieth century surface fires that removed evidence of some of the trees that had been present in 1879 (Fulé et al. 2002), resulting in over-estimation of change. Nevertheless, the large differences in density and basal area between 1879 and 1997–1998 suggest these North Rim sites should not be considered relict areas. The sites might be relictual with regard to larger trees, but the absence of data on diameter distributions prevents assessing this.

Estimation of change at the protected, unburned Gus Pearson Natural Area gave a very different result for ponderosa pine density: it was 55 times the reconstructed value, with 3,099 pines/ha (1,254 pines/acre) in 1992 (Covington et al. 1997). The result for basal area was similar to the North Rim sites: 1.3 times the reconstructed value, with 34 m^2/ha (148 ft^2/acre; calculated from diameter-class data in Covington et al. 1997). This indicates that the large increase in density of ponderosa pine was due to small-diameter trees (see below).

Changes with Euro-American influence also have been examined by resampling permanent plots. The oldest known plots are ~50 that were established in 1909–1913 in National Forests in Arizona and New Mexico to study forest recovery after cutting (Moore et al. 2004; Bell et al. 2009; Sánchez Meador et al. 2010). Resampling these plots has been insightful for various purposes, but because of cutting they are not directly applicable to determining changes in protected stands.

Resampling of plots dating to 1935 in Grand Canyon National Park has proven useful for this purpose (Vankat 2011). Total density remained constant from 1935 to 2004, as an apparent decrease was not statistically significant. Basal area decreased from 50 to 37 m^2/ha (218 to 161 ft^2/acre). Ponderosa pine was constant in density, but its basal area decreased from 42 to 33 m²/ha (183 to 144 ft²/acre). In addition, Vankat (2011) combined data sets from various dates in Grand Canyon National Park and found evidence suggesting that Ponderosa Pine Forest had increased in sapling density since the late nineteenth century, apparently as a result of fire exclusion. He concluded that, depending on forest subtype and variable being examined, Grand Canyon stands had reached or passed a peak in forest density and basal area. Dry Ponderosa Pine Forest was likely near peak density and basal area when sampled in 2004, because it lacked statistically significant increases or decreases in either parameter. Both mesic and moist Ponderosa Pine Forests appear to have passed peak values, with decreases in total density and/or basal area. Decreases were likely due to the interaction of density-dependent mortality and exogenous factors such as climate change and prescribed fires (Vankat 2011).

When density changes are examined by diameter class, it is clear that diameter distributions have changed, with increases in smaller classes (e.g., Covington et al. 1997; Fulé et al. 1997, 2002; Menzel and Covington 1997; Vankat 2011) and in some cases decreases in larger classes (e.g., Covington et al. 1997; Vankat 2011). The overall increases in density and greater homogenization of diameter classes among stands have decreased structural diversity of stands (Dahms and Geils 1997) and landscapes (Allen et al. 2002).

Data in Covington et al. (1997) enable insight into changes in tree-diameter distribution from 1876 to 1992 at the Gus Pearson Natural Area. The estimated density increase of 55 times stated above, which was based on reconstructed density for 1876, was accounted for by seedling and sapling ponderosa pines <30 cm (12 in.) dbh. Although a major increase in these pines undoubtedly occurred, accurate estimation of the increase is impossible because most small pines present in 1876 would have died and decomposed by 1992, resulting in an underestimate of 1876 values (previous section) and therefore an overestimate of change. Also, it is unknown if the 1876 density of small pines – only 5 pines/ha (2 pine/acre) for trees in the above diameter class – was typical or abnormally low for the site. The dates of prior regeneration pulses are unknown (only establishment pulses are known; cf. Mast et al. 1999), yet the timing of regeneration pulses would affect interpretation of density increases. In contrast to the large increase in seedlings and saplings, the density of larger pines (i.e., dbh ≥30 cm/12 in.) was nearly unchanged, with 52 and 49 pines/ha (21 and 20 pines/acre) in 1876 and 1992, respectively. However, all three diameter classes ≥81 cm (32 in.) decreased in density and combined for a decrease from 25 to 5 pines/ha (10–2 pines/acre). Vankat (2011) also reported that small ponderosa pines increased in density and large ponderosa pines decreased; however, the magnitudes of the changes were much smaller (and the time period of 1935 to 2004 was much shorter).

With the increased density of small trees, tree vigor decreased across diameter classes. Tree growth rates declined in all diameter classes, with increased shade and root competition and decreased moisture and nutrients because of thicker litter (Clary and Ffolliott 1969 in Harrington and Sackett 1990; Biondi 1996). It appears that reduced vigor has been especially damaging to older, larger trees, as earlier predicted by Pearson (1950) and Weaver (1951a). Data suggest that competition from smaller, younger trees reduced the vigor of larger, older trees (Feeney et al. 1998; see also Graybill and Rose 1992). Reduced vigor of older trees is inferred from results of ecophysiological measurements following thinning of smaller, younger trees in north-central Arizona; canopy growth, insect-resistance characteristics, and uptake of water, nitrogen, and carbon by older trees all increased with thinning (Stone et al. 1999; Wallin et al. 2004; Zausen et al. 2005). Circumstantial evidence is that many sites have experienced decreased density of larger trees (e.g., Covington et al. 1997; Vankat 2011; see also Crocker-Bedford et al. 2005b). Elevated mortality rates of large trees in Grand Canyon National Park have been related to older trees being more susceptible to pathogens, drought, and injury because of increased stress through increased competition (Kaufmann and Covington 2001).

In contrast to large changes in forest structure, changes in tree composition in protected areas have been minor, except at relatively high elevations in Ponderosa Pine Forest (see below). Although forest reconstructions have indicated increased relative abundance of Gambel oak and/or New Mexico locust, this could partly reflect the above-mentioned issues in reconstructing small, decay-prone individuals of these species (Fulé et al. 2002; Abella and Fulé 2008a). Remeasurement of historical plots in Grand Canyon National Park found suggestions of increases in Gambel oak and New Mexico locust from 1935 to 2004, but none were statistically significant (Vankat 2011).

Data from the remeasurement of historical plots showed that post-1935 changes in species composition were generally limited to increases of white fir and decreases of quaking aspen (Vankat 2011). Changes were different among dry, mesic, and moist Ponderosa Pine Forest. Dry stands exhibited no statistically significant differences in any species from 1935 to 2004. Mesic stands had an increase in density of white fir in the smallest diameter class (10 to <30 cm / 4 to <12 in.), and the species' relative density increased from <1 to 9 % (all diameter classes combined). Moist stands, which are primarily at high elevations, also had an increase in density of white fir in this diameter class, as well as large decreases in density and basal area of quaking aspen. As a result, white fir increased in relative density from 4 to 24 %, and aspen decreased from 64 to 15 %. Changes in moist stands reflect large changes in forest composition reported for stands transitional with Mixed Conifer Forest (Mast and Wolf 2004; Crocker-Bedford et al. 2005a). Such high-elevation stands historically had seed sources of shade-tolerant, fire-sensitive conifers such as white fir within and near stands. As suggested in Sect. 4.2.1, compositional shifts in these stands have changed them into Mixed Conifer Forest (Dahms and Geils 1997; Swetnam et al. 2001; Mast and Wolf 2004; Evans et al. 2011).

In summary, stands of southwestern Ponderosa Pine Forest greatly changed with the influence of Euro-Americans because of a confluence of livestock grazing, fire management, and one or more regeneration pulses of ponderosa pine. All three factors were necessary. Without livestock grazing, some surface fires would have occurred even with fire management. Without fire management, regeneration pulses would have been thinned by fire as livestock grazing declined. Without regeneration pulses, there would have been no dramatic increase in stand densities. Changes in southwestern Ponderosa Pine Forest include increases in tree density, shifts in diameter distributions toward smaller trees, reductions in tree vigor, and shifts in species composition in higher-elevation stands.

4.4.2 *Understory*

A review of understory vegetation of southwestern Ponderosa Pine Forest concluded that reference conditions are difficult to identify (Korb and Springer 2003). Euro-American impacts have been so widespread that there are no known, broadly representative relict sites (Sect. 4.4.1.1). Also, little nineteenth century understory plant material has persisted, although archeological data have been used in combination with ethnobotanical records (Alcoze and Hurteau 2001) and phytoliths have been studied (Kerns et al. 2003).

Brief historical descriptions of past understory conditions are available:

> …every foot being covered with the finest grass… (Beale 1858, for east of the San Francisco Peaks)

> …we walk nearly waist-deep in fine pasture grasses… (Rusby 1889, for the San Francisco Peaks)

> There is no undergrowth to obstruct the view, and after the rainy season the grass beneath the trees is knee-deep in places, but the growth is sparse on account of the rocky nature of the surface. (Merriam 1890b, for the San Francisco Peaks)

> …the ground was well set with perennial grasses and other herbage… It was not an uncommon thing for the early settlers to cut native hay in the pine forests… (Holsinger 1902 for Arizona)

> …hundreds of tons of hay were cut under the actual spread of the forest trees during the [eighteen] sixties and seventies… (Holsinger 1902 for near the city of Prescott in central Arizona)

> The underbrush is very heavy, chiefly oak brush, choke-cherry, scarlet thorn, and wild rose. (DuBois 1903, for San Juan National Forest in the San Juan Mountains; in Romme et al. 2009)

See similar comments by Ingersoll (1885) for the vicinities of the towns of Pagosa Springs and Ignacio in southwestern Colorado.

Photographic evidence of historical conditions is very limited. The landscapes in Figs. 4.23a, b and 4.24 are too distant to show the understory, and livestock grazing modified understories at early dates (Sect. 4.2.6.1).

There is universal agreement that understory conditions changed with Euro-American influence. The changes have been linked to livestock grazing, fire exclusion, increases in tree density, and increases in litter depth (e.g., Arnold 1950; Cooper 1960; Pase and Brown 1994; Sackett et al. 1996; Korb and Springer 2003; Battaglia and Shepperd 2007; Romme et al. 2009). For example, surface fire – by thinning tree regeneration – generally had favored understory plants by reducing competition, increasing nutrient availability, and changing soil-water relationships (Moir et al. 1997).

Today (and presumably in the past), the understory of southwestern Ponderosa Pine Forest is highly variable (Romme et al. 2009). Factors influencing the variability include soil parent material, soil texture, litter depth, precipitation, elevation, topography, fire history, and canopy cover (Laughlin et al. 2005; Abella and Covington 2006; Laughlin and Abella 2007; Romme et al. 2009). Mean understory cover values on isolated, relatively undisturbed sites on the North Rim are ~25 % for ponderosa pine sites and ~47 % for ponderosa pine-Gambel oak sites at somewhat lower elevation (Laughlin et al. 2005). Cover values for individual plots ranged from ~3 to 77 %. At the regional level, shrub species are usually not abundant in most relatively undisturbed stands in northern and central Arizona and New Mexico, but are more abundant in the southern portions of those states and include various oaks (Moir 1993; Pase and Brown 1994; Barton 2002). Shrubs are also abundant in Utah and southwestern Colorado, where Gambel oak is widespread (Bradley et al. 1992; Romme et al. 2009).

Understory species composition has been quantitatively characterized for specific locations (e.g., Laughlin et al. 2004, 2005). Regional characterizations include Hanks et al. (1983), Alexander et al. (1984, 1987), Youngblood and Mauk (1985), DeVelice et al. (1986), Fitzhugh et al. (1987), Moir (1993), Pase and Brown (1994), and Muldavin et al. (1996).

With the scarcity of direct information, understory dynamics must be inferred from various types of research, including studies of the effects of forest thinning and prescribed burning designed to initiate restoration of historical conditions. Long-term results from thinning and burning are not yet available, and inferences from these and other studies can be problematic. Even research on contemporary understories accounts for only 58 % of the variation in plant cover, 22 % of the variation in composition, and 38 % of diversity in Ponderosa Pine Forest on the North Rim (Laughlin et al. 2005). Moreover, historical understory dynamics are likely to have been complex and to have differed among regions (cf. Laughlin et al. 2005). The following paragraphs assess the dynamics of understory cover, species composition, and diversity. All studies are from north-central Arizona, unless otherwise noted.

Understory cover would have changed with overstory successional dynamics. The finding that understory cover was higher on sites burned by high-severity fire vs. lower-severity fire in east-central Arizona (Kuenzi et al. 2008) suggests understory cover is high in early stages of succession (although management reseeding can complicate findings; cf. Foxx 1996; Kuenzi et al. 2008). As succession continues to stages where trees are present, it is likely that understory cover decreases. This is suggested by findings of negative relationships between understory cover and both canopy cover (Arnold 1950) and ponderosa pine basal area (Laughlin et al. 2005, 2011), as well as between understory production and tree density (e.g., Moore and Deiter 1992). These relationships also suggest that understory cover decreased with increased tree densities during fire exclusion. Additional evidence supporting this dynamic is that understory cover is negatively related to time since surface fire (Laughlin et al. 2005), and understory cover and productivity increase with forest restoration treatments of tree thinning and/or management burning (Fig. 4.26; e.g., Huffman and Moore 2004; Laughlin and Fulé 2008). However, some studies

Fig. 4.26 Well-developed herbaceous layer following manual thinning of trees and seeding of herbs near Flagstaff in north-central Arizona (Photograph by Betty J. Huffman)

have shown little relationship between restoration treatments and understory cover (e.g., Korb et al. 2005; Fulé et al. 2005).

Composition of the understory also would have changed with successional dynamics, but this has received little study, except for increases in invasive species following modern, high-severity crown fire (e.g., Crawford et al. 2001; Griffis et al. 2001). Understory composition also changed following fire exclusion (although findings can be inextricably associated with livestock grazing). Grass cover decreased (Cooper 1960; Covington and Moore 1994b; Covington et al. 1997; Fulé et al. 1997; Kerns et al. 2003). This is supported by forest restoration treatments of tree thinning and/or management burning that favored grasses (Weaver 1951b; Sackett et al. 1996; Griffis et al. 2001; Korb and Springer 2003; Moore et al. 2006). In addition, fire exclusion (along with livestock grazing) appears to have led to increases in shrubs, at least in southern Utah (Battaglia and Shepperd 2007).

Changes in understory diversity during succession also have received little study and can be confounded by management reseeding after fire. Understory species richness can be high early in succession (cf. Crawford et al. 2001). It is likely that richness declines as succession continues to stages dominated by trees, as inferred from a negative relationship of understory richness and ponderosa pine basal area

(Laughlin et al. 2005; Laughlin and Grace 2006) and findings that richness is lower in denser forests in Arizona (Clary 1975) and lowest with greatest overstory cover (Laughlin et al. 2007). These findings also indicate that understory richness decreased with increases in tree densities during fire exclusion. Additional evidence of this dynamic is that richness increases following surface fire (Laughlin et al. 2004), is negatively related to time since surface fire (Laughlin et al. 2005; Laughlin and Grace 2006), and decreases without fire (Laughlin et al. 2011). But some studies have found little difference following restoration thinning and burning (e.g., Fulé et al. 2005; Laughlin and Fulé 2008). The decline in understory richness with time since fire can be attributed to post-fire stimulation of seed germination in understory plants, followed by declining understory abundance and increasing ponderosa pine basal area (Laughlin and Grace 2006).

In summary, understory dynamics include increases in understory cover with high-severity disturbance, decreases during succession, and decreases as tree density increased during fire exclusion. Species composition also is dynamic, with decreases in abundance of grasses during succession and with fire exclusion. Species richness increases with high-severity disturbance and decreases during succession and with fire exclusion.

4.5 Conceptual Models

The following nested, three-tiered set of conceptual models is based on my interpretation of best-available information on Ponderosa Pine Forest of the American Southwest. The models summarize and synthesize material on drivers (Sect. 4.2) and processes (Sect. 4.3) of vegetation dynamics, and they account for historical changes (Sect. 4.4). Their format is explained in Sect. 1.7. The models can be used to facilitate understanding of the past and the present. They also can provide insight into effects of future changes in the environment and outcomes of alternative land-management decisions. Moreover, they can guide the development of quantitative models.

4.5.1 Ecosystem-Characterization Model

The ecosystem-characterization model for southwestern Ponderosa Pine Forest emphasizes *Vegetation* (a biotic component) and *Disturbance* (a driver; Fig. 4.27a, Table 4.2). Key aspects of *Vegetation* are structure, fuel, herbaceous cover, and species composition, and these affect various aspects of *Disturbance*. The primary agents of *Disturbance* are fire, drought, and insect outbreaks, and these affect tree mortality, vegetation structure, fuel, and species composition. The two other biotic components are *Soil System* and *Animals*, including insects that cause disturbance.

A second driver is *Weather & Climate*, which causes fires and drought and influences fire behavior, plant vigor, and fuel and soil moisture. The third driver is *Landscape*, with its primary feature being elevation. It influences weather and climate, as well as spread, pattern, and severity of fire and impact of drought. The model also includes six anthropogenic drivers (Fig. 4.27b, Table 4.2): *Livestock Grazing*, *Fire Management*, *Modern Climate Change*, *Invasive Species*, *Recreation*, and *Nearby Land Use*.

4.5.2 Vegetation-Dynamics Models

Separate models are needed to illustrate the vegetation dynamics of moist, mesic, and dry southwestern Ponderosa Pine Forest. The models have similar structure, but there are important differences in species composition of communities. Each vegetation-dynamics model is accompanied by a bar graph to show estimated relative shifts in abundances of the community types from circa 1870 (prior to widespread Euro-American land use) to circa 1970 (end of strict fire exclusion policy) to the present. The estimates are based on my understanding of vegetation dynamics and should be interpreted as showing relative differences and not interpreted quantitatively (hence the absence of a scale for abundance). Although not shown on the graphs, relative abundances shifted prior to circa 1870 as a result of variation in climate and disturbance. Localized differences also have been and are present, but are beyond the scope of these generalized bar graphs.

4.5.2.1 Moist Ponderosa Pine Forest

The vegetation-dynamics model for moist Ponderosa Pine Forest has four states, nine communities, and three transitions (Fig. 4.28, Table 4.3). All occurred historically. *State A* has been more common than the other states, and it encompasses the three most characteristic moist Ponderosa Pine Forest communities, all of which are dominated by ponderosa pine, white fir, and quaking aspen. Community *A1 Old Growth*, which is maintained by tree regeneration coupled with surface fire, was the most common community circa 1870 and included a wide range of stand densities. With reduced frequency of surface fires, community *A1* forms *A2 Old Growth with Dense Understory*. Therefore, *A2* greatly increased with past fire exclusion as ponderosa pine and white fir became more abundant in the understory in the absence of thinning by frequent surface fire. Surface fire can return community *A2* to *A1*, and consequently *A2* decreased since circa 1970 because of management fires. Continued reduced frequency of surface fire results in the maturation of understory trees and changes community *A2* into community *A3 Denser Old Growth*. Moderate tree mortality, as for example with some insect outbreaks, changes community *A3* into *A1* or *A2*.

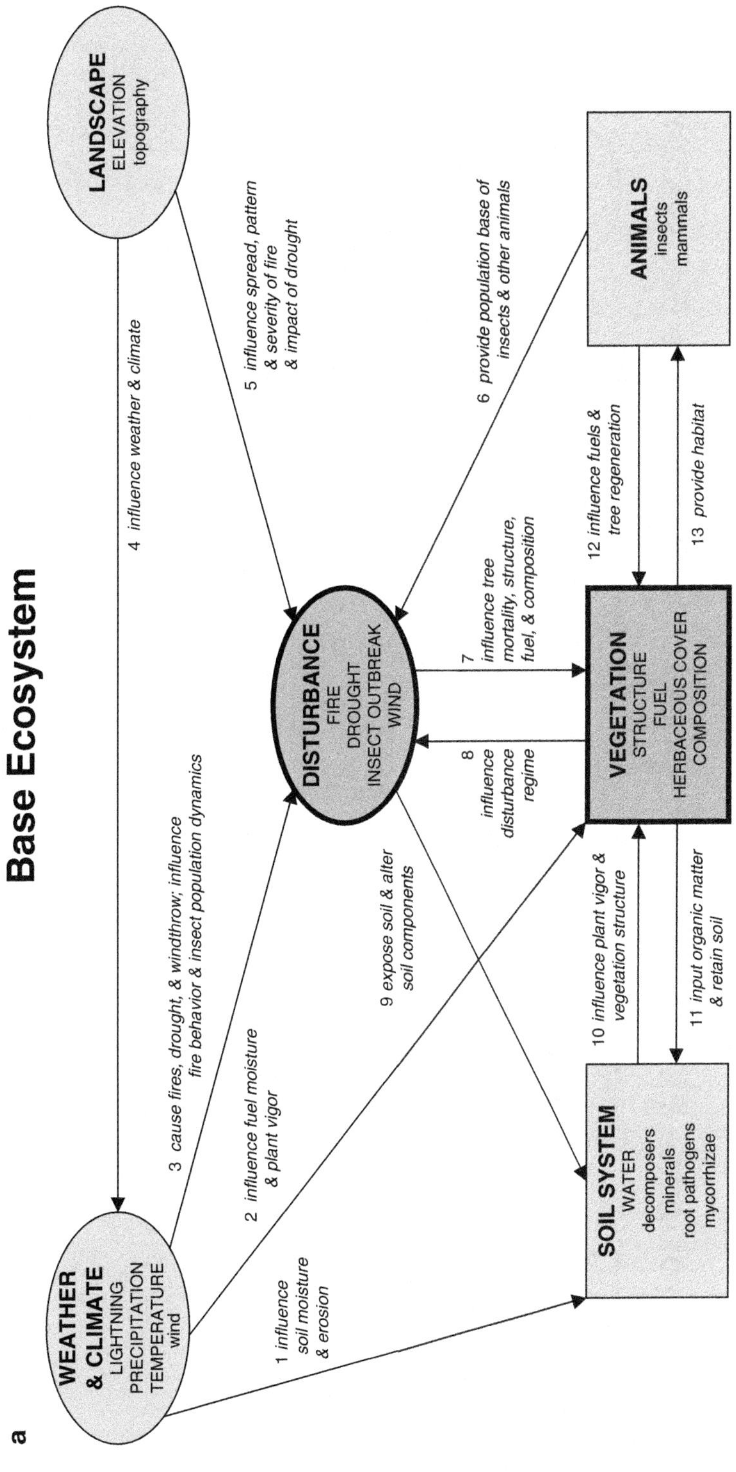
a
Base Ecosystem
WEATHER & CLIMATE
LIGHTNING
PRECIPITATION
TEMPERATURE
wind
LANDSCAPE
ELEVATION
topography
DISTURBANCE
FIRE
DROUGHT
INSECT OUTBREAK
WIND
VEGETATION
STRUCTURE
FUEL
HERBACEOUS COVER
COMPOSITION
SOIL SYSTEM
WATER
decomposers
minerals
root pathogens
mycorrhizae
ANIMALS
insects
mammals
1 influence soil moisture & erosion
2 influence fuel moisture & plant vigor
3 cause fires, drought, & windthrow; influence fire behavior & insect population dynamics
4 influence weather & climate
5 influence spread, pattern & severity of fire & impact of drought
6 provide population base of insects & other animals
7 influence tree mortality, structure, fuel, & composition
8 influence disturbance regime
9 expose soil & alter soil components
10 influence plant vigor & vegetation structure
11 input organic matter & retain soil
12 influence fuels & tree regeneration
13 provide habitat

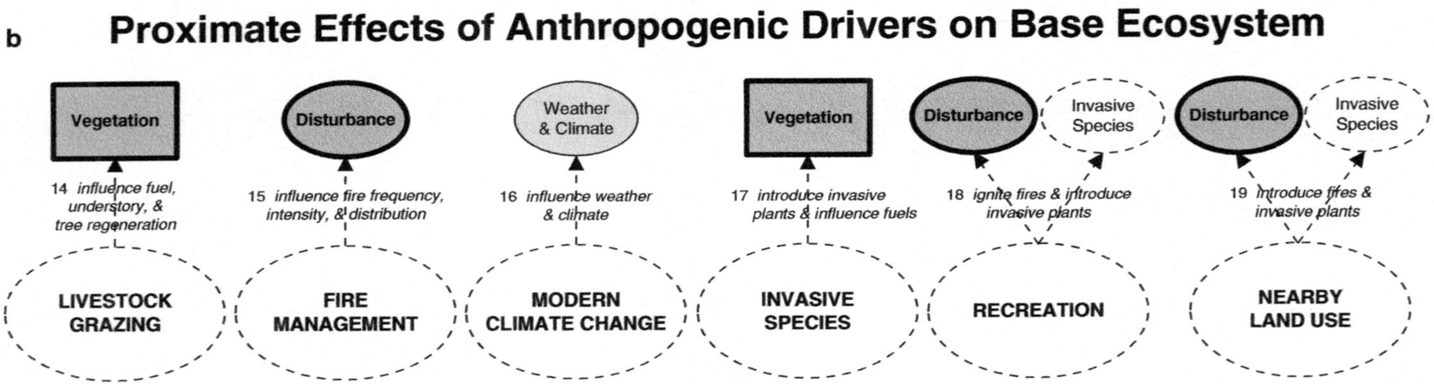

Fig. 4.27 (**a**, **b**) Ecosystem-characterization model for Ponderosa Pine Forest. In the base model (**a**), biotic components are shown as *rectangles* and drivers as *solid ovals*. Text within them identifies major features in *upper case* and other important features in *lower case*. Disturbance and Vegetation are *highlighted*, because they are the basis for the vegetation-dynamics models (Figs. 4.28, 4.29, and 4.30). In the anthropogenic supplement to the base model (**b**), drivers are shown as *dashed ovals*. Relationships in (**a**) and (**b**) are shown as *solid* and *dashed arrows*, respectively, and are described in Table 4.2

Table 4.2 Relationships in the ecosystem-characterization model of southwestern Ponderosa Pine Forest (Fig. 4.27a, b)

Relationships	Circa 1870	Present
1	Precipitation increases soil moisture and can cause soil erosion. Winter precipitation is critical for recharging soil moisture	Same, but infiltration of precipitation, especially summer monsoonal rain, can be reduced by thicker layer of litter and duff, where unburned. Infiltration is reduced and soil erosion can be increased where this layer has been lost in high-severity fires. In addition, climate change increases the severity and duration of drought
2	Precipitation and temperature largely determine fuel moisture and are important to plant vigor	Same, but climate change increases the severity and duration of drought
3	Lightning ignites fires. Consistently low precipitation and high temperature cause drought and affect insect population dynamics. Wind influences fire behavior	Same, but climate change increases drought and insect outbreaks
4	Topography influences weather and climate	Same, but the effects on forest microclimate can be reduced with dense tree cover
5	Fire patterns are influenced by elevation and topography	Same, but the effects of elevation and topography are reduced in areas with higher fuel loadings and more continuous fuels
6	Increases in populations of bark beetles can result in outbreaks	Same
7	Frequent surface fires thin tree regeneration and keeps fuel loadings low. Crown fires kill trees in all size classes. Bark beetles kill scattered large trees, producing coarse fuels. Drought increases stress in trees, surface fire frequency, probability of patchy crown fires, and outbreaks of bark beetles	Same, except past fire exclusion resulted in increased tree densities, fuel loadings, fuel continuity, and homogeneity of vegetation patterns. Surface fires are more likely to crown, and crown fires are more likely to burn across landscapes. Crown fires kill trees in all size classes. Drier climate results in expansive stands of stressed trees and increased probability of fires and bark beetle outbreaks
8	Horizontally continuous fine fuels, light fuel loadings, and vertically discontinuous fuels result in fire regime dominated by frequent, low-severity surface fires. Crown fires occur in areas of heterogeneous topography and steep slopes	Increased fuel loadings and more vertically continuous fuels result in greater probability of extensive crown fires

9	Surface fires partially and patchily expose soil surface and alter soil properties	Past fire exclusion resulted in thicker, more extensive litter and duff and likely altered other soil properties. Hotter, larger fires likely have greater effects on soil properties
10	Water and minerals in soil affect plant vigor and growth and forest structure	Same, but less water is available because of drought, higher temperatures, and thicker litter and duff. Minerals can be less available because of greater sequestering in organic matter
11	Plants add organic matter to soil and stabilize soil, except in areas of crown fires	Same, except areas of crown fires are larger
12	Herbivory influences fuels and tree regeneration	Same, but livestock and increased deer and elk populations have greater effects
13	Vegetation is habitat for animals	Same, but animal populations are changed where habitat is changed
14	Not applicable	Late nineteenth century livestock grazing reduced herb cover and likely influenced tree regeneration. The reduction in herb cover reduced the frequency of surface fires. Impacts of twenty-first century grazing are poorly known
15	Not applicable	Past fire management attempted to suppress fires throughout most of the twentieth century. This resulted in large changes in forest structure and composition. Beginning in the late twentieth century, fire management included prescribed and lightning-ignited management fires
16	Not applicable	Modern climate change affects weather and climate, raising temperature and increasing the frequency and extent of drought
17	Not applicable	Invasive plants have potential to modify the understory and change fire patterns
18	Not applicable	Recreationists ignite fires and introduce and spread invasive plants
19	Not applicable	Nearby land use that initiates fires and colonization by invasive plants can introduce fires and invasives into Ponderosa Pine Forest

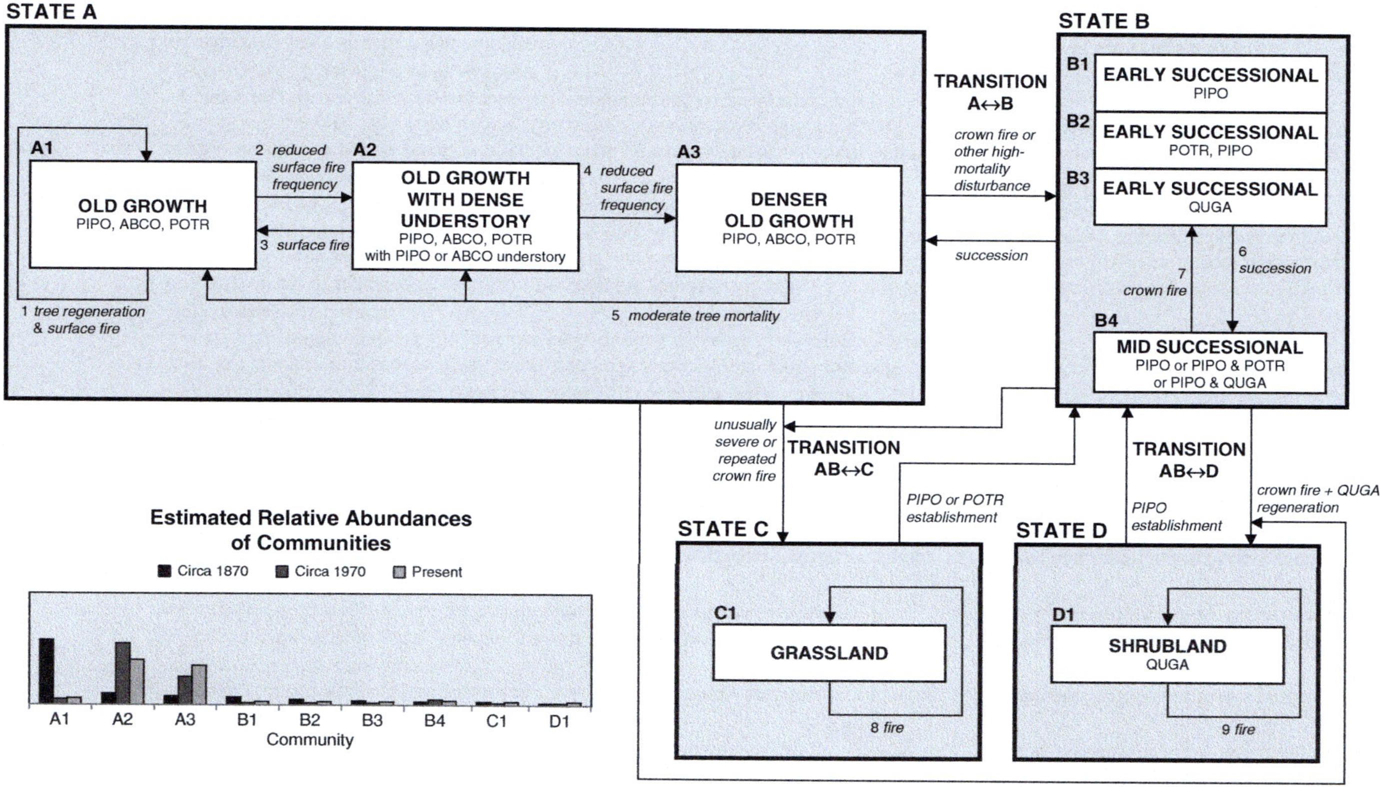

Fig. 4.28 Vegetation-dynamics model for moist Ponderosa Pine Forest. Model provides details on the vegetation-disturbance portion of the ecosystem-characterization model (Fig. 4.27a, b) by showing communities (*smaller*, *open rectangles*) within states (*larger*, *shaded rectangles*), processes that form the communities, and transitions between states. Text along *arrows* identifies pathways that are described in Table 4.3. *Bar graph* shows estimated relative shifts in abundances of the community types from circa 1870 (see Sect. 4.5.2). Species: *ABCO* white fir, *PIPO* ponderosa pine, *POTR* quaking aspen, *QUGA* Gambel oak

Table 4.3 Relationships and transitions in the vegetation-dynamics model of moist Ponderosa Pine Forest (Fig. 4.28)

Relationships and transitions	Circa 1870	Present
1	Periodic tree regeneration is thinned by surface fire, which maintains old growth structure and composition	Same, but surface fire is less frequent
2	Periods of reduced frequency of surface fire leave tree understory unthinned, producing old growth with dense understory	Same and surface fire is less frequent
3	Surface fire thins tree understory, changing stands into old growth without a dense understory	Same, but surface fire is less frequent
4	Periods of reduced frequency of surface fire allow understory trees to grow into the canopy, changing stands into denser old growth	Same and surface fire is less frequent
5	Moderate tree mortality, such as with some insect outbreaks, reduces density of canopy trees, changing stands into old growth or old growth with dense understory	Same and insect outbreaks are more frequent
A ↔ B	Crown fire or other high-mortality disturbance changes old growth communities of State A into early successional communities B1, B2, or B3, depending on the abundance and vigor of quaking aspen and Gambel oak in the old growth stands. Reversion to State A requires succession	Same and crown fire is more common
6	Succession changes early successional stands into mid successional community	Same
7	Crown fire changes mid successional stands into early successional communities B1, B2, or B3, depending on the abundance and vigor of quaking aspen and Gambel oak in the mid successional stands	Same and crown fire is more common
AB ↔ C	Unusually severe or repeated crown fire converts some stands from States A and B into State C. Reversion to State B requires invasion and establishment of ponderosa pine or quaking aspen to form early successional communities B1 or B2, depending on the invading species. Reversion to State A has to be via State B	Same, but unusually severe and repeated crown fire is more common. Larger size of crown fire limits invasion of ponderosa pine and quaking aspen
8	Fire maintains stand structure and composition	Same, but fire is less frequent
AB ↔ D	Crown fire followed by regeneration of Gambel oak converts some stands from States A and B into State D. Reversion to State B requires invasion and establishment of ponderosa pine to form early successional community B3 transitioning into B4. Reversion to State A has to be via State B	Same, but crown fire is more common. Larger crown fire limits invasion of ponderosa pine
9	Fire maintains stand structure and composition	Same, but fire is less frequent

Transition A↔B changes *State A* to *State B* with crown fire or other disturbance that results in high mortality of trees. One of three *Early Successional* communities is formed: *B1* with ponderosa pine, *B2* with quaking aspen and ponderosa pine, or *B3* with shrubs of Gambel oak. The community formed largely depends on the pre-fire abundance and vitality of aspen and Gambel oak. Succession changes *B1*, *B2*, and *B3* into community *B4 Mid Successional* dominated by either (a) ponderosa pine, (b) ponderosa pine and quaking aspen, or (c) ponderosa pine and Gambel oak. Crown fire changes *B4* into *B1*, *B2*, or *B3*, also depending on the pre-fire abundance and vitality of aspen and Gambel oak. *Transition A↔B* can be reversed by succession.

Unusually severe or repeated crown fire can change *States A* and *B* into *State C* via *Transition AB↔C*. *State C* is made up of community *C1 Grassland*, which is maintained by fire. *Transition AB↔C* can be reversed by invasion and establishment of ponderosa pine and/or aspen, converting *C1* into either *B1* or *B2* depending on the invading species. Reversion to State A occurs only via State B.

Crown fire followed by regeneration of Gambel oak can change *States A* and *B* into *State D* via *Transition AB↔D*. *State D* consists of community *D1 Shrubland*, which is dominated by Gambel oak and maintained by fire (Gambel oak sprouts after fire). *Transition AB↔D* can be reversed by invasion and establishment of ponderosa pine, converting *D1* into *B3* transitioning into *B4*. Reversion to State A occurs only via State B.

4.5.2.2 Mesic Ponderosa Pine Forest

The vegetation-dynamics model for mesic Ponderosa Pine Forest has four states, eight communities (one fewer than moist and dry Ponderosa Pine Forest), and three transitions (Fig. 4.29, Table 4.4). All occurred historically. *State A* has been more common than the other states, and it encompasses the three most characteristic mesic Ponderosa Pine Forest communities, all of which are dominated by ponderosa pine. Community *A1 Old Growth*, which is maintained by tree regeneration coupled with surface fire, was the most common community circa 1870 and included a wide range of stand densities. With reduced frequency of surface fires, community *A1* forms *A2 Old Growth with Dense Understory*. Therefore, *A2* greatly increased with fire exclusion as ponderosa pine became more abundant in the understory in the absence of thinning by frequent surface fire. Surface fire can return community *A2* to *A1*, and consequently *A2* decreased since circa 1970 because of management fires. Continued reduced frequency of surface fire results in the maturation of understory trees and changes community *A2* into community *A3 Denser Old Growth*. Moderate tree mortality, as for example with some insect outbreaks, changes community *A3* into *A1* or *A2*.

Transition A↔B changes *State A* to *State B* with crown fire or other disturbance that results in high mortality of trees. One of two *Early Successional* communities is formed: *B1* with ponderosa pine or *B2* with shrubs of Gambel oak. The community formed largely depends on the pre-fire abundance and vitality of Gambel oak.

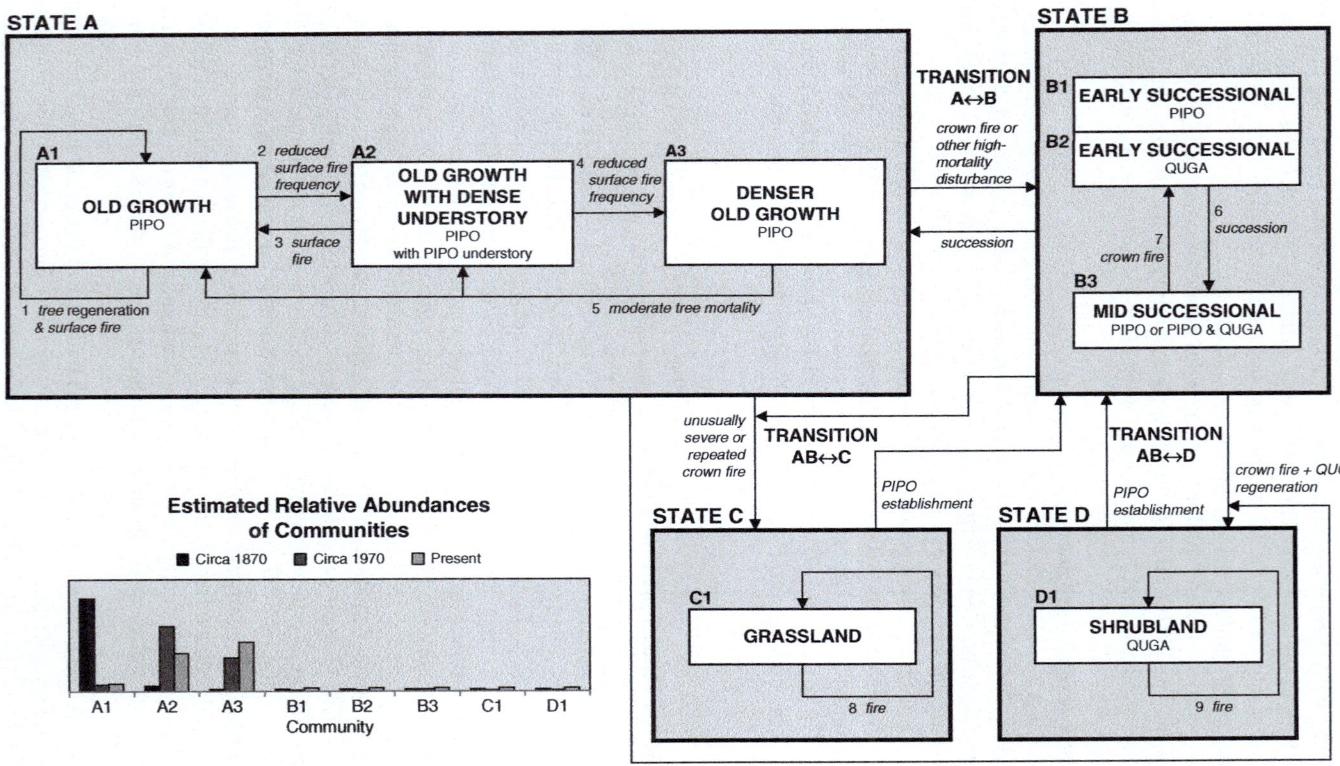

Fig. 4.29 Vegetation-dynamics model for mesic Ponderosa Pine Forest. Model provides details on the vegetation-disturbance portion of the ecosystem-characterization model (Fig. 4.27a, b) by showing communities (*smaller*, *open rectangles*) within states (*larger*, *shaded rectangles*), processes that form the communities, and transitions between states. Text along *arrows* identifies pathways that are described in Table 4.4. *Bar graph* shows estimated relative shifts in abundances of the community types from circa 1870 (see Sect. 4.5.2). Species: *PIPO* ponderosa pine, *QUGA* Gambel oak

Table 4.4 Relationships and transitions in the vegetation-dynamics model of mesic Ponderosa Pine Forest (Fig. 4.29)

Relationships and transitions	Circa 1870	Present
1	Periodic tree regeneration is thinned by surface fire, maintaining stand structure and composition	Same, but surface fire is less frequent
2	Periods of reduced frequency of surface fire leave tree understory unthinned, producing old growth with dense understory	Same and surface fire is less frequent
3	Surface fire thins tree understory, changing stands into old growth without a dense understory	Same, but surface fire is less frequent
4	Periods of reduced frequency of surface fire allow understory trees to grow into the canopy, changing stands into denser old growth	Same and surface fire is less frequent
5	Moderate tree mortality, such as with some insect outbreaks, reduces density of canopy trees, changing stands into old growth or old growth with dense understory	Same and insect outbreaks are more frequent
A ↔ B	Crown fire or other high-mortality disturbance changes old growth communities of State A into early successional communities B1, B2, or B3, depending on the abundance and vigor of quaking aspen and Gambel oak in the old growth stands. Reversion to State A requires succession	Same and crown fire is more common
6	Succession changes early successional stands into mid successional community	Same
7	Crown fire changes mid successional stands into early successional communities B1 or B2, depending on the abundance and vigor of Gambel oak in the mid successional stands	Same and crown fire is more common
AB ↔ C	Unusually severe or repeated crown fire converts some stands from States A and B into State C. Reversion to State B requires invasion and establishment of ponderosa pine to form early successional community B1. Reversion to State A has to be via State B	Same, but unusually severe and repeated crown fire is more common. Larger size of crown fire limits invasion of ponderosa pine
8	Fire maintains stand structure and composition	Same, but fire is less frequent
AB ↔ D	Crown fire followed by regeneration of Gambel oak converts some stands from States A and B into State D. Reversion to State B requires invasion and establishment of ponderosa pine to form early successional community B2 transitioning into B3. Reversion to State A has to be via State B	Same, but crown fire is more common. Larger crown fire limits invasion of ponderosa pine
9	Fire maintains stand structure and composition	Same, but fire is less frequent

Succession changes *B1* and *B2* into community *B3 Mid Successional* dominated by either ponderosa pine or ponderosa pine and Gambel oak. Crown fire changes *B3* into *B1* or *B2*, also depending on the pre-fire abundance and vitality of Gambel oak. *Transition A↔B* can be reversed by succession.

Unusually severe or repeated crown fire can change *States A* and *B* into *State C* via *Transition AB↔C. State C* is made up of community *C1 Grassland*, which is maintained by fire. *Transition AB↔C* can be reversed by invasion and establishment of ponderosa pine, converting *C1* into *B1*. Reversion to State A occurs only via State B.

Crown fire followed by regeneration of Gambel oak can change *States A* and *B* into *State D* via *Transition AB↔D. State D* consists of community *D1 Shrubland*, which is dominated by Gambel oak and maintained by fire (Gambel oak root sprouts after fire). *Transition AB↔D* can be reversed by invasion and establishment of ponderosa pine, converting *D1* into *B2* transitioning into *B3*. Reversion to State A occurs only via State B.

4.5.2.3 Dry Ponderosa Pine forest

The vegetation-dynamics model for dry Ponderosa Pine Forest has four states, nine communities, and three transitions (Fig. 4.30, Table 4.5). All occurred historically. *State A* has been more common than the other states, and it encompasses the three most characteristic dry Ponderosa Pine Forest communities, all of which are dominated by ponderosa pine with pinyons and junipers in the understory. Community *A1 Old Growth*, which is maintained by tree regeneration coupled with surface fire, was the most common community circa 1870 (its range of stand densities was likely narrower than for moist and mesic Ponderosa Pine Forest). With reduced frequency of surface fires, community *A1* forms *A2 Old Growth with Dense Understory*. Therefore, *A2* greatly increased with past fire exclusion as ponderosa pine became more abundant in the understory in the absence of thinning by frequent surface fire. Surface fire can return community *A2* to *A1*, and consequently *A2* decreased since circa 1970 because of management fires. Continued reduced frequency of surface fire results in the maturation of understory trees and changes community *A2* into community *A3 Denser Old Growth*. Moderate tree mortality, as for example with some insect outbreaks, changes community *A3* into *A1* or *A2*.

Transition A↔B changes *State A* to *State B* with crown fire or other disturbance that results in high mortality of trees. One of three *Early Successional* communities is formed: *B1* with ponderosa pine, *B2* with pinyons and junipers, or *B3* with oak shrubs or trees. Succession changes *B1*, *B2*, and *B3* into community *B4 Mid Successional* dominated by either (a) ponderosa pine, (b) ponderosa pine, pinyons, and junipers, or (c) ponderosa pine and oak. Crown fire changes *B4* into *B1*, *B2*, or *B3*. *Transition A↔B* can be reversed by succession.

Unusually severe or repeated crown fire can change *States A* and *B* into *State C* via *Transition AB↔C. State C* is made up of community *C1 Grassland*, which is maintained by fire. *Transition AB↔C* can be reversed by invasion and establishment

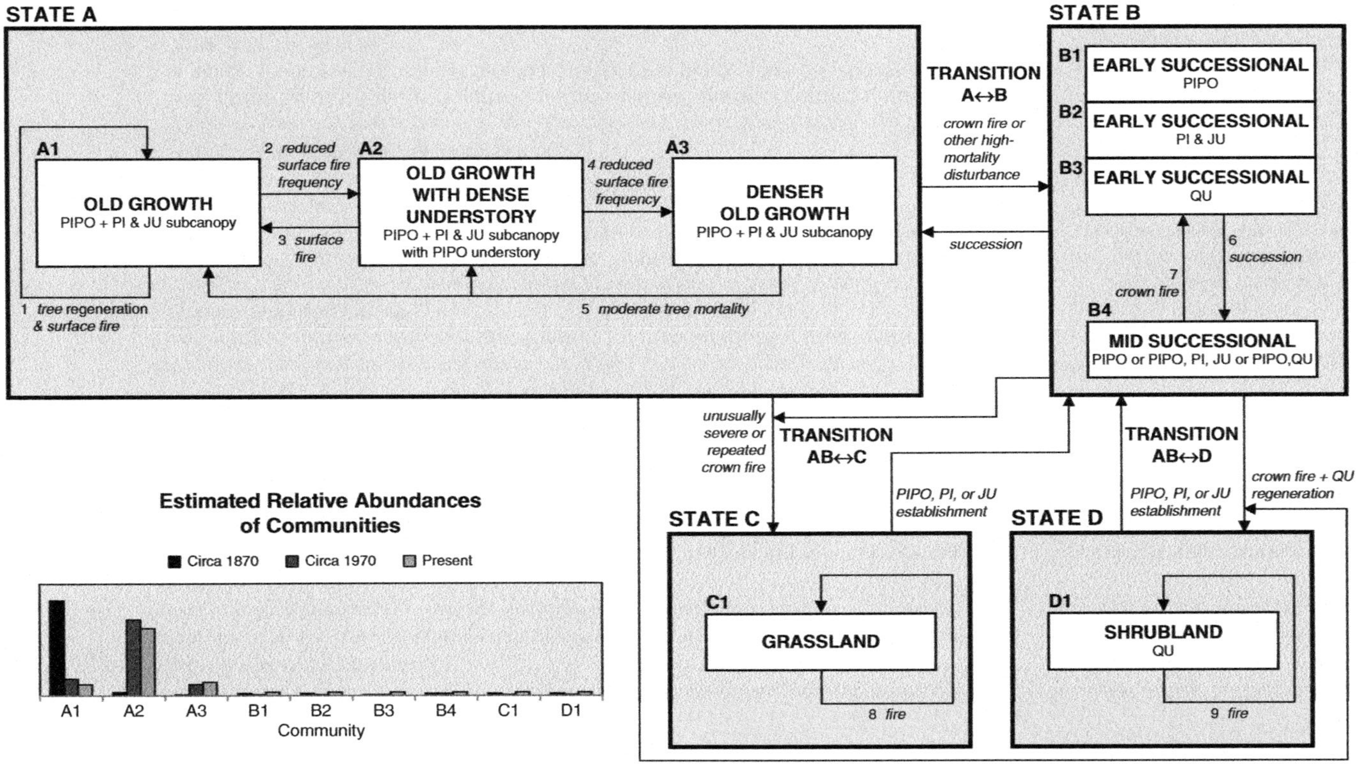

Fig. 4.30 Vegetation-dynamics model for dry Ponderosa Pine Forest. Model provides details on the vegetation-disturbance portion of the ecosystem-characterization model (Fig. 4.27a, b) by showing communities (*smaller*, *open rectangles*) within states (*larger*, *shaded rectangles*), processes that form the communities, and transitions between states. Text along *arrows* identifies pathways that are described in Table 4.5. *Bar graph* shows estimated relative shifts in abundances of the community types from circa 1870 (see Sect. 4.5.2). Species: *JU* junipers, *PI* pinyons, *PIPO* ponderosa pine, and *QU* oaks

Table 4.5 Relationships and transitions in the vegetation-dynamics model of dry Ponderosa Pine Forest (Fig. 4.30)

Relationships and transitions	Circa 1870	Present
1	Periodic tree regeneration is thinned by surface fire, which maintains old growth structure and *composition*	Same, but surface fire is less frequent
2	Periods of reduced frequency of surface fire leave tree understory unthinned, producing old growth with dense understory	Same and surface fire is less frequent
3	Surface fire thins tree understory, changing stands into old growth without a dense understory	Same, but surface fire is less frequent
4	Periods of reduced frequency of surface fire allow understory trees to grow into the canopy, changing stands into denser old growth	Same and surface fire is less frequent
5	Moderate tree mortality, such as with some insect outbreaks, reduces density of canopy trees, changing stands into old growth or old growth with dense understory	Same and insect outbreaks are more frequent
A↔B	Crown fire or other high-mortality disturbance changes old growth communities of State A into early successional communities B1, B2, or B3, depending on the abundance and vigor of oak in the old growth stands and which conifers establish after fire. Reversion to State A requires succession	Same and crown fire is more common
6	Succession changes early successional stands into mid successional community	Same
7	Crown fire changes mid successional stands into early successional communities B1, B2, or B3, depending on the abundance and vigor of oak in the mid successional stands and which conifers establish after fire	Same and crown fire is more common
AB↔C	Unusually severe or repeated crown fire converts some stands from States A and B into State C. Reversion to State B requires invasion and establishment of ponderosa pine, pinyon, or juniper to form early successional communities B1 or B2, depending on the invading species. Reversion to State A has to be via State B	Same, but unusually severe and repeated crown fire is more common. Larger size of crown fire limits invasion of ponderosa pine, pinyon, and juniper
8	Fire maintains stand structure and composition	Same, but fire is less frequent
AB↔D	Crown fire followed by regeneration of oaks converts some stands from States A and B into State D. Reversion to State B requires invasion and establishment of ponderosa pine, pinyon, or juniper to form early successional community B3 transitioning into B4. Reversion to State A has to be via State B	Same, but crown fire is more common. Larger crown fire limits invasion of ponderosa pine, pinyon, and juniper
9	Fire maintains stand structure and composition	Same, but fire is less frequent

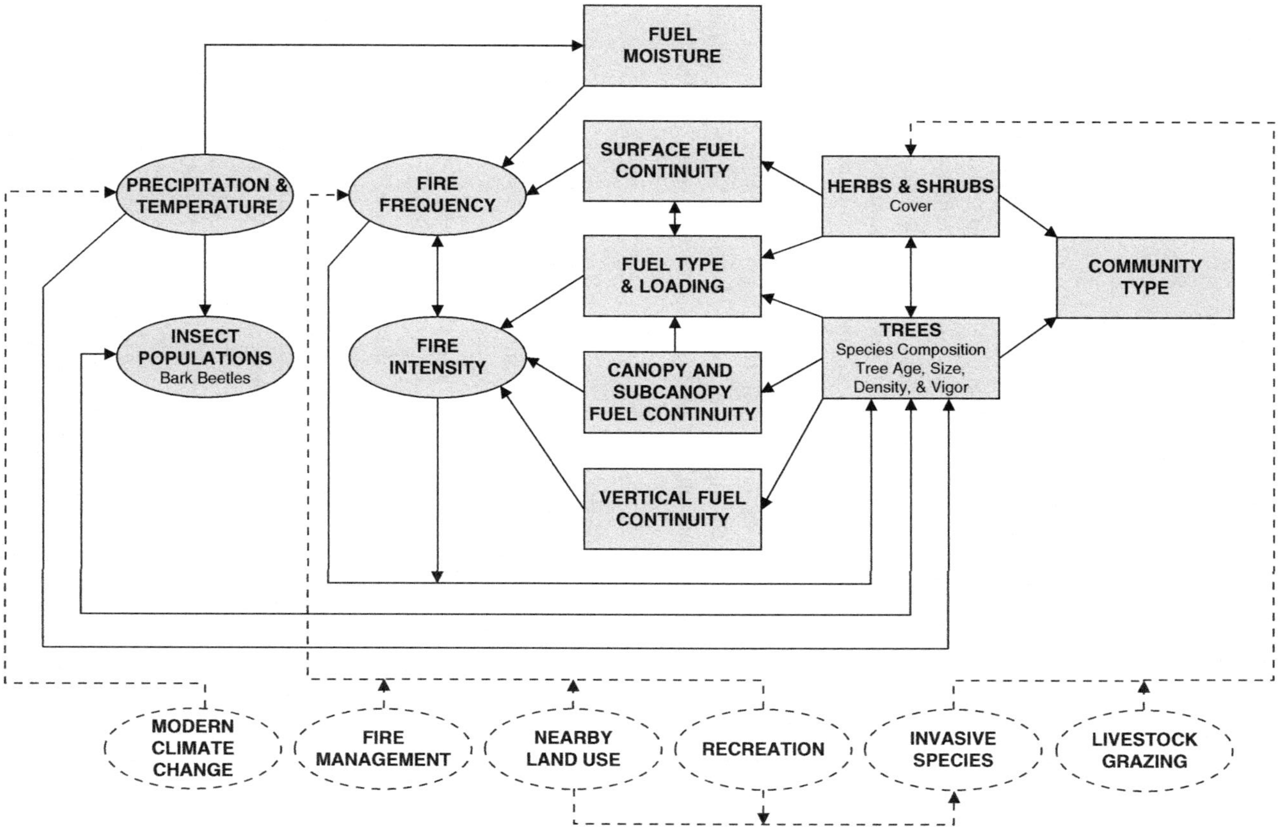

Fig. 4.31 Mechanistic model for Ponderosa Pine Forest. Model provides details on the mechanisms involved in the vegetation-dynamics models (Figs. 4.28, 4.29, and 4.30). *Rectangles* are biotic components (including fuels), *ovals* are drivers, and *dashed ovals* are anthropogenic drivers

of ponderosa pine, pinyons, and/or junipers, converting *C1* into either *B1* or *B2* depending on the invading species. Reversion to State A occurs only via State B.

Crown fire followed by regeneration of oak can change *States A* and *B* into *State D* via *Transition AB↔D*. *State D* consists of community *D1 Shrubland*, which is dominated by oak and maintained by fire. *Transition AB↔D* can be reversed by invasion and establishment of ponderosa pine, pinyons, and/or junipers, converting *D1* into *B3* transitioning into *B4*. Reversion to State A occurs only via State B.

4.5.3 Mechanistic Model

All three vegetation-dynamics models are explained by the same mechanistic model (Fig. 4.31). It has eight biotic components on the right side of the figure (including five aspects of fuels), four drivers on the left side, and six anthropogenic factors at the bottom. In general, *Herbs & Shrubs*, *Trees*, and *Precipitation & Temperature* affect the five fuel characteristics. The fuel characteristics combined with *Fire Intensity*, *Fire Management*, *Nearby Land Use*, and *Recreation* influence *Fire Frequency*. Also, *Fire Frequency*, *Fire Intensity*, *Precipitation & Temperature*, and *Insect Populations* influence characteristics of *Trees*, such as species composition and tree age, size, density, and vigor. *Trees* and *Herbs & Shrubs* determine *Community Type* (of the eight/nine appearing in the vegetation-dynamics models).

4.6 Conclusions and Challenges

Ponderosa Pine Forest is the most thoroughly studied vegetation on southwestern mountains and plateaus. Nevertheless, there are many challenges for researchers and managers. Historical stand structure has been characterized as generally open and park-like; however, some evidence indicates greater variation in stand structure prior to Euro-American settlement. Additional research is needed. A historical fire regime of frequent, low-severity fires is widely documented, but research has indicated that the fire regime also included mixed- and high-severity fire. This too requires additional study, with attention to identifying historical mixed- and high-severity fires, their spatial extent, and factors associated with their spatial and temporal distributions. Fire regimes have been documented mostly by mean fire intervals, but the length of fire-free periods possibly provides insight into differences in species composition among stands. Questions have been raised about methods of forest reconstruction. Comparison of reconstructions by tree rings vs. land surveys in the same landscape could provide insight into the advantages and disadvantages of both methods. Although useful information has been obtained from forest reconstructions based on tree rings, future studies need to report data for diameter distributions and for multiple twentieth century dates to enable comparison of findings with other studies. Increasing stand densities during the twentieth century are well-known, but a more detailed understanding is important. For example, research is needed on the timing of past regeneration pulses of ponderosa pine. Were regeneration levels in late nineteenth century abnormally low, thereby biasing our perspective of historical conditions? Also, have old-growth stands reached or surpassed peak density and basal area, as reported for Grand Canyon National Park? Anthropogenic disturbances have had greater impact on Ponderosa Pine Forest than on other vegetation types on southwestern mountains and plateaus. Key factors needing study and subsequent management planning and action include the direct impacts of air pollution, the spread and control of invasive plants, and the impacts and regulation of recreation. Most importantly, the long-term viability of southwestern Ponderosa Pine Forest is threatened by the combination of climate change, human use, and landscape-scale crown fires. Is conversion to grasslands and shrublands following crown fires – especially repeated crown fires – the fate of southwestern Ponderosa Pine Forest? Well-founded, ecologically based management plans for regional restoration of Ponderosa Pine Forest must be developed and implemented. A major challenge is to develop fire-management programs that achieve forest structure and function that are sustainable during climate change.

References

Abella SR (2008) A unique old-growth ponderosa pine forest in northern Arizona. J Ariz-Nev Acad Sci 40:1–11

Abella SR, Covington WW (2006) Forest ecosystems of an Arizona *Pinus ponderosa* landscape: multifactor classification and implications for ecological restoration. J Biogeogr 33:1368–1383

Abella SR, Denton CW (2009) Spatial variation in reference conditions: historical tree density and pattern on a *Pinus ponderosa* landscape. Can J For Res 39:2391–2403

Abella SR, Fulé PZ (2008a) Changes in Gambel oak densities in southwestern ponderosa pine forests since Euro-American settlement. U.S. Department of Agriculture, Forest Service Research Note RMRS-RN-36, Rocky Mountain Research Station, Fort Collins, CO

Abella SR, Fulé PZ (2008b) Fire effects on Gambel oak in southwestern ponderosa pine-oak forests. U.S. Department of Agriculture, Forest Service Research Note RMRS-RN-34, Rocky Mountain Research Station, Fort Collins, CO

Abella SR, Engel EC, Springer JD, Covington WW (2012) Relationships of exotic plant communities with native vegetation, environmental factors, disturbance, and landscape ecosystems of *Pinus ponderosa* forests, USA. For Ecol Manag 271:65–74

Adams HD, Kolb TE (2005) Tree growth response to drought and temperature in a mountain landscape in northern Arizona, USA. J Biogeor 32:1629–1640

Alcoze T, Hurteau M (2001) Implementing the archaeo-environmental reconstruction technique: rediscovering the historic ground layer of three plant communities in the Greater Grand Canyon region. In: Egan D, Howell EA (eds) The historical ecology handbook: a restorationist's guide to reference ecosystems. Island Press, Washington, DC

Alexander BG Jr, Ronco F Jr, Fitzhugh EL, Ludwig JA (1984) A classification of forest habitat types of the Lincoln National Forest, New Mexico. U.S. Department of Agriculture, Forest Service General Technical Report RM-104, Rocky Mountain Forest and Range Experiment Station, Fort Collins, CO

Alexander BG Jr, Fitzhugh EL, Ronco F Jr, Ludwig JA (1987) A classification of forest habitat types of the northern portion of the Cibola National Forest, New Mexico. U.S. Department of Agriculture, Forest Service General Technical Report RM-143, Rocky Mountain Forest and Range Experiment Station, Fort Collins, CO

Alexander ME, Hawksworth FG (1975) Wildland fires and dwarf mistletoes: a literature review of ecology and prescribed burning. U.S. Department of Agriculture, Forest Service General Technical Report RM-14, Rocky Mountain Forest and Range Experiment Station, Fort Collins, CO

Allen CD (1989) Changes in the landscape of the Jemez Mountains, New Mexico. Dissertation, University of California, Berkeley

Allen CD (2002) Lots of lightning and plenty of people: an ecological history of fire in the upland Southwest. In: Vale TR (ed) Fire, native peoples, and the natural landscape. Island Press, Covelo

Allen CD (2007) Interactions across spatial scales among forest dieback, fire, and erosion in northern New Mexico landscapes. Ecosystems 10:797–808

Allen CD, Breshears DD (1998) Drought-induced shift of a forest-woodland ecotone: rapid landscape response to climate variation. Proc Nat Acad Sci USA 95:14839–14842

Allen CD, Touchan R, Swetnam TW (1995) Landscape-scale fire history studies support fire management action at Bandelier. Park Sci 15:18–19

Allen CD, Savage M, Falk DA, Suckling KF, Swetnam TW, Schulke T, Stacey BP, Morgan P, Hoffman M, Klingel JT (2002) Ecological restoration of southwestern ponderosa pine ecosystems: a broad perspective. Ecol Appl 12:1418–1433

Allen CD, Anderson RS, Jass RB, Toney JL, Baisan CH (2008) Paired charcoal and tree-ring records of high-frequency Holocene fire from two New Mexico bog sites. Int J Wildland Fire 17:115–130

Andrews SR, Daniels JP (1960) A survey of dwarf mistletoes in Arizona and New Mexico. U.S. Department of Agriculture, Forest Service Station Paper No. 49, Rocky Mountain Forest and Range Experiment Station, Fort Collins, CO

Arnold JF (1950) Changes in ponderosa pine-bunchgrass ranges in Arizona resulting from pine regeneration and grazing. J For 48:118–126

Arundel TR (2000) Using General Land Office survey records to determine pre-settlement forest conditions in north-central Arizona, 1878–1879. Thesis, Northern Arizona University, Flagstaff

Bailey VO (1904) Unpublished physiography report – Gallinas Mountains. Transcribed and annotated by William deBuys. On file at U.S. Geological Survey, Jemez Mountains Field Station, Los Alamos, NM

Baker WL (2006) Fire history of ponderosa pine landscapes in Grand Canyon National Park: is it reliable enough for management and restoration? Int J Wildland Fire 15:433–437

Baker WL (2009) Fire ecology in Rocky Mountain landscapes. Island Press, Washington, DC

Bakker JD, Moore MM (2007) Controls on vegetation structure in southwestern ponderosa pine forests, 1941 and 2004. Ecology 88:2305–2319

Barrows JS (1978) Lightning fires in southwestern forests. Final report to USDA Forest Service, Northern Forest Fire Laboratory, Intermountain Forest and Range Experiment Station, under cooperative agreement 16–568 CA with Rocky Mountain Forest and Range Experiment Station, Fort Collins, CO

Barton AM (2002) Intense wildfire in southeastern Arizona: transformation of a Madrean oak-pine forest to oak woodland. For Ecol Manag 165:205–212

Bartos DL (2007) Aspen. In: Hood SM, Miller M (eds) Fire ecology and management of the major ecosystems of southern Utah. U.S. Department of Agriculture, Forest Service General Technical Report RMRS-GTR-202, Rocky Mountain Research Station, Fort Collins, CO

Bates ST (2006) A preliminary checklist of Arizona macrofungi. Canotia 2:47–78

Battaglia MA, Shepperd WD (2007) Ponderosa pine, mixed conifer, and spruce-fir forests. In: Hood SM, Miller M (eds) Fire ecology and management of the major ecosystems of southern Utah. U.S. Department of Agriculture, Forest Service General Technical Report RMRS-GTR-202, Rocky Mountain Research Station, Fort Collins, CO

Beale EF (1858) Wagon road from Fort Defiance to the Colorado River. 35th Congress, 1st Session, Senate Executive Document 124, Washington, DC

Bell DM, Parysow PF, Moore MM (2009) Assessing the representativeness of the oldest permanent inventory plots in northern Arizona ponderosa pine forests. Restor Ecol 17:369–377

Bennett DD, Schmid JM, Mata SA, and Edminster CB (1987) Growth impact of the North Kaibab pandora moth outbreak. U.S. Department of Agriculture, Forest Service Research Note RM-474, Rocky Mountain Forest and Range Experiment Station, Fort Collins, CO

Beschta RL (n.d.) Climatology of the ponderosa pine type in central Arizona. Agricultural Experiment Station Technical Bulletin 228, College of Agriculture, University of Arizona, Tucson [publication is undated, but often cited as 1976]

Biondi F (1996) Decadal-scale dynamics at the Gus Pearson Natural Area: evidence for inverse (a)symmetric competition? Can J For Res 26:1397–1406

Bowman CB (2003) Potential for ozone stress on vegetation at Grand Canyon National Park. Unpublished report, Science Center, Grand Canyon National Park, Coconino, AZ

Bradley AF, Noste NV, Fischer WC (1992) Fire ecology of forests and woodlands in Utah. U.S. Department of Agriculture, Forest Service General Technical Report INT-287, Intermountain Research Station, Ogden, UT

Breece CR, Kolb TE, Dickson BG, McMillin JD, Clancy KM (2008) Prescribed fire effects on bark beetle activity and tree mortality in southwestern ponderosa pine forests. For Ecol Manag 255:119–128

Brown PM, Wu R (2005) Climate and disturbance forcing of episodic tree recruitment in a southwestern ponderosa pine landscape. Ecology 86:3030–3038

Brown PM, Kaye MW, Huckaby LS, Baisan CH (2001) Fire history along environmental gradients in the Sacramento Mountains, New Mexico: influences of local patterns and regional processes. Ecoscience 8:115–126

Brown PM, Heyerdahl EK, Kitchen SG, Weber MH (2008) Climate effects on historical fires (1630–1900) in Utah. Int J Wildland Fire 17:28–39

Chambers CL, Holthausen RS (2000) Montane ecosystems used as rangelands. In: Jemison R, Raish C (eds) Livestock management in the American Southwest: ecology, society, and economics. Elsevier, New York

Clary WP (1975) Range management and its ecological basis in the ponderosa pine type of Arizona: the status of our knowledge. U.S. Department of Agriculture, Forest Service Research Paper RM-RP-158, Rocky Mountain Forest and Range Experiment Station, Fort Collins, CO

Clary WP, Ffolliott PF (1969) Water holding capacity of ponderosa pine forest floor layers. J Soil Water Conserv 24:22–23

Cocke AE, Fulé PZ, Crouse JE (2005) Forest change on a steep mountain gradient after extended fire exclusion: San Francisco Peaks, Arizona, USA. J Appl Ecol 42:814–823

Cooper CF (1960) Changes in vegetation, structure, and growth of southwestern pine forests since white settlement. Ecol Monogr 30:129–164

Cooper CF (1961) Pattern in ponderosa pine forests. Ecology 42:493–499

Covington WW (2003) The evolutionary and historical context. In: Friederici P (ed) Ecological restoration of southwestern ponderosa pine forests. Island Press, Washington, DC

Covington WW, Moore MM (1994a) Postsettlement changes in natural fire regimes and forest structure: ecological restoration of old-growth ponderosa pine forest. J Sustain For 2:153–181

Covington WW, Moore MM (1994b) Southwestern ponderosa forest structure: changes since Euro-American settlement. J For 92:39–47

Covington WW, Sackett SS (1984) The effect of a prescribed burn in southwestern ponderosa pine on organic matter and nutrients in woody debris and forest floor. For Sci 30:183–192

Covington WW, Sackett SS (1986) Effect of periodic burning on soil nitrogen concentrations in ponderosa pine. Soil Sci Soc Am J 50:452–457

Covington WW Sackett SS (1990) Fire effects on ponderosa pine soils and their management implications. In: Krammes JS (tech coord) Effects of fire management of southwestern natural resources. U.S. Department of Agriculture, Forest Service General Technical Report RM-191, Rocky Mountain Forest and Range Experiment Station, Fort Collins, CO

Covington WW, Sackett SS (1992) Soil mineral nitrogen changes following prescribed burning in ponderosa pine. For Ecol Manag 54:175–191

Covington WW, Fulé PZ, Moore MM, Hart SC, Kolb TE, Mast JN, Sackett SS, Wagner MR (1997) Restoring ecosystem health in ponderosa pine forests of the Southwest. J For 95:23–29

Crawford JA, Wahren CHA, Kyle S, Moir WH (2001) Responses of exotic plant species to fires in *Pinus ponderosa* forests in northern Arizona. J Veg Sci 12:261–268

Crocker-Bedford DC, Vankat JL, Bertolette DR, Leatherbury P, McKinnon T, Sipe CL (2005a) Apparent increases in mixed conifer characteristics since 1935 in ponderosa pine – mixed conifer transition forests of Grand Canyon National Park. In: van Riper C III, Mattson DJ (eds) The Colorado Plateau II: biophysical, socioeconomic, and cultural research. Proceedings of the 7th biennial conference of research on the Colorado Plateau. University of Arizona Press, Tucson

Crocker-Bedford DC, Vankat JL, Bertolette DR, Leatherbury P, McKinnon T, Sipe CL (2005b) Apparent reductions in large-diameter trees since 1935 in ponderosa pine forests of Grand Canyon National Park. In: van Riper C III, Mattson DJ (eds) The Colorado Plateau II: biophysical, socioeconomic, and cultural research. Proceedings of the 7th biennial conference of research on the Colorado Plateau. University of Arizona Press, Tucson

Dahms CW, Geils BW (tech eds) (1997) An assessment of forest ecosystem health in the Southwest. U.S. Department of Agriculture, Forest Service General Technical Report RM-GTR-295, Rocky Mountain Forest and Range Experiment Station, Fort Collins, CO

DeByle NV (1985) Animal impacts. In: DeByle NV, Winokur PR (eds) Aspen: ecology and management in the western United States. U.S. Department of Agriculture, Forest Service General Technical Report RM-119, Rocky Mountain Forest and Range Experiment Station, Fort Collins, CO

DeVelice RL, Ludwig JA, Moir WH, Ronco F Jr (1986) A classification of forest habitat types of northern New Mexico and southern Colorado. U.S. Department of Agriculture, Forest Service General Technical Report RM-131, Rocky Mountain Forest and Range Experiment Station, Fort Collins, CO

Douglas LP, Stevens RE (1979) Mountain pine beetle infestation characteristics in ponderosa pine, Kaibab Plateau, Arizona, 1975–1977. U.S. Department of Agriculture, Forest Service Research Note RM-367, Rocky Mountain Forest and Range Experiment Station, Fort Collins, CO

DuBois C (1903) Report on the proposed San Juan Forest Reserve, Colorado. Unpublished report. Supervisor's Office, San Juan National Forest, Durango, CO

Dugan AJ (2012) A landscape fire history and a modern calibration of fire-interval methods in a ponderosa pine forest across the South Rim of Grand Canyon National Park. Thesis, University of Wyoming, Laramie

Dutton CE (1882) The physical geology of the Grand Cañon district. Extract from the Annual Report of the Director of the U.S. Geological Survey – 1880–81. Government Printing Office, Washington, DC

Evans AM, Everett RG, Stephens SL, Youtz JA (2011) Comprehensive fuels treat practices guide for mixed conifer forests: California, central and southern Rockies, and the Southwest. Forest Guild, Santa Fe

Feeney SR, Kolb TE, Covington WW, Wagner MR (1998) Influence of thinning and burning restoration treatments on presettlement ponderosa pines at the Gus Pearson Natural Area. Can J For Res 28:1295–1306

Fitzhugh EL, Moir WH, Ludwig JA, Ronco F Jr (1987) Forest habitat types in the Apache, Gila, and part of the Cibola National Forests. U.S. Department of Agriculture, Forest Service General Technical Report RM-145, Rocky Mountain Forest and Range Experiment Station, Fort Collins, CO

Foxx TS (1996) Vegetation succession after the La Mesa Fire at Bandelier National Monument. In: Allen CD (tech ed) Fire effects in southwestern forests: Proceedings of the Second La Mesa Fire Symposium, Los Alamos, New Mexico, 29–31 Mar 1994. U.S. Department of Agriculture, Forest Service General Technical Report RM-GTR-286, Rocky Mountain Forest and Range Experiment Station, Fort Collins, CO

Fulé PZ, Covington WW, Moore MM (1997) Determining reference conditions for ecosystem management of southwestern ponderosa pine forests. Ecol Appl 7:895–908

Fulé PZ, Heinlein TA, Covington WW, Moore MM (2000) Continuing fire regimes in remote forests of Grand Canyon National Park. In: Cole DN, McCool SF, Borrie WT, O'Loughlin F (compilers) Proceedings: wilderness science in a time of change conference, volume 5: wilderness ecosystems, threats, and management, 23–27 May 1999, Missoula, Montana. U.S. Department

of Agriculture, Forest Service Proceedings RMRS-P-15-VOL-5, Rocky Mountain Research Station, Ogden, UT
Fulé PZ, Covington WW, Moore MM, Heinlein TA, Waltz AEM (2002) Natural variability in forests of the Grand Canyon, USA. J Biogeogr 29:31–47
Fulé PZ, Crouse JE, Heinlein TA, Moore MM, Covington WW, Verkamp G (2003) Mixed-severity fire regime in a high-elevation forest of Grand Canyon, Arizona, USA. Landsc Ecol 18:465–486
Fulé PZ, Crouse JE, Cocke AE, Moore MM, Covington WW (2004) Changes in canopy fuels and potential fire behavior 1880–2040: Grand Canyon, Arizona. Ecol Model 175:231–248
Fulé PZ, Laughlin DC, Covington WW (2005) Pine-oak forest dynamics five years after ecological restoration treatments, Arizona, USA. For Ecol Manag 218:129–145
Gaines EM, Kallander HR, Wagner JA (1958) Controlled burning in southwestern ponderosa pine: results from the Blue Mountain plots, Fort Apache Indian Reservation. J For 56:323–327
Ganey JL, Vojta SC (2011) Tree mortality in drought-stressed mixed-conifer and ponderosa pine forests, Arizona, USA. For Ecol Manag 261:162–168
Gosz JR (1980) Biomass distribution and production budget for a nonaggrading forest ecosystem. Ecology 61:507–514
Graybill DA, Rose MR (1992) Coniferous forests of Arizona and New Mexico. In: Olson RK, Binkley D, Böhm M (eds) The response of western forests to air pollution. Springer, New York
Greenamyre HH (1913) The composite type on the Apache National Forest. U.S. Department of Agriculture, Forest Service, Bulletin 125, Fort Valley Experiment Station, Washington, DC
Griffis KL, Crawford JA, Wagner MR, Moir WH (2001) Understory response to management treatments in northern Arizona ponderosa pine forests. For Ecol Manag 146:239–245
Grissino-Mayer HD, Swetnam TW (1997) Multi-century history of wildfire in the ponderosa pine forests of El Malpais National Monument. In: Mabery K (compiler) Natural history of El Malpais National Monument. Bulletin 156, New Mexico Bureau of Mines & Mineral Resources, Socorro
Grissino-Mayer HD, Swetnam TW (2000) Century-scale climate forcing of fire regimes in the American Southwest. Holocene 10:213–220
Grissino-Mayer HD, Swetnam TW, Adams RK (1997) The rare, old-aged conifers of El Malpais: their role in understanding climatic change in the American Southwest. In: Mabery K (compiler) Natural history of El Malpais National Monument. Bulletin 156, New Mexico Bureau of Mines & Mineral Resources, Socorro
Grissino-Mayer HD, Romme WH, Floyd ML, Hanna DD (2004) Climatic and human influences on fire regimes of the southern San Juan Mountains, Colorado, USA. Ecology 85:1708–1724
Haire SL, McGarigal K (2008) Inhabitants of landscape scars: succession of woody plants after large, severe forest fires in Arizona and New Mexico. Southwest Nat 53:146–161
Haire SL, McGarigal K (2010) Effects of landscape patterns of fire severity on regenerating ponderosa pine forests (*Pinus ponderosa*) in New Mexico and Arizona, USA. Landsc Ecol 25:1055–1069
Hanks JP, Fitzhugh EL, Hanks SR (1983) A habitat type classification system for ponderosa pine forests of northern Arizona. U.S. Department of Agriculture, Forest Service Technical Report RM-97, Rocky Mountain Forest and Range Experiment Station, Fort Collins, CO
Harper KT, Wagstaff FJ, Kunzler LM (1985) Biology and management of the Gambel oak vegetative type: a literature review. U.S. Department of Agriculture, Forest Service General Technical Report INT-179, Intermountain Forest and Range Experiment Station, Ogden, UT
Harrington MG (1982) Stand, fuel, and potential fire behavior characteristics in an irregular southeastern Arizona ponderosa pine stand. U.S. Department of Agriculture, Forest Service Research Note RM-418, Rocky Mountain Forest and Range Experiment Station, Fort Collins, CO
Harrington MG, Hawksworth FG (1990) Interactions of fire and dwarf mistletoe on mortality of southwestern ponderosa pine. In: Krammes JS (tech coord) Effects of fire management of southwestern natural resources. U.S. Department of Agriculture, Forest Service General Technical Report RM-191, Rocky Mountain Forest and Range Experiment Station, Fort Collins, CO

Harrington MG, Sackett SS (1990) Using fire as a management tool in southwestern ponderosa pine forests. In: Krammes JS (tech coord) Effects of fire management of southwestern natural resources. U.S. Department of Agriculture, Forest Service General Technical Report RM-191, Rocky Mountain Forest and Range Experiment Station, Fort Collins, CO

Harris GR, Covington WW (1983) The effect of a prescribed fire on nutrient concentration and standing crop of understory vegetation in ponderosa pine. Can J For Res 13:501–507

Hendricks DM (1985) Arizona Soils. University of Arizona Press, Tucson

Heyerdahl EK, Brown PM, Kitchen S, Weber MH (2011) Multicentury fire and forest histories at 19 sites in Utah and eastern Nevada. U.S. Department of Agriculture, Forest Service General Technical Report RMRS-GTR-261WWW, Rocky Mountain Research Station, Fort Collins, CO. http://www.fs.fed.us/rm/pubs/rmrs_gtr261.pdf. Accessed 12 Jan 2011

Hoffman C, Mathiasen R, Sieg CH (2007) Dwarf mistletoe effects on fuel loadings in ponderosa pine forests in northern Arizona. Can J For Res 37:662–670

Holden ZA, Morgan P, Crimmins MA, Steinhorst RK, Smith AMS (2007a) Fire season precipitation variability influences fire extent and severity in a large southwestern wilderness area, United States. Geophys Res Lett 34:1–5

Holden ZA, Morgan P, Rollins MG, Kavanagh K (2007b) Effects of multiple wildland fires on ponderosa pine stand structure in two southwestern Wilderness Areas, USA. Fire Ecol 3:18–33

Holsinger SJ (1902) The boundary line between the desert and the forest. For Irrig 8:21–27

Huffman DW, Moore MM (2004) Responses of Fendler ceanothus to overstory thinning, prescribed fire, and drought in an Arizona ponderosa pine forest. For Ecol Manag 198:105–115

Huffman DW, Fulé PZ, Pearson KM, Crouse JE (2008) Fire history of pinyon-juniper woodlands at upper ecotones with ponderosa pine forests in Arizona and New Mexico. Can J For Res 38:2097–2108

Ingersoll E (1885) The crest of the continent: a record of a summer's ramble in the Rocky Mountains and beyond. Lakeside Press, Chicago

Iniguez JM, Swetnam TW, Baisan CH (2009) Spatially and temporally variable fire regime on Rincon Peak, Arizona, USA. Fire Ecol 5:3–21

Integrated Taxonomic Information System (2012) http://www.itis.gov. Accessed 1 Apr 2012

Ireland KB, Stan AB, Fulé PZ (2012) Bottom-up control of a northern Arizona ponderosa pine forest fire regime in a fragmented landscape. Landsc Ecol 27:983–997

Ives JC (1861) Report upon the Colorado River of the West, explored in 1857 and 1858. Government Printing Office, Washington, DC

Jenkins SE, Sieg CH, Anderson DE, Kaufman DS, Pearthree PA (2011) Late Holocene geomorphic record of fire in ponderosa pine and mixed-conifer forests, Kendrick Mountain, northern Arizona, USA. Int J Wildland Fire 20:125–141

Karnosky DF, Mankovska B, Percy K, Dickson RE, Podila GK, Sober J, Noormets A, Hendrey G, Coleman MD, Kubiske M, Pregitzer KS, Isebrands JG (1999) Effects of tropospheric O_3 on trembling aspen and interaction with CO_2: results from an O_3-gradient and a FACE experiment. Water Air Soil Pollut 116:311–322

Kaufmann GA, Covington WW (2001) Effect of prescribed burning on mortality of presettlement ponderosa pines in Grand Canyon National Park. In: Vance RK, Edminster CB, Covington WW, Blake JA (compilers) Ponderosa pine ecosystems restoration and conservation: steps toward stewardship, 25–27 Apr 2000, Flagstaff, Arizona. U.S. Department of Agriculture, Forest Service Proceedings RMRS-P-22, Rocky Mountain Research Station, Ogden, UT

Kaufmann MR, Ryan KC, Fulé PZ, Romme WH (2005) Restoration of ponderosa pine forests in the interior western U.S. after logging, grazing, and fire suppression. In: Stanturf JA, Madsen P (eds) Restoration of boreal and temperate forests. CRC Press, Boca Raton

Kaufmann MR, Binkley D, Fulé PZ, Johnson M, Stephens SL, Swetnam TW (2007) Defining old growth for fire-adapted forests of the western United States. Ecol Soc 12(2):15. http://www.ecologyandsociety.org/vol12/iss2/art15/. Accessed 20 Feb 2012

Kearns HSJ, Jacobi WR, Johnson DW (2005) Persistence of pinyon pine snags and logs in southwestern Colorado. West J Appl For 20:247–252

Keeley JE, Zedler PH (1998) Evolution of life histories in *Pinus*. In: Richardson DM (ed) Ecology and biogeography of *Pinus*. Cambridge University Press, New York

Kellogg RS (1902) Forest conditions in southern Arizona. For Irrig 8:501–505

Kerns BK, Moore MM, Timpson ME, Hart SC (2003) Soil properties associated with vegetation patches in a *Pinus ponderosa*-bunchgrass mosaic. West N Am Nat 63:452–462

Klemmedson JO, Smith EL (1979) Vegetation-soil relations of forests, woodlands, and grasslands of Arizona and New Mexico. In: Youngberg CT (ed) Forest soils and land use. Department of Forest and Wood Sciences, Colorado State University, Fort Collins

Koepke DF, Kolb TE, Adams HD (2010) Variation in woody plant mortality and dieback from severe drought among soils, plant groups, and species within a northern Arizona ecotone. Oecologia 163:1079–1090

Kolb TE, Fulé PZ, Wagner MR, Covington WW (2001) Six-year changes in mortality and crown condition of old-growth ponderosa pines in ecological restoration treatments at the G.A. Pearson Natural Area. In: Vance RK, Edminster CB, Covington WW, Blake JA (compilers) Ponderosa pine ecosystems restoration and conservation: steps toward stewardship,25–27 Apr 2000, Flagstaff, Arizona. U.S. Department of Agriculture, Forest Service Proceedings RMRS-P-22, Rocky Mountain Research Station, Ogden, UT

Korb JE, Springer JD (2003) Understory vegetation. In: Friederici P (ed) Ecological restoration of southwestern ponderosa pine forests. Island Press, Washington, DC

Korb JE, Johnson NC, Covington WW (2003) Arbuscular mycorrhizal propagule densities respond rapidly to ponderosa pine restoration treatments. J Appl Ecol 40:101–110

Korb JE, Springer JD, Powers SR, Moore MM (2005) Soil seed banks in *Pinus ponderosa* forests in Arizona: clues to site history and restoration potential. Appl Veg Sci 8:103–112

Kuenzi AM, Fulé PZ, Sieg CH (2008) Effects of fire severity and pre-fire stand treatment on plant community recovery after a large wildfire. For Ecol Manag 255:855–865

Lang DK, Stewart SS (1910) Reconnaissance of the Kaibab National Forest. U.S. Department of Agriculture, Forest Service, unpublished report. http://archive.library.nau.edu/cdm4/document.php?CISOROOT=/cpa&CISOPTR=57641&CISOSHOW=57583. Accessed 7 July 2011

Laughlin DC, Abella SR (2007) Abiotic and biotic factors explain independent gradients of plant community composition in ponderosa pine forests. Ecol Model 205:231–240

Laughlin DC, Fulé PZ (2008) Wildland fire effects on understory plant communities in two fire-prone forests. Can J For Res 38:133–142

Laughlin DC, Grace JB (2006) A multivariate model of plant species richness in forested systems: old-growth montane forests with a long history of fire. Oikos 114:60–70

Laughlin DC, Bakker JD, Stoddard MT, Daniels ML, Springer JD, Gildar CN, Green AM, Covington WW (2004) Toward reference conditions: wildfire effects on flora in an old-growth ponderosa pine forest. For Ecol Manag 199:137–152

Laughlin DC, Bakker JD, Fulé PZ (2005) Understory plant community structure in lower montane and subalpine forests, Grand Canyon National Park, USA. J Biogeogr 32:2083–2102

Laughlin DC, Abella SR, Covington WW, Grace JB (2007) Species richness and soil properties in *Pinus ponderosa* forests: a structural equation modeling analysis. J Veg Sci 18:231–242

Laughlin DC, Moore MM, Fulé PZ (2011) A century of increasing pine density and associated shifts in understory plant strategies. Ecology 92:556–561

Layser EF, Schubert GH (1979) Preliminary classification for the coniferous forest and woodland series of Arizona and New Mexico. U.S. Department of Agriculture, Forest Service Research Paper RM-208, Rocky Mountain Forest and Range Experiment Station, Fort Collins, CO

Lewis DB (2003) Fire regimes of Kipuka forests in El Malpais National Monument, New Mexico. Thesis, University of Tennessee, Knoxville

Longwell CR (1958) Clarence Edward Dutton 1841–1912: a biographical memoir. National Academy of Sciences, Washington, DC. http://books.nap.edu/html/biomems/cdutton.pdf. Accessed 8 Feb 2012

Madany MH, West NE (1980) Fire history of two montane forest areas of Zion National Park. In: Stokes MA, Dieterich JH (tech coords) Proceedings of the fire history workshop, 20–24 Oct 1980, Tucson, Arizona. U.S. Department of Agriculture, Forest Service General Technical Report RM-81, Rocky Mountain Forest and Range Experiment Station, Fort Collins, CO

Madany MH, West NE (1983) Livestock grazing: fire regime interactions within montane forests of Zion National Park, Utah. Ecology 64:661–667

Madany MH, West NE (1984) Vegetation of two relict mesas in Zion National Park. J Range Manag 37:456–461

Maker HJ, Saugherty LA (1986) Soils. In: Williams JL (ed) New Mexico in maps, 2nd edn. University of New Mexico Press, Albuquerque

Margolis EQ, Balmat J (2009) Fire history and fire-climate relationships along a fire regime gradient in the Santa Fe Municipal Watershed, NM, USA. For Ecol Manag 258:2416–2430

Mast JN, Wolf JJ (2004) Ecotonal changes and altered tree spatial patterns in lower mixed-conifer forests, Grand Canyon National Park, Arizona, U.S.A. Landsc Ecol 19:167–180

Mast JN, Fulé PZ, Moore MM, Covington WW, Waltz AEM (1999) Restoration of presettlement age structure of an Arizona ponderosa pine forest. Ecol Appl 9:228–239

McGlone CM, Egan D (2009) The role of fire in the establishment and spread of nonnative plants in Arizona ponderosa pine forests: a review. J Ariz-Nev Acad Sci 41:75–86

McGlone CM, Springer JD, Covington WW (2009a) Cheatgrass encroachment on a ponderosa pine forest ecological restoration project in northern Arizona. Ecol Restor 27:37–46

McGlone CM, Springer JD, Laughlin DC (2009b) Can pine forest restoration promote a diverse and abundant understory and simultaneously resist nonnative invasion? For Ecol Manag 258:2638–2646

McGlone CM, Sieg CH, Kolb TE (2011) Invasion resistance and persistence: established plants win, even with disturbance and high propagule pressure. Biol Invasions 13:291–304

McGlone CM, Sieg CH, Kolb TE, Nietupsky T (2012a) Established native perennial grasses outcompete an invasive annual grass regardless of soil water and nutrient availability. Plant Ecol 213:445–457

McGlone CM, Stoddard MT, Springer JD, Daniels ML, Fulé PZ, Covington WW (2012b) Nonnative species influence vegetative response to ecological restoration: two forests with divergent restoration outcomes. For Ecol Manag 285:195–203

McMillin J, Wagner MR (1993) Influence of stand characteristics and site quality on sawfly population dynamics. In: Wagner MR, Raffa KF (eds) Sawfly life history adaptations to woody plants. Academic, San Diego

Menzel JP, Covington WW (1997) Changes from 1876 to 1994 in a forest ecosystem near Walnut Canyon, northern Arizona. In: van Riper C III, Deshler ET (eds) Proceedings of the 3rd biennial conference of research on the Colorado Plateau. U.S. Department of the Interior, National Park Service Transactions and Proceedings Series NPS/NRNAU/NRTP-97/12

Merriam CH (1890a) Forest trees of the San Francisco Mountain region, Arizona. In: Merriam CH, Stejneger L. Results of a biological survey of the San Francisco Mountain region and desert of the Little Colorado, Arizona. U.S. Department of Agriculture, Division of Ornithology and Mammalogy, North American Fauna No. 3, Government Printing Office, Washington, DC

Merriam CH (1890b) Part I–General results of a biological survey of the San Francisco Mountain region in Arizona, with special reference to the distribution of species. In: Merriam CH, Stejneger L. Results of a biological survey of the San Francisco Mountain region and desert of the Little Colorado, Arizona. U.S. Department of Agriculture, Division of Ornithology and Mammalogy, North American Fauna No. 3, Government Printing Office, Washington, DC

Milchunas DG (2006) Responses of plant communities to grazing in the southwestern United States. U.S. Department of Agriculture, Forest Service General Technical Report RMRS-GTR-169, Rocky Mountain Research Station, Fort Collins, CO

Miller JM, Keen FP (1960) Biology and control of the western pine beetle: a summary of the first fifty years of research. U.S. Department of Agriculture, Forest Service Miscellaneous Publication No. 800. Washington, DC

Miller KK, Wagner MR (1989) Effect of pandora moth (Lepidoptera: Saturniidae) defoliation on growth of ponderosa pine in Arizona. J Econ Entomol 82:1682–1686

Moir WH (1993) Alpine tundra and coniferous forest. In: Dick-Peddie WA New Mexico vegetation: past, present, and future. University of New Mexico Press, Albuquerque

Moir WH, Geils B, Benoit MA, Scurlock D (1997) Ecology of southwestern ponderosa pine forests. In: Block WM, Finch DM (tech eds) Songbird ecology in southwestern ponderosa pine forests:

a literature review. U.S. Department of Agriculture, Forest Service General Technical Report RM-GTR-292, Rocky Mountain Forest and Range Experiment Station, Fort Collins, CO
Möllhausen B (1858) Diary of a journey from the Mississippi to the coasts of the Pacific with a United States Government expedition, vol II (trans: Sinnett P). Longman, Brown, Green, Longmans, & Roberts, London
Moore MM, Deiter DA (1992) Stand density index as a predictor of forage production in northern Arizona pine forests. J Range Manag 45:267–271
Moore MM, Huffman DW, Fulé PZ, Covington WW, Crouse JE (2004) Comparison of historical and contemporary forest structure and composition on permanent plots in southwestern ponderosa pine forests. For Sci 50:162–176
Moore MM, Casey CA, Bakker JD, Springer JD, Fulé PZ, Covington WW, Laughlin DC (2006) Herbaceous vegetation responses (1992–2004) to restoration treatments in a ponderosa pine forest. Rangel Ecol Manag 59:135–144
Mueggler WF (1985) Forage. In: DeByle NV, Winokur PR (eds) Aspen: ecology and management in the western United States. U.S. Department of Agriculture, Forest Service General Technical Report RM-119, Rocky Mountain Forest and Range Experiment Station, Fort Collins, CO
Muldavin EH, DeVelice RL, Ronco F Jr (1996) A classification of forest habitat types: southern Arizona and portions of the Colorado Plateau. U.S. Department of Agriculture, Forest Service General Technical Report RM-GTR-287, Rocky Mountain Forest and Range Experiment Station, Fort Collins, CO
National Park Service (2002) Air quality in the National Parks, 2nd edn. U.S. Department of the Interior, National Park Service, Air Resources Division, Washington, DC
Negrón JF, McMillin JD, Anhold JA, Coulson D (2009) Bark beetle-caused mortality in a drought-affected ponderosa pine landscape in Arizona, USA. For Ecol Manag 257:1353–1362
Pase CP, Brown DE (1994) Rocky Mountain (Petran) and Madrean montane conifer forests. In: Brown DE (ed) Biotic communities: southwestern United States and northwestern Mexico. University of Utah Press, Salt Lake City
Paulsen HA Jr (1975) Range management in the central and southern Rocky Mountains: a summary of the status of our knowledge by range ecosystems. U.S. Department of Agriculture, Forest Service Research Paper RM-154, Rocky Mountain Forest and Range Experiment Station, Fort Collins, CO
Pearson GA (1923) Natural reproduction of western yellow pine in the Southwest. U.S. Department of Agriculture Bulletin 1105. Washington, DC
Pearson GA (1942) Herbaceous vegetation a factor in natural regeneration of ponderosa pine in the southwest. Ecol Monogr 12:315–338
Pearson GA (1950) Management of ponderosa pine in the Southwest, as developed by research and experimental practice. U.S. Department of Agriculture Agricultural Monograph 6, Washington, DC
Pinchot G (1947) Breaking new ground. Harcourt, Brace & Company, New York. 1998 Commemorative Edition reprint by Island Press, Washington, DC
Plummer FG (1904) Forest conditions in the Black Mesa Forest Reserve, Arizona. U.S. Geological Survey Series H, Forestry, 8, Professional Paper No. 23, Washington, DC
Pollock MM, Suckling K (1997) Presettlement conditions of ponderosa pine forests in the American Southwest. Southwest Center for Biological Diversity, Tucson
Potter LD, Krenetsky JC (1967) Plant succession with released grazing on New Mexico range lands. J Range Manag 20:145–151
Powell JW (1890) The non-irrigable lands of the arid region. Century Mag 39:915–922
Prior-Magee JS, Boykin KG, Bradford DF, Kepner WG, Lowry JH, Schrupp DL, Thomas KA, Thompson BC (2007) Ecoregional gap analysis of the southwestern United States: the Southwest regional gap analysis project final report. http://fws-nmcfwru.nmsu.edu/swregap/report/SWReGAP%20Final%20Report.pdf. Accessed 11 July 2012
Raish C, Yong W, Marzluff J (1997) Contemporary human use of southwestern ponderosa pine forests. In: Block WM, Finch DM (tech eds) Songbird ecology in southwestern ponderosa pine forests: a literature review. U.S. Department of Agriculture, Forest Service General

Technical Report RM-GTR-292, Rocky Mountain Forest and Range Experiment Station, Fort Collins, CO

Reynolds HG (1969) Aspen grove use by deer, elk, and cattle in southwestern coniferous forests. U.S. Department of Agriculture, Forest Service Research Note RM-138, Rocky Mountain Forest and Range Experiment Station, Fort Collins, CO

Roccaforte JP, Fulé PZ, Chancellor WW, Laughlin DC (2012) Woody debris and tree regeneration dynamics following severe wildfires in Arizona ponderosa pine forests. Can J For Res 42:593–604

Rollins MG, Morgan P, Swetnam T (2002) Landscape-scale controls over 20th century fire occurrence in two large Rocky Mountain (USA) wilderness areas. Landsc Ecol 17:539–557

Romme WH, Jamieson DW, Redders JS, Bigsby G, Lindsey JP, Kendall D, Cowen R, Kreykes T, Spencer AW, Ortega JC (1992) Old-growth forests of the San Juan National Forest in southwestern Colorado. In: Kauffman MR, Moir WH, Bassett RL (tech coords) Old growth forests in the Southwest and Rocky Mountain regions, proceedings of a workshop, 9–13 Mar 1992, Portal, Arizona. U.S. Department of Agriculture, Forest Service General Technical Report RM-213, Rocky Mountain Forest and Range Experiment Station, Fort Collins, CO

Romme WH, Floyd ML, Hanna D, Grissino-Mayer HD, Green D, Redders JS (2009) Ponderosa pine forests. In: Romme WH, Floyd ML, Hanna D (eds) Historical range of variability and current landscape condition analysis: south central highlands section, southwestern Colorado & northwestern New Mexico. Colorado Forest Restoration Institute, Colorado State University and Region 2, U.S. Forest Service, Washington, DC

Roos CI, Swetnam TW (2012) A 1416-year reconstruction of annual, multidecadel, and centennial variability in area burned for ponderosa pine forests of the southern Colorado Plateau region, Southwest USA. Holocene 22:281–291

Roos CI, Sullivan AP III, McNamee C (2010) Paleoecological evidence for systematic indigenous burning in the upland Southwest. In: Dean RM (ed) The archaeology of anthropogenic environments. Center for Archaeological Investigations, Occasional Paper No. 37, Southern Illinois University, Carbondale

Rothrock JT (1875) Preliminary botanical report, with remarks upon the general topography of the region traversed in New Mexico and Arizona; its climatology, forage-plants, timber, irrigation, sanitary conditions, &c. In: Wheeler GM Annual report upon the geographical explorations and surveys west of the one hundredth meridian, in California, Nevada, Utah, Arizona, Colorado, New Mexico, Wyoming, and Montana. Appendix FF of the Annual Report of the Chief of Engineers for 1874, Government Printing Office, Washington, DC

Rusby HH (1889) General floral features of the San Francisco and Mogollon Mts. of Arizona and New Mexico, and their adjacent regions. Trans N Y Acad Sci 8:76–81

Sackett SS, Haase SM, Harrington MG (1996) Lessons learned from fire use for restoring southwestern ponderosa pine ecosystems. In: Covington WW, Wagner PK (tech coords) Conference on adaptive ecosystem restoration and management: restoration of Cordilleran conifer landscapes of North America. U.S. Department of Agriculture, Forest Service General Technical Report RM-GTR-278, Rocky Mountain Forest and Range Experiment Station, Fort Collins, CO

Sánchez Meador AJ, Moore MM, Bakker JD, Parysow PF (2009) 108 years of change in spatial pattern following selective harvest of a *Pinus ponderosa* stand in northern Arizona, USA. J Veg Sci 20:79–90

Sánchez Meador AJ, Parysow PF, Moore MM (2010) Historical stem-mapped permanent plots increase precision of reconstructed reference data in ponderosa pine forests of northern Arizona. Restor Ecol 18:224–234

Sánchez-Martínez G, Wagner MR (2002) Bark beetle community structure under four ponderosa pine forest stand conditions in northern Arizona. For Ecol Manag 170:145–160

Savage M (1991) Structural dynamics of a southwestern pine forest under chronic human influence. Ann Assoc Am Geogr 81:271–289

Savage M, Mast JN (2005) How resilient are southwestern ponderosa pine forests after crown fires? Can J For Res 35:967–977

Savage M, Brown PM, Feddema J (1996) The role of climate in a pine forest regeneration pulse in the southwestern United States. Ecoscience 3:310–318

Schmid JM, Bennett DD (1988) The North Kaibab pandora moth outbreak, 1978–1984. U.S. Department of Agriculture, Forest Service General Technical Report RM-153, Rocky Mountain Forest and Range Experiment Station, Fort Collins, CO

Schubert GH (1974) Silviculture of southwestern ponderosa pine: the status of our knowledge. U.S. Department of Agriculture, Forest Service Research Paper RM-123, Rocky Mountain Forest and Range Experiment Station, Fort Collins, CO

Scurlock D, Finch DM (1997) A historical review. In: Block WM, Finch DM (tech eds) Songbird ecology in southwestern ponderosa pine forests: a literature review. U.S. Department of Agriculture, Forest Service General Technical Report RM-GTR-292, Rocky Mountain Forest and Range Experiment Station, Fort Collins, CO

Sieg CH, Phillips BG, Moser LP (2003) Exotic invasive plants. In: Friederici P (ed) Ecological restoration of southwestern ponderosa pine forests. Island Press, Washington, DC

Stephens SL, Fulé PZ (2005) Western pine forests with continuing frequent fire regimes: possible reference sites for management. J For 103:357–362

Stoddard MT, McGlone CM (2008) Effects of ecological restoration alternative treatments on nonnative plant species establishment. In: Olberding SD, Moore MM (tech coords) Fort Valley Experimental Forest: a century of research 1908–2008. U.S. Department of Agriculture, Forest Service Proceedings RMRS-P-53CD, Rocky Mountain Research Station, Fort Collins, CO

Stoddard MT, McGlone CM, Fulé PZ, Laughlin DC, Daniels ML (2011) Native plants dominate understory vegetation following ponderosa pine forest restoration treatments. West N Am Nat 71:206–214

Stone JE, Kolb TE, Covington WW (1999) Effects of restoration thinning on presettlement *Pinus ponderosa* in northern Arizona. Restor Ecol 7:172–182

Strom BA, Fulé PZ (2007) Pre-wildfire fuel treatments affect long-term ponderosa pine forest dynamics. Int J Wildland Fire 16:128–138

Swetnam TW, Baisan CH (1996) Historical fire regime patterns in the southwestern United States since AD 1700. In: Allen CD (tech ed) Fire effects in southwestern forests: Proceedings of the Second La Mesa Fire Symposium, Los Alamos, New Mexico, 29–31 Mar 1994. U.S. Department of Agriculture, Forest Service General Technical Report RM-GTR-286, Rocky Mountain Forest and Range Experiment Station, Fort Collins, CO

Swetnam TW, Baisan CH (2003) Tree-ring reconstructions of fire and climate history in the Sierra Nevada and southwestern United States. In: Veblen TT, Baker WL, Montenegro G, Swetnam TW (eds) Fire and climatic change in temperate ecosystems of the western Americas. Springer, New York

Swetnam TW, Betancourt JL (1990) Fire-Southern Oscillation relations in the southwestern United States. Science 249:1017–1020

Swetnam TW, Betancourt JL (1998) Mesoscale disturbance and ecological response to decadal climatic variability in the American Southwest. J Clim 11:3128–3147

Swetnam TW, Baisan CH, Kaib JM (2001) Forest fire histories of the sky islands of La Frontera. In: Webster GL, Bahre CJ (eds) Changing plant life of La Frontera: observations on vegetation in the US/Mexico borderlands. University of New Mexico Press, Albuquerque

Touchan R, Allen CD, Swetnam TW (1996) Fire history and climatic patterns in ponderosa pine and mixed-conifer forests of the Jemez Mountains, northern New Mexico. In: Krammes JS (tech coord) Effects of fire management of southwestern natural resources. U.S. Department of Agriculture, Forest Service General Technical Report RM-191, Rocky Mountain Forest and Range Experiment Station, Fort Collins, CO

U.S. Forest Service (2010) 2010 tornado event: large scale event recovery plan. U.S. Department of Agriculture, Forest Service, Southwest Region, Coconino National Forest, Flagstaff/Mogollon Ranger Districts. http://a123.g.akamai.net/7/123/11558/abc123/forestservic.download.akamai.com/11558/www/nepa/75678_FSPLT2_032314.pdf. Accessed 15 Feb 2012

U.S. Forest Service (2011a) Environmental assessment: Post-tornado resource protection and recovery project, Coconino National Forest. U.S. Department of Agriculture, Forest Service, Southwestern Region. http://a123.g.akamai.net/7/123/11558/abc123/forestservic.download.akamai.com/11558/www/nepa/75678_FSPLT2_035459.pdf. Accessed 15 Feb 2012

U.S. Forest Service (2011b) Forest insect and disease conditions in the Southwestern Region, 2010. U.S. Department of Agriculture, Forest Service Southwestern Region Forestry and Forest Health PR-R3-16-7. http://www.fs.usda.gov/Internet/FSE_DOCUMENTS/stelprdb5316384.pdf. Accessed 5 Mar 2012

Vankat JL (2010) Forest changes in Grand Canyon National Park and implications for management. Report to U.S. Department of the Interior, National Park Service, Grand Canyon National Park, Coconino, Arizona

Vankat JL (2011) Post-1935 changes in forest vegetation of Grand Canyon National Park, Arizona, USA: Part 1—Ponderosa pine forest. For Ecol Manag 261:309–325

Villanueva-Díaz J, McPherson GR (1995) Forest stand structure in mountains of Sonora, Mexico and New Mexico, USA. In: DeBano LF, Ffolliott PF, Ortega-Rubio A, Gottfried GJ, Hamre RH, Edminster CB (tech coords) Biodiversity and management of the Madrean archipelago: the sky islands of southwestern United States and northwestern Mexico. U.S. Department of Agriculture, Forest Service, General Technical Report RM-GTR-264, Rocky Mountain Forest and Range Experiment Station, Fort Collins, CO

Wallin KF, Kolb TE, Skov KR, Wagner MR (2004) Seven-year results of thinning and burning restoration treatments on old ponderosa pines at the Gus Pearson Natural Area. Restor Ecol 12:239–247

Weaver H (1951a) Fire as an ecological factor in the southwestern ponderosa pine forests. J For 49:93–98

Weaver H (1951b) Observed effects of prescribed burning on perennial grasses in the ponderosa pine forests. J For 49:93–98

Weaver H (1959) Ecological changes in the ponderosa pine forest of the Warm Springs Indian Reservation in Oregon. J For 57:15–20

Weng C, Jackson ST (1999) Late Glacial and Holocene vegetation history and paleoclimate of the Kaibab Plateau, Arizona. Palaeogeogr Palaeoclimatol Palaeoecol 153:179–201

White AS (1985) Presettlement regeneration patterns in a southwestern ponderosa pine stand. Ecology 66:589–594

White CS (1986) Effects of prescribed fire on rates of decomposition and nitrogen mineralization in a ponderosa pine ecosystem. Biol Fertil Soils 2:87–95

White CS (1996) The effects of fire on nitrogen cycling processes within Bandelier National Monument, NM. In: Allen CD (tech ed) Fire effects in southwestern forests: proceedings of the second La Mesa fire symposium, Los Alamos, New Mexico, 29–31 Mar 1994. U.S. Department of Agriculture, Forest Service General Technical Report RM-GTR-286, Rocky Mountain Forest and Range Experiment Station, Fort Collins, CO

Williams MA, Baker WL (2011) Testing the accuracy of new methods for reconstructing historical structure of forest landscapes using GLO survey data. Ecol Monogr 81:63–88

Williams MA, Baker WL (2012) Spatially extensive reconstructions show variable-severity fire and heterogeneous structure in historical western United States dry forests. Glob Ecol Biogeogr 21:1042–1052

Williams MA, Baker WL (2013) Variability of historical forest structure and fire across ponderosa pine landscapes of the Coconino Plateau and South Rim of Grand Canyon National Park, Arizona, USA. Landsc Ecol 28:297–310

Wilson JL, Tkacz BM (1996) Historical perspectives on forest insects and pathogens in the Southwest: implications for restoration of ponderosa pine and mixed conifer forests. In: Covington WW, Wagner PK (tech coords) Conference on adaptive ecosystem restoration and management: restoration of Cordilleran conifer landscapes of North America. U.S. Department of Agriculture, Forest Service General Technical Report RM-GTR-278, Rocky Mountain Forest and Range Experiment Station, Fort Collins, CO

Wood RE (1983) Mortality caused by root diseases and associated pests on six National Forests in Arizona and New Mexico. U.S. Department of Agriculture, Forest Service Southwestern Region Forest Pest Management Report R3-83-13, Albuquerque, NM

Woolsey TS Jr (1911) Western yellow pine in Arizona and New Mexico. U.S. Department of Agriculture, Forest Service Bulletin 101, Washington, DC

Youngblood AP, Mauk RL (1985) Coniferous forest habitat types of central and southern Utah. U.S. Department of Agriculture, Forest Service General Technical Report INT-187, Intermountain Research Station, Ogden, UT

Zausen GL, Kolb TE, Bailey JD, Wagner MR (2005) Long-term impacts of stand management on ponderosa pine physiology and bark beetle abundance in northern Arizona: a replicated landscape study. For Ecol Manag 218:291–305

Zegler TJ, Moore MM, Fairweather ML, Ireland KB, Fulé PZ (2012) *Populus tremuloides* mortality near the southwestern edge of its range. For Ecol Manag 282:196–207

Part II
Woodland, Savanna, Grassland, and Shrublands

Chapter 5
Pinyon-Juniper Vegetation

Abstract Pinyon-Juniper vegetation covers more area on the mountains and plateaus of the American Southwest than all other vegetation types combined. It occurs in an elevational band below Ponderosa Pine Forest and above desert shrublands and semi-desert grasslands. Stands are dominated by pinyons and/or junipers of shorter height than forest tree species at higher-elevation. Stands have diverse structure, composition, and ecology and are categorized as persistent woodland, wooded shrubland, and savanna. The key driver of Pinyon-Juniper vegetation is variation in climate, which interacts with other drivers such as insect outbreaks to affect tree regeneration and mortality. Persistent woodland also experiences infrequent, high-intensity, stand-replacing fires. Anthropogenic disturbances include livestock grazing, fire management, modern climate change, invasive species, recreation, and nearby land use. Vegetation dynamics have been dominated by infill, expansion, and succession. Historical conditions are poorly known, but trees have increased in density in many stands since the late nineteenth century (infill) and have invaded adjacent grasslands (expansion). Understories have been altered by livestock grazing and increased tree densities. Vegetation dynamics are illustrated in a nested, three-tiered set of conceptual models. Key conclusions and challenges for researchers and land managers are summarized.

5.1 Introduction

Pinyon-Juniper (*Pinus-Juniperus*) vegetation is dominated by relatively short trees typically only 3–10 m (10–33 ft) in height (Figs. 5.1 and 5.2). Other names for Pinyon-Juniper vegetation are Piñon-Juniper woodland and pygmy conifer woodland. It occurs at elevations below Ponderosa Pine Forest (*Pinus ponderosa*) and above desert shrublands and semi-desert grasslands. Pinyon-Juniper vegetation covers more area on the mountains and plateaus of the American Southwest than all other vegetation types combined, with approximately 153,100 km^2 (59,100

J.L. Vankat, *Vegetation Dynamics on the Mountains and Plateaus of the American Southwest*, Plant and Vegetation 8, DOI 10.1007/978-94-007-6149-0_5,

Fig. 5.1 Pinyon-Juniper vegetation near Sedona in central Arizona (Photograph by Betty J. Huffman)

Fig. 5.2 Pinyon-Juniper vegetation in central Utah (Photograph by author)

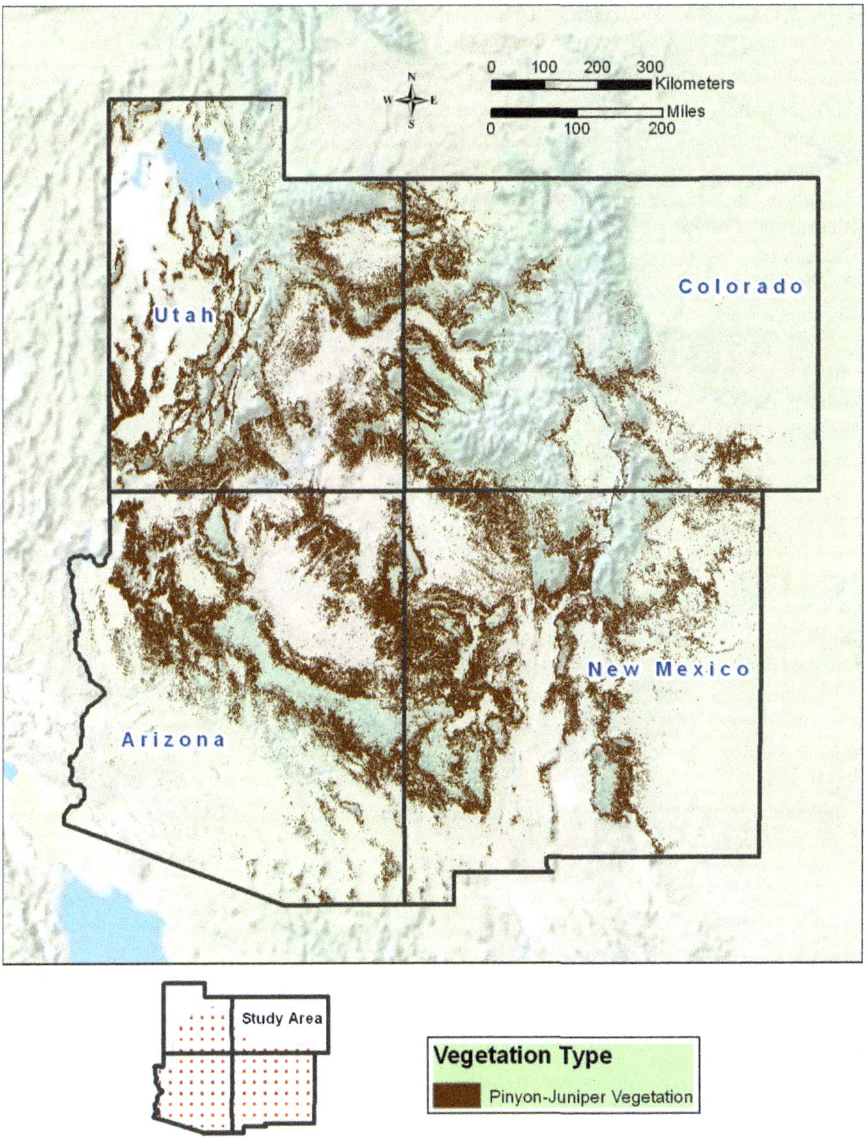

Fig. 5.3 Distribution of Pinyon-Juniper vegetation in the American Southwest. The map shows all of the states of Arizona, New Mexico, Colorado, and Utah, and the American Southwest is illustrated in *red* on the small map. (Source: U.S. Geological Survey National Gap Analysis Program 2005 Southwest Regional Gap Analysis Project, RS/GIS Laboratory, College of Natural Resources, Utah State University, Logan, Utah. Map prepared by Monica Swihart)

miles^2), which is 19.6 % of the American Southwest (Fig. 5.3; calculations based on Prior-Magee et al. 2007). Unlike other southwestern vegetation types that show latitudinal or longitudinal variation in area, Pinyon-Juniper is approximately equally extensive throughout the American Southwest. Pinyon-Juniper vegetation is also

Table 5.1 Common and scientific names of species in this chapter. Primary sources: Integrated Taxonomic Information System (2012) for plants and animals and Bates (2006) for fungi

Plants	
Alligator juniper	*Juniperus deppeana* Steud.
Antelope bitterbrush	*Purshia tridentata* (Pursh) DC.
Big bluegrass	*Poa secunda* J. Presl
Big sagebrush	*Artemisia tridentata* Nutt.
Bluebunch wheatgrass	*Pseudoroegneria spicata* (Pursh) Á. Löve
Blue grama	*Bouteloua gracilis* (Kunth) Lag. ex Griffiths
Canada thistle	*Cirsium arvense* (L.) Scop.
Cheatgrass	*Bromus tectorum* L.
Colorado pinyon	*Pinus edulis* Engelm.
Fir	*Abies* P. Mill.
Gambel oak	*Quercus gambelii* Nutt.
Goosefoot	*Chenopodium* L.
Grama	*Bouteloua* Lag.
Indiangrass	*Sorghastrum nutans* (L.) Nash
Juniper	*Juniperus* L.
Juniper mistletoe	*Phoradendron juniperinum* Engelm. ex Gray
Mexican pinyon	*Pinus cembroides* Zucc.
Mountain mahogany	*Cercocarpus montanus* Raf.
Musk thistle	*Carduus nutans* L.
Oneseed juniper	*Juniperus monosperma* (Engelm.) Sarg.
Pinyon dwarf mistletoe	*Arceuthobium divaricatum* Engelm.
Pinyon	*Pinus* L.
Ponderosa pine	*Pinus ponderosa* Douglas ex P. Lawson & C. Lawson
Sagebrush	*Artemisia* L.
Singleleaf pinyon	*Pinus monophylla* Torr. & Frém.
Spruce	*Picea* A. Dietr.
Sunflower	*Helianthus* L.
Utah juniper	*Juniperus osteosperma* (Torr.) Little
Utah serviceberry	*Amelanchier utahensis* Koehne
Western juniper	*Juniperus occidentalis* Hook.
Western wheatgrass	*Pascopyrum smithii* (Rydb.) Barkworth & D.R. Dewey
Animals	
Buffalo	*Bison bison* Linnaeus, 1758
Deer	*Odocoileus* Rafinesque, 1832
Elk	*Cervus elaphus* Linnaeus, 1758
Pinyon bark beetle	*Ips confusus* LeConte, 1876
Pronghorn antelope	*Antilocapra americana* Ord, 1815
Fungi	
Armillaria root rot	*Armillaria* (Fr.:Fr.) Staude
Black stain root disease	*Leptographium wageneri* var. *wageneri* (W.B. Kendr.) M.J. Wingf., 1985

widespread in other areas of western North America, especially in the Great Basin of the United States and in Mexico.

Southwestern Pinyon-Juniper vegetation is highly diverse in its physical environments and biological components (e.g., Springfield 1976). There are regional differences in dominant tree species associated with differences in the proportion of summer (monsoonal) precipitation. The most widespread tree species are Colorado pinyon (*Pinus edulis*) and Utah juniper (*Juniperus osteosperma*). From northwest to southeast, Colorado pinyon is replaced by singleleaf pinyon (*P. monophylla*), which in turn is replaced by Mexican pinyon (*P. cembroides*) along a gradient of increasing proportion of summer precipitation (Fig. 1.17; see tree distribution maps in Romme et al. 2009a). Similarly, Utah juniper is replaced by oneseed juniper (*J. monosperma*), which in turn is replaced by alligator juniper (*J. deppeana*) along this gradient. Some stands are dominated by either pinyon or juniper, with juniper dominance common at lower elevations and on drier sites.

There are large regional and local differences in shrubs and herbaceous plants. Shrubs tend to be more abundant in central and northern areas and include big sagebrush (*Artemisia tridentata*) and other sagebrush species at low elevations and Gambel oak (*Quercus gambelii*) and mountain-mahogany (*Cercocarpus montanus*) at high elevations. Understories, particularly in the south and east, are often dominated by warm-season grasses such as blue grama (*Bouteloua gracilis*) and other gramas. Biological soil crusts composed of mosses, lichens, microfungi, algae, and cyanobacteria can be present. Crusts occur on and slightly below the soil surface in sites where tree canopies are open and thick litter is absent (Belnap and Lange 2001).

Southwestern Pinyon-Juniper vegetation also has great variation in structure. Despite common use of the term "woodland", stand structure ranges from closed-canopy forest to open-canopy woodland, shrubland, and savanna. Different authors classify Pinyon-Juniper stands differently (e.g., Moir and Carleton 1987; Dick-Peddie 1993a, b; Romme et al. 2003, 2007, 2008, 2009a; Jacobs et al. 2008). The classification used in this chapter divides Pinyon-Juniper into three broadly variable types – persistent woodland, wooded shrubland, and savanna – based on differences in canopy structure, understory characteristics, and historical disturbance regimes (Romme et al. 2008, 2009a).

Persistent woodland (Fig. 5.4) is the most widespread Pinyon-Juniper type in the American Southwest, and stands occur on sites where soils and climate are consistently favorable to pinyons and/or junipers (Romme et al. 2008, 2009a). Canopy cover and tree size range from a closed or nearly closed canopy of relatively tall trees on more productive sites to a very open canopy of small trees on poor substrates. Persistent woodlands occur on various topographic sites and substrates, but are usually on uplands with shallow, poorly developed, rocky soils (Jacobs et al. 2008), including sites where trees grow in cracks in exposed bedrock. The relationship with uplands is less pronounced northward where winter moisture dominates (Jacobs 2011). Understory cover is influenced by canopy cover and soil, but is frequently sparse and consists of shrubs, subshrubs, forbs, and grasses. Biological soil crusts are present, but bare soil or rock can dominate between trees, and litter dominates below trees. Some upper-elevation, more-mesic sites have greater tree cover and have been considered Pinyon-Juniper forest (Moir and Carleton 1987; Dick-Peddie 1993b). This book includes these sites with persistent woodlands.

Wooded shrubland (Fig. 5.5) has a well-developed shrub layer below a moderately open to sparse tree canopy. Dominant shrubs include big sagebrush, other

Fig. 5.4 Stand of Pinyon-Juniper persistent woodland in South Rim region of Grand Canyon National Park, north-central Arizona (Photograph by Betty J. Huffman)

Fig. 5.5 Stand of Pinyon-Juniper wooded shrubland in northeastern Arizona (Photograph by Betty J. Huffman)

Fig. 5.6 Stand of Pinyon-Juniper savanna in north-central Arizona (Photograph by author)

sagebrush species, Gambel oak, and mountain-mahogany. The herbaceous cover of grasses and forbs differs among stands. As with persistent woodlands, wooded shrublands occur on various topographic sites and substrates. These range from mountain slopes with rocky, shallow soils to valleys with deep, finer soils. Wooded shrublands occur throughout the American Southwest, but most commonly in Utah where winter precipitation exceeds summer precipitation (Romme et al. 2008, 2009a).

Savanna (i.e., wooded grassland; Fig. 5.6) has a nearly continuous understory of grasses and forbs below a moderately open to open tree canopy. Shrubs are sometimes present, but usually are not abundant. The most common trees are Colorado pinyon and/or oneseed or alligator juniper. Savannas occur where conditions are favorable for grasses, including gentle uplands and valleys where soils are moderately deep and fine- to coarse-textured. They are especially common in basins and on foothills of central and southern Arizona and New Mexico, where summer rainfall predominates. Pinyon-Juniper savanna can be considered a low elevation, dry variant of woodland (Moir and Carleton 1987) that is ecotonal with grassland (Dick-Peddie 1993b), but savanna also can occur at higher elevations adjacent to Ponderosa Pine Forest.

There is extensive literature on Pinyon-Juniper vegetation. This chapter is almost entirely based on research from the American Southwest and review papers applicable to the region.

5.2 Drivers

Key drivers of southwestern Pinyon-Juniper vegetation are landscape, climate, soil, animals, natural disturbance, and anthropogenic disturbance. Each driver is important in vegetation dynamics, and therefore all are incorporated in the conceptual models (Sect. 5.5).

5.2.1 Landscape

Pinyon-Juniper vegetation occurs at approximately 1,300–2,200 m (4,300–7,200 ft) elevation in the mid-latitudes of its distribution in the American Southwest (Fig. 5.7). Transitions with Ponderosa Pine Forest at high elevations, desert shrublands and semi-desert grasslands at low elevations, and Interior Chaparral Shrubland at all elevations often occur as mosaics of stands. Alternatively, the transition with Ponderosa Pine Forest can be gradual where pinyons and junipers occur as subcanopy species in Ponderosa Pine Forest and where understory species also overlap. The disturbance regimes of these adjacent vegetation types, along with competitive

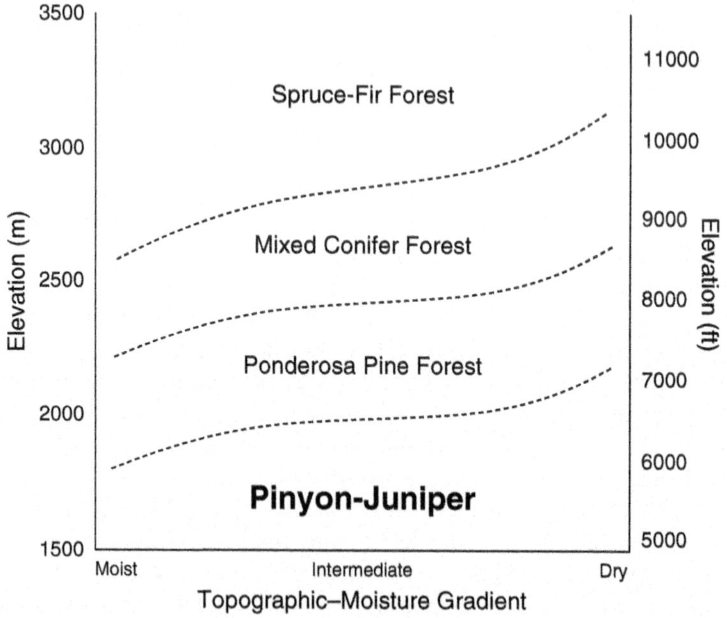

Fig. 5.7 Ecological distribution of Pinyon-Juniper vegetation on the mountains and plateaus of the American Southwest along gradients in elevation and topographic-moisture. Elevations are approximate and generally representative of mid-latitudes of the region, i.e., landscapes in northern Arizona and northern New Mexico. Lower elevations have desert shrublands and semi-desert grasslands

Fig. 5.8 Monsoonal precipitation falling on Pinyon-Juniper vegetation on the South Rim region of Grand Canyon National Park, north-central Arizona (Photograph by Mark Jefferson)

interactions among species, likely have been important in determining ecotones (Johnsen 1962; Allen and Breshears 1998; Chambers et al. 1999; Jacobs 2008).

As mentioned in Sect. 5.1, topography can be important in the distribution of Pinyon-Juniper vegetation. Persistent woodlands are most commonly on uplands, wooded shrublands are on mountain slopes and in valleys, and savannas are in valley basins and on foothills. Dense persistent woodlands are often associated with rugged slopes, canyons, and mesa tops where barriers prevent the spread of fire. Pinyons are usually more abundant than junipers at higher elevations; vice-versa at lower elevations. This is due to differential tree mortality rather than differential seedling establishment (Martens et al. 2001). The differences in mortality between pinyons and junipers are thought to be related to species' rooting patterns that affect acquisition of water. The differences in rooting patterns increase with tree size/age and thus affect tree mortality more than seedling establishment.

5.2.2 *Climate*

Mean annual precipitation is 25–55 cm (10–22 in.), with some winter precipitation falling as snow, yet little snow accumulates (Paulsen 1975; Springfield 1976; Brown 1994; Milchunas 2006; Ffolliott and Gottfried 2008). The seasonal pattern is typical of the American Southwest, with dry springs and early summers, monsoonal precipitation in July-August (Fig. 5.8), moderate precipitation in early fall, and high amounts of frontal precipitation in late fall and winter (Sect. 1.2.2). Amounts and seasonality of precipitation vary by latitude, elevation, and local topography (Springfield 1976; Milchunas 2006). For example, the monsoon season accounts for ~50 % of annual precipitation in most of New Mexico and southern Arizona (Fig. 1.17), and this decreases with increasing latitude to ~30 % in western and northern Arizona and southwestern Colorado and to ~20–30 % in southern and central Utah. Precipitation amounts, especially winter precipitation, also vary on annual and decadal time scales, as related to the El Niño-Southern Oscillation (ENSO; Sect. 1.2.2; Gottfried et al. 1995), and this variation in climate has major effects on the dynamics of southwestern Pinyon-Juniper vegetation (Sect. 5.2.5.3).

Summer temperatures are warm, with high temperatures typically reaching ~32 ° C (90 ° F; Ffolliott and Gottfried 2008). Cool to cold winter temperatures result in freezing temperatures on 150 or more days a year and result in a growing season averaging 120 days in central Utah (Price and Evans 1937) and longer in the south and at lower elevations. Wind has important effects on evapotranspiration and soil erosion (Miller 2005).

Modern climate change is described in Sect. 1.6.3 and is treated as an anthropogenic disturbance in Sect. 5.2.6.3.

5.2.3 Soil

Pinyon-Juniper vegetation occurs on a wide range of soils derived from a variety of parent materials (Paulsen 1975), but soils are generally rocky and thin (Brown 1994). Soils have a mesic temperature regime and an ustic moisture regime (Klemmedson and Smith 1979). Major soil suborders are Ustolls, Ustalfs, Orthids, Orthents, Argids, and Usterts (Klemmedson and Smith 1979). Differences based on parent material are more important than in less-arid vegetation (Miller 2005). Surface rocks can limit tree cover (Harper et al. 2003). Soils have major effects on vegetation (such as influencing canopy cover and shrub cover), as is apparent from research on relict sites (Mason et al. 1967; Thatcher and Hart 1974).

Important biological components of soils include fungal pathogens, especially black stain root disease (*Leptographium wageneri* var. *wageneri*) and Armillaria root rot (*Armillaria* spp.). Their roles as agents of natural disturbance are described in Sect. 5.2.5.2.

5.2.4 Animals

Native animals affecting the dynamics of Pinyon-Juniper vegetation include pinyon bark beetle (*Ips confusus*), which has important impacts (Sect. 5.2.5.2). Common large native ungulates include deer (*Odocoileus* spp.) and elk (*Cervus elaphus*), as well as pronghorn antelope (*Antilocapra americana*) in open sites and adjacent grassland patches. Impacts of these native ungulates on vegetation dynamics are unknown.

5.2.5 Natural Disturbance

The natural disturbance regime of Pinyon-Juniper vegetation in the American Southwest is dominated by the interactions of fire, insects, and drought. In contrast to forests at higher elevation (e.g., see Sect. 2.2.5.2), there is little evidence that wind is an important disturbance factor acting directly on Pinyon-Juniper vegetation. This is likely because of the relatively short height and small crowns of the trees. Nevertheless, windthrow of pinyons does occur (Romme et al. 2009b), and wind importantly influences the spread of fire.

5.2.5.1 Fire

Fire as a factor in southwestern Pinyon-Juniper vegetation was recognized as early as the mid nineteenth century:

> Again, we spotted the scene of a [juniper] forest, the trees of which all had died at the same time, namely I assume as a result of an extended fire… (Möllhausen in Shaw HG 2006, for northwestern Arizona in 1858)

Eight percent of all lightning fires in tree-dominated vegetation on National Forests of Arizona and New Mexico occur in woodlands, including 21 % of lightning fires reaching at least 4 ha (10 acres; Barrows 1978, for 1960–1974). Although much insight into fire regimes has been obtained since the beginning of the twenty-first century, the fire regime of southwestern Pinyon-Juniper vegetation continues to be a subject of speculation.

It appears that Pinyon-Juniper vegetation has a range of fire regimes reflective of diverse tree densities, understory structure and composition, and site conditions (cf. Baker and Shinneman 2004). Studies of Pinyon-Juniper fire regimes are challenging. Although pinyons and junipers can form fire scars (Baker and Shinneman 2004), fires frequently kill pinyons rather than scar them. Also, junipers often have false and missing rings (Milchunas 2006) as a result of weather. Therefore, it is difficult to cross-date fires and determine fire-return intervals. Pinyon-Juniper fire history usually is reconstructed using indirect methods, such as dating the age structure of homogeneous patches of trees adjacent to burned snags to determine the minimum time since the last fire.

Formerly, Pinyon-Juniper persistent woodlands of the American Southwest were thought to have had a historical fire regime of frequent, low-severity surface fires (see examples in Baker and Shinneman 2004). Moreover, it was said this fire regime was altered, first by livestock grazing that reduced herbaceous fuels that formerly carried surface fires and later by fire suppression policies of land-management agencies. This scenario is similar to historical changes in the fire regime of Ponderosa Pine Forest (see Sects. 4.2.6.1 and 4.2.6.2). This view is challenged by observations that (a) continuous fuels are lacking, (b) pinyons and junipers are thin-barked and therefore relatively fire-intolerant, (c) trees with multiple fire scars are rare on most sites (Baker and Shinneman 2004; Romme et al. 2009a), and (d) many trees have low branches that would ladder fires into tree crowns. Instead, the presence of charred snags evidences high-intensity, stand-replacing fires in the past, as noted in the above-quoted historical statement by Möllhausen. Stand-replacing fires also occur today (Fig. 5.9a, b).

Research on the historical fire regime of Pinyon-Juniper persistent woodlands in Mesa Verde National Park (Floyd et al. 2000, 2004), Glen Canyon National Recreation Area in south-central Utah (Floyd et al. 2008), and the Uncompahgre Plateau of west-central Colorado (Shinneman and Baker 2009) has concluded that low-severity surface fires are unlikely to have been important in the historical fire regime. Instead, stand age structures (especially maximum tree ages) across the landscapes indicated stand-replacing fires with turnover times of 400–600 years or longer. Such fires were not limited by ignitions, because scattered burned snags evidence lighting ignitions that did not spread (Floyd et al. 2008). Limiting factors for stand-replacing fires appear to be weather related, because these fires generally have followed dry winters and/or springs (Floyd et al. 2004) and require strong winds for spreading (Floyd et al. 2008). Soils (which influence shrub cover) and topography also can affect the distribution of high-severity fires (Arnold et al. 1964; Floyd et al. 2008). The frequency and extent of high-severity fires in Mesa Verde National Park near the turn of the twenty-first century much exceeded findings for 1700–1900, but that was attributed to weather, not fire exclusion (Floyd et al. 2004). Indeed, large fires have become increasingly frequent in western North America as climate warming has lengthened fire seasons (Westerling et al. 2006).

It is possible that persistent woodlands elsewhere in the Southwest have variations or alternatives to a regime of infrequent high-severity fires. For example, it has been suggested that persistent woodlands at the ecotone between Pinyon-Juniper and Ponderosa Pine Forest have a surface fire regime (e.g., Allen 1989). Baker and Shinneman (2004) reviewed the evidence and found only two studies that had cross-dated fires in order to determine if trees had burned in the same year: Allen (1989) and Baisan and Swetnam (1997). However, neither study had confirmed that unscarred trees had survived the fires and were not of more recent origin. A more recent study of ecotones in north-central Arizona and north-central New Mexico (Huffman et al. 2008a) addressed the survival issue. It was determined that surface fires had not spread far from Ponderosa Pine Forest into Pinyon-Juniper woodland (the terminology "Pinyon-Juniper woodland" here and hereafter follows its use by the authors of the study cited; in many cases it is equivalent to "Pinyon-Juniper persistent woodland"). Rather, there was an abrupt difference in fire regimes at the ecotone, likely because of differences in microclimates and soils that affected vegetation cover and fuel characteristics. In addition, they found that patches of Colorado pinyons with similar maximum ages were small and the maximum ages differed among patches. This indicated multiple, patchy, stand-replacing fires (Despain and Mosley 1990 also described patchy historical fires in ecotonal stands in north-central Arizona). This patchiness suggests fuel discontinuities between patches, and this was supported by an associated study in which management-ignited fires failed to spread (Huffman et al. 2009).

Fig. 5.9 (**a**) Crown fire in Pinyon-Juniper persistent woodland and (**b**) aftermath of fire in Gila National Forest, west-central New Mexico (Photographs by Kari Greer/U.S. Forest Service, Gila National Forest)

Another possible variation in the fire regimes of persistent woodlands was described for two sites in southwestern Texas (Poulos et al. 2009). Although outside of this book's geographical range, the findings are possibly applicable to the American Southwest. Fires were cross-dated, and ages of selected unscarred trees were sampled. The presence of trees with multiple fire scars and stands with multiple cohorts of Mexican pinyon indicated that low-severity surface fires had been common. Mean fire-return intervals at the two sites were 11 and 37 years (10 % scarring). Frequent fires were said to have resulted from grassy fuels on large expanses of the study areas and steep slopes that facilitated spread of fires from lowland grassland into upland Pinyon-Juniper vegetation.

The fire regime of southwestern Pinyon-Juniper wooded shrublands is less-well to poorly known (Romme et al. 2009a, b). The historical and current fire regime is thought to be similar to the infrequent, high-severity fire regime of persistent woodlands (Romme et al. 2009a), because wooded shrublands also lack continuous fine fuels to carry surface fires. Instead, the primary fuels are the crowns of trees and shrubs, and strong winds likely are needed to spread ignitions. Fire rotations appear to be very long, but there is little evidence on fire sizes and severities (Romme et al. 2009a).

The historical fire regime of southwestern Pinyon-Juniper savanna is also poorly known (Baker and Shinneman 2004; Romme et al. 2009a, b). The well-developed herbaceous layer likely supports spreading surface fires during dry seasons (Baker and Shinneman 2004; Romme et al. 2009a), perhaps especially following periods of above average precipitation when herbaceous fuels accumulate (e.g., Sect. 4.2.5.1). Historical fire intervals were likely short, but little is known about past fire sizes and severities (Romme et al. 2009a). Fire regimes possibly were influenced by proximity to vegetation that burned frequently, such as grasslands and Ponderosa Pine Forest. The fire regime likely thinned tree populations in savannas (Baker and Shinneman 2004). For example, fire in a savanna in south-central New Mexico killed 14 % of Colorado pinyons and 24 % of oneseed junipers (Dwyer and Pieper 1967). Small oneseed juniper trees are approximately twice as likely to be killed by fire than are large trees (Jameson 1962).

5.2.5.2 Biotic Agents

The vegetation dynamics of many Pinyon-Juniper persistent woodlands are often driven more by biotic disturbance agents and climate than by fire (Romme et al. 2009a). Important biotic disturbance agents in the American Southwest include pinyon bark beetle, fungal pathogens such as Armillaria root rot and black stain root disease, and pinyon dwarf mistletoe (*Arceuthobium divaricatum*). These biotic agents are continuously present in landscapes, but not always at levels influencing vegetation dynamics. Their effects are better known for persistent woodlands than for wooded shrublands and savannas. In general, more biotic agents affect and have greater impacts on pinyons than on junipers (e.g., Romme et al. 2009b).

Fig. 5.10 Pinyons killed by pinyon bark beetles near Flagstaff in north-central Arizona. Living trees are junipers (Photograph by Joel McMillin, Forest Health, Arizona Zone, U.S. Forest Service)

Pinyon bark beetles have been the most impactful biotic disturbance agent affecting southwestern Pinyon-Juniper persistent woodlands since at least the mid twentieth century (Fig. 5.10). Major outbreaks and subsequent mortality of pinyons occurred in the 1950s (Allen 1989; Betancourt et al. 1993; Swetnam et al. 1999), as well as in the 1990s (Swetnam and Betancourt 1998; Gottfried and Pieper 2000), and in the early 2000s (Breshears et al. 2005; Shaw JD 2006; Floyd et al. 2009). Each of these outbreaks was associated with regional drought during which water stress reduced the ability of pinyons to resist attack by bark beetles. Major outbreaks are generally limited to a few years – 3 to 4 years in the case of the early 2000s – but the effects of outbreaks on vegetation dynamics last decades or longer (Santos and Whitman 2010). For example, loss of cone-bearing pinyons can affect trajectories of stand recovery (Romme et al. 2009b). However, because outbreaks cause little mortality of junipers, they do not produce geographical or elevational range contractions of Pinyon-Juniper vegetation or conversion to other types of vegetation (Shaw JD 2006; Witt and Shaw 2010).

Pinyon mortality is best known from the early 2000s, when it peaked in Arizona and New Mexico at 7,747 km^2 (2,991 miles^2) in 2003 (based on incomplete aerial surveys; U.S. Forest Service 2004). Over 50 million pinyons died in New Mexico alone in 2001–2005 (Eager 2008). Mortality was highly variable among sites, ranging from 0 to nearly 100 % (Shaw et al. 2005). Even where pinyon mortality was low, it showed a large increase, e.g., from 0.8 % before drought to 7.1 % with drought in Grand Staircase-Escalante National Monument in south-central Utah (Witt and Shaw 2010), an increase of nearly an order of magnitude.

The degree of pinyon mortality from pinyon bark beetle outbreaks in the American Southwest has been related to numerous factors. Broad categories of factors affecting pinyon mortality include regional conditions, site environmental conditions, site biological factors, and pinyon tree characteristics. These are explored in the following four paragraphs.

Regional conditions related to greater pinyon mortality include high-severity drought (Wilson and Tkacz 1992; Breshears et al. 2005), because water stress is the key driving force of pinyon mortality (Breshears et al. 2005; Mueller et al. 2005). Rising temperatures are also important because they amplify water stress, increase the number of beetle generations in a year (Breshears et al. 2005; Bentz et al. 2009), and increase trees' respiratory losses of carbon, which has been identified as the primary mechanism of pinyon mortality (McDowell et al. 2008; Adams et al. 2009).

Evidence that site environmental conditions are important in pinyon mortality caused by bark beetles includes the finding that sites with high mortality in the 1990s also had high mortality in the early 2000s (Santos and Whitham 2010). Site conditions related to greater pinyon mortality from bark beetles are those that affect water availability and/or temperature, including low moisture availability (Allen 1989; Peterman et al. 2012), shallow soils (Gitlin et al. 2006), low elevation (Allen 1989; Shaw JD 2006; Santos and Whitham 2010), south-facing slope aspects (Ogle et al. 2000; Gitlin et al. 2006), steeper slopes (Santos and Whitham 2010), and length of drought (Peterman et al. 2012).

Site biological factors related to greater pinyon mortality from bark beetles include later successional stage (i.e., time since disturbance; Shaw JD 2006), greater and lesser pinyon dominance of stand basal area (Witt and Shaw 2010), and greater pinyon density (Negrón and Wilson 2003; Bentz et al. 2009; Santos and Whitham 2010). Stand density has been found to be unrelated to pinyon mortality (Floyd et al. 2009). Additional biological factors related to high pinyon mortality are presence of understory shrubs (Santos and Whitham 2010), position in vegetation interspaces (in the case of pinyon seedlings; Mueller et al. 2005), and greater infestation by pinyon dwarf mistletoe (Negrón and Wilson 2003) and black stain root disease (J. Lundquist, personal communication cited in Romme et al. 2009b).

Pinyon tree characteristics related to greater mortality from bark beetles are slower growth rates (Huffman et al. 2008b), greater sensitivity to past climate variation (Ogle et al. 2000), older age (Swetnam and Betancourt 1998; Huffman et al. 2008b), greater height (Huffman et al. 2008b), and larger size (Negrón and Wilson 2003; Mueller et al. 2005; Huffman et al. 2008b; Floyd et al. 2009; Santos and Whitham 2010). Nevertheless, some studies found no relationship to pinyon diameter (Shaw JD 2006) or that the relationship is largely limited to small trees (Mueller et al. 2005). Severe drought can override some of these relationships (Floyd et al. 2009).

Both Armillaria root rot and black stain root disease can kill pinyons (Rogers 1995; Kearns and Jacobi 2005), but black stain root disease has received more attention from plant ecologists. This root pathogen has infected pinyons in southwestern

Fig. 5.11 Dead pinyons in a mortality center of black stain root disease in southwestern Colorado (Photograph by William R. Jacobi, Colorado State University)

Colorado, southeastern Utah, and New Mexico (Shaw et al. 2005; Fairweather et al. 2006). It causes greatest mortality following years of consistent, heavy summer precipitation (J. Worrall, personal communication cited in Romme et al. 2009b). Pinyon mortality from black stain root disease affects vegetation dynamics (Romme et al. 2009b).

Black stain root disease has limited dispersal within stands of Pinyon-Juniper vegetation, and its patchy areas of impacts are referred to as mortality centers (Kearns and Jacobi 2005) or infection centers (Fairweather et al. 2006); see Fig. 5.11. Dead pinyons occupy the middle of these centers and trees of different states of decline are toward the perimeter (Fairweather et al. 2006). Thirty newly formed centers studied in southwestern Colorado had a mean area of 0.3 ha (0.7 acres; Kearns and Jacobi 2005). The mean annual radial expansion of these mortality centers was 1.1 m (3.6 ft) and was unrelated to any site data collected, including pinyon density. Sixty-eight percent of the pinyons within the centers were dead, 76 % were affected by black stain fungi, and 70 % had evidence of pinyon bark beetle. There was no statistically significant difference in density of pinyon regeneration inside vs. outside these centers.

Pinyons with black stain root disease can be more susceptible to bark beetle attack (Kearns and Jacobi 2005) and thereby likely sustain bark beetle populations during moist periods (J. Lundquist, personal communication in Romme et al. 2009b). Black stain root disease in the Mesa Verde region interacts so closely with pinyon bark beetle that they can be considered an integrated agent of pinyon mortality (Romme et al. 2009b).

Pinyon dwarf mistletoe is a parasitic plant that is especially abundant on Colorado and singleleaf pinyons but can occur on other pinyons (Shaw et al. 2005). Juniper mistletoe (*Phoradendron juniperinum*) is parasitic on junipers. Mistletoes cause host trees to accumulate resins and form dense clusters of branches ("witches brooms") that can facilitate crown scorching of trees (Conklin and Fairweather 2010; Sect. 4.2.5.3). Pinyon dwarf mistletoe reduces tree growth and vigor, predisposing host trees to attack by other biological agents (Shaw et al. 2005). Infected trees are more susceptible to and perhaps preferred by pinyon bark beetles (Negrón and Wilson 2003). Pinyon dwarf mistletoe also can kill pinyons, especially smaller trees (Shaw et al. 2005).

5.2.5.3 Climate Variation

Decadal-scale climate variation is a major driver of Pinyon-Juniper vegetation dynamics. It is linked to episodic tree recruitment and mortality events (Betancourt et al. 1993; Swetnam and Betancourt 1998; Barger et al. 2009; Romme et al. 2009a; Shinneman and Baker 2009; Jacobs 2011). Consequently, Pinyon-Juniper vegetation is an oscillating system with inextricable links among climate, fire, and insect outbreaks.

While fire and insects are proximate causes of tree mortality, drought can be considered the ultimate cause. This is evidenced by fires being linked to dry conditions (Sect. 5.2.5.1), as well as by pinyon mortality having occurred without a bark beetle outbreak in some areas of north-central Arizona during the 1990s drought (Mueller et al. 2005). Droughts have long impacted Pinyon-Juniper stands. For example, current stands – even old stands – rarely have living trees predating 1600, apparently as a result of extreme drought in the late sixteenth century (Swetnam and Brown 1992; Swetnam and Betancourt 1998). The next drought of similar or greater intensity was in the 1950s. It also greatly affected some stands of Pinyon-Juniper, e.g., more than 90 % of single leaf pinyon alive at a site in central New Mexico in 1940 died by 1956 (Betancourt et al. 1993).

The drought of the early 2000s differed from the 1950s by having warmer temperatures (Breshears et al. 2005). This appears to be critical because experimental evidence indicated that higher temperatures substantially shorten the time for drought-induced mortality of Colorado pinyon (Adams et al. 2009; Sect. 5.2.6.3). Consequently, drought in the early 2000s resulted in greater, more geographically extensive mortality of pinyon (Breshears et al. 2005; see Sect. 5.2.5.2).

Junipers appear more resistant to drought than pinyons (e.g., Nowak et al. 1999). Evidence includes junipers exhibiting less mortality than pinyons during drought (Shaw JD 2006; Floyd et al. 2009; Koepke et al. 2010). The background mortality rate for junipers for 5 years across the range of Pinyon-Juniper is <0.1 % (Witt and Shaw 2010). Although this rate increased by an order of magnitude with drought to <1 %, that percentage is well below even the background rate of pinyons. This is possibly related to juniper being affected by fewer biotic agents, leaving drought as the only important cause of juniper mortality besides fire (Floyd et al. 2009).

Climate variation also includes periodic wet periods. These are associated with pulses of tree regeneration and establishment (Swetnam et al. 1999; Romme et al. 2009a; Shinneman and Baker 2009), as well as mortality from black stain root disease (see previous section). See Sect. 5.3 for discussion of tree regeneration.

5.2.6 *Anthropogenic Drivers*

People have lived in southwestern Pinyon-Juniper vegetation for over 10,000 years (Cartledge and Propper 1993) – longer than any other vegetation type on the mountains and plateaus of the American Southwest. Ancestral Native Americans were so connected with Pinyon-Juniper vegetation that its distribution is a proxy for the distribution of their settlements (Schlanger and Larralde 2008). They used Pinyon-Juniper vegetation as a source of food through the hunting of game and gathering of pinyon nuts, as well as sites for agricultural crops. Local losses of vegetation cover occurred, primarily through burning to create agricultural fields (Kohler and Matthews 1988; Kohler 1992; note: these publications referred to forest, but the studies were on Pinyon-Juniper vegetation; see also Vale 2002). In addition, the vegetation was a source of wood for construction of dwellings and other structures, as well as fuel for cooking, heating, and pottery firing (Gottfried and Pieper 2000; Schlanger and Larralde 2008).

Human uses of Pinyon-Juniper vegetation increased with Spanish and Euro-American settlement. Included were livestock grazing, as well as cutting trees for mining supports, production of charcoal for smelting, ties for railroads, and fuel for locomotives (Ernst and Pieper 1996). Anthropogenic drivers that have affected relatively undisturbed stands include livestock grazing, fire management, modern climate change, invasive species, recreation, and nearby land use. Each of these can be important in vegetation dynamics, and therefore all are incorporated in the conceptual models (Sect. 5.5).

5.2.6.1 Livestock Grazing

Livestock grazing in the American Southwest began with the introduction of livestock by the Spanish in 1540 and greatly expanded in Spanish, Native-American, and eventually Euro-American populations over the following three centuries (Gottfried and Pieper 2000; Sect. 1.6.1; Fig. 5.12). Because Spanish and Native-American settlements were in or near Pinyon-Juniper vegetation, livestock grazing has impacted this vegetation longer than higher-elevation vegetation types (Milchunas 2006). Moreover, with moderate winter weather usually enabling year-around accessibility, grazing impacts on Pinyon-Juniper vegetation occurred throughout the year (Romme et al. 2009b).

Livestock grazing has been considered a likely cause of increased tree density in Pinyon-Juniper vegetation (see discussion in Romme et al. 2009a). Grazing can

Fig. 5.12 Sheep grazing in 1909 on Kaibab National Forest, north-central Arizona (Photograph by G. A. Pearson, courtesy of Fort Valley Experimental Forest Historic Images, U.S. Forest Service)

favor tree regeneration by reducing species that compete with tree seedlings and by increasing shrubs that can act as nurse plants for tree seedlings (see Sect. 5.3). In addition, grazing has been said to have decreased fire frequency by reducing the amount and connectivity of herbaceous fuels, as in Mixed Conifer and Ponderosa Pine Forests (Sects. 3.2.6.1 and 4.2.6.1). However, with the possible exception of reducing fire frequencies in savannas, it is unlikely that livestock grazing significantly impacted the historical fire regimes of southwestern Pinyon-Juniper vegetation (Baker and Shinneman 2004; Romme et al. 2009a). For example, research indicates that the fire regimes of persistent woodlands in Mesa Verde National Park and Glen Canyon National Recreation Area have been little affected by livestock grazing (Floyd et al. 2000, 2004, 2008).

Other impacts of grazing have been studied by comparing vegetation structure and composition of grazed areas vs. lightly or ungrazed areas such as relict sites. Impacts of grazing can depend on soil type, tree cover, and pre-historical grazing by large ungulates such as buffalo (*Bison bison*), but can be overshadowed by increases in tree cover (Milchunas 2006). Comparison of a grazed and a relict site in south-central Utah indicated that grazing did not affect tree abundance in persistent woodlands (Harris et al. 2003; Barger et al. 2009), although another study of the same sites indicated somewhat greater cover of Colorado pinyon on the grazed site

(Guenther et al. 2004). Also, research in southwestern Colorado suggested that livestock grazing approximately triples tree seedlings and sapling densities through increases in pinyon (Shinneman and Baker 2009). In addition, a modeling study that utilized field data from across the American Southwest indicated that single-seed juniper had invaded nearby grasslands in areas with a high proportion of summer rainfall and that the invasion was most likely related to livestock grazing having reduced competition and removed fuels for fires (Jacobs 2011).

Grazed sites also can differ from ungrazed sites in terms of shrubs, herbs, and biological soil crusts. Grazed sites have greater abundance of shrubs (Harris et al. 2003; Guenther et al. 2004; Tausch and Hood 2007), but can have lower shrub species richness (Guenther et al. 2004). With regard to herbs, grazed sites have lower production (Pieper 1968) and differences in species composition (Jameson et al. 1962; Guenther et al. 2004) such as fewer palatable species (Milchunas 2006) and either less grass cover (Potter and Krenetsky 1967; Beymer and Klopatek 1992) or no differences in grass cover (Guenther et al. 2004). Grazed sites also have greater abundance of forbs (Beymer and Klopatek 1992; Harris et al. 2003) and more annuals (Milchunas 2006). Species richness can be unaffected (Guenther et al. 2004). In addition, grazed sites (because of trampling) have less cover of biological soil crusts (Beymer and Klopatek 1992; Guenther et al. 2004) and more bare ground (Guenther et al. 2004). The effects of grazing can be exacerbated by drought (Jacobs 2011).

An important indirect effect of livestock grazing has been mechanical removal of trees by chaining or cabling to attempt to increase forage (Fig. 5.13a, b). Mechanical removal became common after World War II (Ernst and Pieper 1996; Gottfried and Pieper 2000). Trees were uprooted by dragging a heavy chain or cable strung between two powerful bulldozer tractors. By the early twenty-first century, over 700 such treatments had been applied to approximately 2,835 km^2 (1,095 $miles^2$) on the Colorado Plateau (Peters and Cobb 2008), an area that makes up much of the American Southwest. A tree canopy often regenerated where understory seedlings and saplings were not removed (Romme et al. 2009b). Mechanical removal of trees is not included in the Pinyon-Juniper conceptual models because it – like logging in forests – results in stands that are not in a near-natural state.

5.2.6.2 Fire Management

Fire management possibly preceded Spanish settlement, as it has been said that Native Americans burned Pinyon-Juniper vegetation to increase and protect wild resources such as Indiangrass (*Sorghastrum nutans*), sunflower (*Helianthus*), goosefoot (*Chenopodium*), and pinyons (including their edible "nuts"; Sullivan 1996 in Raish 2004; B. Pikyavit, personal communication in Alcoze 2003). Fire management of Pinyon-Juniper vegetation by non-natives was first attempted early in Spanish settlement with the establishment of ranching regulations termed the Mexican Mesta (Weckmann 1992), which included punishment for burning savannas or fields (Dusenberry 1963). Evidence of ecological impacts attributable to this regulation is lacking.

Fig. 5.13 (**a**, **b**) Removal of pinyon and juniper trees in the 1960s in Dixie National Forest in southwest Utah by chaining. (**a**) A large chain strung between two bulldozer tractors was dragged to uproot and topple trees. (**b**) After uprooted trees were removed, the area was reseeded to increase forage for livestock (Photographs courtesy of U.S. Forest Service via Special Collections, Sherratt Library, Southern Utah University)

Although modern fire management began in the early twentieth century, it is unlikely that fire suppression had large impacts on the historical fire regimes of southwestern Pinyon-Juniper persistent woodlands (Baker and Shinneman 2004; Romme et al. 2009a). For example, the fire regimes of persistent woodlands in Mesa Verde National Park and Glen Canyon National Recreation Area have been little affected by fire suppression (Floyd et al. 2000, 2004, 2008). In contrast, Poulos et al. (2009) reported that fire suppression was the likely cause of twentieth century changes in the fire regime at two Pinyon-Juniper sites in southwestern Texas (Sect. 5.2.5.1). The fire regimes of wooded shrublands and savannas are insufficiently known to reach conclusions about effects of fire suppression, but evidence of broad-scale effects has yet to be documented.

Twenty-first century fire management in some areas of the American Southwest has not focused on Pinyon-Juniper vegetation. For example, Grand Canyon National Park has an active program of management fires in Ponderosa Pine and Mixed Conifer Forests, but no such management occurs in Pinyon-Juniper vegetation below the canyon rim, and fire management in Pinyon-Juniper vegetation above the canyon rim is largely limited to manually thinning stands to protect park infrastructure (Grand Canyon National Park 2012).

5.2.6.3 Modern Climate Change

Die-off of Colorado pinyon in the American Southwest in the early 2000s was greater and extended into moister sites than the 1950s die-off (Breshears et al. 2005; Sect. 5.2.5.3). Precipitation amounts were similar during the two droughts, but higher temperatures occurred in the early 2000s, apparently as the result of modern climate change (Breshears et al. 2005). The higher temperatures were critical, as evidenced by the finding that an increase of approximately 4 °C (7 °F) reduces the time for drought-induced mortality in Colorado pinyon by 28 % to 18 weeks (Adams et al. 2009). Modeling has suggested a future fivefold increase in the frequency of regional-scale die-offs of Colorado pinyon (Adams et al. 2009) and large contractions of its distribution in most of the American Southwest (Cole et al. 2008). However, a field-based study to examine effects of climate variability on National Park Service lands in southeastern Utah found no change in Colorado pinyon cover and increases in cover of Utah juniper and shrubs (Munson et al. 2011, for 1989–2008)

Modern climate change has been linked to increased atmospheric CO_2 concentrations (Intergovernmental Panel on Climate Change 2007), and it has been suggested that those concentrations also have directly impacted the dynamics of Pinyon-Juniper vegetation. For example, elevated CO_2 concentrations have been implicated in late twentieth century increases in western juniper (*Juniperus occidentalis*) in woodlands in Oregon (Soulé et al. 2004). Nevertheless, supporting evidence, at least in the American Southwest, is insufficient to consider CO_2 concentration to be directly important in tree dynamics in Pinyon-Juniper vegetation (Romme et al. 2009a). Experimental research has indicated that increased CO_2 concentrations could have increased productivity and fuel loadings of the invasive cheatgrass (*Bromus tectorum*), with subsequent potential effects on the fire regime where cheatgrass is abundant (Ziska et al. 2005; see following section).

5.2.6.4 Invasive Species

Invasive plant species can be uncommon in southwestern Pinyon-Juniper vegetation (Gottfried and Pieper 2000; Guenther et al. 2004), and little information is available on most exotics. One study found that greater numbers of non-native species were positively correlated with richness of native species, indicating that habitats with high biodiversity are at greater risk (Floyd et al. 2006). But not all exotic plant species are aggressive invasives and not all native plant species are adversely affected by the presence of exotics (Ott et al. 2001).

The invasive of primary concern in Pinyon-Juniper vegetation is cheatgrass, an annual species with no native ecological equivalent. As summarized by Sieg et al. (2003), its broad ecological amplitude allows it to grow and reproduce in a variety of environments. This and dispersal attached to animals, including humans, give it the ability to expand its already wide geographic distribution. Cheatgrass is a cool-season species that can capture resources before most native species. Its growth can produce extensive cover of fine fuels every year, giving it the potential to shorten fire intervals.

The abundance of cheatgrass can increase rapidly. It had been uncommon on the mesa tops of Mesa Verde National Park until dense stands abruptly appeared in 2003 (Floyd et al. 2006). Cheatgrass joined musk thistle (*Carduus nutans*) and Canada thistle (*Cirsium arvense*) as the invasives of most concern following disturbance by large, stand-replacing fires in the Park's older woodlands. The spread of cheatgrass into these burned areas has the potential to change the fire regime from infrequent stand-replacing fires occurring only during extreme dry periods to frequent surface fires disconnected from drought. Such a change in fire regime would inhibit successional recovery of Pinyon-Juniper vegetation and could result in a vegetation type conversion to grassland. Similar concerns have been expressed for Pinyon-Juniper vegetation elsewhere in the American Southwest, including Glen Canyon National Recreation Area (Floyd et al. 2008) and southern Utah (Tausch and Hood 2007).

As mentioned in the previous section, increased CO_2 concentration has the potential to increase the productivity and fuel loadings of cheatgrass (Ziska et al. 2005). Moreover, cheatgrass increases after fire. Not only has this occurred in Mesa Verde National Park (see above), but a study in central Utah determined its cover averaged 17, 19, and 22 % in three unburned sites and was 29, 55, and 70 % in paired burned areas (Ott et al. 2001). The researchers concluded that cheatgrass made it difficult to restore Pinyon-Juniper communities.

5.2.6.5 Recreation

Impacts of recreation on southwestern Pinyon-Juniper vegetation are understudied, but are likely significant because of accessibility and use by recreationists.

Some impacts are spatially limited (e.g., construction and use of roads and campsites), but others are more extensive, such as the introduction and spread invasive plants. Pinyon-Juniper invasives have aggressively moved into sites of both dispersed and developed recreation in Arizona (Fenner 2008). In addition, use of off-highway vehicles negatively impacts Pinyon-Juniper understories and soils, including reducing biological soil crusts.

5.2.6.6 Nearby Land Use

Nearby land use is an important anthropogenic driver. Pinyon-Juniper vegetation borders more adjacent lands, especially disturbed lands, and has more cities, towns, and other human developments embedded within it than vegetation at higher elevations. Although under-studied, this proximity facilitates the spread of invasive species and fires into southwestern Pinyon-Juniper vegetation.

5.3 Processes

The vegetation dynamics of southwestern Pinyon-Juniper are dominated by three processes: infill, expansion, and succession (each with a corresponding inverse process: infill & thinning, expansion & contraction, and succession & disturbance). These processes play key roles in the conceptual models (Sect. 5.5).

Infill, expansion, and succession require regeneration of pinyons and/or junipers, which can be problematical in dry environments. Seeds of pinyons and junipers are dispersed by birds and mammals – pinyon seeds by scatter-hoarding species and juniper seeds by frugivorous species that ingest seeds and defecate (Chambers et al. 1999). Pinyon seeds, but not juniper seeds, must be buried by their dispersers (Chambers et al. 1999; Chambers 2001).

Seedling survival for both species is facilitated by amelioration of environmental conditions such as by shading (Meagher 1943), but environmental modification is necessary for survival of pinyon seedlings (Chambers 2001). Studies in north-central Arizona have found strong spatial associations between small Colorado pinyons and either nurse plants or other structures that provided greater soil moisture availability and/or shade (Pearson and Theimer 2004; Mueller et al. 2005). Even junipers can facilitate pinyon growth (Landis and Bailey 2005; Haire and McGarigal 2008), at least on drier sites (Jacobs 2008). Establishment pulses of Colorado pinyon are associated with wet periods (Shinneman and Baker 2009).

Like pinyons, junipers often establish below nurse plants, but junipers also can establish in interspaces, especially in regions with a high proportion of summer rainfall (Chambers et al. 1999). Interspaces are characterized by stressful conditions (Breshears et al. 1998) that are likely to be ameliorated by summer rainfall. Establishment pulses of Utah juniper roughly coincide with periods of severe, prolonged drought, possibly because mortality of pinyon favors juniper establishment (Shinneman and Baker 2009).

5.3.1 *Infill and Expansion*

Infill refers to increasing tree density and cover in stands that are already Pinyon-Juniper vegetation. It has occurred in many stands since the mid nineteenth century, as evidenced by comparison of historical photographs with current scenes and by tree age structures that reflect an increase of trees. Examples where infill has been documented in the American Southwest include persistent woodlands and possibly savannas in central, north-central, and northwestern Arizona (Johnsen and Elson 1979; Gordon et al. 1992; Ffolliott and Gottfried 2002; Landis and Bailey 2005; Shaw HG 2006), persistent woodlands and wooded shrublands in Mesa Verde National Park (Floyd et al. 2004), and savannas in southwestern and south-central New Mexico (Miller 1999; Fuchs 2001, 2002). Infill has not been universal, as some Pinyon-Juniper stands have remained stable or decreased in tree density and cover (Romme et al. 2009a).

The inverse of infill is thinning, i.e., decrease of tree density and cover in Pinyon-Juniper stands. This has been observed at a regional scale in the American Southwest, with extensive die-off of pinyons during the 1950s, 1990s, and early 2000s (Sect. 5.2.5.2).

Expansion refers to pinyons and junipers establishing in areas lacking trees, such as shrublands and grasslands (Figs. 5.14a, b and 5.15a, b). Expansion since the mid nineteenth century has been documented through examination of historical photographs and tree age structures and has varied among stands. Examples where expansion has been documented in the American Southwest include shrublands in Mesa Verde National Park (Floyd et al. 2004) and former grasslands in northwestern Arizona (Shaw HG 2006), north-central New Mexico (Jacobs et al. 2008), and southwestern and south-central New Mexico (Miller 1999; Fuchs 2001, 2002). Expansion is usually followed by infill as early invading trees modify the environment by reducing herbaceous cover, altering soil properties, and producing shade (cf. Tausch 1999; Jacobs 2008). Infill likely accelerates as early invading trees mature and begin to produce seeds.

The opposite of expansion is contraction, which is the elimination of pinyon and juniper tree cover by natural processes. Contraction is not as well-studied as expansion in the American Southwest, but elevational contractions and expansions tied to Holocene climate variations have been documented for pinyon-juniper vegetation in the Great Basin (Miller and Wigand 1994). Presumably, climate shifts have had similar effects in the American Southwest. Indeed, this is implied by the expansion of pinyons and junipers into areas where they formerly existed (cf. Sallach 1986), i.e., where contraction is thought to have preceded the observed expansion. The extensive die-offs of pinyons described in Sect. 5.2.5.2 illustrate contraction of pinyons, but not necessarily of Pinyon-Juniper vegetation, because junipers have persisted in many of these areas (e.g., Shaw JD 2006; Koepke et al. 2010; Witt and Shaw 2010).

Additional differences between infill and expansion include sources of seeds – largely within stand sources for infill and outside stand sources for expansion. There also are differences in regeneration sites; by definition, trees (which are potential nurse plants) are present where infill occurs but absent where expansion occurs.

Fig. 5.14 (**a**) Increase of oneseed juniper from 1899 to (**b**) 2008 in area between the Enchanted Mesa and Acoma Pueblo (photograph point) in west-central New Mexico ((**a**) by W.H. Jackson/U.S. Geological Survey; (**b**) by Roger L. Rainwater)

Potential causes of infill and/or expansion include recovery from past disturbance, natural range expansion, livestock grazing, fire exclusion, climatic variability, and elevated CO_2 (Romme et al. 2008, 2009a). Each of these is discussed below. Identifying the causes of infill and expansion is critical to understanding whether the increases in tree density observed since the mid nineteenth century are due to natural factors or are related to Euro-American settlement and subsequent land use.

Fig. 5.15 (**a**) Area of juniper expansion into sagebrush shrubland in central Utah (Photograph by author). (**b**) Area of juniper expansion into grassland in northern New Mexico in middle ground of photograph; also note living junipers in immediate foreground and dead pinyons killed by bark beetles behind them (Photograph by Brian Jacobs)

Recovery from past disturbance is one cause of apparent infill and expansion (e.g., Huffman et al. 2012). Past disturbances include naturally occurring events such as stand-replacing fire, where subsequent increases in tree density reflect succession (see the following section). Such sites are often evident because of charred snags and downed logs, which in the case of juniper can persist for many decades because of decay-resistant wood. Sites where humans have removed trees are not always apparent today, and this can obscure determining post-disturbance recovery vs. range expansion. For example, sites where Native Americans cleared Pinyon-Juniper vegetation surrounding their settlements centuries ago are not readily evident today without research, e.g., areas of northwestern New Mexico (Betancourt and Van Devender 1981; Samuels and Betancourt 1982) and southwestern Colorado (Kohler and Matthews 1988). Sites of recent clearance of Pinyon-Juniper for fuel wood, enhancement of livestock forage, etc. are usually apparent from cut stumps and other tree remains (cf. Shaw HG 2006).

Natural range expansion also accounts for some expansion and infill. This is evidenced by growth of young trees on little-disturbed sites that lack snags and logs of larger, older trees evidencing past disturbance. Range expansion of pinyons and junipers in the American Southwest has been correlated with favorable changes in climate and therefore can be a natural process (e.g., Betancourt 1987; Davis 1987; for regions near the Southwest, see for example Miller and Wigand 1994; Gray et al. 2006).

Livestock grazing has long been considered a likely cause of infill and expansion, as both it and increasing tree density became widespread in the second half of the nineteenth century. Grazing potentially favors tree regeneration by reducing competition for tree seedlings and by increasing shrubs that can act as nurse plants for tree seedlings. Research has been insufficient to judge the importance of grazing as a direct factor in the processes of infill and expansion (Romme et al. 2009a). More recently, a modeling study based on field data concluded that the invasion of grasslands by single-seed juniper in areas with summer rainfall was likely related to livestock grazing (Jacobs 2011). Another study also found grazing affected tree recruitment (Shinneman and Baker 2009), but others found no effect of grazing on tree abundance (Harris et al. 2003; Barger et al. 2009). See Sect. 5.2.6.1 for additional details. Range expansion in grasslands degraded by livestock grazing in north-central New Mexico appears primarily correlated with soil moisture and has occurred most commonly at lower elevations and on sites with relatively high productivity and deep soils (Jacobs et al. 2008).

Fire exclusion, which has resulted in increased tree densities in forests at higher elevations of the American Southwest (e.g., Sect. 4.4.1.2), also has been postulated as a cause of infill and expansion of Pinyon-Juniper vegetation. However, fires were infrequent in persistent woodlands (Sect. 5.2.5.1), and therefore fire exclusion cannot be a primary mechanism for their infill (Romme et al. 2009a). If fires had been frequent in wooded shrublands, shrublands, savannas, and grasslands, fire exclusion could have resulted in infill and expansion of pinyons and/or junipers. But this is hypothetical. Few data are available on the fire regimes of wooded shrublands, shrublands, savannas, and grasslands because of the paucity of fire scars (Baker and Shinneman 2004; Romme et al. 2009a; Sect. 5.2.5.1).

Variation in nineteenth century climate also correlates temporally with infill and expansion. Pinyon-Juniper vegetation, particularly its trees, is especially sensitive to changes in precipitation and temperature (Sect. 5.2.5.3). Evidence from southwestern Colorado supports an important role for such changes affecting tree density (Shinneman and Baker 2009). Extensive dieback of Colorado pinyon during drought (Breshears et al. 2005; Shaw et al. 2005) indicates the importance of climate change on population dynamics of trees.

Elevated CO_2 has been postulated as a cause of infill and expansion (Sect. 5.2.6.3). Nevertheless, supporting evidence is insufficient to consider it directly important in tree population dynamics in the American Southwest (Romme et al. 2009a).

5.3.2 Succession

Succession occurs following stand-replacing crown fire in persistent woodlands (re-establishment of trees after thinning by insects, disease, and drought is treated as infill). Succession is influenced by such factors as climate, soils, age of the stand burned, and other pre-fire conditions, as well as residual seeds and seed dispersal, residual coarse woody debris, fire size and severity, livestock grazing, and invasive species (Barney and Frischknecht 1974; Dick-Peddie 1993b; Ott et al. 2001). Succession in southwestern Pinyon-Juniper vegetation appears to have been described first by Leopold (1924), who noted vegetation changes in areas with charred junipers in southern and central Arizona.

Multiple studies have focused on changes in plant species composition during succession. The general pattern is that burned sites are initially dominated by annual herbs that are soon followed by perennial grasses and forbs. Within a few decades, shrubs dominate. That dominance can persist for several decades, but tree seedlings establish within the shrubs and gradually overtop them. Junipers usually establish before pinyons. Trees increase in density and cover and woodland is formed in 100–300 years (Leopold 1924; Arnold et al. 1964; Erdman 1970; Barney and Frischknecht 1974; Floyd 1982; Tress and Klopatek 1987; Dick-Peddie 1993b; Paysen et al. 2000; Miller and Tausch 2001; Romme et al. 2003).

The specific species involved in succession differ by region and site. For example, Gambel oak, Utah serviceberry (*Amelanchier utahensis*), mountain-mahogany, and antelope bitterbrush (*Purshia tridentata*) are dominant shrubs in Mesa Verde National Park (Erdman 1970), and big sagebrush is dominant in Grand Canyon National Park (Jameson et al. 1962; Rowlands and Brian 2001). Succession is unlikely to lead to equilibrium conditions because of episodic mortality and recruitment related to ongoing variations in climate (Betancourt et al. 1993). In fact, succession usually takes longer than the cycle of climate change that causes large-scale mortality and regeneration (Sects. 5.2.5.3 and 5.3.1).

Another approach to elucidating Pinyon-Juniper succession focused on structural parameters along a chronosequence of 13 sites in north-central Arizona that ranged from 3 to 370 years since fire (Fig. 5.16a–c; Huffman et al. 2012).

Fig. 5.16 (**a–c**) Three stands from a chronosequence of 13 stands in Pinyon-Juniper persistent woodland in north-central Arizona studied by Huffman et al. (2012). Time since fire in these stands is 3, 85, and 340 years for **a**, **b**, and **c**, respectively (Photographs by David W. Huffman)

Fig. 5.16 (continued)

Although chronosequences do not replicate succession (because of differences in site conditions), they can provide insight into successional changes. Tree species present included Colorado pinyon, Utah juniper, and oneseed juniper. Total shrub cover was highly variable throughout the chronosequence and unrelated to time since fire. Tree seedlings first occurred 6 years into the chronosequence, but their density also was highly variable throughout the chronosequence and unrelated to time since fire. Juniper was likely to establish earlier than pinyon. Trees >1.37 m (4.5 ft) height first occurred 30 years into the chronosequence. Juniper and total live tree density both had positive linear relationships with time since fire, but pinyon did not. Aboveground live-tree biomasses of pinyons, junipers, and both species combined exhibited positive curvilinear relationships with time since fire, and that for both species combined approached an asymptote at about 250 years. All sites had diameter distributions weighted to smaller trees, except for three of the six sites that were 130 years or older, which had relatively greater numbers of mid-diameter trees. Total snag density decreased to a minimum density at about 240 years and then increased. Pinyon snag density exhibited no relationship and juniper snag density had a negative linear relationship with time since fire. Rotten downed logs exhibited a positive linear relationship with time since fire, but sound logs and total logs had no relationship. An index of overall structural complexity had a positive linear relationship with time since fire.

5.4 Historical Changes

5.4.1 Overstory

5.4.1.1 Pre-Euro-American Settlement

The earliest, useful insights on the overstory of southwestern Pinyon-Juniper vegetation come from written descriptions dating to the mid nineteenth century. They indicated variable tree density, ranging from open to dense stands:

> Every high hill, every dell, every vale or knook [sic] seemed thickly coated with a living green of rich grass and set about with [junipers] from 12 to 25 feet high like an old orchard. (Pratt for central Utah in 1851; Stanley and Camp 1935)

> A species of [juniper]...is found in most localities, and frequently gives the landscape the appearance of an old apple orchard. (Tidball for north-central Arizona in 1854; Shaw HG 2006)

> Part of the road had to be cut, the timber...was so thick. (Sherburne for north-central Arizona in 1854; Gordon 1988)

> ...we struck heavy and thick timber, of pine, [juniper], and piñon, where we were detained hours without being able to get through it; and it is barely possible to pass it on foot. (Aubry for northwestern or north-central Arizona in 1854; Wyman 1932,)

See other descriptions in Christensen and Johnson (1964), Gordon et al. (1992), and Shaw HG (2006).

Photographs showing Pinyon-Juniper vegetation date back to at least 1867 (Figs. 5.17 and 5.18). They indicate vegetation distribution as well as stand structure. As with early descriptions, the sites shown in early photographs are too widely dispersed and Pinyon-Juniper vegetation is too varied to permit generalizations beyond the specific sites. In addition, scenes of low tree density can be challenging to interpret, if no evidence of past fire is apparent in the photograph. Is the tree density low because the photograph shows savanna or wooded shrubland, because it shows persistent woodland in early or mid-succession, or because there has been tree removal by Native Americans, the Spanish, or early Euro-Americans? On-site visits sometime resolve such questions, because of the persistence of juniper snags, stumps, and logs.

Quantitative insight on historical conditions has been sought through study of relict areas little influenced by Euro-Americans and likely by Native Americans as well. Most relict areas are the tops of mesas and buttes that have been isolated from livestock grazing, tree cutting, and recreation by surrounding steep, rocky slopes. Such sites are relatively common in the American Southwest (Van Pelt and Tuhy 1991; Van Pelt and Johnson 1993), where even partial surveys have identified over 120 sites on National Park Service lands (Rowlands and Brian 2001). Relict sites with studies of southwestern Pinyon-Juniper vegetation include Spy Mesa in northwestern Arizona (Thatcher and Hart 1974), both Fishtail Mesa (Jameson et al. 1962; Rowlands and Brian 2001) and Shiva Temple (Beymer and Klopatek 1992) in Grand Canyon National Park, Wide Rock Butte in Canyon de Chelly National Monument in northeastern Arizona (Schmutz et al. 1976), No Man's Mesa in Grand

Fig. 5.17 Pinyon-Juniper vegetation in 1867 along Tecalote Creek in north-central New Mexico (Photograph by Alexander Gardner (or associates) courtesy of Boston Public Library)

Fig. 5.18 Pinyon-Juniper vegetation in 1867 near Cygnus Mountain (now called Mount Hope) in west-central Arizona (Photograph by Alexander Gardner (or associates) courtesy of Boston Public Library)

Staircase-Escalante National Monument (Mason et al. 1967; Harris et al. 2003; Guenther et al. 2004; Barger et al. 2009), and Greatheart and Church Mesas in Zion National Park in southwestern Utah (Madany and West 1984).

Tree cover recorded for these relict sites has ranged from <1 to 30 % for Colorado pinyon and from 0 to 10 % for Utah juniper (Jameson et al. 1962; Schmutz et al. 1976; Madany and West 1984; Rowlands and Brian 2001; Guenther et al. 2004). Tree density on No Man's Mesa was reported as 220 canopies/ha (89 canopies/acre) for Colorado pinyon and Utah juniper combined (Harris et al. 2003) and 282 trees/ha (114 trees/acre) for Colorado pinyon (Barger et al. 2009). Pinyon basal area on the same site was 9 m^2/ha (39 ft^2/ha; Barger et al. 2009).

Research findings from these relict sites have limitations. Some sites, including the one most frequently studied (i.e., No Man's Mesa), had been grazed for a period in the past, and the effects of that grazing are unknown but possibly significant. In addition, many sites are small in area (those listed above average 224 ha/554 acres), and Pinyon-Juniper vegetation often occupies only a portion of the area. Moreover, the tops of isolated mesas and buttes can be unique environments, lacking a full suite of species and having greater rock cover and drier conditions, as well as different soils, fire histories, geology, slopes, aspects, and vegetation than surrounding areas (Van Pelt and Johnson 1993; Guenther et al. 2004; Milchunas 2006). Consequently, findings from relict studies likely are not widely applicable, especially to the broad range of diversity that is characteristic of southwestern Pinyon-Juniper vegetation.

Another method that has been used to quantitatively describe historical conditions in southwestern Pinyon-Juniper vegetation is stand reconstruction. Reconstruction techniques were originally developed and have been used more frequently in Ponderosa Pine Forest (Sect. 4.4.1). This method usually involves analyzing rings of living trees and using decay classes to estimate ages of snags and downed logs currently on the site to determine/estimate which were alive and at what dbh on a selected date in the past. Reconstructions underestimate historical density and basal area when evidence of trees has been lost by combustion or decomposition (Fulé et al. 2002, 2003) or is otherwise missing.

Reconstructions to 1860 stand conditions in Pinyon-Juniper woodlands on three soils in central Arizona reported historical total tree densities of 110–246 trees/ha (45–100 trees/acre; Landis and Bailey 2005; unstated lower diameter limit). Densities of Colorado pinyon were 3–29 trees/ha (1–12 trees/acre) and Utah juniper were 94–217 trees/ha (38–88 trees/acre). Historical total basal area values were 5.0–10.6 m^2/ha (21.8–46.2 ft^2/acre). Basal areas of Colorado pinyon were 0–0.3 m^2/ha (0–1.3 ft^2/acre) and Utah juniper were 4.7–10.6 m^2/ha (20.5–46.2 ft^2/acre).

Another study reconstructed 1875 stand conditions for two Pinyon-Juniper woodland sites in northwestern Arizona (Huffman et al. 2008b). It determined similar historical total tree densities of 104 and 261 trees/ha (42 and 106 trees/acre) for trees 1.37 m (4.5 ft) height. Densities of Colorado pinyon were 41 and 65 trees/ha (17 and 26 trees/acre) and Utah juniper were 63 and 196 trees/ha (25 and 79 trees/acre). Basal area values were not provided.

Several concerns can be raised about these reconstruction studies. Neither provided data on individual diameter classes. This prevents direct comparison of results between studies as well as inference of nineteenth century stand dynamics. In addition, neither study reported data for intermediate dates that would provide insight into twentieth century dynamics. Moreover, evidence of nineteenth century trees undoubtedly was lost by the twenty-first century date of sampling. Neither study mentioned evidence of fire, so loss of evidence by combustion appears not to be an issue; however, loss of pinyons by decomposition is an issue. Study of Colorado pinyon snags and logs in southwestern Colorado determined that they do not persist beyond 25 years (Kearns et al. 2005; the study was done within infection centers of black stain root disease, but this fungus does not cause decay and dies with its host, according to Fairweather et al. 2006). Therefore, the above reconstruction studies done in 2002 likely missed pinyons dying between 1860/1875 and ~1977, a period that includes possible extensive pinyon mortality during the 1950s drought. The loss of pinyon to decomposition results in substantial underestimation of pinyon and total nineteenth century densities and basal areas.

In conclusion, the lack of details in historical accounts, the uncertain usefulness of data from relict sites, and the apparent inaccuracy of reconstructions combine with great diversity of vegetation to limit understanding of historical conditions of Pinyon-Juniper vegetation in the American Southwest.

5.4.1.2 Post-Euro-American Settlement

Despite decades of research, it remains challenging to understand changes in southwestern Pinyon-Juniper vegetation since Euro-American settlement. Reviews by Romme et al. (2008, 2009) have provided much-needed framework for greater clarity, but many challenges remain.

Among several reasons for the difficulties in understanding changes is the diversity of vegetation. All the divisions of Spruce-Fir (*Picea-Abies*), Mixed Conifer, and Ponderosa Pine Forests have similar structure: forest. In contrast, the divisions of Pinyon-Juniper vegetation have different structures: woodland, wooded shrubland, and savanna.

Another reason is that multiple major processes are involved in Pinyon-Juniper vegetation dynamics: infill, expansion, and succession and their paired, inverse processes of thinning, contraction, and disturbance. Six processes and three types of vegetation produce an 18-cell matrix. To what degree has infill occurred in persistent woodlands? Has contraction occurred in wooded shrublands? Has stand-replacing disturbance occurred in savanna? But even this matrix is oversimplified, because the structural diversity of pinyon-juniper vegetation only hints at the actual diversity affecting vegetation changes. Factors such as regional differences in climate and species, as well as local differences in soils, landforms, etc. also influence vegetation changes (e.g., Romme et al. 2009a; Jacobs 2011), as of course do differences in land uses.

A third reason that vegetation changes since post-Euro-American settlement are challenging to understand is the period of human land use. Not only is it much longer than in other types of upland southwestern vegetation (Sect. 5.2.6), but it also overlaps with periods of natural infill-thinning, expansion-contraction, and succession-disturbance. This overlap makes it difficult to disentangle human impacts and natural processes.

Fig. 5.19 Pinyon-Juniper vegetation of mostly junipers in 1871 near Truxton in northwestern Arizona (Photograph by Timothy H. O'Sullivan courtesy of National Archives and Records Administration)

The following several paragraphs review what different approaches – such as use of historical photographs and descriptions, relict sites, reconstructions, and repeat samplings – have shown about changes in southwestern Pinyon-Juniper vegetation since Euro-American settlement.

Early descriptions and photographs provide insight into vegetation changes in some specific areas (e.g., Figs. 5.19 and 5.20). A review of historical descriptions of valleys in central Utah led to the conclusion that junipers had infilled stands on hillsides and expanded into grasslands and sagebrush shrublands since the mid nineteenth century (Christensen and Johnson 1964; Rogers 1982). Repeat photography of images taken at different locations in north-central and northwestern Arizona in 1867 determined that pinyons and junipers had infilled in seven of the nine sites of photographs showing Pinyon-Juniper vegetation and had thinned in the other two sites (Gordon et al. 1992; sites had no evidence of tree removals). A later study in the same region that reviewed historical descriptions, revisited described sites, and included repeat photography (Shaw HG 2006) concluded:

> The picture that emerges of the general study area, for the mid-19th century, is one of a dry short grass prairie intermixed with stands of juniper. It may have been more savanna-like than it is now. Woodlands now seem denser, mainly on ridges in areas where they existed in 1854, but I see no evidence that they have greatly extended their range into the larger valleys…

Fig. 5.20 Pinyon-Juniper vegetation of mostly pinyons in 1891 in the San Francisco volcanic field, north-central Arizona. Note the variation in tree density in landscape (Photograph by G.K. Gilbert courtesy of U.S. Geological Survey)

Comparison of relict sites to paired non-relict sites potentially reveals changes that have taken place with human influences such as livestock grazing. Such studies have produced no clear evidence of changes in overstories, but their broader applicability is possibly limited (see previous section).

Another approach has been comparison of reconstructed nineteenth century tree densities and basal areas to recent samples from the same sites. Both of the reconstruction studies described in the previous section indicated large increases in trees since the nineteenth century on all sites studied. Mean percentage increases from 1860 to 2002 for three sites with different soils in central Arizona were 428 % in total stand density and 487 % in total stand basal area (Landis and Bailey 2005; unstated lower diameter limit). Respective percentage increases were 1,456 and 851 % for Colorado pinyon and 314 and 480 % for Utah juniper. Mean percentage increases from 1875 to 2002 for two sites in northwestern Arizona were 403 % in total stand density, 796 % in Colorado pinyon density, and 242 % increase in Utah juniper density for trees 1.37 m (4.5 ft) height (Huffman et al. 2008b; basal area values were not provided).

These values of changes in pinyon based on reconstructed nineteenth century values are inaccurate. Colorado pinyons dying by ~1977 would not have been evident by 2002 because of decomposition (see previous section). As a result, pinyon and total tree densities and basal areas in 1860/1875 were greatly underestimated and percentage increases to 2002 greatly overestimated. Moreover, understanding these reported increases in tree density and basal area is possibly further complicated by succession or by expansion followed by infill. Either of these processes is suggested by the lack of trees established before 1800 at one site examined by Huffman et al. (2008b), but they interpreted canopy changes at the site as infill of savanna.

Resampling historical study plots in southwestern Pinyon-Juniper vegetation potentially offers insight into canopy changes, but study plots older than several decades are uncommon. A qualitative study compared vegetation descriptions for 1870 survey lines in central Utah with vegetation observed almost a century later (Christensen and Johnson 1964). It was reported that juniper had expanded into grasslands as well as sagebrush shrublands and had infilled areas of expansion and old stands of juniper. Similar conclusions were reached by examination of historical accounts from the mid nineteenth century, as stated above. In contrast, no change in cover of Colorado pinyon or Utah juniper occurred from 1958 to 1996 on permanent plots located in Pinyon-Juniper vegetation on Fishtail Mesa, a relict site in Grand Canyon National Park (Jameson et al. 1962; Rowlands and Brian 2001). Both tree species had increased in sagebrush shrubland, apparently as part of post-fire succession.

Regional-scale insights into canopy changes in Pinyon-Juniper vegetation have been provided by determining the presence/absence of trees preceding Euro-American settlement at over 1,000 sites in the American Southwest (Jacobs 2011). Results indicated single-seed juniper had invaded nearby grasslands in areas with a relatively high percentage of summer rainfall and that the invasion was most likely related to livestock grazing.

In conclusion, the lack of accurate details on overstories of southwestern Pinyon-Juniper vegetation prior to Euro-American settlement (previous section) and the great diversity of conditions encompassed by the vegetation complicate region-wide understanding of overstory changes in the American Southwest. Nevertheless, research results using different approaches to studying overstory changes over the last century provide an overview. Findings include infill in some established stands, expansion into grassland, and increases during succession following stand-replacing fires. However, findings differ among sites, as some stands have experienced little change in the overstory.

5.4.2 Understory

There are no historical data on the composition and structure of the understory of Pinyon-Juniper vegetation of the American Southwest, but there are general descriptions, such as:

> …a country of shrub [junipers]…and richly clothed in bunch grass. (Pratt for central Utah in 1850; Christensen and Johnson 1964)

Such descriptions led to the conclusion that the pre-Euro-American settlement vegetation of central Utah included foothills "covered with grasses, scattered junipers…, and sagebrush", with bluebunch wheatgrass (*Pseudoroegneria spicata*) and big bluegrass (*Poa secunda*) as the most abundant grasses, along with western wheatgrass (*Pascopyrum smithii*) in level areas scattered among the foothills (Christensen and Johnson 1964). Later expansion and infill by Utah juniper and increases in sagebrush resulted in greatly reduced grass cover in the foothills (Christensen and Johnson 1964). There are regional differences in historical descriptions. For example, historical statements did not describe dense grass cover in Pinyon-Juniper vegetation in northwestern Arizona (although it was indicated for nearby vegetation; cf. Shaw HG 2006).

These differences in historical understories of central Utah and northwestern Arizona appear to represent Pinyon-Juniper savanna versus persistent woodlands. They also underscore the variability within Pinyon-Juniper vegetation, a characteristic apparent in overstory density but even more evident in understory structure and composition. Understory differences are affected by many factors such as grazing history, fire history, tree density, and soil features such as soil type, texture, and moisture (Paulsen 1975; Rowlands and Brian 2001; Milchunas 2006).

Historical changes in the understory of pinyon-juniper vegetation have been driven by livestock grazing, changes in tree densities, introduction of invasive species, fire and fire exclusion, climate variation, etc. Comparison of grazed and ungrazed sites has yielded information on the effects of livestock grazing on understories. These are detailed in Sect. 5.2.6.1, but in general grazing increased shrubs and decreased shrub richness. In addition, grazing affected herbs by reducing production, as well as altered species composition by reducing grasses and increasing forbs and annuals. Livestock also reduced cover of biological crusts by trampling.

Historical changes in understory vegetation due to increased tree densities likely included reduced production (Jameson 1962) and changed species composition (Pieper 1990). These are potentially reversed with reductions in tree density (cf. Clary 1971; Brockway et al. 2002). Invasive species are discussed in Sect. 5.2.6.4. Changes with succession following stand-replacing fire are outlined in Sect. 5.3.2. Understory changes due to climate variation and other factors are poorly known, in part because of interactive effects.

One attempt at greater clarity considered changes in both overstories and understories in highly impacted stands and classified stands into four degraded states, each with altered ecosystem processes: invaded, annualized, woody dominated, and severely eroded (Miller 2005). *Invaded* stands have functionally important invasive exotic plant species, but ecosystem processes (fire, etc.) are relatively little changed. *Annualized* stands have dominance by weedy annuals such as cheatgrass, and vegetation structure and ecosystem processes are greatly altered. *Woody-dominated* stands have persistent increased abundance of woody plants, and ecosystem processes such as fire can be affected. *Severely eroded* stands occur with the erosion of soils and resultant changes in resources.

In conclusion, the lack of data on historical structure and composition of southwestern Pinyon-Juniper understory prevents full understanding of change

(Romme et al. 2009b). Moreover, understanding is complicated by the same factors that challenge understanding of overstory changes, i.e., diversity of vegetation, multiple processes of vegetation dynamics, and long history of human land use (Sect. 5.4.1.2).

5.5 Conceptual Models

The following nested, three-tiered set of conceptual models is based on my interpretation of best-available information on Pinyon-Juniper vegetation of the American Southwest. The models summarize and synthesize material on drivers (Sect. 5.2) and processes (Sect. 5.3) of vegetation dynamics, and they account for historical changes (Sect. 5.4). Their format is explained in Sect. 1.7. The models can be used to facilitate understanding of the past and the present. They also can provide insight into effects of future changes in the environment and outcomes of alternative land-management decisions. Moreover, they can guide the development of quantitative models.

5.5.1 Ecosystem-Characterization Model

The ecosystem-characterization model for southwestern Pinyon-Juniper vegetation emphasizes *Vegetation* (a biotic component) and *Disturbance* (a driver; Fig. 5.21a, Table 5.2). Key aspects of *Vegetation* are structure, fuel, and tree, shrub, and herb cover, and these affect various aspects of *Disturbance*. The primary agents of *Disturbance* are drought, insect outbreaks, and fire, and these affect tree mortality and vegetation structure. The two other biotic components are *Soil System*, including fungal pathogens, and *Animals*, including insects that cause disturbance. A second driver is *Weather & Climate*, which causes drought and fires and influences soil moisture, fuel moisture, species recruitment, plant vigor, fire behavior, and insect population dynamics. Thus, climate drives patterns of infill & thinning and expansion & contraction. The third driver is *Landscape*, with its primary features being position and topography. It influences weather and climate, as well as patterns of disturbances. The model also includes six anthropogenic drivers (Fig. 5.21b, Table 5.2): *Livestock Grazing*, *Fire Management*, *Modern Climate Change*, *Invasive Species*, *Recreation*, and *Nearby Land Use*.

5.5.2 Vegetation-Dynamics Models

Separate models are needed to illustrate the vegetation dynamics of persistent woodlands, wooded shrublands, and savannas of southwestern Pinyon-Juniper vegetation.

Each vegetation-dynamics model is accompanied by a bar graph to show estimated relative shifts in abundances of the community types from circa 1870 (prior to widespread Euro-American land use) to circa 1970 (end of strict fire exclusion policy) to the present. The estimates are based on my understanding of vegetation dynamics and should be interpreted as showing relative differences and not interpreted quantitatively (hence the absence of a scale for abundance). Although not shown on the graphs, relative abundances shifted prior to circa 1870 as a result of variation in climate and disturbance. Localized differences also have been and are present, but are beyond the scope of these generalized bar graphs.

5.5.2.1 Pinyon-Juniper Persistent Woodland

The vegetation-dynamics model for Pinyon-Juniper persistent woodland has two states, five communities, and one transition (Fig. 5.22, Table 5.3). All five communities occurred historically. *State A* encompasses the most characteristic community, *A1 Woodland*, which is dominated by pinyons, junipers, and shrubs. It has been and continues to be the most common community. Community *A1* forms community *A2 Denser Woodland* (dominated by the same species) when tree recruitment results in infill, and *A2* forms *A1* when tree mortality results in thinning. Both of these communities can be impacted by stand-replacing fire and form community *A3 Early Successional* dominated by forbs, grasses, or shrubs, depending on local site conditions and time since the previous fire. With succession, trees establish in the community, forming community *A4 Mid Successional* dominated by various proportions of grasses, forbs, shrubs, junipers, and pinyons. Additional succession changes community *A4* into *A1*.

Transition A↔B changes *State A* to *State B* (community *B1 Grassland*) with stand-replacing fire followed by colonization by grasses that can include the invasive cheatgrass. Community *B1* is maintained by frequent fire. *Transition A↔B* can be reversed by invasion and establishment of junipers and/or pinyons.

5.5.2.2 Pinyon-Juniper Wooded Shrubland

Little is known about the vegetation dynamics of Pinyon-Juniper wooded shrublands, so this model is largely hypothetical. The model has two states, four communities, and one transition (Fig. 5.23, Table 5.4). All four communities occurred historically. The core and most common community was and is *A2 Wooded Shrubland*, dominated by shrubs, pinyons, and junipers. Tree mortality changes *A2* into community *A1 Shrubland* and appears as contraction of Pinyon-Juniper vegetation. *A1* forms *A2* with tree establishment and appears as expansion of Pinyon-Juniper vegetation. Additional tree establishment infills community *A2* and forms community *A3 Woodland*, which is dominated by pinyons, junipers, and shrubs. Tree mortality results in thinning and changes *A3* into community *A2*. In addition, *A3* and *A2* can change into *A1* by stand-replacing fire and regeneration of shrubs.

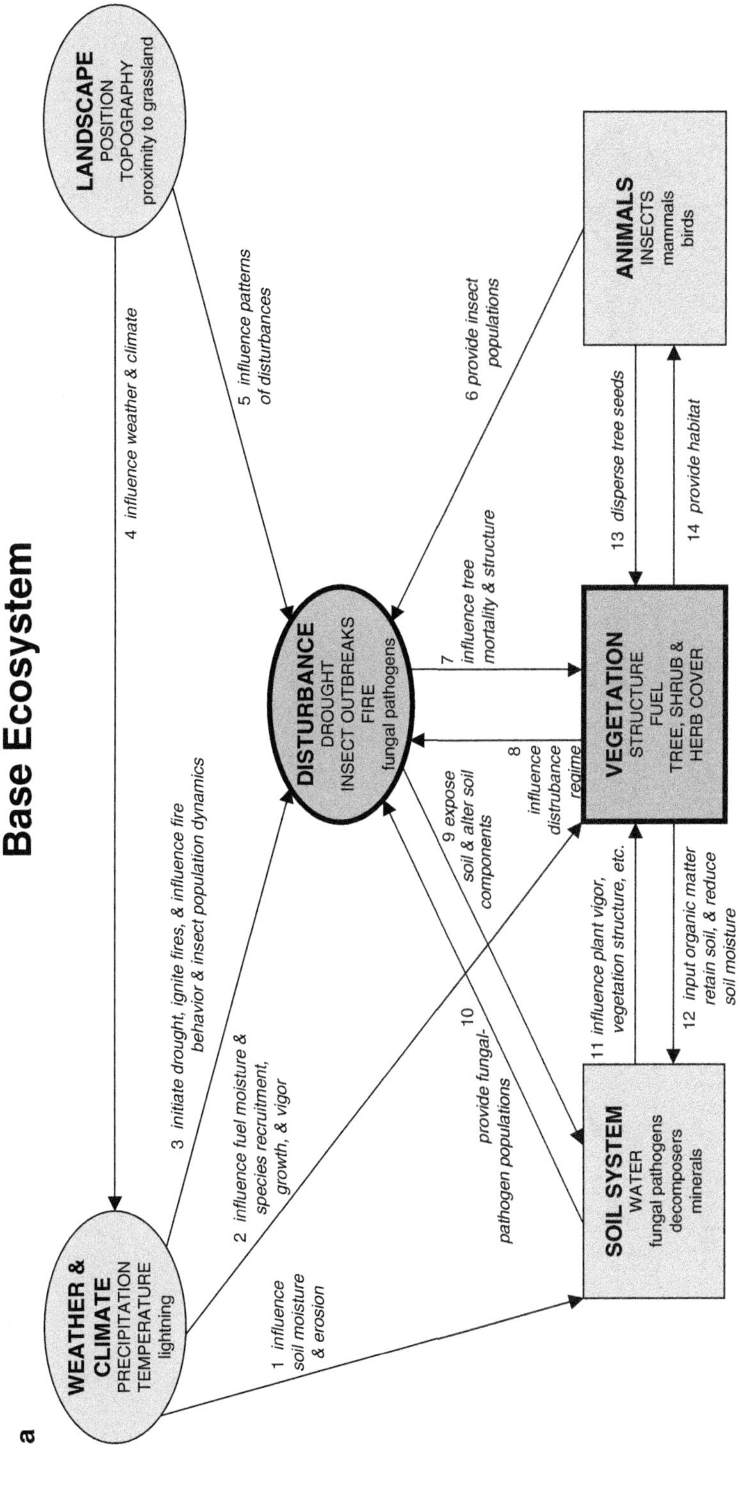
a
Base Ecosystem
LANDSCAPE
POSITION
TOPOGRAPHY
proximity to grassland
WEATHER & CLIMATE
PRECIPITATION
TEMPERATURE
lightning
DISTURBANCE
DROUGHT
INSECT OUTBREAKS
FIRE
fungal pathogens
ANIMALS
INSECTS
mammals
birds
VEGETATION
STRUCTURE
FUEL
TREE, SHRUB & HERB COVER
SOIL SYSTEM
WATER
fungal pathogens
decomposers
minerals
1 influence soil moisture & erosion
2 influence fuel moisture & species recruitment, growth, & vigor
3 initiate drought, ignite fires, & influence fire behavior & insect population dynamics
4 influence weather & climate
5 influence patterns of disturbances
6 provide insect populations
7 influence tree mortality & structure
8 influence distrubance regime
9 expose soil & alter soil components
10 provide fungal-pathogen populations
11 influence plant vigor, vegetation structure, etc.
12 input organic matter retain soil, & reduce soil moisture
13 disperse tree seeds
14 provide habitat

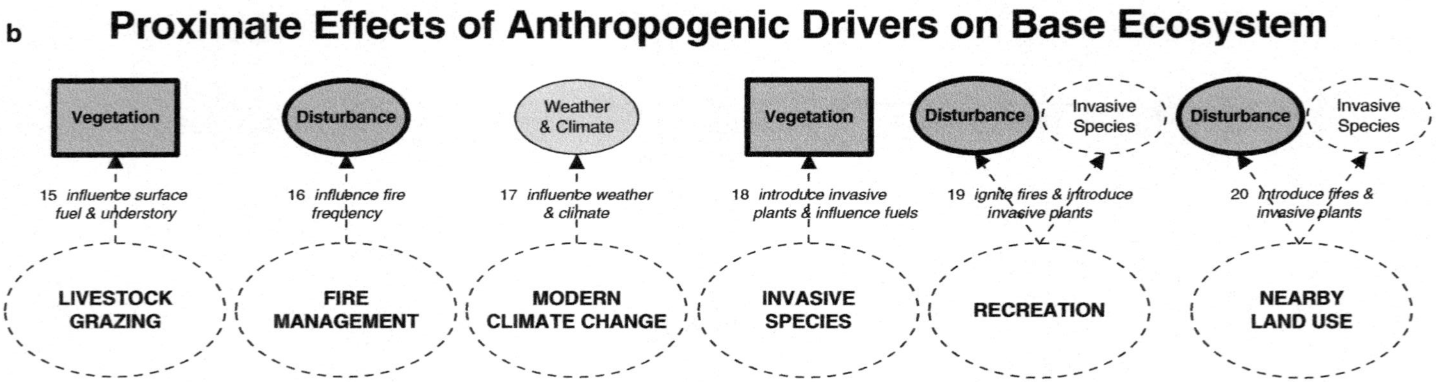

Fig. 5.21 (**a**, **b**) Ecosystem-characterization model for Pinyon-Juniper vegetation. In the base model (**a**), biotic components are shown as *rectangles* and drivers as *solid ovals*. Text within them identifies major features in *upper case* and other important features in *lower case*. Disturbance and Vegetation are *highlighted*, because they are the basis for the vegetation-dynamics models (Figs. 5.22, 5.23, and 5.24). In the anthropogenic supplement to the base model (**b**), drivers are shown as *dashed ovals*. Relationships in (**a**) and (**b**) are shown as *solid* and *dashed arrows*, respectively, and are described in Table 5.2

Table 5.2 Relationships in the ecosystem-characterization model of Pinyon-Juniper vegetation (Fig. 5.21a, b)

Relationships	Circa 1870	Present
1	Precipitation and temperature affect soil moisture. Winter precipitation is critical for recharging soil moisture. Precipitation and wind increase soil erosion	Same, but climate change increases the severity and duration of drought
2	Precipitation and temperature largely determine fuel moisture and are important in plant recruitment, growth, and vigor	Same, but climate change increases the severity and duration of drought
3	Consistently low precipitation and high temperature cause drought and affect insect population dynamics. Lightning ignites fires, and wind influences fire behavior	Same, but climate change increases temperatures, which increases drought. Drought increases the frequency and size of insect outbreaks
4	Topography influences weather and climate	Same
5	Patterns of fires and insect outbreaks are influenced by position within the landscape and topography	Same
6	Increases in populations of bark beetles can result in outbreaks	Same
7	Stand-replacing fires kill trees of persistent woodland and possibly wooded savanna. Frequent surface fires possibly thin tree populations in savanna. Bark beetles can kill trees in all three Pinyon-Juniper types. Drought increases stress in trees and outbreaks of bark beetles	Same, except that fire exclusion possibly reduces fire frequencies in savanna, and climate change increases temperatures, which increases drought. Drought increases the frequency and size of insect outbreaks
8	In persistent woodland and possibly in wooded shrubland, continuous fuels combine with strong wind (and lightning) to produce stand-replacing fires. In savanna, continuous understory fuels and discontinuous canopy fuels likely result in frequent, low-severity surface fires	Same
9	Fires partially and patchily expose soil surface and alter soil properties	Same
10	Increases in populations of fungal root pathogens result in outbreaks	Same
11	Water and minerals in soil affect plant vigor and growth and vegetation structure	Same, but less water is available because of drought and higher temperatures
12	Plants add organic matter to soil, stabilize soil, and reduce soil moisture, except in areas of crown fire	Same
13	Animals are important in dispersing seeds of trees	Same
14	Vegetation is habitat for animals	Same

15	Livestock grazing of savanna near areas of human settlement likely reduced fuel continuity by reducing herb cover	Same, but livestock grazing is more widespread and possibly more intensive
16	Apparently ineffective and of limited area	Past fire exclusion likely reduced fire frequency in savanna
17	Not applicable	Modern climate change affects weather and climate, raising temperature and increasing the frequency and extent of drought
18	Not applicable	Invasive plants can modify the herbaceous layer, change fuels, and change fire patterns
19	Not applicable	Recreationists ignite fires and introduce and spread invasive plants
20	Not applicable	Nearby land use that initiates fires and colonization by invasive plants can introduce fires and invasives into Pinyon-Juniper vegetation

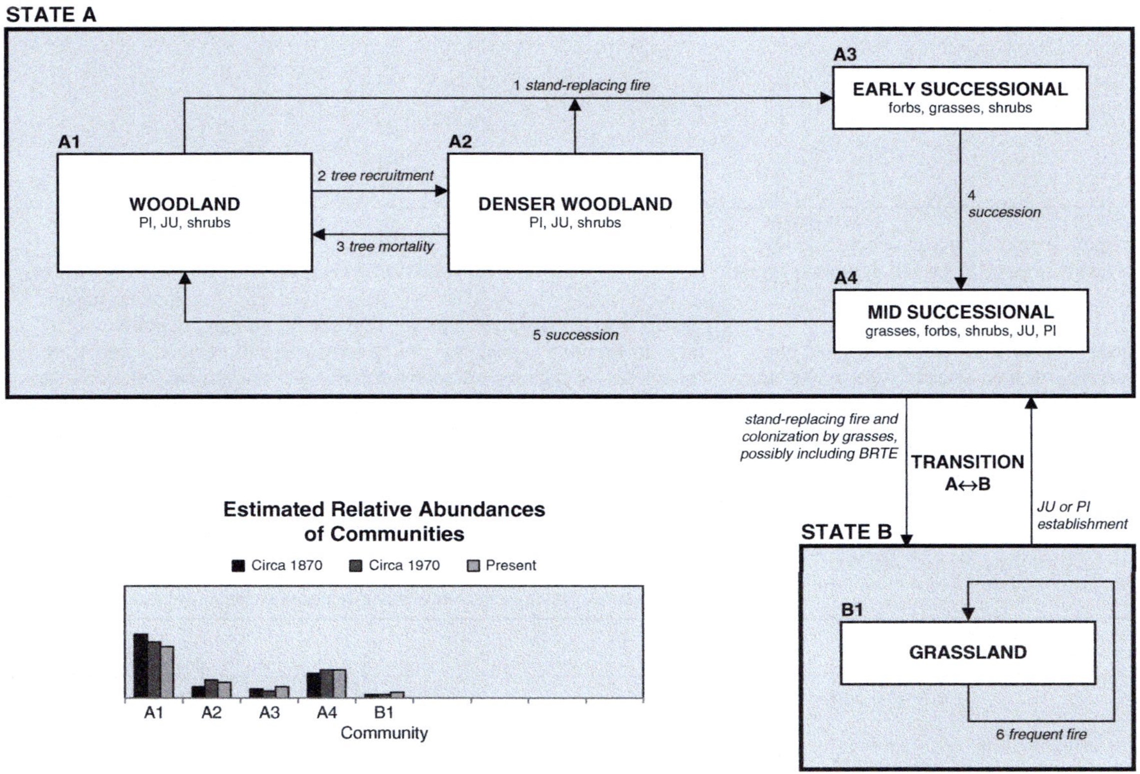

Fig. 5.22 Vegetation-dynamics model for Pinyon-Juniper persistent woodland. Model provides details on the vegetation-disturbance portion of the ecosystem-characterization model (Fig. 5.21a, b) by showing communities (*smaller, open rectangles*) within states (*larger, shaded rectangles*), processes that form the communities, and transitions between states. Text along *arrows* identifies pathways that are described in Table 5.3. *Bar graph* shows estimated relative shifts in abundances of the community types from circa 1870 (see Sect. 5.5.2). Species: *BRTE* cheatgrass, *JU* junipers, *PI* pinyons

Table 5.3 Relationships and transition in the vegetation-dynamics model of Pinyon-Juniper persistent woodland (Fig. 5.22)

Relationships and transition	Circa 1870	Present
1	Stand-replacing fire changes woodland and denser woodland into early successional community	Same
2	Tree recruitment infills woodland, producing denser woodland	Same
3	Tree mortality from drought and insects thins denser woodland, producing woodland	Same, but climate change increases drought, and drought increases insect outbreaks
4	Succession changes early successional stands into mid successional stands dominated by herbs and shrubs with scattered junipers and pinyons	Same
5	Succession changes mid successional stands into woodland	Same
A↔B	Stand-replacing fire followed by colonization of grasses changes woodland or denser woodland stands of State A into grassland community of State B. Reversion to State A requires invasion and establishment of junipers and/or pinyons	Same, but invasive species such as cheatgrass can be present and increase the frequency of stand-replacing fire
6	Frequent fire maintains grassland structure and composition	Same, but fire frequency is possibly increased by invasive species such as cheatgrass and decreased by fire management

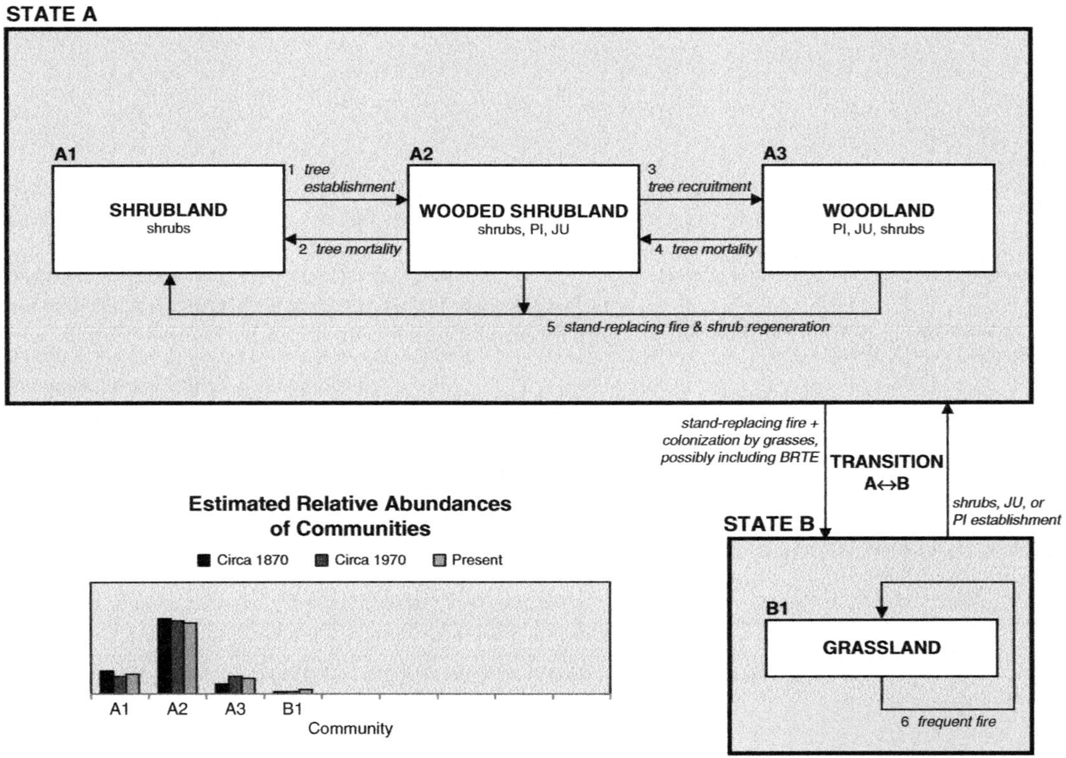

Fig. 5.23 Vegetation-dynamics model for Pinyon-Juniper wooded shrubland. Model provides largely hypothetical details on the vegetation-disturbance portion of the ecosystem-characterization model (Fig. 5.21a, b) by showing communities (*smaller, open rectangles*) within states (*larger*, *shaded rectangles*), processes that form the communities, and transitions between states. Text along *arrows* identifies pathways that are described in Table 5.4. *Bar graph* shows estimated relative shifts in abundances of the community types from circa 1870 (see Sect. 5.5.2). Species: *BRTE* cheatgrass, *JU* junipers, *PI* pinyons

Table 5.4 Relationships and transition in the vegetation-dynamics model for Pinyon-Juniper wooded shrubland (Fig. 5.23)

Relationships	Circa 1870	Present
1	Tree invasion and establishment change shrubland into wooded shrubland	Same
2	Tree mortality from drought and insect outbreaks changes wooded shrubland into shrubland	Same, but climate change increases drought, and drought increases insect outbreaks
3	Tree recruitment infills wooded shrubland, producing woodland	Same
4	Tree mortality from drought and insect outbreaks thins woodland, producing wooded shrubland	Same, but climate change increases drought, and drought increases insect outbreaks
5	Stand-replacing fire followed by shrub regeneration changes wooded shrubland and woodland into shrubland	Same
A↔B	Stand-replacing fire followed by colonization of grasses changes stands of State A into grassland of State B. Reversion to State A requires invasion and establishment of shrubs, junipers, and/or pinyons	Same, but invasive species such as cheatgrass can be present and increase the frequency of stand-replacing fire
6	Frequent fire maintains grassland structure and composition	Same, but fire frequency is possibly increased by invasive species such as cheatgrass and decreased by fire management

Because little is known about the dynamics of Pinyon-Juniper wooded shrublands, the model is largely hypothetical

Transition A↔B changes any of the communities of *State A* to *State B* (community *B1 Grassland*) with stand-replacing fire followed by colonization by grasses that can include the invasive cheatgrass. Community *B1* is maintained by frequent fire. At least theoretically, *Transition A↔B* can be reversed by invasion and establishment of shrubs, junipers, and/or pinyons.

5.5.2.3 Pinyon-Juniper Savanna

Little is known about the vegetation dynamics of Pinyon-Juniper savanna, so this model is largely hypothetical. The model has one state, three communities, and no transitions (Fig. 5.24, Table 5.5). All three communities occurred historically. The core and most common community was and is *A2 Savanna*, dominated by grasses, forbs, shrubs, junipers, and pinyons. Frequent fire maintains *A2*, but tree mortality changes *A2* into community *A1 Grassland* and appears as contraction of Pinyon-Juniper vegetation. Frequent fire maintains *A1*. Alternatively, *A1* forms *A2* with tree establishment and this appears as expansion of Pinyon-Juniper vegetation. Additional tree recruitment infills community *A2* and forms community *A3 Woodland*, which is dominated by pinyons, junipers, and shrubs. Tree mortality results in thinning and changes *A3* into community *A2*. *A3* can also change into *A1* by stand-replacing fire. Grasses can include the invasive cheatgrass.

5.5.3 Mechanistic Models

Differences among southwestern Pinyon-Juniper persistent woodlands, wooded shrublands, and savanna require two mechanistic models (pending full characterization of fire regimes).

5.5.3.1 Pinyon-Juniper Persistent Woodland and Wooded Shrubland

The vegetation-dynamics models for Pinyon-Juniper persistent woodland and wooded shrubland are explained by the same mechanistic model (Fig. 5.25). It has six biotic components on the right side of the figure (including three aspects of fuels), three drivers on the left side, and four anthropogenic factors at the bottom. In general, *Herbs & Shrubs*, *Trees*, and *Precipitation & Temperature* affect the three fuel characteristics. All three fuel components, *Nearby Land Use*, and *Recreation* influence *Crown Fire*. Also, *Crown Fire*, *Precipitation & Temperature*, and *Insect Populations* influence characteristics of *Trees*, such as tree size, density, cover, and vigor. *Trees* and *Herbs & Shrubs* determine *Community Type* (of the five/four appearing in the vegetation-dynamics models). *Modern Climate Change* influences *Precipitation & Temperature. Nearby Land Use* and *Recreation* are sources of *Invasive Species*, which influence the species composition and cover of the herbs in *Herbs & Shrubs*.

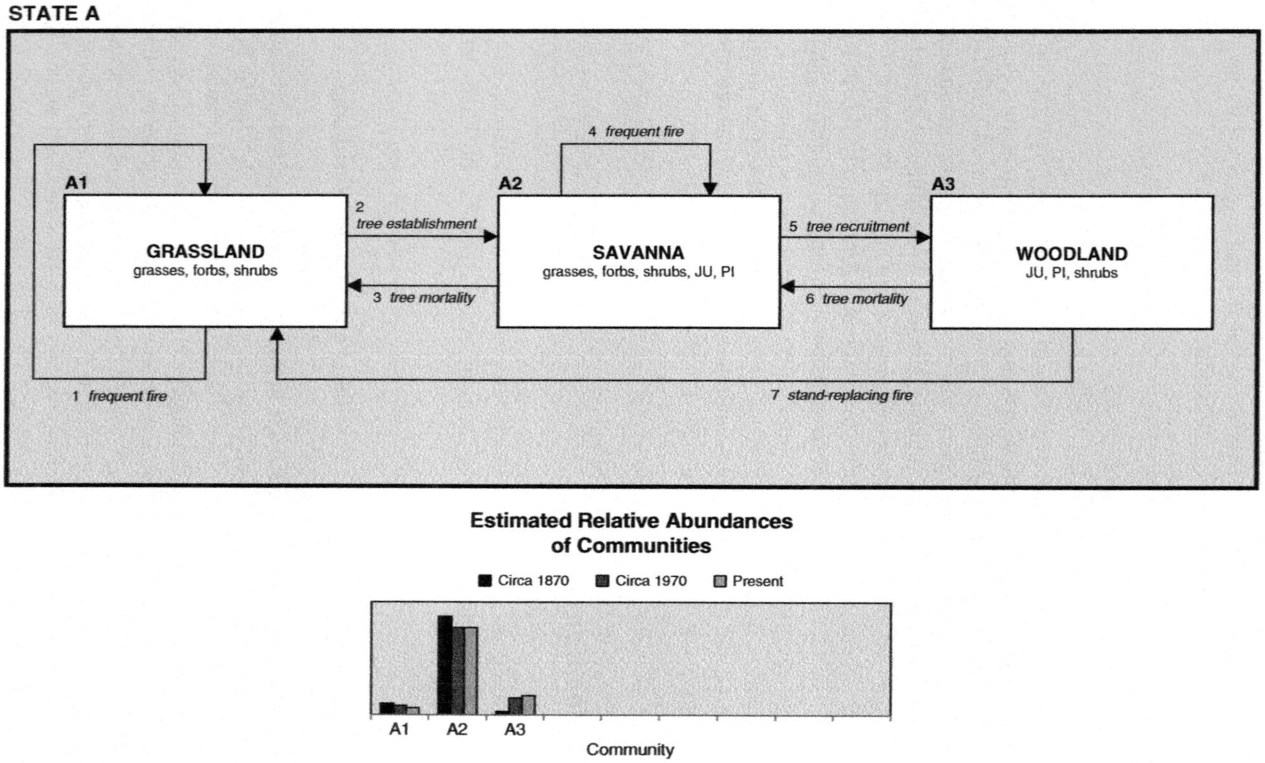

Fig. 5.24 Vegetation-dynamics model for Pinyon-Juniper savanna. Model provides largely hypothetical details on the vegetation-disturbance portion of the ecosystem-characterization model (Fig. 5.21a, b) by showing communities (*smaller*, *open rectangles*) within states (*larger*, *shaded rectangles*), processes that form the communities, and transitions between states. Text along *arrows* identifies pathways that are described in Table 5.5. *Bar graph* shows estimated relative shifts in abundances of the community types from circa 1870 (see Sect. 5.5.2). Grasses can include the invasive cheatgrass. Species: *JU* junipers, *PI* pinyons

Table 5.5 Relationships in the vegetation-dynamics model of Pinyon-Juniper savanna (Fig. 5.24)

Relationships	Circa 1870	Present
1	Frequent fire maintains stand structure and composition	Same, but fire frequency is possibly altered by invasive species such as cheatgrass and by fire management
2	Tree invasion and establishment changes grassland into savanna	Same, but can be facilitated by more widespread and intensive livestock grazing. Also is possibly affected by alterations in the fire regime related to invasive species such as cheatgrass and fire management
3	Tree mortality from drought, insects, and fire changes savanna into grassland	Same, but climate change increases drought, and drought increases insect outbreaks. Also is possibly affected by alterations in the fire regime related to invasive species such as cheatgrass and fire management
4	Frequent fire maintains stand structure and composition	Same, but fire frequency is possibly altered by invasive species such as cheatgrass and by fire management
5	Tree recruitment infills savanna, producing woodland	Same, but can be facilitated by more widespread and intensive livestock grazing. Also is possibly affected by alterations in the fire regime related to invasive species such as cheatgrass and fire management
6	Tree mortality from drought and insects thins woodland, producing savanna	Same, but climate change increases drought, and drought increases insect outbreaks
7	Stand-replacing fire changes woodland into grassland	Same

Because little is known about the dynamics of Pinyon-Juniper savanna, the model is largely hypothetical

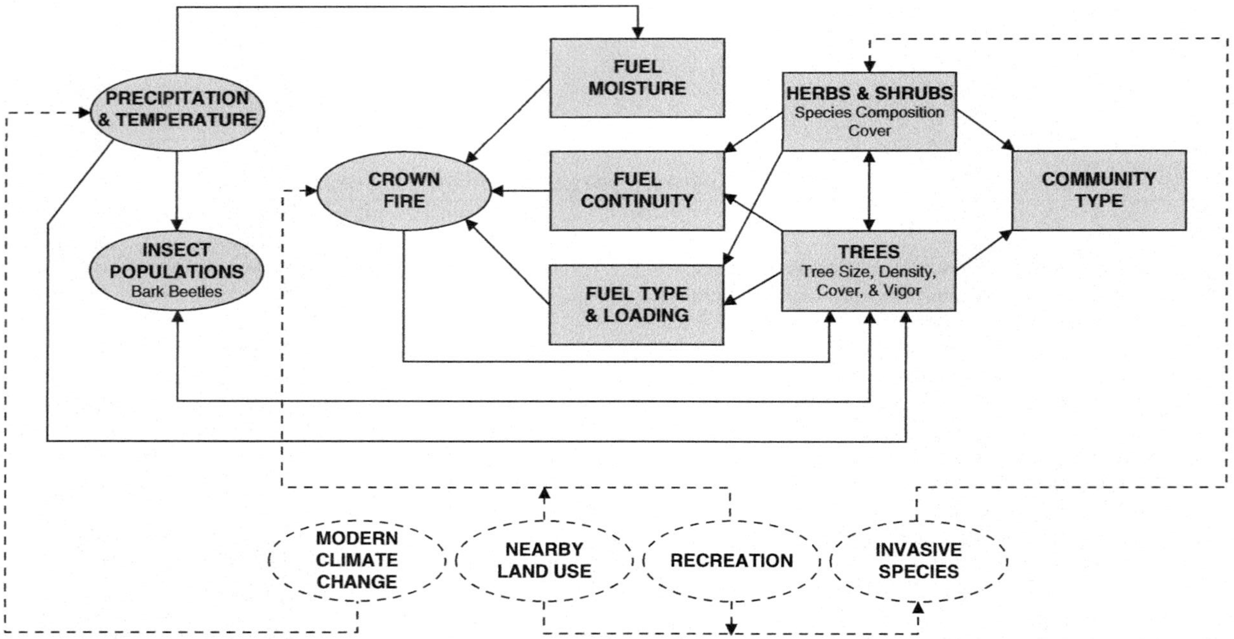

Fig. 5.25 Mechanistic model for Pinyon-Juniper persistent woodland and wooded shrubland. Model provides details on the mechanisms involved in the vegetation-dynamics models (Figs. 5.22 and 5.23). *Rectangles* are biotic components (including fuels), *ovals* are drivers, and *dashed ovals* are anthropogenic drivers

5.5.3.2 Pinyon-Juniper Savanna

The vegetation-dynamics model for Pinyon-Juniper savanna is explained in the second mechanistic model (Fig. 5.26). It also has six biotic components on the right side of the figure (including three aspects of fuels) and three drivers on the left side, but has six anthropogenic factors at the bottom (two more than in the previous mechanistic model). In general, *Herbs & Shrubs*, *Trees*, and *Precipitation & Temperature* affect the three fuel characteristics. All three fuel components, *Fire Management*, *Nearby Land Use*, and *Recreation* influence *Surface Fire*. Also, *Surface Fire*, *Precipitation & Temperature*, and *Insect Populations* influence characteristics of *Trees*, such as tree density and vigor. *Trees* and *Herbs & Shrubs* determine *Community Type* (of the three appearing in the vegetation-dynamics models). *Modern Climate Change* influences *Precipitation & Temperature. Nearby Land Use* and *Recreation* are sources of *Invasive Species. Invasive Species* and *Livestock Grazing* influence the species composition and cover of the herbs in *Herbs & Shrubs*.

5.6 Conclusions and Challenges

Despite much research, Pinyon-Juniper vegetation is as poorly understood as any vegetation type covered in this book – an unfortunate situation considering it covers the majority of the land area on mountains and plateaus of the American Southwest. Pinyon-Juniper vegetation appears to have a broad range of fire regimes, including infrequent, high-severity crown fire in persistent woodlands. However, the fire regimes of wooded shrublands and savannas are largely unknown and challenging to study. The lack of clear understanding of fire regimes severely inhibits development of ecologically based land-management practices. But fire is likely secondary, as vegetation dynamics are driven more by biotic disturbance agents such as pinyon bark beetle and by climate variation. Fortunately, more is known about these disturbance agents. Another area of great need is research on anthropogenic drivers. Given the lengthy inhabitation of Pinyon-Juniper vegetation by humans, it is important to better disentangle anthropogenic effects from natural effects. More specifically, additional research is needed on past and present impacts of livestock grazing, effects of fire exclusion, present and future impacts of climate change, spread and effects of invasive species, impact and management of recreation, and effects of nearby land use. Research is also needed on processes of vegetation dynamics, including the under-studied process of contraction, causes of infill, effects of grazing and fire exclusion on infill, and possible role of elevated CO_2. More information is also needed on historical conditions, but management should not necessarily consider nineteenth century conditions to be objectives for vegetation restoration. Instead, management needs to recognize Pinyon-Juniper vegetation as an oscillating system with links to climate variations that drive the processes of expansion & contraction and infill & thinning and affect succession & disturbance.

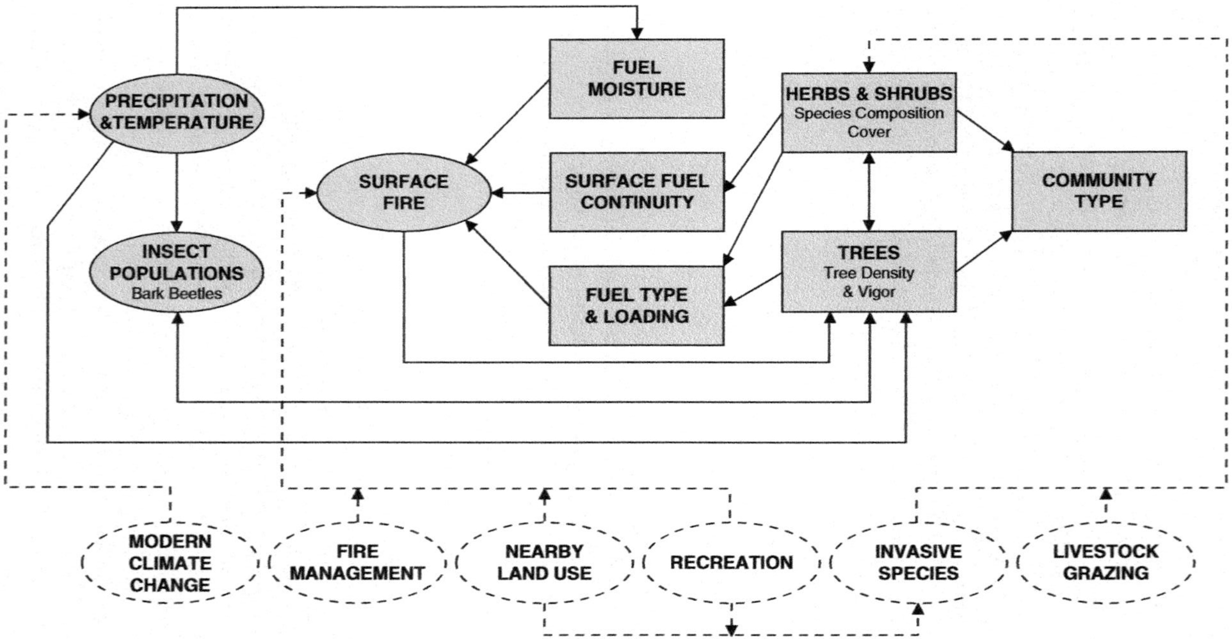

Fig. 5.26 Mechanistic model for Pinyon-Juniper savanna. Model provides details on the mechanisms involved in the vegetation-dynamics model (Fig. 5.24). *Rectangles* are biotic components (including fuels), *ovals* are drivers, and *dashed ovals* are anthropogenic drivers

Therefore, land managers should not over-react to fluctuations that appear to be within the broad natural boundaries of the system. Instead, managers should rely on site-specific studies, at least until the variability of Pinyon-Juniper vegetation is better understood and more detailed generalities are available.

References

Adams HD, Guardiola-Claramonte M, Barron-Gafford GA, Villegas JC, Breshears DD, Zou CB, Troch PA, Huxman TE (2009) Temperature sensitivity of drought-induced tree mortality portends increased regional die-off under global change-type drought. Proc Nat Acad Sci USA 106:7063–7066

Alcoze T (2003) First peoples in the pines: historical ecology of humans and ponderosas. In: Friederici P (ed) Ecological restoration of southwestern ponderosa pine forests. Island Press, Washington, DC

Allen CD (1989) Changes in the landscape of the Jemez Mountains, New Mexico. Dissertation, University of California, Berkeley

Allen CD, Breshears DD (1998) Drought-induced shift of a forest-woodland ecotone: rapid landscape response to climate variation. Proc Nat Acad Sci USA 95:14839–14842

Arnold JF, Jameson DA, Reid EH (1964) The pinyon-juniper type of Arizona: effects of grazing, fire, and tree control. U.S. Department of Agriculture, Forest Service Production Research Report No. 84, Washington, DC

Baisan CH, Swetnam TW (1997) Interactions of fire regimes and land use in the central Rio Grande Valley. U.S. Department of Agriculture, Forest Service Research Paper RM-RP-330, Rocky Mountain Forest and Range Experiment Station, Fort Collins, CO

Baker WL, Shinneman DJ (2004) Fire and restoration of piñon-juniper woodlands in the western United States: a review. For Ecol Manag 189:1–21

Barger NN, Adams HD, Woodhouse C, Neff JC, Asner GP (2009) Influence of livestock grazing and climate on pinyon pine (*Pinus edulis*) dynamics. Rangel Ecol Manag 62:531–539

Barney MA, Frischknecht NC (1974) Vegetation changes following fire in the pinyon-juniper type of west-central Utah. J Range Manag 27:91–96

Barrows JS (1978) Lightning fires in southwestern forests. Final report to USDA Forest Service, Northern Forest Fire Laboratory, Intermountain Forest and Range Experiment Station, under cooperative agreement 16–568 CA with Rocky Mountain Forest and Range Experiment Station, Fort Collins, CO

Bates ST (2006) A preliminary checklist of Arizona macrofungi. Canotia 2:47–78

Belnap J, Lange OL (eds) (2001) Biological soil crusts: structure, function, and management. Springer, New York

Bentz B, Logan J, MacMahon J, Allen CD, Ayres M, Berg E, Carroll A, Hansen M, Hicke J, Joyce L, Macfarlane W, Munson S, Negrón J, Paine T, Powell J, Raffa K, Régnière J, Reid M, Romme B, Seybold SJ, Six D, Tomback D, Vandygriff J, Veblen T, White M, Witcosky J, Wood D (2009) Bark beetle outbreaks in western North America: causes and consequences. University of Utah Press, Salt Lake City

Betancourt JL (1987) Paleoecology of pinyon-juniper woodlands: summary. In: Everett RL (compiler) Proceedings – Pinyon-juniper conference, 13–16 Jan 1986, Reno, Nevada. U.S. Department of Agriculture, Forest Service General Technical Report INT-215, Intermountain Research Station, Ogden, UT

Betancourt JL, Van Devender TR (1981) Holocene vegetation in Chaco Canyon, New Mexico. Science 214:656–658

Betancourt JL, Pierson EA, Rylander KA, Fairchild-Parks JA, Dean JS (1993) Influence of history and climate on New Mexico piñon-juniper woodlands. In Aldon EF, Shaw DW (tech coords) Managing piñon-juniper ecosystems for sustainability and social needs: proceedings of the symposium, 26–30 Apr 1993, Santa Fe, New Mexico. U.S. Department of Agriculture, Forest Service General Technical Report RM-236, Rocky Mountain Research Station, Fort Collins, CO

Beymer RJ, Klopatek JM (1992) Effects of grazing on cryptogamic crusts in pinyon-juniper woodlands in Grand Canyon National Park. Am Midl Nat 127:139–148

Breshears DD, Nyhan JW, Heil CE, Wilcox BP (1998) Effects of woody plants on microclimate in a semiarid woodland: soil temperature and evaporation in canopy and intercanopy patches. Int J Plant Sci 159:1010–1017

Breshears DD, Cobb NS, Rich PM, Price KP, Allen CD, Balice RG, Romme WH, Kastens JH, Floyd ML, Belnap J, Anderson JJ, Myers OB, Meyer CW (2005) Regional vegetation die-off in response to global-change-type drought. Proc Natl Acad Sci 102:15144–15148

Brockway DG, Gatewood RG, Paris RB (2002) Restoring grassland savannas from degraded pinyon-juniper woodlands: effects of mechanical overstory reduction and slash treatment alternatives. J Environ Manag 64:179–197

Brown DE (1994) Great Basin conifer woodland. In: Brown DE (ed) Biotic communities: southwestern United States and northwestern Mexico. University of Utah Press, Salt Lake City

Cartledge TR, Propper JG (1993) Piñon-juniper ecosystems through time: information and insights from the past. In Aldon EF, Shaw DW (eds) Managing piñon-juniper ecosystems for sustainability and social needs: proceedings of the symposium, 26–30 Apr 1993, Santa Fe, New Mexico. U.S. Department of Agriculture, Forest Service General Technical Report RM-236, Rocky Mountain Research Station, Fort Collins, CO

Chambers JC (2001) *Pinus monophylla* establishment in an expanding *Pinus-Juniperus* woodland: environmental conditions, facilitation and interacting factors. J Veg Sci 12:27–40

Chambers JC, Vander Wall SB, Schupp EW (1999) Seed and seedling ecology of piñon and juniper species in the pygmy woodlands of western North America. Bot Rev 65:1–38

Christensen EM, Johnson HB (1964) Presettlement vegetation and vegetational change in three valleys in central Utah. Brigh Young Univ Sci Bull Biolog Ser 4(4):1–16

Clary WP (1971) Effects of Utah juniper removal on herbage yields from Springerville soils. J Range Manag 24:373–378

Cole KL, Ironside KE, Arundel ST, Duffy P, Shaw J (2008) Modeling future plant distributions on the Colorado Plateau: an example using *Pinus edulis*. In: van Riper C III, Sogge MK (eds) The Colorado Plateau III: Integrating research and resources management for effective conservation. Proceedings of the 8th biennial conference of research on the Colorado Plateau. University of Arizona Press, Tucson

Conklin DA, Fairweather ML (2010) Dwarf mistletoes and their management in the southwest. U.S. Department of Agriculture, Forest Service, Southwestern Region, R3-FH-10-01. http://www.fs.fed.us/r3/publications/documents/dm_management.pdf. Accessed 12 Jan 2012

Davis OK (1987) Palynological evidence for historic juniper invasion in central Arizona: a late-Quaternary perspective. In: Everett RL (compiler) Proceedings – Pinyon-juniper conference, 13–16 Jan 1986, Reno, Nevada. U.S. Department of Agriculture, Forest Service General Technical Report INT-215, Intermountain Research Station, Ogden, UT

Despain DW, Mosley JC (1990) Fire history and stand structure of a pinyon-juniper woodland at Walnut Canyon National Monument, Arizona. U.S. Department of the Interior, National Park Service Technical Report No. 34. Cooperative National Park Resources Studies Unit, University of Arizona, Tucson

Dick-Peddie WA (1993a) Ecology and diversity on piñon-juniper woodland in New Mexico. In Aldon EF, Shaw DW (eds) Managing piñon-juniper ecosystems for sustainability and social needs: proceedings of the symposium, 26–30Apr 1993, Santa Fe, New Mexico. U.S. Department of Agriculture, Forest Service General Technical Report RM-236, Rocky Mountain Research Station, Fort Collins, CO

Dick-Peddie WA (1993b) New Mexico vegetation: past, present, and future. University of New Mexico Press, Albuquerque

Dusenberry WH (1963) The Mexican Mesta: the administration of ranching in Colonial Mexico. University of Illinois Press, Urbana
Dwyer DD, Pieper RD (1967) Fire effects on blue grama-pinyon-juniper rangeland in New Mexico. J Range Manag 20:359–362
Eager T (2008) Piñon mortality from 2001 to 2005: causes and management strategies. In: Gottfried GJ, Shaw JD, Ford PL (compilers) Ecology, management, and restoration of piñon-juniper and ponderosa pine ecosystems: combined proceedings of the 2005 St. George, Utah and 2006 Albuquerque, New Mexico workshops. U.S. Department of Agriculture, Forest Service Proceedings RMRS-P-51, Rocky Mountain Research Station, Fort Collins, CO
Erdman JA (1970) Pinyon-juniper succession after natural fires on residual soils of Mesa Verde, Colorado. Brigh Young Univ Sci Bull Biolog Ser 11(2):1–26
Ernst R, Pieper RD (1996) Changes in pinyon-juniper vegetation: a brief history. Rangel 18:14–16
Fairweather ML, McMillin J, Rogers T, Conklin D, Fitzgibbon B (2006) Field guide to insects and diseases of Arizona and New Mexico forests. U.S. Department of Agriculture, Forest Service Southwestern Region MR-R3-16-3, Albuquerque, NM
Fenner P (2008) Effects of invasive plants on public land management of pinyon-juniper woodlands in Arizona. In: Gottfried GJ, Shaw JD, Ford PL (compilers) Ecology, management, and restoration of piñon-juniper and ponderosa pine ecosystems: combined proceedings of the 2005 St. George, Utah and 2006 Albuquerque, New Mexico workshops. U.S. Department of Agriculture, Forest Service Proceedings RMRS-P-51, Rocky Mountain Research Station, Fort Collins, CO
Ffolliott PF, Gottfried GJ (2002) Dynamics of a pinyon-juniper stand in northern Arizona: a half-century history. U.S. Department of Agriculture, Forest Service Research Paper RMRS-RP-35, Rocky Mountain Research Station, Fort Collins, CO
Ffolliott PF, Gottfried GJ (2008) Plant communities and associations. In: Ffolliott PF, Davis OK (eds) Natural environments of Arizona: from deserts to mountains. University of Arizona Press, Tucson
Floyd ME (1982) The interaction of piñon pine and Gambel oak in plant succession near Dolores, Colorado. Southwest Nat 27:143–147
Floyd ML, Romme WH, Hanna DD (2000) Fire history and vegetation pattern in Mesa Verde National Park, Colorado, U.S.A. Ecol Appl 10:1666–1680
Floyd ML, Hanna DD, Romme WH (2004) Historical and recent fire regimes in piñon-juniper woodlands on Mesa Verde, Colorado, U.S.A. For Ecol Manag 198:269–289
Floyd ML, Hanna D, Romme WH, Crews TE (2006) Predicting and mitigating weed invasions to restore natural post-fire succession in Mesa Verde National Park, Colorado, USA. Int J Wildland Fire 15:247–259
Floyd ML, Romme WH, Hanna DD, Winterowd M, Hanna D, Spence J (2008) Fire history of piñon-juniper woodlands on Navajo Point, Glen Canyon National Recreation Area. Nat Areas J 28:26–36
Floyd ML, Clifford M, Cobb NS, Hanna D, Delph R, Ford P, Turner D (2009) Relationship of stand characteristics to drought-induced mortality in three southwestern piñon-juniper woodlands. Ecol Appl 19:1223–1230
Fuchs EH (2001) The changing faces of Lincoln County's watersheds. Water Resources Research Institute, 46th New Mexico Annual Water Conference Proceedings 323:69–82. http://wrri.nmsu.edu/publish/watcon/proc46/fuchs.pdf. Accessed 11 Apr 2012
Fuchs EH (2002) Historic increases in woody vegetation in Lincoln County, New Mexico. Vanguard Printing, Albuquerque
Fulé PZ, Covington WW, Moore MM, Heinlein TA, Waltz AEM (2002) Natural variability in forests of the Grand Canyon, USA. J Biogeogr 29:31–47
Fulé PZ, Crouse JE, Heinlein TA, Moore MM, Covington WW, Verkamp G (2003) Mixed-severity fire regime in a high-elevation forest of Grand Canyon, Arizona, USA. Landsc Ecol 18:465–486
Gitlin AR, Sthultz CM, Bowker MA, Stumpf S, Paxton KL, Kennedy K, Muñoz A, Bailey JK, Whitham TG (2006) Mortality gradients within and among dominant plant populations as barometers of ecosystem change during extreme drought. Conserv Biol 20:1477–1486

Gordon MM (ed) (1988) Through Indian country to California: John P. Sherburne's diary of the Whipple Expedition, 1853–1854. Stanford University Press, Stanford, CA

Gordon BR, Parrott GP, Smith JB (1992) Vegetation changes in northern Arizona: the Alexander Gardner photos. Rangel 14:308–320

Gottfried GJ, Pieper RD (2000) Pinyon-juniper rangelands. Jemison R, Raish C (eds) Livestock management in the American Southwest: ecology, society, and economics. Elsevier, New York

Gottfried GJ, Swetnam TW, Allen CD, Betancourt JL, Chung-MacCoubrey AL (1995) Pinyon-juniper woodlands. In: Finch DM, Tainter JA (tech coords) Ecology, diversity, and sustainability of the Middle Rio Grande Basin. U.S. Department of Agriculture, Forest Service, General Technical Report RM-GTR-268, Rocky Mountain Forest and Range Experiment Station, Fort Collins, CO

Grand Canyon National Park (2012) Grand Canyon National Park Fire Management Plan. U.S. Department of the Interior, National Park Service, Grand Canyon National Park, Grand Canyon, AZ. http://www.nps.gov/grca/parkmgmt/upload/GRCA_FMP.pdf. Accessed 30 Apr 2012

Gray ST, Betancourt JL, Jackson ST, Eddy RG (2006) Role of multidecadal climate variability in a range extension of pinyon pine. Ecology 87:1124–1130

Guenther D, Stohlgren TJ, Evangelista P (2004) A comparison of a near-relict site and a grazed site in a pinyon-juniper community in the Grand Staircase-Escalante National Monument, Utah. In: van Riper CIII, Cole KL (eds) The Colorado Plateau: culture, biological, and physical research. University of Arizona Press, Tucson

Haire SL, McGarigal K (2008) Inhabitants of landscape scars: succession of woody plants after large, severe forest fires in Arizona and New Mexico. Southwest Nat 53:146–161

Harper KT, Sanderson SC, McArthur ED (2003) Pinyon-juniper woodlands in Zion National Park, Utah. West N Am Nat 63:189–202

Harris AT, Asner GP, Miller ME (2003) Changes in vegetation structure after long-term grazing in pinyon-juniper ecosystems: integrating imaging spectroscopy and field studies. Ecosystems 6:368–383

Huffman DW, Fulé PZ, Pearson KM, Crouse JE (2008a) Fire history of pinyon-juniper woodlands at upper ecotones with ponderosa pine forests in Arizona and New Mexico. Can J For Res 38:2097–2108

Huffman DW, Stoddard MT, Fulé PZ, Covington WW, Smith HB (2008b) A demonstration project to test ecological restoration of a pinyon-juniper ecosystem. In: Gottfried GJ, Shaw JD, Ford PL (compilers) Ecology, management, and restoration of piñon-juniper and ponderosa pine ecosystems: combined proceedings of the 2005 St. George, Utah and 2006 Albuquerque, New Mexico workshops. U.S. Department of Agriculture, Forest Service Proceedings RMRS-P-51, Rocky Mountain Research Station, Fort Collins, CO

Huffman DW, Fulé PZ, Crouse JE, Pearson KM (2009) A comparison of fire hazard mitigation alternatives in pinyon-juniper woodlands of Arizona. For Ecol Manag 257:628–635

Huffman DW, Crouse JE, Chancellor WW, Fulé PZ (2012) Influence of time since fire on pinyon-juniper woodland structure. For Ecol Manag 274:29–37

Integrated Taxonomic Information System (2012) http://www.itis.gov. Accessed 3 Apr 2012

Intergovernmental Panel on Climate Change (2007) Summary for policymakers. In: Solomon S, Qin D, Manning M, Chen Z, Marquis M, Averyt KB, Tignor M, Miller HL (eds) Climate Change 2007: The Physical Science Basis. Contribution of Working Group I to the Fourth Assessment Report of the Intergovernmental Panel on Climate Change. Cambridge University Press, Cambridge, UK/New York

Jacobs BF (2008) Southwestern U.S. juniper savanna and piñon-juniper woodland communities: ecological history and natural range of variability. In: Gottfried GJ, Shaw JD, Ford PL (compilers) Ecology, management, and restoration of piñon-juniper and ponderosa pine ecosystems: combined proceedings of the 2005 St. George, Utah and 2006 Albuquerque, New Mexico workshops. U.S. Department of Agriculture, Forest Service Proceedings RMRS-P-51, Rocky Mountain Research Station, Fort Collins, CO

Jacobs BF (2011) Spatial patterns and ecological drivers of historic piñon-juniper woodland expansion in the American southwest. Ecography 34:1085–1095

Jacobs BF, Romme WH, Allen CD (2008) Mapping "old" vs. "young" piñon-juniper stands with a predictive topo-climatic model. Ecol Appl 18:1627–1641

Jameson DA (1962) Effects of burning on a galleta-black grama range invaded by juniper. Ecology 43:760–763

Jameson DA, Williams JA, Wilton EW (1962) Vegetation and soils of Fishtail Mesa, Arizona. Ecology 43:403–410

Johnsen TN Jr (1962) One-seed juniper invasion of northern Arizona grasslands. Ecol Monogr 32:187–207

Johnsen TN Jr, Elson JW (1979) Sixty years of change on a central Arizona grassland-juniper woodland ecotone. U.S. Department of Agriculture, Agricultural Research, Western Region, Science and Education Administration Agricultural Reviews and Manuals ARM-W-7, Oakland, CA

Kearns HSJ, Jacobi WR (2005) Impacts of black stain root disease in recently formed mortality centers in the piñon-juniper woodlands of southwestern Colorado. Can J For Res 35:461–471

Kearns HSJ, Jacobi WR, Johnson DW (2005) Persistence of pinyon pine snags and logs in southwestern Colorado. West J Appl For 20:247–252

Klemmedson JO, Smith EL (1979) Vegetation-soil relations of forests, woodlands, and grasslands of Arizona and New Mexico. In: Youngberg CT (ed) Forest soils and land use. Department of Forest and Wood Sciences, Colorado State University, Fort Collins

Koepke DF, Kolb TE, Adams HD (2010) Variation in woody plant mortality and dieback from severe drought among soils, plant groups, and species within a northern Arizona ecotone. Oecologia 163:1079–1090

Kohler TA (1992) Prehistoric human impact on the environment in the upland North American southwest. Popul Environ 13:255–268

Kohler TA, Matthews MH (1988) Long-term Anasazi land use and forest reduction: a case study from southwest Colorado. Am Antiq 53:537–564

Landis AG, Bailey JD (2005) Reconstruction of age structure and spatial arrangement of piñon-juniper woodlands and savannas of Anderson Mesa, northern Arizona. For Ecol Manag 204:221–236

Leopold A (1924) Grass, brush, timber, and fire in southern Arizona. J For 22:1–10

Madany MH, West NE (1984) Vegetation of two relict mesas in Zion National Park. J Range 37:456–461

Martens SN, Breshears DD, Barnes FJ (2001) Development of species dominance along an elevational gradient: population dynamics of *Pinus edulis* and *Juniperus monosperma.* Int J Plant Sci 162:777–783

Mason LR, Andrews HM, Carley JA, Haacke ED (1967) Vegetation and soils of No Man's Land Mesa relict area, Utah. J Range Manag 20:45–49

McDowell N, Pockman WT, Allen CD, Breshears DD, Cobb N, Kolb T, Plaut J, Sperry J, West A, Williams DG, Yepez EA (2008) Mechanisms of plant survival and mortality during drought: why do some plants survive while others succumb to drought? New Phytol 178:719–739

Meagher GS (1943) Reaction of pinon and juniper seedlings to artificial shade and supplemental watering. J For 41:480–482

Milchunas DG (2006) Responses of plant communities to grazing in the southwestern United States. U.S. Department of Agriculture, Forest Service General Technical Report RMRS-GTR-169, Rocky Mountain Research Station, Fort Collins, CO

Miller ME (1999) Use of historic aerial photography to study vegetation change in the Negrito Creek watershed, southwestern New Mexico. Southwest Nat 44:121–137

Miller ME (2005) The structure and function of dryland ecosystems – conceptual models to inform long-term ecological monitoring. U.S. Geological Survey, Biological Resources Discipline, Scientific Investigations Report 2005–5197, Denver

Miller RF, Tausch RJ (2001) The role of fire in juniper and pinyon woodlands: a descriptive analysis. In: Galley KEM, Wilson TP (eds) Proceedings of the invasive species workshop: the role of fire in the spread and control of invasive species. Fire Conference 2000: the First National Congress on Fire Ecology, Prevention and Management. Miscellaneous publication no. 11, Tall Timbers Research Station, Tallahassee, FL

Miller RF, Wigand PE (1994) Holocene changes in semiarid pinyon-juniper woodlands. BioScience 44:465–474

Moir WH, Carleton JP (1987) Classification of pinyon-juniper (P-J) sites on the National Forests of the Southwest. In: Everett RL (compiler) Proceedings – Pinyon-juniper conference, 13–16 Jan 1986, Reno, Nevada. U.S. Department of Agriculture, Forest Service General Technical Report INT-215, Intermountain Research Station, Ogden, UT

Mueller RC, Scudder CM, Porter ME, Trotter RT III, Gehring CA, Whitham TG (2005) Differential tree mortality in response to severe drought: evidence for long-term vegetation shifts. J Ecol 93:1085–1093

Munson SM, Belnap J, Schelz CD, Moran M, Carolin TW (2011) On the brink of change: plant responses to climate on the Colorado Plateau. Ecosphere 2(6):art68

Negrón JF, Wilson JL (2003) Attributes associated with probability of infestation by the piñon ips, *Ips confusus* (Coleoptera: Scolytidae), in piñon pine, *Pinus edulis*. West N Am Nat 63:440–451

Nowak RS, Moore DJ, Tausch RJ (1999) Ecophysiological patterns of pinyon and juniper. In: Monsen SB, Stevens R (compilers) Proceedings: ecology and management of pinyon-juniper communities within the Interior West. U.S. Department of Agriculture, Forest Service Proceedings RMRS-P-9, Rocky Mountain Research Station, Ogden, UT

Ogle K, Whitham TG, Cobb NS (2000) Tree-ring variation in pinyon predicts likelihood of death following severe drought. Ecology 81:3237–3243

Ott JE, McArthur ED, Sanderson SC (2001) Plant community dynamics of burned and unburned sagebrush and pinyon-juniper vegetation in west-central Utah. In McArthur ED, Fairbanks DL (compilers) Shrubland ecosystem genetics and biodiversity: proceedings, 13–15 June 2000, Provo, Utah. U.S. Department of Agriculture, Forest Service Proceedings RMRS-P-21, Rocky Mountain Research Station, Ogden, UT

Paulsen HA Jr (1975) Range management in the central and southern Rocky Mountains: a summary of the status of our knowledge by range ecosystems. U.S. Department of Agriculture, Forest Service Research Paper RM-154, Rocky Mountain Forest and Range Experiment Station, Fort Collins, CO

Paysen TE, Ansley RJ, Brown JK, Gottfried GJ, Haase SM, Harrington MG, Narog MG, Sackett SS, Wilson RC (2000) Fire in western shrubland, woodland, and grassland ecosystems. In: Brown JK, Smith JK (eds) Wildland fire in ecosystems: effects of fire on flora. U.S. Department of Agriculture, Forest Service General Technical Report RMRS-GTR-42-vol. 2, Rocky Mountain Forest and Range Experiment Station, Ogden, UT

Pearson KM, Theimer TC (2004) Seed-caching responses to substrate and rock cover by two *Peromyscus* species: implications for pinyon pine establishment. Oecologia 141:76–83

Peterman W, Waring RH, Seager T, Pollock WL (2012) Soil properties affect pinyon pine-juniper response to drought. Ecohydrol (published online; DOI: 10.1002/eco.1284)

Peters M, Cobb NS (2008) Removal of pinyon-juniper woodlands on the Colorado Plateau. In: Gottfried GJ, Shaw JD, Ford PL (compilers) Ecology, management, and restoration of piñon-juniper and ponderosa pine ecosystems: combined proceedings of the 2005 St. George, Utah and 2006 Albuquerque, New Mexico workshops. U.S. Department of Agriculture, Forest Service Proceedings RMRS-P-51, Rocky Mountain Research Station, Fort Collins, CO

Pieper RD (1968) Comparison of vegetation on grazed and ungrazed pinyon-juniper grassland sites in southcentral New Mexico. J Range Manag 21:51–53

Pieper RD (1990) Overstory-understory relations in pinyon-juniper woodlands in New Mexico. J Range Manag 43:413–415

Potter LD, Krenetsky JC (1967) Plant succession with released grazing on New Mexico range lands. J Range Manag 20:145–151

Poulos HM, Gatewood RG, Camp AE (2009) Fire regimes of the piñon-juniper woodlands of Big Bend National Park and the Davis Mountains, west Texas, USA. Can J For Res 39:1236–1246

Price R, Evans RB (1937) Climate of the west front of the Wasatch Plateau in central Utah. Mon Weather Rev 65:291–301

Prior-Magee JS, Boykin KG, Bradford DF, Kepner WG, Lowry JH, Schrupp DL, Thomas KA, Thompson BC (2007) Ecoregional gap analysis of the southwestern United States: the Southwest regional gap analysis project final report. http://fws-nmcfwru.nmsu.edu/swregap/report/SWReGAP%20Final%20Report.pdf. Accessed 11 July 2012

Raish C (2004) Historic and contemporary land use in southwestern grassland ecosystems. In: Finch DM (ed) Assessment of grassland ecosystem conditions in the southwestern United States, vol 1. U.S. Department of Agriculture, Forest Service General Technical Report RMRS-GTR-135-vol 1, Rocky Mountain Research Station, Fort Collins, CO

Rogers GG (1982) Then and now: a photographic history of vegetation changes in the central Great Basin desert. University of Utah Press, Salt Lake City

Rogers TJ (1995) Insect and disease associates of the piñon-juniper woodlands. In Aldon EF, Shaw DW (tech coords) Managing piñon-juniper ecosystems for sustainability and social needs: proceedings of the symposium, 26–30 Apr 1993, Santa Fe, New Mexico. U.S. Department of Agriculture, Forest Service General Technical Report RM-236, Rocky Mountain Research Station, Fort Collins, CO

Romme WH, Floyd-Hanna L, Hanna DD (2003) Ancient piñon-juniper forests of Mesa Verde and the West: a cautionary note for forest restoration programs. In: Omi PN, Joyce LA (tech eds) Fire, fuel treatments, and ecological restoration: Conference proceedings. U.S. Department of Agriculture, Forest Service, Rocky Mountain Research Station, RMRS-P-29, Fort Collins, CO

Romme W, Allen C, Bailey J, Baker W, Bestelmeyer B, Brown P, Eisenhart K, Floyd-Hanna L, Huffman D, Jacobs B, Miller R, Muldavin E, Swetnam T, Tausch R, Weisberg P (2007) Historical and modern disturbance regimes of piñon-juniper vegetation in the western U.S. Colorado Forest Restoration Institute and The Nature Conservancy. http://jornada.nmsu.edu/bibliography/08-026.pdf. Accessed 11 Apr 2012

Romme WH, Allen CD, Bailey JD, Baker WL, Bestelmeyer BT, Brown PM, Eisenhart KS, Floyd-Hanna L, Huffman DW, Jacobs BF, Miller RF, Muldavin EH, Swetnam TW, Tausch RJ, Weisberg PJ (2008) Historical and modern disturbance regimes, stand structures, and landscape dynamics in piñon-juniper vegetation of the western U.S. Colorado Forest Restoration Institute. http://warnercnr.colostate.edu/docs/frs/cfri/PJSynthesis.pdf. Accessed 11 Apr 2012

Romme WH, Allen CD, Bailey JD, Baker WL, Bestelmeyer BT, Brown PM, Eisenhart KS, Floyd ML, Huffman DW, Jacobs BF, Miller RF, Muldavin EH, Swetnam TW, Tausch RJ, Weisberg PJ (2009a) Historical and modern disturbance regimes, stand structures, and landscape dynamics in piñon-juniper vegetation of the western United States. Rangel Ecol Manag 62:203–222

Romme WH, Redders JS, Floyd-Hanna L, Hanna DD (2009b) Other vegetation types. In: Romme WH, Floyd ML, Hanna D (eds) Historical range of variability and current landscape condition analysis: south central highlands section, southwestern Colorado & northwestern New Mexico. Colorado Forest Restoration Institute, Colorado State University and Region 2, U.S. Forest Service, Washington, DC

Rowlands PG, Brian NJ (2001) Fishtail Mesa: a vegetation resurvey of a relict area in Grand Canyon National Park, Arizona. West N Am Nat 61:159–181

Sallach BK (1986) Vegetation changes in New Mexico documented by repeat photography. Thesis, New Mexico State University, Las Cruces

Samuels ML, Betancourt JL (1982) Modeling the long-term effects of fuelwood harvests on pinyon-juniper woodlands. Environ Manag 6:505–515

Santos MJ, Whitham TG (2010) Predictors of *Ips confusus* outbreaks during a record drought in southwestern USA: implications for monitoring and management. Environ Manag 45:239–249

Schlanger S, Larralde S (2008) Why is cultural resource site density high in the piñon-juniper woodland? In: Gottfried GJ, Shaw JD, Ford PL (compilers) Ecology, management, and restoration of piñon-juniper and ponderosa pine ecosystems: combined proceedings of the 2005 St. George,

Utah and 2006 Albuquerque, New Mexico workshops. U.S. Department of Agriculture, Forest Service Proceedings RMRS-P-51, Rocky Mountain Research Station, Fort Collins, CO
Schmutz EM, Dennis AE, Harlan A, Hendricks D, Zauderer J (1976) An ecological survey of Wide Rock Butte in Canyon de Chelly National Monument, Arizona. J Ariz Acad Sci 11:114–125
Shaw HG (2006) Wood plenty, grass good, water none: vegetation changes in Arizona's upper Verde River watershed from 1850 to 1997. U.S. Department of Agriculture, Forest Service General Technical Report RMRS-GTR-177, Rocky Mountain Research Station, Fort Collins, CO
Shaw JD (2006) Population-wide changes in pinyon-juniper woodlands caused by drought in the American Southwest: effects on structure, composition, and distribution. In: Lafortezza R, Sanesi G (eds) Patterns and processes in forest landscapes: consequences of human management. Proceedings of the 4th meeting of the International Union of Forest Research Organizations Working Party 8.01.03, 26–29 Sept 2006, Locorotondo, Bari, Italy
Shaw JD, Steed BE, DeBlander LT (2005) Forest inventory and analysis (FIA) annual inventory answers the question: what is happening to pinyon-juniper woodlands? J For 103:280–285
Shinneman DJ, Baker WL (2009) Historical fire and multidecadal drought as context for piñon–juniper woodland restoration in western Colorado. Ecol Appl 19:1231–1245
Sieg CH, Phillips BG, Moser LP (2003) Exotic invasive plants. In: Friederici P (ed) Ecological restoration of southwestern ponderosa pine forests. Island Press, Washington, DC
Soulé PT, Knapp PA, Grissino-Mayer HD (2004) Human agency, environmental drivers, and western juniper establishment during the Late Holocene. Ecol Appl 14:96–112
Springfield HW (1976) Characteristics and management of Southwestern pinyon-juniper ranges: the status of our knowledge. U.S. Department of Agriculture, Forest Service Research Paper RM-160, Rocky Mountain Forest and Range Experiment Station, Fort Collins, CO
Stanley RH, Camp CL (1935) A Mormon mission to California in 1851 from the diary of Parley Parker Camp. Calif Hist Soc Q 14:59–73
Sullivan AP III (1996) Risk, anthropogenic environments, and Western Anasazi subsistence. In: Tainter JA, Tainter BB (eds) Evolving complexity and environmental risk in the prehistoric Southwest. Proceedings of the workshop on resource stress, economic uncertainty, and human response in the prehistoric Southwest, 25–29 Feb 1992, Santa Fe, New Mexico. Proceedings Vol 24, Santa Fe Institute Studies in the Sciences of Complexity, Addison-Wesley, Reading, MA
Swetnam TW, Betancourt JL (1998) Mesoscale disturbance and ecological response to decadal climatic variability in the American Southwest. J Clim 11:3128–3147
Swetnam TW, Brown PM (1992) Oldest known conifers in the southwestern United States: temporal and spatial patterns of maximum age. In: Kauffman MR, Moir WH, Bassett RL (tech coords) Old growth forests in the Southwest and Rocky Mountain regions, proceedings of a workshop, 9–13 Mar 1992, Portal, Arizona. U.S. Department of Agriculture, Forest Service General Technical Report RM-213, Rocky Mountain Forest and Range Experiment Station, Fort Collins, CO
Swetnam TW, Allen CD, Betancourt JL (1999) Applied historical ecology: using the past to manage for the future. Ecol Appl 9:1189–1206
Tausch RJ (1999) Historic pinyon and juniper woodland development. In: Monsen SB, Stevens R (compilers) Proceedings: ecology and management of pinyon-juniper communities within the Interior West. U.S. Department of Agriculture, Forest Service Proceedings RMRS-P-9, Rocky Mountain Research Station, Ogden, UT
Tausch RJ, Hood S (2007) Pinyon/juniper woodlands. In: Hood SM, Miller M (eds) Fire ecology and management of the major ecosystems of southern Utah. U.S. Department of Agriculture, Forest Service General Technical Report RMRS-GTR-202, Rocky Mountain Research Station, Fort Collins, CO
Thatcher AP, Hart VL (1974) Spy Mesa yields better understanding of pinyon-juniper in range ecosystem. J Range Manag 27:354–357
Tress JA Jr, Klopatek JM (1987) Successional changes in community structure of pinyon-juniper woodlands of north-central Arizona. Everett RL (compiler) Proceedings: Pinyon-juniper conference, 13–16 Jan 1986, Reno, Nevada. U.S. Department of Agriculture, Forest Service General Technical Report INT-215, Intermountain Research Station, Ogden, UT

U.S. Forest Service (2004) Forest insect and disease conditions in the Southwestern Region, 2003. U.S. Department of Agriculture, Forest Service Southwestern Region Forestry and Forest Health R3-04-02. http://www.fs.usda.gov/Internet/FSE_DOCUMENTS/stelprdb5238439.pdf. Accessed 21 Apr 2012

Vale TR (2002) The pre-European landscape of the United States: pristine or humanized? In: Vale TR (ed) Fire, native peoples, and the natural landscape. Island Press, Covelo

Van Pelt NS, Johnson DW (1993) Isolated butte and mesa summits of the Colorado Plateau. In Aldon EF, Shaw DW (tech coords) Managing piñon-juniper ecosystems for sustainability and social needs: proceedings of the symposium, 26–30 Apr 1993, Santa Fe, New Mexico. U.S. Department of Agriculture, Forest Service General Technical Report RM-236, Rocky Mountain Research Station, Fort Collins, CO

Van Pelt N, Tuhy JS (1991) Relict vegetation sites: urgent inventory need for desert parks. Park Sci 11(3):20

Weckmann L (1992) The medieval heritage of Mexico. López-Morillas FM (translator). Fordham University Press, New York

Westerling AL, Hidalgo HG, Cayan DR, Swetnam TW (2006) Warming and earlier spring increase western U.S. forest wildfire activity. Science 313:940–943

Wilson JL, Tkacz BM (1992) Pinyon ips outbreak in pinyon juniper woodlands in northern Arizona: a case study. In Ffolliott PF, Gottfried GJ, Bennett DA, Hernandez VM, Ortega-Rubio A, Hamre RH (tech eds) Ecology and management of oak and associated woodlands: perspectives in the southwestern United States and northern Mexico. U.S. Department of Agriculture, Forest Service General Technical Report RM-218, Rocky Mountain Forest and Range Experiment Station, Fort Collins, CO

Witt C, Shaw JD (2010) Characteristics of pinyon-juniper woodlands in Grand Staircase-Escalante National Monument: changes since monument establishment and prospects for future monitoring. In: Proceedings, Learning from the land 2006 science symposium, 12–14 Sept 2006, Southern Utah University, Cedar City, Utah. U.S. Department of the Interior, Bureau of Land Management, Grand Staircase-Escalante National Monument, UT. http://www.treesearch.fs.fed.us/pubs/36936. Accessed 11 Apr 2012

Wyman WD (1932) F.X. Aubrey: Santa Fé freighter, pathfinder and explorer. N M Hist Rev 7:1–31

Ziska LH, Reeves JB III, Blank B (2005) The impact of recent increases in atmospheric CO_2 on biomass production and vegetative retention of cheatgrass (*Bromus tectorum*): implications for fire disturbance. Glob Chang Biol 11:1325–1332

Chapter 6
Subalpine-Montane Grassland

Abstract Stands of southwestern Subalpine-Montane Grassland occur as scattered, small to large openings within Spruce-Fir, Mixed Conifer, and Ponderosa Pine Forests. Most stands are dominated by bunchgrasses. Stands are divided into moist and mesic-dry grasslands. Natural disturbances are primarily fire, herbivory, and extreme weather. Historically, moist stands likely burned infrequently and during drought years. Mesic-dry stands likely burned more often, with fire frequency correlated with that of the surrounding forest vegetation and with the dryness of the grassland. The most important anthropogenic disturbance is livestock grazing; others are fire management, modern climate change, invasive species, recreation, and nearby land use. Vegetation dynamics are dominated by three processes: rapid regrowth by resprouting from below-ground structures, succession, and tree encroachment. Where tree encroachment occurs, it appears related to increased precipitation, warmer temperatures, reduced fire, and effects of ungulate herbivory. Specifics of historical conditions are poorly known. Past livestock grazing reduced vegetation cover, altered species composition, and increased invasive plants. Vegetation dynamics are illustrated in a nested, three-tiered set of conceptual models. Key conclusions and challenges for researchers and land managers are summarized.

6.1 Introduction

Subalpine-Montane Grassland on the mountains and plateaus of the American Southwest is dominated by herbaceous plants of various heights; shrubs are also present. Stands generally occur as openings scattered within Spruce-Fir (*Picea-Abies*), Mixed Conifer, and Ponderosa Pine (*Pinus ponderosa*) Forests (Figs. 6.1 and 6.2), making Subalpine-Montane Grassland the most naturally fragmented of the vegetation types covered in this book (cf. Fletcher and Robbie 2004). Stands are also referred to as meadows or parks. Stands are in the range of small openings under 1 ha (2.5 acres) to large expanses of more than 1,000 ha (2,500 acres). With its patchy distribution, Subalpine-Montane Grassland occupies only a small percentage

J.L. Vankat, *Vegetation Dynamics on the Mountains and Plateaus of the American Southwest*, Plant and Vegetation 8, DOI 10.1007/978-94-007-6149-0_6,

Fig. 6.1 Subalpine-Montane Grassland on the San Francisco Peaks in north-central Arizona. Note other stands of Subalpine-Montane Grassland in the area, including a large stand partially obscured by the tree in the middle of the photograph (Photograph by Betty J. Huffman)

Fig. 6.2 Subalpine-Montane Grassland in a valley in the North Rim region of Grand Canyon National Park, north-central Arizona (Photograph by Betty J. Huffman)

Table 6.1 Common and scientific names of species in this chapter. Primary source: Integrated Taxonomic Information System (2012)

Plants	
Arizona fescue	*Festuca arizonica* Vasey
Douglas-fir	*Pseudotsuga menziesii* (Mirbel) Franco
Engelmann spruce	*Picea engelmannii* Parry ex Engelm.
Fescue	*Festuca* L.
Fir	*Abies* P. Mill.
Intermediate wheatgrass	*Thinopyrum intermedium* (Host) Barkworth & D.R. Dewey
Juniper	*Juniperus* L.
Kentucky bluegrass	*Poa pratensis* L.
Mountain muhly	*Muhlenbergia montana* (Nutt.) Hitchc.
Oatgrass	*Danthonia* DC.
Orchardgrass	*Dactylis glomerata* L.
Pinyon	*Pinus* L.
Ponderosa pine	*Pinus ponderosa* Douglas ex P. Lawson & C. Lawson
Quaking aspen	*Populus tremuloides* Michx.
Screwleaf muhly	*Muhlenbergia virescens* (Kunth) Trin.
Sedge	*Carex* L.
Smooth brome	*Bromus inermis* Leyss.
Southwestern white pine	*Pinus strobiformis* Engelm.
Spruce	*Picea* A. Dietr.
Subalpine fir	*Abies lasiocarpa* var. *lasiocarpa* (Hook.) Nutt.
Thurber fescue	*Festuca thurberi* Vasey
Timothy	*Phleum pratense* L.
White fir	*Abies concolor* (Gord. & Glend.) Lindl. Ex Hildebr.
Animals	
Cattle	*Bos taurus* Linnaeus, 1758
Deer	*Odocoileus* Rafinesque, 1832
Elk	*Cervus elaphus* Linnaeus, 1758
Grasshopper	suborder Caelifera
Northern pocket gopher	*Thomomys talpoides* Richardson, 1928
Sheep	*Ovis aries* Linnaeus, 1758

of most landscapes, e.g., 2 % of the forested portion of the landscape of Grand Canyon National Park in north-central Arizona (Warren et al. 1982).

Subalpine-Montane Grassland covers approximately 10,400 km^2 (4,000 miles^2) of the American Southwest, which is 1.3 % of the region's area (Fig. 6.3; calculations based on Prior-Magee et al. 2007). This is the second smallest area of the vegetation types covered in this book. The maximum area of Subalpine-Montane Grassland in the region occurs in central Utah, where for example it occupies most of the subalpine zone of the Wasatch Plateau (Ellison 1954). By definition, Subalpine-Montane Grassland is below treeline. Above treeline is alpine tundra, and although many of its stands include graminoid species, its ecology is different from grasslands.

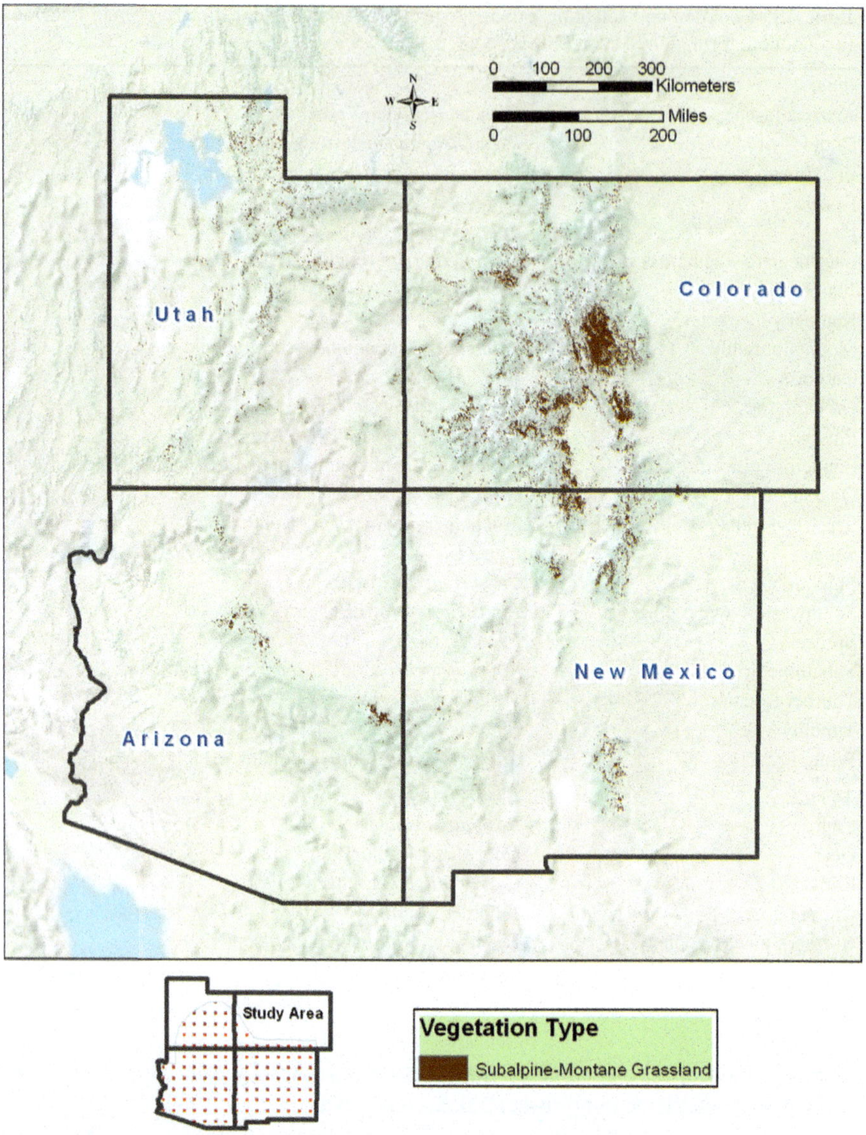

Fig. 6.3 Distribution of Subalpine-Montane Grassland in the American Southwest. The map shows all of the states of Arizona, New Mexico, Colorado, and Utah, and the American Southwest is illustrated in *red* on the small map (Source: U.S. Geological Survey National Gap Analysis Program 2005 Southwest Regional Gap Analysis Project, RS/GIS Laboratory, College of Natural Resources, Utah State University, Logan, Utah. Map prepared by Monica Swihart)

The occurrence of Subalpine-Montane Grassland as patches within forests has led to discussions of their origin. Two hypotheses proposed early in the twentieth century continue to be applicable today: (a) site conditions limit forest trees and (b) disturbance such as stand-replacing fire replaced forest with grassland (Pearson 1913). Anecdotal evidence supporting site conditions as the cause is the association of some stands with particular habitats, including basins with microclimates involving cold air drainage and soils contrasting with those of surrounding forest (Pearson 1913). Evidence of stand-replacing fire being a causal factor is that large stands of Subalpine-Montane Grassland on mountain slopes occur on dry, expansive slopes where (a) clumps of trees are generally restricted to the upslope and lee side of topographic breaks, (b) ridgetops are often sites of abrupt transition from grassland on drier slopes to forests on mesic slopes, and (c) fire-scarred trees are present in grassland-forest ecotones (Allen 1984). In addition, some twentieth and twenty-first century forest fires have produced stands of grasses and other herbs with little or no evidence of tree invasion decades later (Savage and Mast 2005). Other disturbances, such as interaction of drought with insect and disease outbreaks, also can create some patches of Subalpine-Montane Grassland (Fletcher and Robbie 2004).

It appears that both hypotheses on the origin of stands of Subalpine-Montane Grassland (i.e., limiting site conditions and forest fire) are correct and that different factors dominate on different sites. In addition, the two factors can act in combination: limiting site conditions appear to account for Subalpine-Montane Grassland in the center of some basins and drainages, whereas fire had determined the grassland-forest boundary at the perimeter.

The specific site conditions limiting to forest have been subject to much speculation. Proposed factors include (a) soil factors such as poor drainage, fine texture, less moisture, and frost heaving; (b) climate factors such as cold air drainage, severe frosts, winter desiccation, and seasonal drought/long-term precipitation patterns; and (c) biotic factors such as thick sod, allelopathy, absence of necessary mycorrhizal symbionts, competition from herbs, and herbivory (Pearson 1913, 1931; Merkle 1962; Moir 1967; Rietveld 1975; Moir and Ludwig 1979; Allen 1984, 1989; Dyer and Moffett 1999; Fletcher and Robbie 2004; Moore and Huffman 2004; Coop and Givnish 2007a, b, 2008). Few experimental data are available, but research in Subalpine-Montane Grassland of the Valles Caldera National Preserve in north-central New Mexico examined the roles of soil texture, minimum temperature, grass competition, and ungulate herbivory (Coop and Givnish 2008). It was concluded that differential establishment of ponderosa pine seedlings was driven by low minimum temperatures and that these temperatures likely had interacted with fire to determine historical grassland-forest boundaries.

Southwestern Subalpine-Montane Grassland encompasses a diverse array of stands. Species distributions appear to be influenced primarily by soil texture, soil moisture, elevation, site exposure (e.g., ridges), and disturbance (Merkle 1953; Dick-Peddie 1993; Brown 1994a). This chapter divides stands into moist and mesic-dry types, largely based on topographic position (cf. McHenry 1933; Klemmedson and Smith 1979; Chambers and Holthausen 2000).

Moist Subalpine-Montane Grassland (Fig. 6.4) occurs in bottoms of shallow valleys and other wet areas. They are sometimes called cienegas. Most stands are small and dominated by sedges (*Carex* spp.; e.g., Patton and Judd 1970; Dick-Peddie 1993). Stands have lower species richness than mesic-dry stands and higher herbaceous species richness than adjacent forest (Patton and Judd 1970). Moist stands generally have a discontinuous distribution and are relatively small, occupying <100 ha (250 acres; Rasmussen 1941; Dick-Peddie 1993; Brown 1994a).

Mesic-dry Subalpine-Montane Grassland (Fig. 6.5) occurs upslope and adjacent to moist stands, as well as on other relatively dry sites. Mesic-dry stands in most of the American Southwest are dominated by fescue (*Festuca* spp.) bunchgrasses of up to 1 m (3 ft) height. Perennial and annual forbs, sod-grasses, and sedges occur in bunchgrass interspaces. However, stands in central Utah were dominated by tall perennial forbs before extensive livestock grazing began, and grasses and sedges were also present. In general, upper-elevation stands are associated with Spruce-Fir Forest and moist-mesic Mixed Conifer Forest and are dominated by Thurber fescue (*Festuca thurberi*). Common associated species include Arizona fescue (*Festuca arizonica*), and forbs are often abundant. Lower-elevation stands cover greater total land area (Fletcher and Robbie 2004) and occur within dry-mesic Mixed Conifer Forest and Ponderosa Pine Forest. Associated species in lower-elevation stands include mountain muhly (*Muhlenbergii montanus*), screwleaf muhly (*M. virescens*), and oatgrasses (*Danthonia* spp.). Shrubs tend to be more abundant than at higher elevations. Mesic-dry stands have higher herbaceous species richness than adjacent forest (Patton and Judd 1970). Stands range in size from as small as moist stands to much larger.

Both moist and mesic-dry Subalpine-Montane Grassland exhibit seasonal variation (cf. Fletcher and Robbie 2004) and disturbance-related variation (see Sect. 6.4.2). Regional variation is also important, especially between central Utah and other regions of the Southwest, e.g., compare Ellison (1954) and Gill (2007) to Dick-Peddie (1993) and Romme et al. (2009). Species lists are available (e.g., Ellison 1954; Dick-Peddie 1993; Brown 1994a, c; White 2002). Few classifications have been published. Upper-elevation grasslands of the Apache and Sitgreaves National Forests in east-central Arizona have been divided into four community types (White 2002). Most species in all four of these types are C_3, perennial, and native. Grasses dominate in cover and percentage composition. Subalpine stands have higher cover of cryptogams and grasses, as well as relatively more species of perennials. Montane stands have higher cover of annuals and relatively more species of forbs, biennials, and annuals.

Although regions outside of the American Southwest also have subalpine-montane grasslands (e.g., the Rocky Mountains of central Colorado; Fig. 6.3), there are differences in climate and species composition. Therefore, this chapter is based on findings from the American Southwest, despite little quantitative information being available (cf. White 2002).

Fig. 6.4 Moist Subalpine-Montane Grassland in central Utah (Photograph by author)

Fig. 6.5 Mesic-Dry Subalpine-Montane Grassland near the San Francisco Peaks in north-central Arizona (Photograph by Betty J. Huffman)

6.2 Drivers

Key drivers of southwestern Subalpine-Montane Grassland are landscape, climate, soil, animals, natural disturbance, and anthropogenic disturbance. Each driver is important in vegetation dynamics, and therefore all are incorporated in the conceptual models (Sect. 6.5).

6.2.1 Landscape

Subalpine-Montane Grassland occurs nearly throughout the elevational range of coniferous forests in the American Southwest, from approximately 1,800 to 3,500 m (6,200 to 11,500 ft) in the mid-latitudes of its distribution in the American Southwest (Fig. 6.6). As described in the previous section, elevation and topography are correlated with differences in species composition. Many stands, especially in the

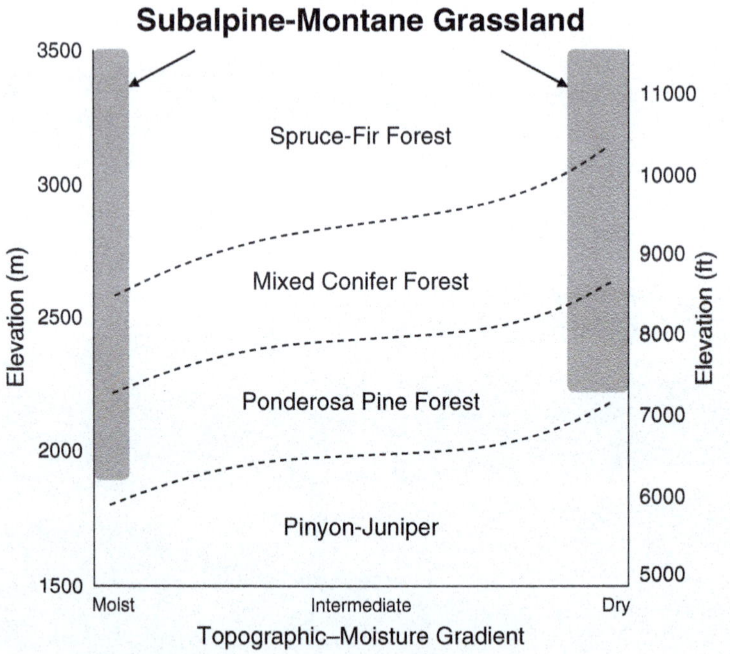

Fig. 6.6 Ecological distribution of Subalpine-Montane Grassland (*shaded area*) on the mountains and plateaus of the American Southwest along gradients in elevation and topographic-moisture. Elevations are approximate and generally representative of mid-latitudes of the region, i.e., landscapes in northern Arizona and northern New Mexico

montane zone, occur in shallow, relatively flat depressions that are only slightly lower in elevation than forests (Pearson 1913). In deeper depressions and drainages, moist stands can be present in the bottoms and mesic-dry stands occur on slopes that lead to the surrounding forests. Mesic-dry stands also occur elsewhere on a variety of slope aspects and inclinations. For example, sites include large, little-dissected, relatively steep, south-facing mountain slopes of the Jemez Mountains in north-central New Mexico (Allen 1984, 1989). On a more local scale, topographic variations and their associated microclimates can be keys to species' distributions (e.g., Ellison 1954). For example, Kentucky bluegrass (*Poa pratensis*) occurs in narrow valley bottoms in Apache and Sitgreaves National Forests in east-central Arizona (Laing et al. 1987).

6.2.2 Climate

Few long-term climate data have been published for Subalpine-Montane Grassland in the American Southwest. In general, summers are warm to cool, and winters are cold. The range for mean annual temperature is 1–5 °C (34–41 °F) for subalpine grasslands and 4–8 °C (39–46 °F) for montane grasslands in the Apache and Sitgreaves National Forests (White 2002). Mean annual precipitation has been estimated as 50–100 cm (20–39 in.) for grassland in the subalpine zone (Milchunas 2006), and precipitation in the montane zone likely averages near the lower end of that range. Values in Apache and Sitgreaves National Forests are 66–78 cm (26–31 in.) with 150–170 cm (59–67 in.) of snow for subalpine grasslands and 50–62 cm (20–24 in.) with 120–150 cm (47–59 in.) of snow for montane grasslands (White 2002). In general, the percentage of precipitation falling as snow is 50–75 % in the subalpine zone, where snow commonly covers sites from October through May (Ford et al. 2004). The percentage and time of coverage by snow decrease toward lower elevations, and snow buildup is much less for stands on south aspects (Allen 1984). The growing season is short, especially in the subalpine zone where it is often <100 days (Brown 1994a), and frosts occasionally occur during the summer. The average number of frost-free days per year is 70–90 in subalpine grasslands and 90–100 in montane grasslands in the Apache and Sitgreaves National Forests (White 2002).

Subalpine-Montane Grassland in valleys, even shallow valleys, often has a microclimate different from that of adjacent forests. Data collected over a 4-year period at the Fort Valley Experimental Forest in north-central Arizona indicated maximum temperatures were similar to adjacent forested sites, with annual means of 15.1 °C in Subalpine-Montane Grassland vs. 14.6 °C in adjacent Ponderosa Pine Forest (59.2 vs. 58.3 °F; Pearson 1913). However, minimum temperatures were lower in the Subalpine-Montane Grassland throughout the year, with annual means of −5.2 °C in contrast to −1.6 °C in the Ponderosa Pine Forest (22.7 vs. 29.1 °F). Monthly mean minimum temperatures in the Subalpine-Montane Grassland were above freezing only in July–September. Except during cloudy weather, the diurnal range of temperatures was much greater in Subalpine-Montane Grassland than in Ponderosa Pine Forest, especially in winter. Subalpine-Montane Grassland also experienced more wind and had greater evaporation than adjacent forest.

Modern climate change is described in Sect. 1.6.3 and is treated as an anthropogenic disturbance in Sect. 6.2.6.3.

6.2.3 Soil

Stands of southwestern Subalpine-Montane Grassland in shallow valley bottoms are on finely-textured alluvial or colluvial soils (Peet 2000). Many stands, especially in the montane zone, occur in shallow, relatively flat depressions with deep clay or clay-loam soils (Klemmedson and Smith 1979). Soils are variable, but most are deep, well-developed, and well- to poorly drained (Warren et al. 1982; Brown 1994a, c). Soil moisture regimes are udic and ustic, and soil temperature regimes are cryic and frigid (Ford et al. 2004). Soils resemble prairie soils with a deep, dark, organic A horizon (Moir 1967). Most are Mollisols, and some are Alfisols and weakly developed Entisols (Laing et al. 1987; Maker and Saugherty 1986; Miller et al. 1995). The Mollisols of mesic-dry Subalpine-Montane Grassland on slopes of the Jemez Mountains have deep profiles of >1.5 m (5 ft; Allen 1984).

Subalpine-Montane Grassland soils change upslope in shallow valleys. Soils of valley bottoms in the North Rim region of Grand Canyon National Park tend to be relatively poorly drained loams or clay loams, while slopes closer to surrounding forests are well-drained, moderately deep to deep gravelly loam to clay loam (Moore and Huffman 2004). These well-drained soils tend to be dry.

Soils of Subalpine-Montane Grassland also differ by elevation. Soils of subalpine stands with Thurber fescue have udic soil moisture regimes and cryic soil temperature regimes. Soils of montane stands with Arizona fescue have ustic soil moisture regimes and frigid soil temperature regimes (W.A. Robbie, personal communication).

Subalpine-Montane Grassland soils can be similar or different from forest soils. In Ponderosa Pine Forest, small stands are usually on the same kind of soil as the forest, but larger stands can be on different kinds of soil than forest (Klemmedson and Smith 1979). The Mollisols of mesic-dry Subalpine-Montane Grassland on slopes of the Jemez Mountains sharply differ from the Alfisols and Inceptisols of adjacent forests (Allen 1984). Subalpine-Montane Grassland soils on the Wasatch Plateau of central Utah tend to be deeper and less rocky than soils supporting Spruce-Fir Forest (Ellison 1954).

6.2.4 Animals

Southwestern Subalpine-Montane Grassland is important habitat for wildlife. Common animals influencing vegetation dynamics include ungulates such as deer

(*Odocoileus* spp.) and elk (*Cervus elaphus*; Merkle 1953; Brown 1994a, b; Wolters 1996), although these species spend more time in adjacent forests (Patton and Judd 1970). Many rodent species are present (Brown 1994a, b), including northern pocket gopher (*Thomomys talpoides*), which affects plant species composition by loosening heavy soils (Ellison and Aldous 1952; Merkle 1953). Such burrowing animals also provide sites of exposed mineral soil where competition is low – sites that can be utilized by seedlings. Herbivory by grasshoppers (suborder Caelifera) also can be significant. Animals can cause substantial seedling mortality of ponderosa pines encroaching on Subalpine-Montane Grassland (Coop and Givnish 2008).

6.2.5 Natural Disturbance

The natural disturbance regime of Subalpine-Montane Grassland in the American Southwest is dominated by fire, herbivory, and extreme weather. In addition, wind importantly influences the spread of fire.

6.2.5.1 Fire

The historical fire regime of southwestern Subalpine-Montane Grassland is poorly known. Moist stands likely burned only infrequently during drought years when plants and litter dried. Mesic-dry stands likely burned more often and fire frequency was probably correlated with that of the surrounding forest vegetation and to the dryness of the stand (Fig. 6.7a, b). Therefore, fires were less frequent in stands within Spruce-Fir Forest and moist-mesic Mixed Conifer Forest, i.e., forests that burned less frequently (Sects. 2.2.5.1 and 3.2.5.1), and more frequent in stands within dry-mesic Mixed Conifer Forest and Ponderosa Pine Forest (Sects. 3.2.5.1 and 4.2.5.1; Dick-Peddie 1993; Romme et al. 2009). Some mesic-dry stands typically do not burn during the primary fire season (Bradley et al. 1992). Indeed, larger stands can act as fire breaks during mid-summer, but can burn in late summer and early fall, as well as in early spring before vegetation green-up begins (Bradley et al. 1992). Mean fire return intervals in Subalpine-Montane Grassland dominated by Thurber fescue are thought to be 18–22 years (U.S. Forest Service 2012b). Mesic-dry Subalpine-Montane Grassland on slopes of the Jemez Mountains is estimated to have burned every 15 years or less, including both lightning- and possible human-ignited fires (Allen 1984).

Fire can alter species composition, and its effects likely depend on post-fire weather (cf. Milchunas 2006). Fire reduces litter, thereby increasing both nutrient availability and soil temperature, but reducing soil moisture. Historical fires apparently restricted encroachment of trees (e.g., Allen 1984; Moore and Huffman 2004; Romme et al. 2009; Sect. 6.3.3).

Fig. 6.7 (**a**) Fire in mesic-dry Subalpine-Montane Grassland in Gila National Forest, west-central New Mexico (Photograph by Chris Ader/U.S. Forest Service, Gila National Forest). (**b**) Burn mosaic (*brown patches* are burned, and *green* and *tan patches* are unburned) in mesic-dry Subalpine-Montane Grassland in Fort Apache Reservation, east-central Arizona (Photograph by Kari Greer/U.S. Forest Service)

Fig. 6.8 Elk grazing in Subalpine-Montane Grassland in Valles Caldera National Preserve, north-central New Mexico. Smoky haze is from nearby forest fire (Photograph by Jayson Coil)

6.2.5.2 Herbivory

Herbivory by deer and elk (Fig. 6.8) reduces total plant cover and affects species composition by reducing palatable species and increasing less palatable ones (e.g., Wolters 1996). Secondary impacts of this herbivory likely include increased soil temperature and decreased soil moisture. Additional impacts can parallel those of livestock grazing (Sect. 6.2.6.1). Browsing by unusually large populations of deer has been a factor restricting encroachment of quaking aspen (*Populus tremuloides*) in Subalpine-Montane Grassland in the North Rim region of Grand Canyon National Park (Moore and Huffman 2004).

6.2.5.3 Extreme Weather

Weather extremes can have several effects on southwestern Subalpine-Montane Grassland. These include intensive rainfall events that result in incisement (down-cutting, gullying) of drainages and lowering of water tables. Subsequent drying slowly changes moist stands into mesic-dry stands of Subalpine-Montane Grassland as species adapted to wet conditions are gradually replaced over months and years by species adapted to drier conditions. Extreme drought can similarly change stands from moist to mesic-dry. Drought also can result in mortality of encroaching trees and in overall reduction of vegetation cover. Unusually cold temperatures can combine with cold-air drainage to produce unseasonable frosts that cause tree mortality (Moore and Huffman 2004; Coop and Givnish 2007b).

6.2.6 Anthropogenic Disturbance

Land Use by Native Americans in the American Southwest was concentrated in Pinyon-Juniper vegetation (*Pinus-Juniperus*; Sects. 1.5.1.1 and 5.2.6), but Subalpine-Montane Grassland was likely used for hunting and possibly other purposes such as agriculture, especially at low elevations. Little information is available on anthropogenic drivers affecting southwestern Subalpine-Montane Grassland, except for livestock grazing. Other anthropogenic drivers related primarily to Euro-American land use are fire management, modern climate change, invasive species, recreation, and nearby land use. Each of these can be important in vegetation dynamics, and therefore all are incorporated in the conceptual models (Sect. 6.5).

6.2.6.1 Livestock Grazing

Southwestern Subalpine-Montane Grassland has been grazed by cattle (*Bos taurus*) and sheep (*Ovis aries*) since at least the mid nineteenth century (Fig. 6.9). Although herbivory by deer and elk is natural, historical livestock grazing was different because of the large numbers of livestock. Livestock grazing was especially intensive from approximately 1880 to the early twentieth century:

> ...between 1888 and 1905, the Wasatch [Plateau] was a vast dust bed, grazed, trampled, and burned to the utmost. The timber cover was reduced, the brush thinned, the weeds and grass cropped to the roots, and such sod as existed was broken and worn. (Reynolds 1911)

Livestock grazing affected the understories of adjacent forests (Sects. 2.2.6.1, 3.2.6.1, and 4.2.6.1), but livestock utilized meadows proportionately more than forests because of greater accessibility, more forage, and abundant palatable species (Patton and Judd 1970). Livestock grazing continues in many stands today (Fig. 6.10), but levels are much reduced from the early twentieth century.

Livestock grazing has increased invasive species, because invasives are spread attached to animal coats and in their feces. Moreover, grazing that reduces vegetation cover can facilitate colonization by invasives. Although invasives can decline within a few years after grazing is reduced, recovery is incomplete (Dick-Peddie 1993; Wolters 1996). Land managers have seeded invasives such as Kentucky bluegrass, intermediate wheatgrass (*Thinopyrum intermedium*), orchardgrass (*Dactylis glomerata*), smooth brome (*Bromus inermis*), and timothy (*Phleum pratense*) as forage for livestock and for revegetation of disturbed areas (Merkle 1953; Romme et al. 2009).

See Sect. 6.4.2 for more on impacts of livestock grazing.

Fig. 6.9 Cattle grazing in Subalpine-Montane Grassland circa 1910 in Fort Valley, north-central Arizona (Photograph by A.G. Varela, courtesy of Fort Valley Experimental Forest Historic Images, U.S. Forest Service)

Fig. 6.10 Sheep grazing in mesic-dry Subalpine-Montane Grassland in southwestern Utah (Photograph by Betty J. Huffman)

6.2.6.2 Fire Management

The history of fire management in southwestern Subalpine-Montane Grassland parallels that of fire management in adjacent forests (Sects. 2.2.6.2, 3.2.6.2, and 4.2.6.2), except prescribed fires rarely have been intended for grassland. Fire management likely has had little effect on moist Subalpine-Montane Grassland; however, fire suppression focused on adjacent forests likely contributed to tree encroachment into some mesic-dry stands (Allen 1984; Merola-Zwartjes 2004; Moore and Huffman 2004; see Sect. 6.4.2).

6.2.6.3 Modern Climate Change

Effects of modern climate change on southwestern Subalpine-Montane Grassland have been little studied. Climate change likely has contributed to tree encroachment into mesic-dry stands (Dyer and Moffett 1999; Merola-Zwartjes 2004; Zier and Baker 2006; Sect. 6.3.3). One aspect of modern climate change is increased frequency of extreme events (Sect. 1.6.3), and these can have a variety of effects on Subalpine-Montane Grassland (see Sect. 6.2.5.3).

6.2.6.4 Invasive Species

Invasive plants are abundant in Subalpine-Montane Grassland of the American Southwest, so much so that they dominate many moist stands in the Jemez Mountains (Allen 1989) and elsewhere. For example, the invasive Kentucky bluegrass characterizes one of the three major Subalpine-Montane Grassland communities described by Romme et al. (2009) for north-central New Mexico and southwestern Colorado. This community occurs across subalpine and montane zones, and includes other invasives as associated species. The community persists because of grazing, either by livestock (Fletcher and Robbie 2004; Romme et al. 2009) or native ungulates (Wolters 1996). Compaction of moist Subalpine-Montane Grassland by livestock trampling inhibits water infiltration, which results in drier soil on the site, changes the area from moist to mesic-dry Subalpine-Montane Grassland, and favors expansion by Kentucky bluegrass (Fletcher and Robbie 2004). Other invasive plants such as intermediate wheatgrass, orchardgrass, smooth brome, and timothy also dominate some sites (Romme et al. 2009). Floristic study of Subalpine-Montane Grassland in the Apache and Sitgreaves National Forests determined that 28 of 199 species (14 %) were non-native (White 2002).

6.2.6.5 Recreation

No studies have explicitly addressed effects of recreation on southwestern Subalpine-Montane Grassland, but the presence of invasive species in recreational areas suggests recreation facilitates their spread and establishment (cf. Romme et al. 2009). Recreation is also a source of fires. Moreover, recreation that reduces vegetation cover, such as off-highway vehicle use, can increase erosion.

6.2.6.6 Nearby Land Use

Land use in Mixed Conifer and Ponderosa Pine Forests is considered to be an anthropogenic driver of southwestern Subalpine-Montane Grassland (Romme et al. 2009). This nearby land use is a source of fires and invasive plants. Land use in Spruce-Fir Forest is less extensive and intensive; therefore, it is less of a driver of Subalpine-Montane Grassland.

6.3 Processes

Important processes in southwestern Subalpine-Montane Grassland are rapid regrowth following light to moderate disturbance, succession following more severe disturbance, and tree encroachment with shifts in the environment. These three processes play key roles in the conceptual models (Sect. 6.5).

6.3.1 Rapid Regrowth

The primary process of vegetation dynamics in southwestern Subalpine-Montane Grassland is rapid regrowth, which occurs within weeks to months following disturbances such as fire and herbivory (Fig. 6.11). The key to rapid regrowth is that grasses and sedges sprout and regrow from meristems (zones of dividing cells) at or below the soil surface where they are little damaged by fire or herbivory. Moreover, many herbs have storage structures at shallow soil depth, where they are protected from disturbance and provide carbohydrates for rapid regrowth. Also, grassland species are relatively small in stature and therefore individuals rapidly regrow to full size.

Little is known about the fire ecology of the dominant bunchgrasses, Thurber fescue and Arizona fescue. Root crowns of Thurber fescue appear to be able to survive fire (Bradley et al. 1992), and plants likely recover 2–3 years after fire (U.S. Forest Service 2012b). Arizona fescue has meristems in the soil where they are protected from heat produced by fires, and thereby plants survive and rapidly recover following most fires (U.S. Forest Service 2012a).

Other species also rapidly recover following fire by sprouting from below-ground structures such as roots, rhizomes, and bulbs The greater the depth of these structures in the soil, the more likely the species will be undamaged by fire and regenerate (Antos et al. 1983). Other species can regenerate from soil seed banks or by dispersal of seeds from outside the burned area. Therefore, the rate and degree of post-fire

Fig. 6.11 Rapid regrowth of small stand of mesic-dry Subalpine-Montane Grassland following crown fire in the North Rim region of Grand Canyon National Park, north-central Arizona (Photograph by Fire and Aviation Management, Grand Canyon National Park)

recovery are influenced by species composition (reflecting different modes and rates of regeneration). Additional factors likely to influence recovery include severity of fire, time since previous fire (accumulation of litter influences fuel loadings), season of fire (species dormant at the time of fire are more likely to survive and regenerate), and patchiness of fire (unburned patches facilitate more rapid recovery of the stand; Antos et al. 1983; Baker 2009).

The other common disturbance, herbivory, occurs annually, but herbivory by deer and elk is typically a less-severe disturbance than fire. Rapid recovery from light to moderate herbivory largely parallels that described above for fire, but can be more complete. Intensive, repeated herbivory, such as with livestock grazing from the late nineteenth century into the twentieth century, has greater impacts. For example, Thurber fescue and Arizona fescue can resprout following light to moderate herbivory by utilizing carbohydrates stored in roots, but intensive repeated herbivory can exhaust carbohydrate reserves and result in mortality. Neither of these common native bunchgrasses produces and spreads by rhizomes (underground stems). Therefore, self-replacement following mortality requires regrowth from seed, a slow process that additional herbivory can prevent. Consequently, Thurber fescue and Arizona fescue are often replaced by species spreading below-ground by rhizomes. There are no common, native rhizomatous grasses in Southwestern Subalpine-Montane Grassland, so replacement is often by invasive rhizomatous species such as Kentucky bluegrass and smooth brome (Paulson and Baker 2006).

6.3.2 Succession

Succession occurs in southwestern Subalpine-Montane Grassland following high-severity disturbances such as soil erosion and long, intensive herbivory. Rapid regrowth can play an important role early in succession, except where the severity of the disturbance causes substantial mortality of plants (including their belowground structures). Chronic disturbance, such as long-term livestock grazing, can similarly limit the role of rapid regrowth in succession.

Little is known about succession in Subalpine-Montane Grassland, especially where information is unavailable on stand structure and composition before grazing began (cf. Sect. 6.4.1) and where invasive plants are present. Decadal variations in species composition in response to weather, grazing, etc. also can obscure successional patterns. Moreover, details of succession are highly species-, site-, region-, and disturbance-specific (e.g., Ellison 1954).

Succession can involve tree encroachment where encroachment is sustained (next section).

6.3.3 Tree Encroachment

The colonization and establishment of trees in Subalpine-Montane Grassland is termed tree encroachment (Fig. 6.12a, b). It can be followed by infill by additional trees, as well as by mortality that reduces or eliminates encroachment. Tree encroachment has been studied in a few locations.

Encroachment into Subalpine-Montane Grassland on the slopes of the Jemez Mountains was mostly by ponderosa pine and Douglas-fir (*Pseudotsuga menziesii*), but it also included quaking aspen and Engelmann spruce (*Picea engelmannii*; Allen 1984, 1989). In the past, invading trees were likely killed by fire and winter desiccation. During the twentieth century, tree establishment was enabled by fire exclusion, and establishment occurred primarily during the period between the end of intensive sheep grazing and recovery of Subalpine-Montane Grassland vegetation.

Initial tree encroachment in the White Mountains of south-central New Mexico is by Douglas-fir, which is a nurse-plant for southwestern white pine (*Pinus strobiformis*; Dyer and Moffett 1999). The lack of evidence implicating livestock grazing or fire exclusion led to the conclusion that encroachment had resulted from increased precipitation and possibly warmer temperatures. Climate was also considered the primary driver of encroachment – mostly by quaking aspen – in small Subalpine-Montane Grassland stands within forests of the San Juan Mountains of southwestern Colorado (Zier and Baker 2006).

Most tree encroachment on the North Rim of Grand Canyon National Park has involved quaking aspen, but other trees – in decreasing order of abundance – are spruce, subalpine fir (*Abies lasiocarpa*), white fir (*Abies concolor*), and ponderosa pine (Moore and Huffman 2004). Both spruce and ponderosa pine appear to

Fig. 6.12 (**a**, **b**) Tree encroachment in mesic-dry Subalpine-Montane Grassland at two elevations on the San Francisco Peaks in north-central Arizona: (**a**) moderately high elevation and (**b**) moderately low elevation; (note small saplings of ponderosa pine in front of *yellow-colored* quaking aspen) (Photographs by Betty J. Huffman)

progressively encroach from the grassland-forest boundary. In contrast, quaking aspen and firs infill among trees. Encroachment occurred during a period of climate change, fire exclusion, and herbivory by livestock and native ungulates, making it challenging to identify a primary driver of encroachment.

In conclusion, tree encroachment – along with infill and mortality – produces spatially dynamic grassland-forest boundaries. Encroachment can involve most tree species that are dominant in adjacent or nearby forests. It is favored by various changes in environmental factors, including increased precipitation, warmer temperatures, reduced fire, reduced ungulate herbivory, and possibly by increased ungulate herbivory that reduces competition. Tree mortality occurs with reduced precipitation, colder temperatures, and fire. Tree seedling mortality can be caused by additional factors such as herbivory (Sects. 6.2.4 and 6.2.5.2).

6.4 Historical Changes

6.4.1 Pre-Euro-American Settlement

Little is known about southwestern Subalpine-Montane Grassland prior to Euro-American influence. Historical descriptions provide qualitative information on cover:

> ...we found luxuriant bunch-grass covering the ground as thickly as it could stand. (Rothrock 1878, for south-central Colorado in 1873)

as well as quantitative information on stand size:

> These parks are of varying extent, from a mere glade of five acres up to tracts embracing 14,000 acres [2 to 5700 ha]... (Leiberg et al. 1904, for the vicinity of the San Francisco Peaks in north-central Arizona)

> Everywhere through the forest we encounter beautiful open parks, from a few acres to several square miles in area. (Rusby 1889, for north-central Arizona in 1883)

and stand height:

> ...we passed successive vales and glades, filled with verdant grass knee high to our mules... (Beale 1858, for east of the San Francisco Peaks)

> ...the grasses are...often nearly two yards [1.8 m] high... (Rusby 1889, for north-central Arizona in 1883)

Historical photographs (Figs. 6.13 and 6.14) also provide little information on stand composition and structure. However, they can document the expansion of Subalpine-Montane Grassland (such as caused by forest fire) as well as their contraction (by tree encroachment).

Nevertheless, it is difficult to obtain a detailed, accurate description of conditions prior to Euro-American settlement (Fletcher and Robbie 2004). Historical data are lacking, relict sites are rare and generally small, and the effects of livestock grazing were both widespread and intensive. In addition, there are few dateable, long-lived plants (i.e., trees).

Fig. 6.13 Subalpine-Montane Grassland in 1871 at the base of the San Francisco Peaks in north-central Arizona. Note the abundance of bunchgrasses (Photograph by Timothy H. O'Sullivan, courtesy of the National Archives and Records Administration)

Intensive livestock grazing altered the vegetation of the Wasatch Plateau so thoroughly and so long ago that there are no historical accounts of it, and researchers in the mid twentieth century found no early inhabitants who remembered what it was like (Ellison 1954; Prevedel et al. 2005). Evidence pieced together from small areas that were much less-intensively grazed and from observations of changes that have occurred in grazing exclosures indicated that mesic-dry Subalpine-Montane Grassland in this region had high species richness of tall perennial forbs along with grasses and sedges; plants and litter likely combined for about 70 % cover (Ellison1954).

Conditions in mesic-dry Subalpine-Montane Grassland in north-central New Mexico and southwestern Colorado (and probably elsewhere in New Mexico and Arizona) are thought to have included high diversity and cover of herbaceous species (Romme et al. 2009). Thurber fescue and Arizona fescue dominated and likely combined for more than 50 % cover. Bunchgrass interspaces had forbs, sod-forming grasses, and sedges. Litter and plant cover combined was probably 80–90 %. Annual forbs and species that increase with disturbance were minor components of most stands. Moist Subalpine-Montane Grassland likely had been more common before Euro-American settlement.

Fig. 6.14 Subalpine-Montane Grassland in 1867 in north-central New Mexico. Bare patches in foreground are likely from livestock grazing. Trees in background are ponderosa pine (Photograph by Alexander Gardner, courtesy of the Boston Public Library)

6.4.2 Post-Euro-American Settlement

The lack of data and paucity of information on pre-Euro-American conditions in southwestern Subalpine-Montane Grassland also limit understanding of changes since then (e.g., Ellison 1954). Nevertheless, it is clear that changes have occurred, and livestock grazing has been a primary driver (Sect. 6.2.6.1). For example, comparison of data collected in the Apache and Sitgreaves National Forests for 1913–1915 to 1997–1998 determined significant decreases in vegetation cover and increases in bare ground that were more related to livestock and elk grazing than to either fire exclusion or variation in precipitation (White 2002). Greater change occurred in montane grassland types than in subalpine grassland types.

The impacts of livestock grazing depend on many variables, including site conditions, intensity and duration of grazing, and whether grazing was by sheep, cattle, or both. General effects of livestock grazing on Subalpine-Montane Grassland of the American Southwest include: (a) changed structure and composition (e.g., reduced cover, altered species composition, increased shrubs, reduced species diversity, and introduction and establishment of invasive plants); (b) changed processes of vegetation dynamics (e.g., altered succession and increased tree invasion); (c) modified site conditions (e.g., reduced likelihood of fire, reduced animal species diversity, altered populations of native animal species, increased soil exposure and compaction, reduced water infiltration, increased erosion, and reduced nutrient cycling); and (d) decreased resilience, decreased connectivity, and greater fragmentation (e.g., Ellison 1954; Fleischner 1994; Fletcher and Robbie 2004; Merola-Zwartjes 2004; Prevedel et al. 2005; Romme et al. 2009). While the general impacts of historical grazing are clear, details of the impacts are not. Even the impacts of more recent livestock grazing have been characterized as poorly known, confusing, and controversial (Merola-Zwartjes 2004).

Changes in plant species composition resulting from livestock grazing have been a focus of research. In general terms, there have been reductions in palatable species and increases in less-palatable or unpalatable species. A quantitative study of Subalpine-Montane Grassland in the Apache and Sitgreaves National Forests determined that almost 95 % of species shared by all four community types were not preferred forage species for livestock (White 2002). In general, bunchgrasses have decreased (Milchunas 2006; Zier and Baker 2006; see Sect. 6.3.1), especially Thurber fescue. Also, the tall perennial forbs that were the primary growth form in central Utah were reduced by livestock grazing (Ellison and Aldous 1952; Ellison 1954). Species that have increased with grazing include unpalatable shrubs, forbs, and uncommon grasses, as well as invasive plants, rhizomatous species, and species from drier habitats (Ellison and Aldous 1952; Ellison 1954; Merola-Zwartjes 2004; Prevedel et al. 2005; Milchunas 2006; Zier and Baker 2006). Sheep and cattle have different effects on species composition. Sheep grazing results in palatable forbs being replaced by grasses, and cattle grazing results in palatable grasses being replaced by unpalatable forbs and shrubs (Ellison 1954; Milchunas 2006). However, the impacts of livestock grazing are not always separable from grazing by deer and elk (cf. Rambo and Faeth 1999).

Overall, species changes in north-central New Mexico and southwestern Colorado (and likely elsewhere in New Mexico and Arizona) were characterized as changes from tall bunchgrass to short sod- or forb-dominated Subalpine-Montane Grassland (Romme et al. 2009). Species changes in Utah involved replacement of tall perennial forbs with shorter forbs, grasses, and shrub species (Ellison 1954; Lewis 1993 in Prevedel et al. 2005). Erosion is thought to have increased with livestock grazing and to have led to stream incisement, which lowered water tables and led to drying of some moist Subalpine-Montane Grassland (Patton and Judd 1970; Bradley et al. 1992; cf. Sect. 6.2.5.3) and replacement of sedges with species of mesic-dry stands. Changes from 1913–1915 to 1997–1998 in the Apache and Sitgreaves National Forests were characterized as negatively affecting soil and site stability, hydrologic function, and biotic community integrity (White 2002).

Changes involving trees are a special case, because of the uncommonness of that growth form in Subalpine-Montane Grassland. Not all stands have experienced encroachment by trees (Allen 1984; Dyer and Moffett 1999; Zier and Baker 2006). Evidence of long-term persistence of treeless Subalpine-Montane Grassland includes large stands on mountains in north-central New Mexico that early Spanish explorers named for having extensive grasslands (Allen 1984), e.g., Cerro Pelon ("bald peak") and Cerro Pelado ("bare peak"). Historical photographs also provide evidence of persistence of some stands (e.g., Zier and Baker 2006).

Nevertheless, other stands of Subalpine-Montane Grassland have decreased in size or have been lost since Euro-American settlement because of tree encroachment (see Sect. 6.3.3 for mechanisms). For example, mesic-dry Subalpine-Montane Grassland on the slopes of the Jemez Mountains has been invaded by trees, reducing the area of grassland by 55 % during 1935–1981 with the disappearance of some small stands and fragmentation of larger stands (Allen 1989). Decrease in size of stands has also been reported for the White Mountains of New Mexico (Dyer and Moffett 1999), the North Rim region of Grand Canyon National Park (Moore and Huffman 2004), and the San Juan Mountains (Zier and Baker 2006).

In addition, changes in adjacent forests, particularly Ponderosa Pine Forest, likely have affected Subalpine-Montane Grassland. Before Euro-American settlement, grassland species were extensive in the understory of open forest stands, resulting in connectivity among grassland patches. As fire exclusion resulted in increased tree densities, the forest understory decreased (Sect. 4.4.2) and thereby the connectivity of grassland species was reduced (Fletcher and Robbie 2004).

6.5 Conceptual Models

The following nested, three-tiered set of conceptual models is based on my interpretation of best-available information on Subalpine-Montane Grassland of the American Southwest. The models summarize and synthesize material on drivers (Sect. 6.2) and processes (Sect. 6.3) of vegetation dynamics, and they account for historical changes (Sect. 6.4). Their format is explained in Sect. 1.7. The models can be used to facilitate understanding of the past and the present. They also can provide insight into effects of future changes in the environment and outcomes of alternative land-management decisions. Moreover, they can guide the development of quantitative models.

6.5.1 Ecosystem-Characterization Model

The ecosystem-characterization model for southwestern Subalpine-Montane Grassland emphasizes *Vegetation* (a biotic component) and *Disturbance* (a driver; Fig. 6.15a, Table 6.2). Key aspects of *Vegetation* are structure, cover, sprouting, species composition and fuel, and these affect *Disturbance*. The primary agents of *Disturbance* are fire, herbivory, and extreme weather, and these cause shoot mortality followed by resprouting of some species. A second biotic component is *Soil System*, the key aspects of which are water, water table, and texture, all of which influence vegetation structure and composition. The third biotic component is *Animals*. They affect vegetation through herbivory, especially when animal populations increase to where herbivory is a disturbance. Animals also affect the soil system by burrowing, which loosens and aerates soils and exposures mineral soil. Larger mammals also can compact soils. A second driver is *Weather & Climate*, which ignites fires, causes weather extremes, and influences fire behavior, fuel moisture, plant vigor, soil moisture, water table depth, and erosion. The third driver is *Landscape*, with its primary features being topography, elevation, proximity to forest, and landscape position. It influences weather, climate, water runoff, deposition of fine soil particles, spread and pattern of fire, and impact of drought. Landscape position combines with the Soil System to determine differences between moist and mesic-dry Subalpine-Montane Grassland. The model also includes six anthropogenic drivers (Fig. 6.15b, Table 6.2): *Livestock Grazing*, *Fire Management*, *Modern Climate Change*, *Invasive Species*, *Recreation*, and *Nearby Land Use*.

6.5.2 Vegetation-Dynamics Model

The same vegetation-dynamics model represents both moist and mesic-dry Subalpine-Montane Grassland. The vegetation-dynamics model is accompanied by a bar graph to show estimated relative shifts in abundances of the community types from circa 1870 (prior to widespread Euro-American land use) to circa 1970 (end of strict fire exclusion policy) to the present. The estimates are based on my understanding of vegetation dynamics and should be interpreted as showing relative differences and not interpreted quantitatively (hence the absence of a scale for abundance). Although not shown on the graph, relative abundances shifted prior to circa 1870 as a result of variation in climate and disturbance. Localized differences also have been and are present, but are beyond the scope of this generalized bar graph.

The vegetation dynamics model of moist and mesic-dry Subalpine-Montane Grassland has three states with five communities and two transitions (Fig. 6.16, Table 6.3). All communities occurred historically. *State A* is composed of community *A1 Moist Grassland*, which is characterized by sedges, forbs, grasses, and shrubs. It rarely burns and is maintained by a high water table. *State B* includes community *B1 Mesic-Dry Grassland*, which is dominated by grasses, forbs, and shrubs. *B1* was the only common community circa 1870 (most stands of *A1 Moist Grassland* were and are small). It can be maintained by fire. Alternatively, it can form community *B2 Degraded Mesic-Dry Grassland* by disturbance such as intensive or long-lasting herbivory. *B2* has been the most common community since circa 1870. It is dominated by forbs, shrubs, and grasses, including invasive plants, and is maintained by ongoing disturbance. Without disturbance, succession can change *B2* into *B1*. Alternatively, tree establishment (encroachment) can change *B2* or *B1* into *B3 Wooded Mesic-Dry Grassland*. *B3* is dominated by the same growth forms as *B1* and *B2*, with invasion and establishment of trees. Tree species are those found in adjacent or nearby forests. *B3* can revert to community *B1* of *B2* with tree mortality.

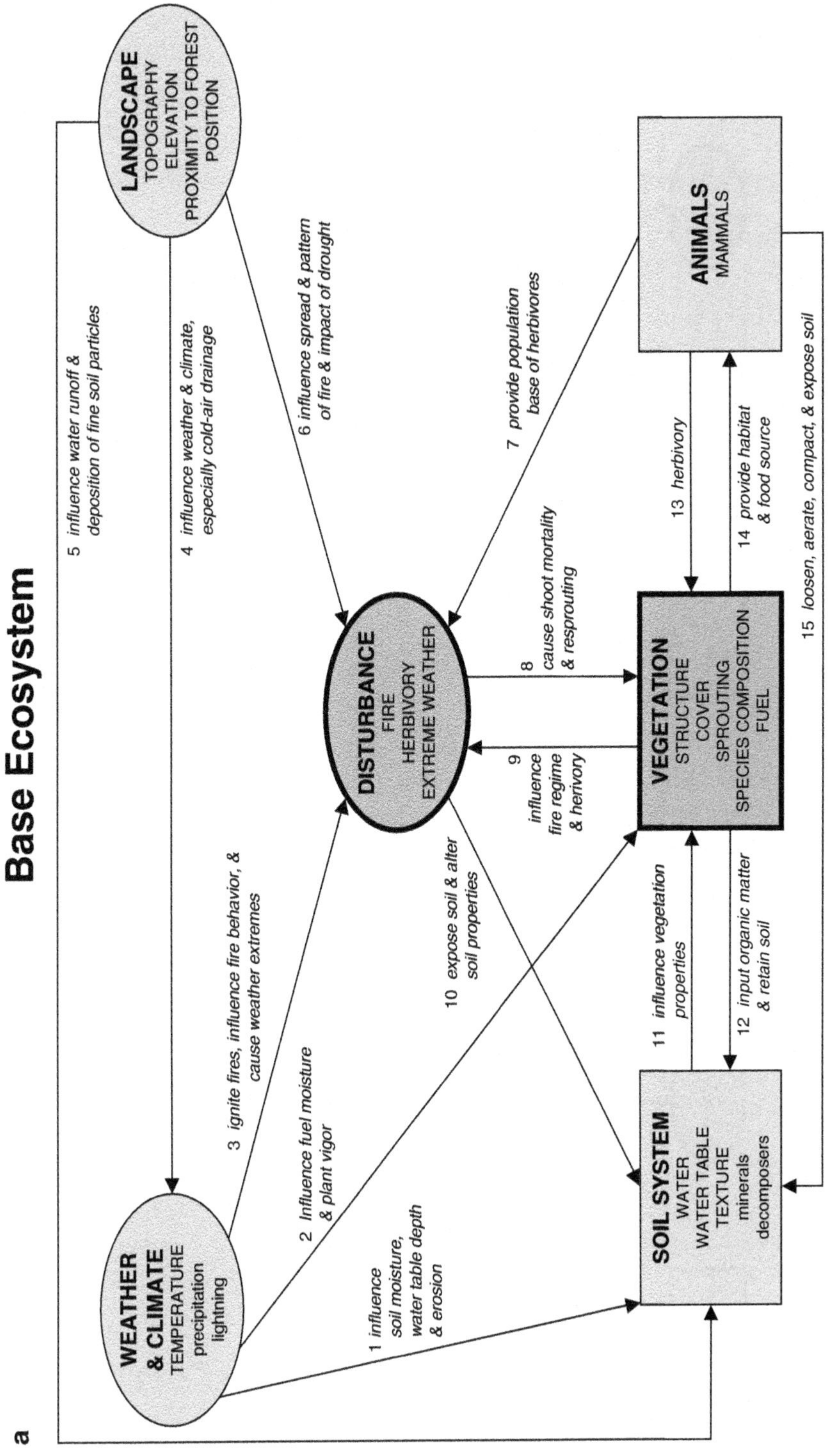
a
Base Ecosystem
LANDSCAPE
TOPOGRAPHY
ELEVATION
PROXIMITY TO FOREST
POSITION
WEATHER & CLIMATE
TEMPERATURE
precipitation
lightning
DISTURBANCE
FIRE
HERBIVORY
EXTREME WEATHER
VEGETATION
STRUCTURE
COVER
SPROUTING
SPECIES COMPOSITION
FUEL
ANIMALS
MAMMALS
SOIL SYSTEM
WATER
WATER TABLE
TEXTURE
minerals
decomposers
1 influence soil moisture, water table depth & erosion
2 Influence fuel moisture & plant vigor
3 ignite fires, influence fire behavior, & cause weather extremes
4 influence weather & climate, especially cold-air drainage
5 influence water runoff & deposition of fine soil particles
6 influence spread & pattern of fire & impact of drought
7 provide population base of herbivores
8 cause shoot mortality & resprouting
9 influence fire regime & herivory
10 expose soil & alter soil properties
11 influence vegetation properties
12 input organic matter & retain soil
13 herbivory
14 provide habitat & food source
15 loosen, aerate, compact, & expose soil

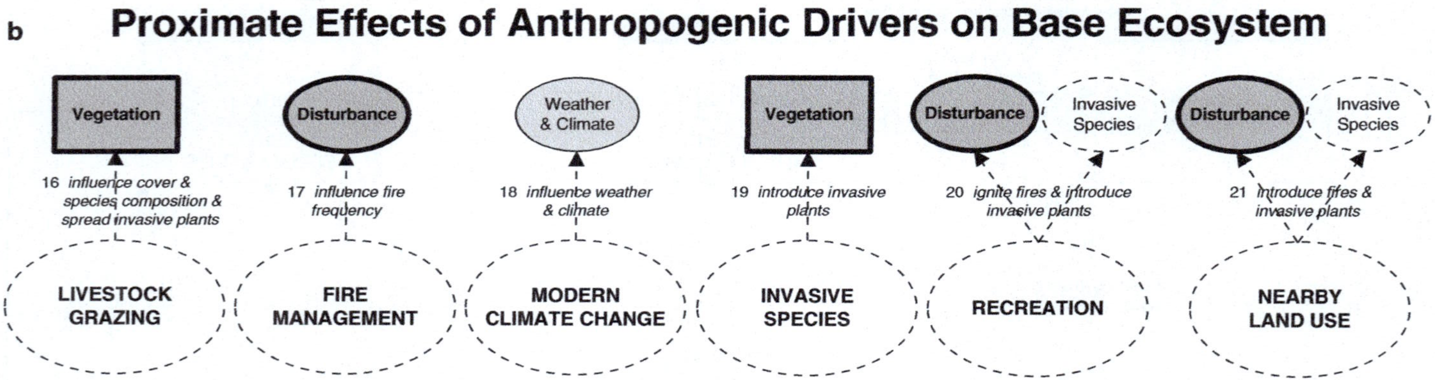

Fig. 6.15 (**a**, **b**) Ecosystem-characterization model for Subalpine-Montane Grassland. In the base model (**a**), biotic components are shown as *rectangles* and drivers as *solid ovals*. Text within them identifies major features in *upper case* and other important features in *lower case*. Disturbance and Vegetation are *highlighted*, because they are the basis for the vegetation-dynamics model (Fig. 6.16). In the anthropogenic supplement to the base model (**b**), drivers are shown as *dashed ovals*. Relationships in (**a**) and (**b**) are shown as *solid* and *dashed arrows*, respectively, and are described in Table 6.2

Table 6.2 Relationships in the ecosystem-characterization model of southwestern Subalpine-Montane Grassland (Fig. 6.15a, b)

Relationships	Circa 1870	Present
1	Precipitation influences soil moisture, water table depth, and erosion	Same, but climate change increases the severity and duration of drought
2	Precipitation and temperature largely determine fuel moisture and are important in plant recruitment, growth, and vigor	Same, but climate change increases the severity and duration of drought
3	Lightning ignites fires, wind influences fire behavior, and weather extremes occur	Same, but weather extremes are more common
4	Topography influences weather and climate, especially cold-air drainage	Same
5	Topography influences water runoff, deposition of fine soil particles	Same
6	Fires can spread from other areas of the landscape (especially from adjacent forests), and position in landscape and topography influence patterns of fire and drought	Same, but fires are probably less frequent
7	Increase in populations of herbivores result in increased herbivory	Same
8	Fires and herbivory cause shoot mortality and lead to resprouting	Same
9	Cover of plants and litter influence fire regime, and species composition influences herbivory	Same
10	Fires expose soil surface and alter soil properties	Same, but fires are probably less frequent
11	Water, depth of water table, etc. influence plant cover and species composition	Same
12	Plants add organic matter to soil and stabilize soil	Same
13	Herbivory influences plant cover and species composition	Same
14	Vegetation is habitat and food for animals	Same
15	Burrowing animals loosen, aerate, and expose soil, and large mammals compact soil	Same
16	Not applicable	Livestock grazing reduces plant cover, alters species composition, and spreads invasive species

17	Not applicable	Fire suppression reduces the frequency and spread of fires
18	Not applicable	Modern climate change increases temperatures, the frequency and extent of drought, and other extreme weather
19	Not applicable	Invasive plants change species composition
20	Not applicable	Recreationists ignite fires and introduce and spread invasive plants
21	Not applicable	Nearby land use that initiates fires and colonization by invasive plants can introduce fires and invasives into Subalpine-Montane Grassland

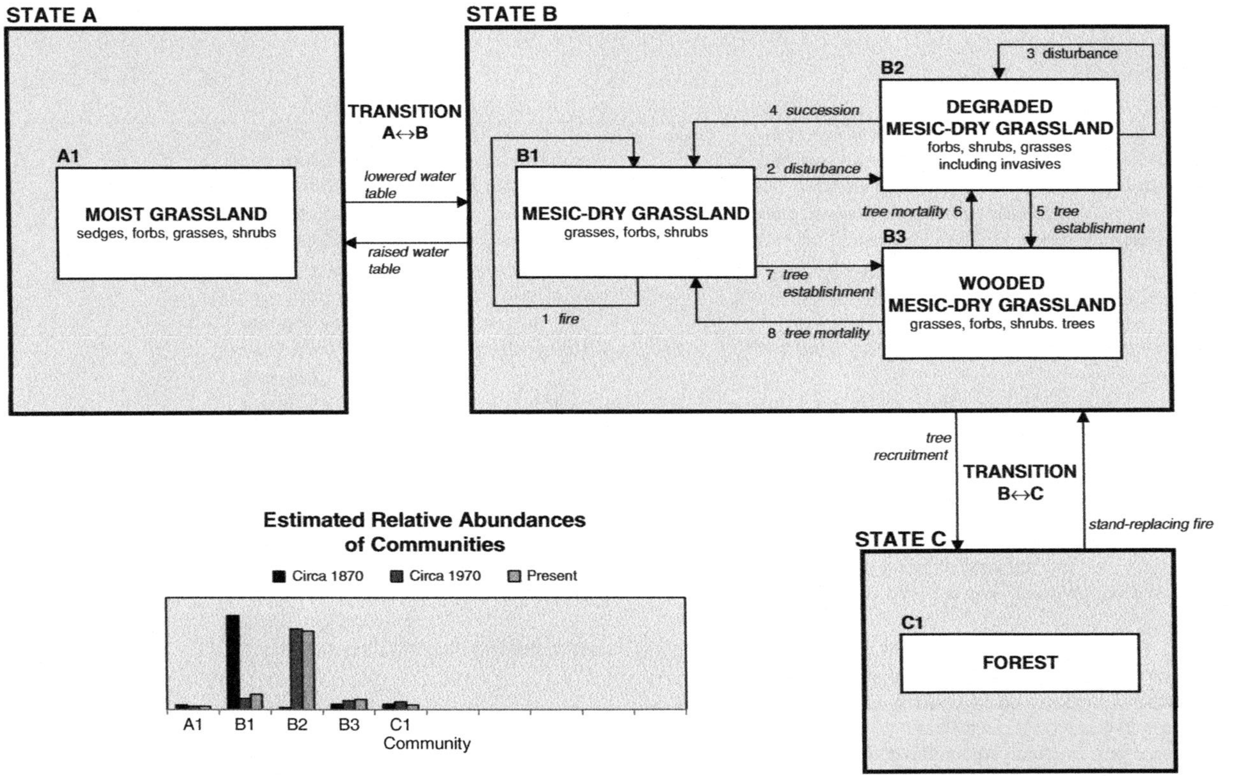

Fig. 6.16 Vegetation-dynamics model for Subalpine-Montane Grassland (both moist and mesic-dry subtypes). Model provides details on the vegetation-disturbance portion of the ecosystem-characterization model (Fig. 6.15a, b) by showing communities (*smaller*, *open rectangles*) within states (*larger*, *shaded rectangles*), processes that form the communities, and transitions between states. Text along *arrows* identifies pathways that are described in Table 6.3. *Bar graph* shows estimated relative shifts in abundances of the community types from circa 1870 (see Sect. 6.5.2)

Table 6.3 Relationships and transitions in the vegetation-dynamics model of Subalpine-Montane Grassland (Fig. 6.16)

Relationships and transitions	Circa 1870	Present
A↔B	Decrease in depth of water table can change moist grassland of State A into mesic-dry grassland of State B. Reversion to State A requires a raised water table	Same, but decreases in depth of water table are more common. Livestock grazing results in greater soil erosion that can initiate stream incisement. Also, modern climate change increases the frequency and extent of extreme weather events such as high precipitation that can result in incisement
1	Frequent fires kill trees, maintaining treeless mesic-dry grassland	Same, but fires are less frequent
2	Disturbance changes mesic-dry grassland into degraded mesic-dry grassland	Same, but disturbance is more common because of livestock grazing and invasive plant species
3	Disturbance maintains degraded mesic-dry grassland	Same, but disturbance is more common
4	Succession changes degraded mesic-dry grassland into mesic-dry grassland	Same, but is possibly affected by invasive species
5	Tree invasion and establishment convert degraded mesic-dry grassland into wooded mesic-dry grassland	Same, but tree establishment is likely more common because fires are less frequent
6	Tree mortality converts wooded mesic-dry grassland into degraded mesic-dry grassland	Same, but is likely less common because fires are less frequent
7	Tree invasion and establishment convert mesic-dry grassland into wooded mesic-dry grassland	Same, but tree establishment is likely more common because fires are less frequent
8	Tree mortality converts wooded mesic-dry grassland into mesic-dry grassland	Same, but is likely less common because fires are less frequent and invasive species are present
B↔C	Infill of trees changes wooded mesic-dry grassland into forest. Reversion to State B requires stand-replacing fire	Same, but transition likely occurs more frequently, because infill increases with infrequent fire and because stand-replacing fire is more common

Transition A↔B changes *State A* to *State B* with lowering of the water table, which is caused by incisement or long drought. *Transition A↔B* can be reversed by rising of the water table.

Transition B↔C changes *State B* (community *B3*) to *State C* with tree recruitment that infills the wooded site. *State C* is composed of community *C1 Forest*. *Transition B↔C* is reversed by stand-replacing fire, forming either community *B1*, *B2*, or *B3*, depending on the colonizing species and survival of trees. See also Sects. 2.5.2, 3.5.2, and 4.5.2 for relationships between forests and Subalpine-Montane Grassland.

6.5.3 *Mechanistic Model*

The mechanistic model (Fig. 6.17) has six biotic components on the right side of the figure (including three aspects of fuels), five drivers on the left side, and six anthropogenic factors at the bottom. In general, *Trees* influence *Fuel Type & Loading*, which influences *Fire Intensity*, which affects the density of *Trees*. *Herbs & Shrubs* affect *Fuel Continuity* and *Fuel Type & Loading*, both of which – along with *Fuel Moisture* and *Fire Frequency* in adjacent forest – affect *Fire Frequency* in grassland. *Fire Frequency* in grassland and *Weather* affect the density of *Trees*. *Weather* also affects the *Water Table* and *Fuel Moisture*. *Weather* and *Water Table* affect the cover and species composition of *Herbs & Shrubs*. *Herbs & Shrubs* and *Trees* determine the community type.

Modern Climate Change affects *Weather*. *Fire Management*, *Nearby Land Use*, and *Recreation* affect *Fire Frequency* in grassland. *Nearby Land Use* and *Recreation* also affect *Invasive Species*. *Invasive Species* and *Livestock Grazing* affect the cover and species composition of *Herbs & Shrubs*.

6.6 Conclusions and Challenges

Subalpine-Montane Grassland of the American Southwest is understudied. Consequently there are relatively few conclusions and many challenges for researchers and managers. Some research needs are basic, such as more information on site conditions, including correlations with elevation and topography. In addition, better understanding of site factors that limit tree growth is important, as is the percentage, ecological distribution, and permanence of stands formed by site conditions vs. historical crown fire. The fire regime is poorly known, and land managers need information on fire frequency. Land managers also need a better understanding of almost all aspects of anthropogenic drivers, including the effects of historical and present-day grazing, fire exclusion, prescribed burning, and modern climate change. Understanding the impacts of recreation and the need for its regulation is important

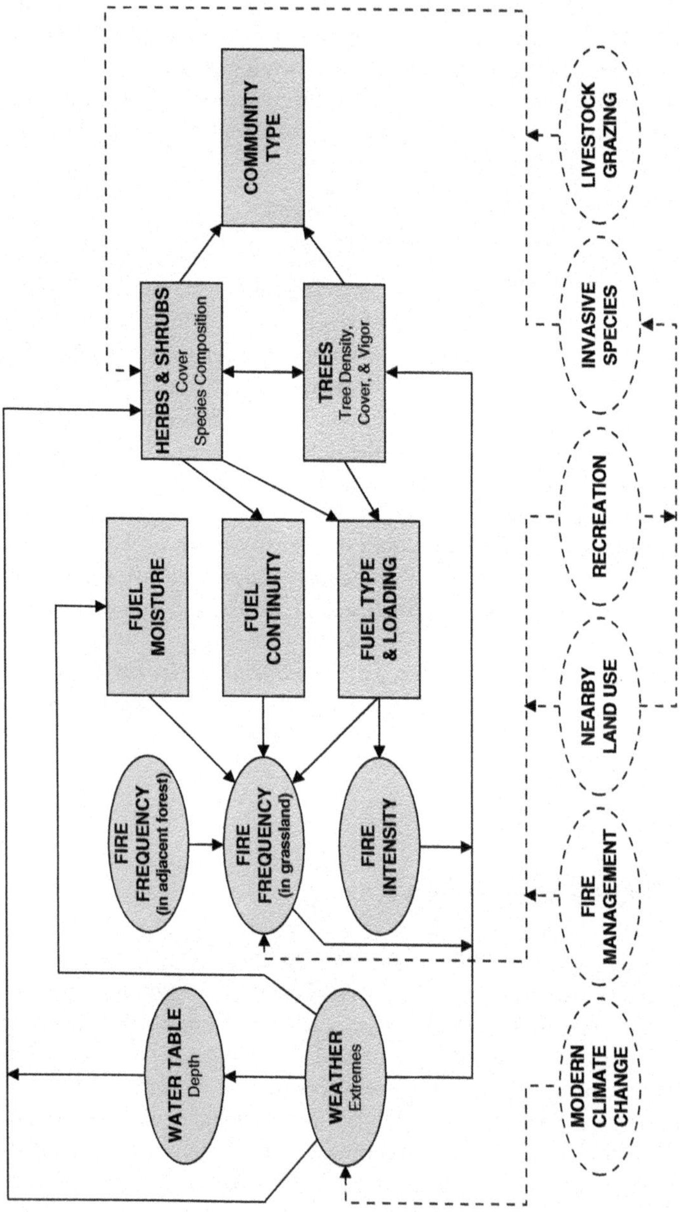

Fig. 6.17 Mechanistic model for Subalpine-Montane Grassland. Model provides details on the mechanisms involved in the vegetation-dynamics model (Fig. 6.16). *Rectangles* are biotic components (including fuels), *ovals* are drivers, and *dashed ovals* are anthropogenic drivers

to managers, as is the effects of land use and management of adjacent forests on Subalpine-Montane Grassland. Stand dynamics involve rapid regrowth, succession, and tree encroachment, none of which are well understood. The fire ecology of species, especially the dominant bunchgrasses, needs study, as do factors influencing successional patterns. Tree encroachment has received more attention, but additional research is needed on its relationship to ungulate herbivory, including that of livestock. Little is known about conditions prior to Euro-American settlement, yet that is essential to more fully understand changes that followed.

References

Allen CD (1984) Montane grassland in the landscape of the Jemez Mountains, New Mexico. Thesis, University of Wisconsin, Madison

Allen CD (1989) Changes in the landscape of the Jemez Mountains, New Mexico. Dissertation, University of California, Berkeley

Antos JA, McCune B, Bara C (1983) The effect of fire on an ungrazed western Montana grassland. Am Midl Nat 110:354–361

Baker WL (2009) Fire ecology in Rocky Mountain landscapes. Island Press, Washington, DC

Beale EF (1858) Wagon road from Fort Defiance to the Colorado River. 35th Congress, 1st Session, Senate Executive Document 124, Washington, DC

Bradley AF, Noste NV, Fischer WC (1992) Fire ecology of forests and woodlands in Utah. U.S. Department of Agriculture, Forest Service General Technical Report INT-287, Intermountain Research Station, Ogden, UT

Brown DE (1994a) Alpine and subalpine grasslands. In: Brown DE (ed) Biotic communities: southwestern United States and northwestern Mexico. University of Utah Press, Salt Lake City

Brown DE (ed) (1994b) Biotic communities: southwestern United States and northwestern Mexico. University of Utah Press, Salt Lake City

Brown DE (1994c) Montane meadow grassland. In: Brown DE (ed) Biotic communities: southwestern United States and northwestern Mexico. University of Utah Press, Salt Lake City

Chambers CL, Holthausen RS (2000) Montane ecosystems used as rangelands. In: Jemison R, Raish C (eds) Livestock management in the American Southwest: ecology, society, and economics. Elsevier, New York

Coop JD, Givnish TJ (2007a) Gradient analysis of reversed treelines and grasslands of the Valles Caldera, New Mexico. J Veg Sci 18:43–54

Coop JD, Givnish TJ (2007b) Spatial and temporal patterns of recent forest encroachment in montane grasslands of the Valles Caldera, New Mexico, USA. J Biogeogr 34:914–927

Coop JD, Givnish TJ (2008) Constraints on tree seedling establishment in montane grasslands of the Valles Caldera, New Mexico. Ecology 89:1101–1111

Dick-Peddie WA (1993) New Mexico vegetation: past, present, and future.. University of New Mexico Press, Albuquerque

Dyer JM, Moffett KE (1999) Meadow invasion from high-elevation spruce-fir forest in south-central New Mexico. Southwest Nat 44:444–456

Ellison L (1954) Subalpine vegetation of the Wasatch Plateau, Utah. Ecol Monogr 24:89–184

Ellison L, Aldous CM (1952) Influence of pocket gophers on vegetation of subalpine grassland in central Utah. Ecology 33:177–186

Fleischner TL (1994) Ecological costs of livestock grazing in western North America. Conserv Biol 8:629–644

Fletcher R, Robbie WA (2004) Historic and current conditions of southwestern grasslands. In: Finch DM (ed) Assessment of grassland ecosystem conditions in the southwestern United

States, vol 1. U.S. Department of Agriculture, Forest Service General Technical Report RMRS-GTR-135-vol 1, Rocky Mountain Research Station, Fort Collins, CO
Ford PL, Potter DU, Pendleton R, Pendleton D, Robbie WA, Gottfried GJ (2004) Southwestern grassland ecology. In: Finch DM (ed) Assessment of grassland ecosystem conditions in the southwestern United States, vol 1. U.S. Department of Agriculture, Forest Service General Technical Report RMRS-GTR-135-vol 1, Rocky Mountain Research Station, Fort Collins, CO
Gill RA (2007) Influence of 90 years of protection from grazing on plant and soil processes in the subalpine of the Wasatch Plateau, USA. Rangel Ecol Manag 60:88–98
Integrated Taxonomic Information System (2012) http://www.itis.gov. Accessed 22 July 2012
Klemmedson JO, Smith EL (1979) Vegetation-soil relations of forests, woodlands, and grasslands of Arizona and New Mexico. In: Youngberg CT (ed) Forest soils and land use. Department of Forest and Wood Sciences, Colorado State University, Fort Collins
Laing L, Ambos N, Subirge T, McDonald C, Nelson C, Robbie W (1987) Terrestrial ecosystems survey of the Apache-Sitgreaves National Forests. USDA Forest Service, Southwestern Region, Albuquerque, NM
Leiberg JB, Rixon TF, Dodwell A (1904) Forest conditions in the San Francisco Mountains Forest Reserve, Arizona. Department of the Interior, U.S. Geological Survey Series H, Forestry, 7, Professional Paper No. 22, Washington, DC
Lewis ME (1993) Successional changes in the tall forb type of the Wasatch Plateau. U.S. Department of Agriculture, Forest Service unpublished report, Intermountain Region, Ogden, UT
Maker HJ, Saugherty LA (1986) Soils. In: Williams JL (ed) New Mexico in maps, 2nd edn. University of New Mexico Press, Albuquerque
McHenry DE (1933) Woodland parks on the North Rim. Gd Canyon Nat Notes 8:195–198
Merkle J (1953) An ecological analysis of the meadows on the Kaibab Plateau, Arizona. Year Book Am Philos Soc 1953:157–160
Merkle J (1962) Plant communities of the Grand Canyon area, Arizona. Ecology 43:698–710
Merola-Zwartjes M (2004) Biodiversity, functional processes, and the ecological consequences of fragmentation in southwestern grasslands. In: Finch DM (ed) Assessment of grassland ecosystem conditions in the southwestern United States, vol 1. U.S. Department of Agriculture, Forest Service General Technical Report RMRS-GTR-135-vol 1, Rocky Mountain Research Station, Fort Collins, CO
Milchunas DG (2006) Responses of plant communities to grazing in the southwestern United States. U.S. Department of Agriculture, Forest Service General Technical Report RMRS-GTR-169, Rocky Mountain Research Station, Fort Collins, CO
Miller G, Ambos N, Boness P, Reyher D, Robertson G, Scalzone K, Steinke R, Subirge T (1995) Terrestrial ecosystems survey of the Coconino National Forest. USDA Forest Service, Southwestern Region, Albuquerque
Moir WH (1967) The subalpine tall grass, *Festuca thurberi*, community of Sierra Blanca, New Mexico. Southwest Nat 12:321–328
Moir WH, Ludwig JA (1979) A classification of spruce-fir and mixed conifer habitat types of Arizona and New Mexico. U.S. Department of Agriculture, Forest Service Research Paper RM-207, Rocky Mountain Forest and Range Experiment Station, Fort Collins, CO
Moore MM, Huffman DW (2004) Tree encroachment on meadows of the North Rim, Grand Canyon National Park, Arizona, U.S.A. Arct Antarct Alp Res 36:474–483
Patton DR, Judd BI (1970) The role of wet meadows as wildlife habitat in the Southwest. J Range Manag 23:272–275
Paulson DD, Baker WL (2006) The nature of southwestern Colorado: recognizing human legacies and restoring natural places. University Press of Colorado, Boulder
Pearson GA (1913) A meteorological study of parks and timbered areas in the western yellow-pine forests of Arizona and New Mexico. Mon Weather Rev 41:1615–1629
Pearson GA (1931) Forest types in the Southwest as determined by climate and soil. Technical Bulletin 247. U.S. Department of Agriculture, Forest Service, Washington, DC
Peet RK (2000) Forests and meadows of the Rocky Mountains. In: Barbour MG, Billings WD (eds) North American terrestrial vegetation, 2nd edn. Cambridge University Press, New York

Prevedel DA, McArthur ED, Johnson CM (2005) Beginnings of range management: an anthology of the Sampson-Ellison photo plots (1913 to 2003) and a short history of the Great Basin Experiment Station. U.S. Department of Agriculture, Forest Service General Technical Report RMRS-GTR-154, Rocky Mountain Research Station, Fort Collins, CO
Prior-Magee JS, Boykin KG, Bradford DF, Kepner WG, Lowry JH, Schrupp DL, Thomas KA, Thompson BC (2007) Ecoregional gap analysis of the southwestern United States: the Southwest regional gap analysis project final report. http://fws-nmcfwru.nmsu.edu/swregap/report/SWReGAP%20Final%20Report.pdf. Accessed 11 July 2012
Rambo JL, Faeth SH (1999) Effect of vertebrate grazing on plant and insect community structure. Conserv Biol 13:1047–1054
Rasmussen DI (1941) Biotic communities of Kaibab Plateau, Arizona. Ecol Monogr 11:229–275
Reynolds RVR (1911) Grazing and floods: a study of conditions in the Manti National Forest, Utah. U.S. Department of Agriculture, Forest Service Bulletin 91, Washington, DC
Rietveld WJ (1975) Phytotoxic grass residues reduce germination and initial root growth of ponderosa pine. U.S. Department of Agriculture, Forest Service Research Paper RM-153, Rocky Mountain Forest and Range Experiment Station, Fort Collins, CO
Romme WH, Redders JS, Floyd-Hanna L, Hanna DD (2009) Other vegetation types. In: Romme WH, Floyd ML, Hanna D (eds) Historical range of variability and current landscape condition analysis: south central highlands section, southwestern Colorado & northwestern New Mexico. Colorado Forest Restoration Institute, Colorado State University and Region 2, U.S. Forest Service, Fort Collins
Rothrock JT (1878) Notes on Colorado in 1873. Reports upon the botanical collections made in portions of Nevada, Utah, California, Colorado, New Mexico, and Arizona during the years of 1871, 1872, 1873, 1874, and 1875. In: Wheeler GM. Report upon United States geographical surveys west of the one hundredth meridian, vol. VI-Botany. U.S. Army, Engineer Department, Government Printing Office, Washington, DC
Rusby HH (1889) General floral features of the San Francisco and Mogollon Mts. of Arizona and New Mexico, and their adjacent regions. Trans N Y Acad Sci 8:76–81
Savage M, Mast JN (2005) How resilient are southwestern ponderosa pine forests after crown fires? Can J For Res 35:967–977
U.S. Forest Service (2012a) *Festuca arizonica*. U.S. Department of Agriculture, Forest Service Fire Effects Information System. http://www.fs.fed.us/database/feis/plants/graminoid/fesari/all.html#FIRE ECOLOGY. Accessed 31 May 2012
U.S. Forest Service (2012b) *Festuca thurberi*. U.S. Department of Agriculture, Forest Service Fire Effects Information System. http://www.fs.fed.us/database/feis/plants/graminoid/festhu/all.html#APPENDIX: FIRE REGIME TABLE. Accessed 31 May 2012
Warren PL, Reichhardt KL, Mouat DA, Brown BT, Johnson RR (1982) Vegetation of Grand Canyon National Park. Technical Report No. 9, Cooperative National Park Resources Study Unit. University of Arizona, Tucson
White MR (2002) Characterization of, and changes in the subalpine and montane grasslands, Apache-Sitgreaves National Forests, Arizona. Dissertation, Northern Arizona University, Flagstaff
Wolters GL (1996) Elk effects on Bandelier National Monument meadows and grasslands. In: Allen CD (tech ed) Fire effects in southwestern forests: proceedings of the second La Mesa fire symposium, Los Alamos, New Mexico, 29–31 Mar 1994. U.S. Department of Agriculture, Forest Service General Technical Report RM-GTR-286, Rocky Mountain Forest and Range Experiment Station, Fort Collins, CO
Zier JL, Baker WL (2006) A century of vegetation change in the San Juan Mountains, Colorado: an analysis using repeat photography. For Ecol Manage 228:251–262

Chapter 7
Gambel Oak Shrubland

Abstract Gambel Oak Shrubland is the second-most abundant shrubland on the mountains and plateaus of the American Southwest. It is more extensive in the northern half of the region. It is bounded at high elevation by forest, primarily Ponderosa Pine Forest in the south and Mixed Conifer Forest in the north. At low elevation, it is bounded by Pinyon-Juniper vegetation or sagebrush shrubland. It is dominated by Gambel oak and other deciduous shrubs. Stands are divided into southern and northern shrublands. Fire is a primary driver of Gambel Oak Shrubland, and the fire regime is dominated by infrequent, high-severity fires that occur during drought. Anthropogenic disturbances include livestock grazing, fire management, modern climate change, invasive species, recreation, and nearby land use. Vegetation dynamics are dominated by rapid regrowth from sprouts following disturbance. Some stands appear to be a seral stage in forest or woodland succession; other stands are more stable in structure and composition. Historical conditions are poorly known, but the regional distribution of southwestern Gambel Oak Shrubland is generally unchanged and stands in many areas have increased in both size and density. Vegetation dynamics are illustrated in a nested, three-tiered set of conceptual models. Key conclusions and challenges for researchers and land managers are summarized.

7.1 Introduction

Gambel Oak Shrubland (*Quercus gambelii*) is the second-most abundant shrubland on the mountains and plateaus of the American Southwest (Figs. 7.1 and 7.2). It is also known as Petran chaparral, deciduous thicket scrub, mountain shrub, mountain brush, mountain mahogany-oak (*Cercocarpus-Quercus*) scrub, and Rocky Mountain bushland (Brown 1994). Gambel Oak Shrubland is dominated by deciduous shrubs 1–4 m (3–13 ft) tall. Gambel oaks in at least north-central New Mexico and southwestern Colorado are generally <2 m (7 ft) tall and <5 cm (2 in.) diameter at breast height (dbh), i.e., at 1.4 m (4.5 ft; Romme et al. 2009). In older stands, Gambel oak reaches 5 m (16 ft) height and 10 cm (4 in.) dbh.

J.L. Vankat, *Vegetation Dynamics on the Mountains and Plateaus of the American Southwest*, Plant and Vegetation 8, DOI 10.1007/978-94-007-6149-0_7,

Fig. 7.1 Gambel Oak Shrubland in summer in central Utah. Note its landscape position below Mixed Conifer Forest (Photograph by author)

Fig. 7.2 Gambel Oak Shrubland in *fall colors* in the LaSal Mountains of east-central Utah. Other vegetation includes *gray*, rock-dominated alpine tundra above *dark-green* Spruce-Fir (*Picea-Abies*) and Mixed Conifer Forests that include *yellow-green* stands of quaking aspen (*Populus tremuloides*) (Photograph by Jay Ross)

Fig. 7.3 Distribution of Gambel Oak Shrubland in the American Southwest. The *map* shows all of the states of Arizona, New Mexico, Colorado, and Utah, and the American Southwest is illustrated in *red* on the small map (Source of data: U.S. Geological Survey National Gap Analysis Program 2005 Southwest Regional Gap Analysis Project, RS/GIS Laboratory, College of Natural Resources, Utah State University, Logan, Utah. Map prepared by Monica Swihart)

Gambel Oak Shrubland covers about 4,600 km^2 (1,800 $miles^2$) of the American Southwest, which is 0.6 % of the region's area and the smallest of the vegetation types covered in this book (Fig. 7.3; calculations based on Prior-Magee et al. 2007). There are other upland shrublands in the region (cf. Dick-Peddie 1993), but only

Table 7.1 Common and scientific names of species in this chapter. Primary source: Integrated Taxonomic Information System (2012)

Plants	
Antelope bitterbrush	*Purshia tridentata* (Pursh) DC.
Bigtooth maple	*Acer grandidentatum* Nutt.
Brome	*Bromus* L.
Cheatgrass	*Bromus tectorum* L.
Chokecherry	*Prunus virginiana* L.
Cliff fendlerbush	*Fendlera rupicola* Gray
Cliff-rose	*Purshia mexicana* var. *stansburyana* (Torr.) S.L. Welsh
Colorado pinyon	*Pinus edulis* Engelm.
Curl-leaf mountain mahogany	*Cercocarpus ledifolius* Nutt.
Douglas-fir	*Pseudotsuga menziesii* (Mirbel) Franco
Fir	*Abies* P. Mill.
Gambel oak	*Quercus gambelii* Nutt.
Juniper	*Juniperus* L.
Kentucky bluegrass	*Poa pratensis* L.
Mountain mahogany (genus)	*Cercocarpus* Kunth
Mountain mahogany	*Cercocarpus montanus* Raf.
Mountain snowberry	*Symphoricarpos oreophilus* A. Gray
New Mexico locust	*Robinia neomexicana* A. Gray
Oak	*Quercus* L.
Pinyon	*Pinus* L.
Ponderosa pine	*Pinus ponderosa* Douglas ex P. Lawson & C. Lawson
Quaking aspen	*Populus tremuloides* Michx.
Rocky Mountain juniper	*Juniperus scopulorum* Sarg.
Sagebrush	*Artemisia* L.
Skunkbush	*Rhus trilobata* Nutt.
Snowberry	*Symphoricarpos* Duhamel
Spruce	*Picea* A. Dietr.
Utah juniper	*Juniperus osteosperma* (Torr.) Little
Utah serviceberry	*Amelanchier utahensis* Koehne
Western serviceberry	*Amelanchier alnifolia* (Nutt.) Nutt. ex M. Roem.
Western wheatgrass	*Pascopyrum smithii* (Rydb.) Barkworth & D.R. Dewey
White fir	*Abies concolor* (Gord. & Glend.) Lindl. Ex Hildebr.
Animals	
Cattle	*Bos taurus* Linnaeus, 1758
Deer	*Odocoileus* Rafinesque, 1832
Elk	*Cervus elaphus* Linnaeus, 1758
Flathead borer	*Agrilus quercicola* Fisher, 1928
Looper	*Lambdina punctata* Hulst, 1899 = *Lambdina vitraria* Grote, 1883
Sheep	*Ovis aries* Linnaeus, 1758

Interior Chaparral Shrubland covers more area and is included in this book. Southwestern Gambel Oak Shrubland is most extensive in southern and central Utah and southwestern Colorado. It covers less area in New Mexico and is uncommon in Arizona. Gambel Oak Shrubland also occurs outside the American

Southwest, especially in northern Utah and central and northern Colorado (see Fig. 7.3).

Gambel Oak Shrubland is bounded at high elevation by forest, primarily Ponderosa Pine Forest (*Pinus ponderosa*) in the southern portion of its range and Mixed Conifer Forest in the northern portion. It is bounded at low elevation by Pinyon-Juniper vegetation (*Pinus-Juniperus*) or sagebrush shrubland (*Artemisia*). Gambel oak occurs in these adjacent vegetation types and can grow as both shrubs and trees in stands of Ponderosa Pine Forest and Pinyon-Juniper vegetation.

Stands occur on a wide variety of sites, and vegetation structure ranges from dense, homogeneous shrub thickets to heterogeneous mixtures of shrub clumps and interspace openings (Brown 1994). Clumps of Gambel oak range in size from 3 to over 150 m^2 (32–1,615 ft^2; Van Epps 1974). Underground, Gambel oak has a shallow lignotuber (swollen root crown), woody rhizomes, and roots. The lignotubers and rhizomes are capable of sprouting (suckering; Engle et al. 1983; Tiedemann et al. 1987). Fire stimulates sprouting (e.g., Floyd et al. 2000). Clumps of Gambel oak reproduce mostly clonally and expand at an average rate of about 10 cm/year (4 in./year; Christensen 1955). Patches of shrubs of the same clone can be separated by over 24 m (80 ft; Van Epps 1974). Clonal growth is likely especially advantageous in marginal environments and where there is competition (Neilson and Wullstein 1983; Harper et al. 1985).

Gambel oak is the only shrub species in some stands, but elsewhere grows in different combinations with other species. Species composition depends on elevation, slope aspect, substrate, and geographic area (Spence et al. 1995). Other shrub species include bigtooth maple (*Acer grandidentatum*), western serviceberry (*Amelanchier alnifolia*), Utah serviceberry (*A. utahensis*), curl-leaf mountain mahogany (*Cercocarpus ledifolius*), mountain mahogany (*C. montanus*), cliff fendlerbush (*Fendlera rupicola*), chokecherry (*Prunus virginiana*), cliff-rose (*Purshia mexicana* var. *stansburyana*), antelope bitterbrush (*Purshia tridentata*), New Mexico locust (*Robinia neomexicana*), and mountain snowberry (*Symphoricarpos oreophilus*).

Scattered trees are present in some stands and include pinyons such as Colorado pinyon (*Pinus edulis*), junipers such as Utah juniper (*Juniperus osteosperma*), ponderosa pine, Douglas-fir (*Pseudotsuga menziesii*), white fir (*Abies concolor*), and bigtooth maple (which can be shrubby or grow as trees). With increased presence of trees, some stands appear to be a seral stage in forest or woodland succession, but other stands are more stable in composition and structure. The ground cover within shrub clumps is dominated by leaf litter, but smaller shrubs and herbs are also present. Interspaces have less litter, more bare soil, and herbs of various species (Brown 1958). A list of species is in Brown (1994).

There is little agreement on subdivisions of Gambel Oak Shrubland (cf. MacMahon 1988). This chapter focuses on two broadly defined types in the American Southwest: southern and northern. *Southern Gambel Oak Shrubland* (Fig. 7.4) occurs primarily in northern Arizona, northern New Mexico, and southwestern Colorado. Gambel oak typically dominates. *Northern Gambel Oak Shrubland* (Fig. 7.5) is most extensive in Utah and extends into western Colorado. Its species composition is variable, but Gambel oak and bigtooth maple commonly codominate.

Gambel Oak Shrubland is the least-researched vegetation in this book. This chapter focuses on findings from the American Southwest, but includes some research results from north-central Utah. The applicability of that research to the Southwest is supported by similarity in species composition between north-central and central Utah (cf. Kunzler et al. 1981).

7.2 Drivers

Key drivers of southwestern Gambel Oak Shrubland are landscape, climate, soil, animals, natural disturbance, and anthropogenic disturbance. Each driver is important in vegetation dynamics, and therefore all are incorporated in the conceptual models (Sect. 7.5).

7.2.1 Landscape

Gambel Oak Shrubland typically occurs at approximately 2,000–2,500 m (6,600–8,200 ft) elevation in the mid-latitudes of its distribution in the American Southwest (Fig. 7.6). Its upper-elevational limits are likely due to competition, cold temperatures, and shorter growing seasons, and its lower limits to water stress (Neilson and Wullstein 1983; Harper et al. 1985). At both upper and lower elevational limits, species and stands integrate with adjacent types of vegetation such as Ponderosa Pine Forest (Floyd 1982; Floyd et al. 2000; Romme et al. 2009), and vegetation type is determined by the relative abundances of shrubs vs. trees (Romme et al. 2009). Gambel Oak Shrubland and Ponderosa Pine Forest can occupy similar sites in areas of their transition (Madany and West 1984; Romme et al. 2009). Transitions can be especially broad where Gambel oak dominates successional or alternative states (communities) in the dynamics of Ponderosa Pine Forest or Pinyon-Juniper vegetation (see Sects. 4.5.2 and 5.5.2).

Topography plays a key role in variation in structure and composition of southwestern Gambel Oak Shrubland. Stands at higher elevation and on steeper sites generally have nearly continuous shrub cover (Hayward 1948). Cover decreases with elevation and on flatter slopes such that stands become open, consisting of scattered clumps of shrubs and a relatively well-developed herbaceous layer between them. These decreases in shrub cover are possibly related to fires spreading and burning less thoroughly and thereby causing less widespread

Fig. 7.4 Stand of southern Gambel Oak Shrubland in summer in Capulin Volcano National Monument in northeastern New Mexico (Photograph by Stephanie E.V. Fitzgerald)

Fig. 7.5 Stand of northern Gambel Oak Shrubland in fall in western Colorado (Photograph courtesy of Agustin Goba)

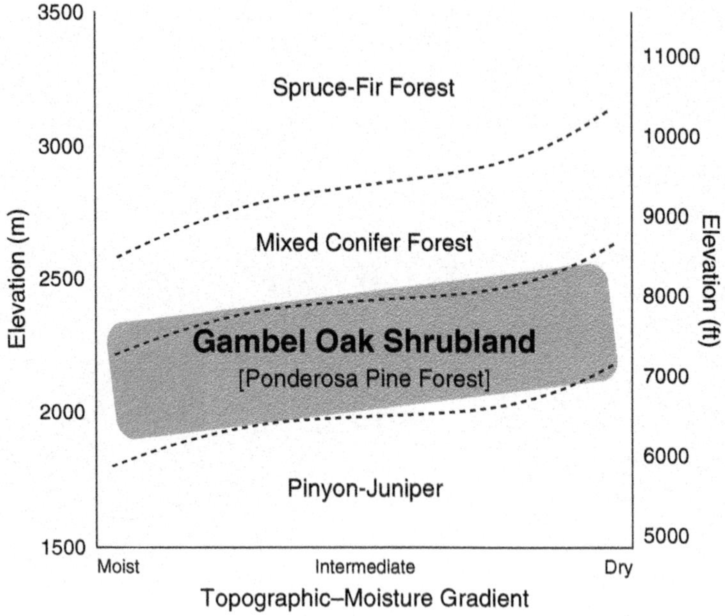

Fig. 7.6 Ecological distribution of Gambel Oak Shrubland on the mountains and plateaus of the American Southwest along gradients in elevation and topographic-moisture. Gambel Oak Shrubland (*shaded area*) is superimposed on Ponderosa Pine Forest. Elevations are approximate and generally representative of mid-latitudes of the region, i.e., landscapes in northern Arizona and northern New Mexico

sprouting (Brown 1958), as also occurs in Interior Chaparral Shrubland (Sect. 8.2.1). In north-central New Mexico and southwestern Colorado, moist higher-elevation sites are dominated by Gambel oak and snowberry (Symphoricarpos spp.) and drier, lower-elevation sites are dominated by mountain mahogany, Utah serviceberry, and cliff fendlerbush (Romme et al. 2009). Slope aspect can be a factor at least in Utah, where stands on south aspects tend to be shorter, are less well-developed, and extend to higher elevations than stands on north aspects (Christensen 1949).

7.2.2 Climate

Few climate data have been published for southwestern Gambel Oak Shrubland (but see Price and Evans 1937; Brown 1958, 1994; Harper et al. 1985). In general, the climate is characterized by cool to cold temperatures that produce moderately long to short growing seasons. Mean monthly maximum temperatures at Mesa Verde National Park in southwestern Colorado range from 4 °C (40 °F) in January to 30 °C (86 °F) in July (Western Regional Climate Center 2012). Mean monthly minimum temperatures range from −7 °C (19 °F) in January to 14 °C (57 °F) in July. The growing season has been reported as 90–136 days (Price and Evans 1937; Brown 1958). Mean annual precipitation is typically 38–56 cm (15–22 in.). It is 46 cm (18 in.) in Mesa Verde, where mean annual snowfall is 203 cm (80 in.). The percentage of precipitation falling in the summer is higher in the south and decreases northward (Fig. 1.17). Winter precipitation is important for recharging soil moisture (Tew 1967). Lightning is a key component of the climate, because it can ignite fires.

Modern climate change is described in Sect. 1.6.3 and is treated as an anthropogenic disturbance in Sect. 7.2.6.3.

7.2.3 Soil

Gambel Oak Shrubland occurs on a variety of soils formed from a variety of parent materials (Christensen 1949; Harper et al. 1985). Soils are generally poorly developed because of steep slopes (Brown 1994), but have a high moisture-holding capacity (Harper et al. 1985). Most soils in stands in southwestern Colorado are Mollisols (Argic Pachic Cryoborolls and Argic Cryoborolls; Steinhoff 1981 in Harper et al. 1985). Such soils tend to be well-drained, moderately deep to deep, and gravelly but fine- to medium-textured (Hendricks 1985). Litter depth can be as much as 8 cm (3 in.; Christensen 1949), but usually is less. Depth of surface soil appears positively correlated with stem size in west-central Colorado (Brown 1958).

7.2.4 Animals

Southwestern Gambel Oak Shrubland is important winter habitat for deer (*Odocoileus* spp.) and elk (*Cervus elaphus*; Hayward 1948; Brown 1958, 1994). Several insect species are associated with Gambel oak (cf. Harper et al. 1985; Cranshaw et al. 1994). These include a looper (tentatively identified as *Lambdina punctata* = *L. vitraria*) that can kill stems after several years of defoliation (Brown 1958). Another insect is a flathead borer (*Agrilus quercicola*) that attacks boles and branches of Gambel oaks stressed by drought, but rarely causes mortality (U.S. Forest Service 2011). No animals have been shown to directly affect vegetation dynamics.

7.2.5 *Natural Disturbance*

Fire and drought are the only natural disturbances that significantly affect southwestern Gambel Oak Shrubland. Nevertheless, there is little information on them.

7.2.5.1 Fire

There are few data on the fire regime of Gambel Oak Shrubland in the American Southwest, in part because Gambel oak and associated shrub species lack fire scars, which are used in most fire-history studies of forests (e.g. Sect. 4.2.5.1). Also, few data are available on fire because several shrub species do not produce clear annual rings (Gambel oak has distinct rings; Floyd et al. 2000; Ex et al. 2011).

The most informative study of the fire regime of southwestern Gambel Oak Shrubland examined fire turnover times in Mesa Verde National Park in southwestern Colorado (cf. Floyd et al. 2000). Past fires were dated and mapped by determining ages of Gambel oak stems that had sprouted after fire. The method recorded fires back to 1840, but likely overlooked many small fires (Floyd et al. 2000). The turnover time during the second half of the nineteenth century, i.e., before significant Euro-American impacts, was about 100 years. Turnover time doubled to 200 years in the first half of the twentieth century, a period when livestock reduced herbaceous fuels and fire suppression began. However, it is unlikely that livestock grazing was important in the reduction of fire, because observations of fire behavior after grazing was stopped in the early twentieth century revealed that fires spread primarily through shrub crowns. Fire suppression also was considered unimportant because it was very limited much of that half-century. That left regionally moist conditions (cf. Swetnam and Betancourt 1998) as the likely cause of the increased turnover time.

In the second half of the twentieth century, turnover time in Mesa Verde returned to the nineteenth century figure of 100 years, despite advanced fire-fighting technologies and a policy of complete fire suppression. Major fires occurred when stands were highly flammable as a result of lengthy drought, dense structure, much leaf litter, and continuous herbaceous fuels in shrub interspaces. Many of these fires began in Pinyon-Juniper vegetation and spread into Gambel Oak Shrubland.

Another data-based study examined the frequency of ignitions and area burned in a large area of Gambel Oak Shrubland in north-central Utah (Wadleigh et al. 1998, for 1973–1997). Lightning-caused ignitions occurred primarily in July and August and averaged two per year. All 50 lightning fires combined burned <0.01 % of the study area (human-caused ignitions were more frequent and burned a much larger area). Even if the spread of lightning-caused fires was reduced by suppression, the fire turnover time was likely several centuries.

Long turnover times suggest that the fire regime of Gambel Oak Shrubland is characterized high fire severities. This was evidenced by the late twentieth century fires in Mesa Verde National Park. In addition, modeling of wildfire hazards in southwestern Colorado indicated high wildfire hazards in Gambel Oak Shrubland (Romme et al. 2006). A fire regime of infrequent but severe fire also is indicated by an overview of fire ecology in Utah (Bradley et al. 1992). The authors stated that burning is usually so unlikely that Gambel Oak Shrubland has been considered a fuel break between more highly flammable vegetation types. Nevertheless, fuel conditions occasionally enabled severe fires, such as when leaves killed by spring frosts or disease remained on the oak shrubs as fine, dry fuels.

Evidence of low-severity fires comes from two relict areas on isolated mesas in Zion National Park in southwestern Utah (Madany and West 1984). Stands of Gambel Oak Shrubland included large stems of fire-sensitive bigtooth maple and chokecherry. The authors hypothesized that the fire regime consisted of patchy, low-severity fires that burned into stands along grassy interspaces and were stopped by moist fuels in clumps of Gambel oak. The broader applicability of this hypothesis is possibly limited. Although Gambel Oak Shrubland covered at least half of both mesas, it appears that stands were interspersed with stands of Ponderosa Pine Forest, and this mosaic possibly affected the fire regime. Also, it is unclear that fire patterns on small relict sites are widely representative (see discussion of relict sites in Sect. 5.4.1.1). Regardless, the occurrence of patchy, low-severity fires does not preclude infrequent, high-severity fires.

In conclusion, at least some areas of southwestern Gambel Oak Shrubland have experienced and continue to experience infrequent, high-severity fires linked to drought. Low-severity fires also can occur, but current evidence suggests they are limited. It is unclear if low-severity fires were more common before livestock grazing (Sect. 7.2.6.1).

7.2.5.2 Drought

Drought is an important disturbance factor in southwestern Gambel Oak Shrubland through its interaction with fire. For example, in Mesa Verde National Park weather and fire size have been strongly related since record-keeping began in 1926 (Omi and Emrick 1980 in Floyd et al. 2000). All large fires in the park occurred with severe early-summer drought, and fires were reduced during moist decades of the twentieth century (Floyd et al. 2000; see previous section).

7.2.6 Anthropogenic Disturbance

Land use by Native Americans in the American Southwest was concentrated in Pinyon-Juniper vegetation (Sects. 1.5.1.1 and 5.2.6), but Gambel Oak Shrubland was likely used for hunting and possibly other purposes. Multiple sources have described effects of anthropogenic factors on southwestern Gambel Oak Shrubland; however, supportive data are generally limited in quantity and restricted spatially. Anthropogenic disturbances considered below and related to Euro-American land use are livestock grazing, fire management, modern climate change, invasive species, recreation, and nearby land use. Each of these can be important in vegetation dynamics, and therefore all are incorporated in the conceptual models (Sect. 7.5).

7.2.6.1 Livestock Grazing

Grazing by cattle (*Bos taurus*) and sheep (*Ovis aries*) has been widespread in southwestern Gambel Oak Shrubland, except for stands that are virtually impenetrable. In general, livestock grazing is thought to have reduced palatable herbs and increased cover of Gambel oak and other shrub species (Harper et al. 1985). Past livestock grazing in Gambel Oak Shrubland in Mesa Verde National Park possibly enhanced twenty-first century abundance of herbaceous plants poisonous to livestock and the abundance of Kentucky bluegrass, a common invasive resistant to grazing (Paulson and Baker 2006). Comparison of relict sites to grazed sites in Zion National Park indicated grazing reduced grasses in shrub interspaces, enabling shrub clones to coalesce and increase in cover (Madany and West 1983). In contrast, other observations indicated grazing reduces Gambel oak (Forsling and Storm 1929; Evans 1936, both in Christensen 1949). Moreover, grazing impacts on species composition can be complex. For example, grazing in north-central Utah was said to have nearly eliminated western wheatgrass (*Pascopyrum smithii*) from Gambel Oak Shrubland (Hayward 1948), yet the species was observed to die out in a grazing exclosure (Nixon 1967). Indirect effects of livestock grazing include mechanical cutting of Gambel oak to increase forage (cf. Harper et al. 1985).

Grazing also has been said to have altered the fire regime. Consumption of fine fuels that potentially carry fires through stands has been implicated in reduced spread of fire (Wadleigh et al. 1998), but the lack of fire scars in Gambel oak makes it challenging to verify this. And as stated above, fires have been observed to spread primarily through shrub canopies, not herbaceous fuels (Floyd et al. 2000). Impacts of grazing on fire possibly are stand-specific, depending on the relative cover of shrubs and herbs. Indirect effects of grazing on the fire regime include stockmen setting fires to reduce shrub cover to enable the movement of livestock and increase grass forage (Reynolds 1911; Brown 1958).

7.2.6.2 Fire Management

Stoppage of burning by Native Americans following widespread Euro-American settlement is said to have affected southwestern Gambel Oak

Shrubland (Brown 1958; see also Wadleigh et al. 1998), but regional-scale ecological impacts of Native Americans on southwestern fire regimes are controversial and undocumented (Sect. 1.5.1.2).

The effects of Euro-American fire management, which was characterized by fire suppression throughout most of the twentieth century, are unclear and possibly disparate. Fire suppression has been considered a factor in successional changes. It has been said to have facilitated replacement of Gambel Oak Shrubland by later-successional Pinyon-Juniper vegetation in Mesa Verde National Park (Erdman 1970), but no clear evidence for this was presented. Fire suppression is also said to have increased bigtooth maple, white fir, and Rocky Mountain juniper (*Juniperus scopulorum*) elsewhere (Harper et al. 1985). In addition, fire suppression has been said to have reduced the spatial extent of fires in north-central Utah (Wadleigh et al. 1998), although no evidence was presented. The observation that advanced fire suppression technology failed to affect major twentieth century fires in Gambel Oak Shrubland in Mesa Verde National Park (Sect. 7.2.5.1) indicates that fire suppression does not influence high-severity fire (Floyd et al. 2000).

7.2.6.3 Modern Climate Change

No research has focused on the effects of modern climate change on southwestern Gambel Oak Shrubland, but it is likely that such effects exist. For example, the conclusion that drought was the primary driving force behind major fires in Mesa Verde National Park in the late twentieth century (Sect. 7.2.5.1) indicates that climate warming and drying can impact fire regimes.

7.2.6.4 Invasive Species

Data on invasive plants in southwestern Gambel Oak Shrubland are limited. Both cheatgrass (*Bromus tectorum*) and Kentucky bluegrass (*Poa pratensis*) were reported as common in stands of central Utah as early as 1981 (Kunzler et al. 1981). Studies have focused on the relationship between invasives and fire. Post-fire resprouting by shrubs and perennial grasses can reduce the influx of invasives (Floyd et al. 2001, 2006). Indeed, comparison of unburned and burned stands in north-central Utah indicated a large decrease in cover of invasives 1 year after fire; nevertheless, the number of invasive species was higher on the burned site (Poreda and Wullstein 1994). Longer-term effects are less well-documented. Cheatgrass can increase following fire in Gambel Oak Shrubland (Kunzler et al. 1981; Wadleigh et al. 1998; Floyd et al. 2006). The number of non-native species after fire was positively correlated with richness of native species in Mesa Verde, indicating that habitats with high biodiversity are at greater risk of invasion (Floyd et al. 2006).

7.2.6.5 Recreation

No studies have explicitly addressed effects of recreation on southwestern Gambel Oak Shrubland, but other research has implicated recreation as an anthropogenic disturbance factor. For example, all recent, large fires in the extensive area in north-central Utah mentioned in Sect. 7.2.5.1 were human-caused and occurred near population centers (Wadleigh et al. 1998). Also, the presence of invasive species in recreational areas suggests recreationists facilitate their spread and establishment.

7.2.6.6 Nearby Land Use

Nearby land use is an important anthropogenic driver of southwestern Gambel Oak Shrubland because of proximity to and intergradations with Ponderosa Pine Forest, Pinyon-Juniper vegetation, and human development. For example, some of the recent large fires mentioned in the previous section occurred in areas adjacent to human developments (Wadleigh et al. 1998). Nearby land use is also a likely source of invasive plants, given their presence in adjacent stands of Ponderosa Pine Forest (Sect. 4.2.6.4), Pinyon-Juniper vegetation (Sect. 5.2.6.4), and sagebrush shrubland, as well as in areas of human development.

7.3 Processes

Important processes in Gambel Oak Shrubland of the American Southwest are rapid regrowth following disturbance and succession following high-severity disturbance. Both of these processes play key roles in the conceptual models (Sect. 7.5).

7.3.1 Rapid Regrowth

The primary process of vegetation dynamics in southwestern Gambel Oak Shrubland is rapid regrowth following disturbance such as fire (Figs. 7.7 and 7.8). Rapid regrowth is also characteristic of Subalpine-Montane Grassland and Interior Chaparral Shrubland (Sects. 6.3.1 and 8.3.1, respectively). Shrubs rapidly regrow because Gambel oak and other species (e.g., Utah serviceberry, mountain snowberry, cliff fendlerbush, and skunkbush (*Rhus trilobata*)) sprout from below-ground structures. The presence of dead stems and stumps indicates regrowth occurs in the same locations as previous clumps (Brown 1958).

Regrowth is so rapid that shrub cover on burned sites is similar to that of unburned sites within a year or two (Poreda and Wullstein 1994; Floyd et al. 2000). Full recovery of shrub height takes longer, typically 15 years in central Utah (range of 6–35 years), with faster recovery at lower elevations and on south and west exposures (Kunzler and Harper 1980). Clumps of shrubs tend to regrow as low, dense

Fig. 7.7 Recently burned stand of Gambel Oak Shrubland in the La Sal Mountains of east-central Utah (Photograph © William Horton, William Horton Photography)

Fig. 7.8 Post-fire sprouts of Gambel oak three years after fire in the La Sal Mountains in east-central Utah. Different patches of *fall colors* suggest different clones of Gambel oak. Fire-killed trees are mostly Douglas-fir (*gray-black*) in Mixed Conifer Forest at high elevation and pinyons and junipers scattered in Gambel Oak Shrubland at mid- to low-elevation (Photograph by Jay Ross)

thickets that become more open as shrubs age (apparently with self-thinning of stems; Brown 1958; Floyd 1982; Bradley et al. 1992). Sprouting also can reduce interspaces, causing clumps of Gambel oak to merge (Brown 1958). Species of other growth forms also regrow after fire, and stands in central Utah return to pre-fire species composition with little if any loss of species (Kunzler et al. 1981).

7.3.2 Succession

The successional status of southwestern Gambel Oak Shrubland has received much attention. Some stands have shown no evidence of substantial, directional changes in species composition and are interpreted to be late-successional (Harper et al. 1985; Bradley et al. 1992). Other stands have been invaded by trees (Harper et al. 1985). Although displacement of Gambel oak has rarely been observed (Harper et al. 1985), many researchers have interpreted the ingrowth of trees as indicating Gambel Oak Shrubland is a seral stage in forest or woodland succession (e.g., Bradley et al. 1992). Alternatively, these changes in species composition are possibly driven by shifts in climate and disturbance regimes, but this has received little attention (see Sect. 7.4.1.2 for possible connections among climate, fire, and livestock grazing that account for twentieth century expansion and infilling of Gambel Oak Shrubland in central and north-central Utah).

Succession begins with rapid regrowth of Gambel oak, herbs, and – if present – other shrub species. This can be followed by colonization and establishment of trees. As these trees grow into the overstory, Gambel oak decreases. Trees continue to reproduce and increasingly dominate the overstory. The tree species differ among sites and regions. Southern Gambel Oak Shrublands have ponderosa pine (Sect. 4.3.2) in upland sites and pinyon and/or juniper, which can use shrubs as nurse plants (Sect. 5.3.2), in lowland sites. Northern stands have bigtooth maple, which both seeds and sprouts, especially on relatively moist sites (Bradley et al. 1992). Maple can be joined by other tree species such as white fir and Douglas-fir in northern stands, but it is unclear if they will eventually dominate (Bradley et al. 1992). Northern stands lack reproduction of Gambel oak by seed (Neilson and Wullstein 1983), but it is unknown if this affects vegetation dynamics.

7.4 Historical Changes

7.4.1 Overstory

7.4.1.1 Pre-Euro-American Settlement

Few specifics are known about the structure and composition of Gambel Oak Shrubland before Euro-American settlement. Shrublands lack the economic value that motivated many early descriptions of forests and grasslands. Nevertheless,

Fig. 7.9 Gambel Oak Shrubland in 1874 in foreground and on hillsides to the *left* and *right* of Horse Lake in Jicarilla Apache Reservation, north-central New Mexico. *Dark* clumps are Gambel oak, *light* areas are herb-dominated interspaces, and trees are ponderosa pine (Photograph by Timothy H. O'Sullivan, courtesy of National Archives and Records Administration)

the writings of early Euro-Americans occasionally mentioned oak-dominated shrublands and described them in terms that match present conditions:

> ...descended [a ridge], breaking through almost impenetrable thickets of chokecherry and dwarf oak... (Escalante in 1776 for north-central Utah, in Bolton 1950)

> The ravines and some of the side hills have groves of oak and [bigtooth] maple on them all of a short shrubby description... (Clyman in 1846 for the Wasatch Mountains of central and north-central Utah, in Christensen 1950)

> ...dense thickets cover all the dry, sandy, and gravelly knolls and foothills below the forest-forming timber trees. ...the vast areas covered by this brush deeply impress the observer. (Sudworth 1900 for west-central Colorado)

Historical photographs suggest that stands were similar in structure, composition, and location to those currently present (Fig. 7.9).

Relict areas possibly provide quantitative insight into historical conditions. Study of two isolated mesa tops in Zion National Park reported total tree density in Gambel Oak Shrubland was 1,666 individuals/ha (674 individuals/acre) for stems >5 cm dbh (Madany and West 1983). Only two species had individuals of that diameter: Gambel oak (85 % relative density) and bigtooth maple (15 %). Eighty percent of these stems were <100 years in age. Cover of Gambel oak was 76 % in this diameter class and 25 % for smaller stems (Madany and West 1984). Cover of bigtooth maple was 15 and 1 %, respectively. Extrapolation of these data to other sites is questionable because of the small size and potentially unique environment of relict sites (Sect. 5.4.1.1) and a small sample size.

Historical conditions of forests have been reconstructed through examination of the ages of current trees, snags, and logs (e.g., Sect. 4.4.1.1), but similar reconstructions have not been done for Gambel Oak Shrubland. The reconstruction approach may not be useful in Gambel Oak Shrubland, because individual stems rarely live more than 200 years and can be prone to rapid decay (Fulé et al. 2002). Examination of the age structure of Gambel oak populations indicates little more than dates of stand origin and subsequent stem recruitment (Sect. 7.2.5.1).

In summary, information on structure and composition of Gambel Oak Shrubland before Euro-American settlement is very limited. Stand locations were similar to areas where Gambel Oak Shrubland occurs today. Stands were abundant and dense in at least some locations, and woody plant composition appears to have been similar to present stands.

7.4.1.2 Post-Euro-American Settlement

Repeat photography has provided insight into changes since Euro-American settlement. New clumps of Gambel oak have established and older clumps have expanded in central and north-central Utah (Christensen 1957; Rogers 1982, George E. Gruell, personal communication, cited in Bradley et al. 1992). These changes were hypothesized to be related to the interaction of climate, fire, and livestock grazing (Rogers 1982), as follows. Unusually low winter temperatures and late spring frosts likely adversely impacted Gambel Oak Shrubland in these areas of Utah in the nineteenth century. Temperatures later moderated and facilitated the expansion of clumps shown in the repeat photographs. Livestock grazing and fires likely prevented formation of new clumps until the first half of the twentieth century when both grazing and fires were reduced, enabling the appearance of new clumps shown in the repeat photographs.

Repeat photography also has shown invasion of small patches of Gambel Oak Shrubland by trees in forested areas of the San Juan Mountains of southwestern Colorado (Zier and Baker 2006). But replacement of shrubland by forest was uncommon and replacement of forest by shrubland was not observed.

Comparison of relict and non-relict areas potentially provides quantitative insight into historical changes. The only such study used the small relict areas (and small sample size) in Zion National Park described in the previous section, albeit compared to a larger, better sampled non-relict area. There was little difference in Gambel oak density between the relict and disturbed area, but the disturbed area had a higher proportion of younger trees (95 % <100 years old) and a much lower density of bigtooth maple (Madany and West 1983). The presence of ponderosa pine and

Rocky Mountain juniper over 100 years in age on the disturbed site indicates the relict and non-relict areas possibly had been ecologically different.

Changes related to anthropogenic disturbances such as grazing and fire management are described in Sects. 7.2.6.1 and 7.2.6.2, respectively. The changes include fewer young stands and greater cover of tree species in some stands.

In addition, Gambel Oak Shrubland is thought to have increased where stand-replacing fires (and logging) have occurred in dry-mesic Mixed Conifer Forest, Ponderosa Pine Forest, and Pinyon-Juniper wooded shrubland (see Sects. 3.5.2.2, 4.5.2, and 5.5.2.2, respectively).

In conclusion, the regional distribution of Gambel Oak Shrubland generally has been stable since Euro-American settlement. Stands have increased in size and density in some areas.

7.4.2 Understory

Little information is available on the herbaceous layer of southwestern Gambel Oak Shrubland prior to Euro-American settlement. Two relict sites in Zion National Park had grass and forb cover of 14 and 17 %, respectively, with grasses dominant in shrub interspaces (Madany and West 1984). It is not known if these data are widely representative (see Sect. 7.4.1.1).

Although the historical characteristics of the herbaceous layer are poorly known, it likely was at least as well-developed as today. Livestock grazing reduced palatable herbs (Harper et al. 1985), especially grasses (Brown 1994). Today, less-palatable forbs are common, as are non-native brome (*Bromus* spp.) grasses and Kentucky bluegrass (Brown 1994). Understory cover is inversely related to overstory density (Christensen 1949) and therefore likely has decreased in areas of increased Gambel oak density.

7.5 Conceptual Models

The following nested, three-tiered set of conceptual models is based on my interpretation of best-available information on Gambel Oak Shrubland of the American Southwest. The models summarize and synthesize material on drivers (Sect. 7.2) and processes (Sect. 7.3) of vegetation dynamics, and they account for historical changes (Sect. 7.4). Their format is explained in Sect. 1.7. The models can be used to facilitate understanding of the past and the present. They also can provide insight into effects of future changes in the environment and outcomes of alternative land-management decisions. Moreover, they can guide the development of quantitative models.

7.5.1 Ecosystem-Characterization Model

The ecosystem-characterization model for southwestern Gambel Oak Shrubland emphasizes *Vegetation* (a biotic component) and *Disturbance* (a driver; Fig. 7.10a, Table 7.2). Key aspects of *Vegetation* are sprouting, structure, and fuel, and these affect *Disturbance*. The primary agents of *Disturbance* are fire and drought, and fire causes stem mortality followed by sprouting. The two other biotic components are *Soil System* and *Animals*. A second driver is *Weather & Climate*, which causes fires and drought and influences fire behavior and fuel and soil moisture. The third driver is *Landscape*, with its primary feature being elevation. It influences weather and climate, as well as spread and pattern of fire and impact of drought. The model also includes six anthropogenic drivers (Fig. 7.10b, Table 7.2): *Livestock Grazing*, *Fire Management*, *Modern Climate Change*, *Invasive Species*, *Recreation*, and *Nearby Land Use*.

7.5.2 Vegetation-Dynamics Models

Two models are needed to illustrate the vegetation dynamics. The first model is for southern Gambel Oak Shrubland. The second model is for northern Gambel Oak Shrubland. The models are similar, but differ in species and number of communities. See Sects. 3.5.2.2, 4.5.2.1, and 4.5.2.2 for relationships between forests and shrublands. Each vegetation-dynamics model is accompanied by a bar graph to show estimated relative shifts in abundances of the community types from circa 1870 (prior to widespread Euro-American land use) to circa 1970 (end of strict fire exclusion policy) to the present. The estimates are based on my understanding of vegetation dynamics and should be interpreted as showing relative differences and not interpreted quantitatively (hence the absence of a scale for abundance). Although not shown on the graph, relative abundances shifted prior to circa 1870 as a result of variation in climate and disturbance. Localized differences also have been and are present, but are beyond the scope of these generalized bar graphs.

7.5.2.1 Southern Gambel Oak Shrubland

The vegetation-dynamics model for southern Gambel Oak Shrubland has one state with five communities (Fig. 7.11, Table 7.3). All communities occurred historically. Community *A1 Young Shrubland* is formed by post-fire sprouting and is dominated by herbs, Gambel oak, and sometimes other shrub species. As sprouts mature, community *A1* changes into community *A2 Shrubland*, which is also dominated by Gambel oak, other shrubs, and herbs. Historically, this has been the most common

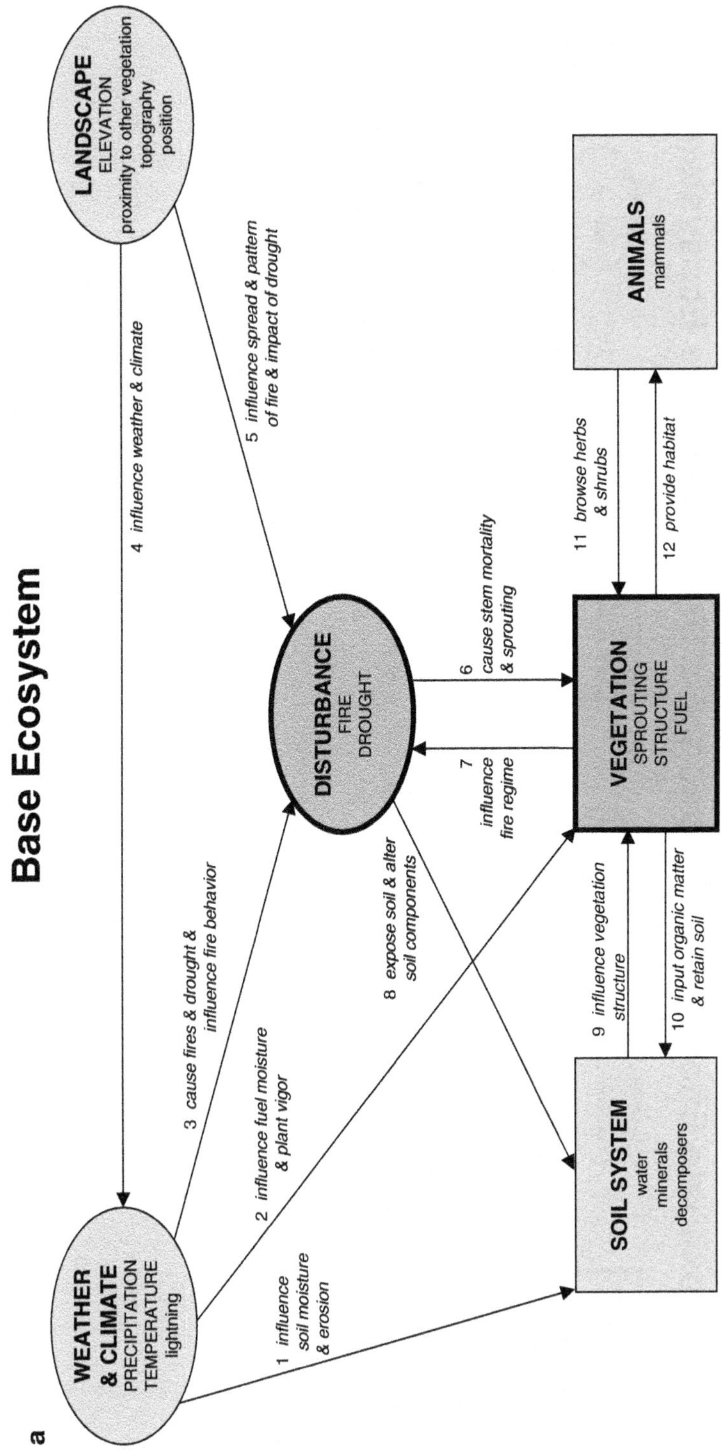
a
Base Ecosystem
LANDSCAPE
ELEVATION
proximity to other vegetation
topography
position
WEATHER & CLIMATE
PRECIPITATION
TEMPERATURE
lightning
DISTURBANCE
FIRE
DROUGHT
VEGETATION
SPROUTING
STRUCTURE
FUEL
SOIL SYSTEM
water
minerals
decomposers
ANIMALS
mammals
1 influence soil moisture & erosion
2 influence fuel moisture & plant vigor
3 cause fires & drought & influence fire behavior
4 influence weather & climate
5 influence spread & pattern of fire & impact of drought
6 cause stem mortality & sprouting
7 influence fire regime
8 expose soil & alter soil components
9 influence vegetation structure
10 input organic matter & retain soil
11 browse herbs & shrubs
12 provide habitat

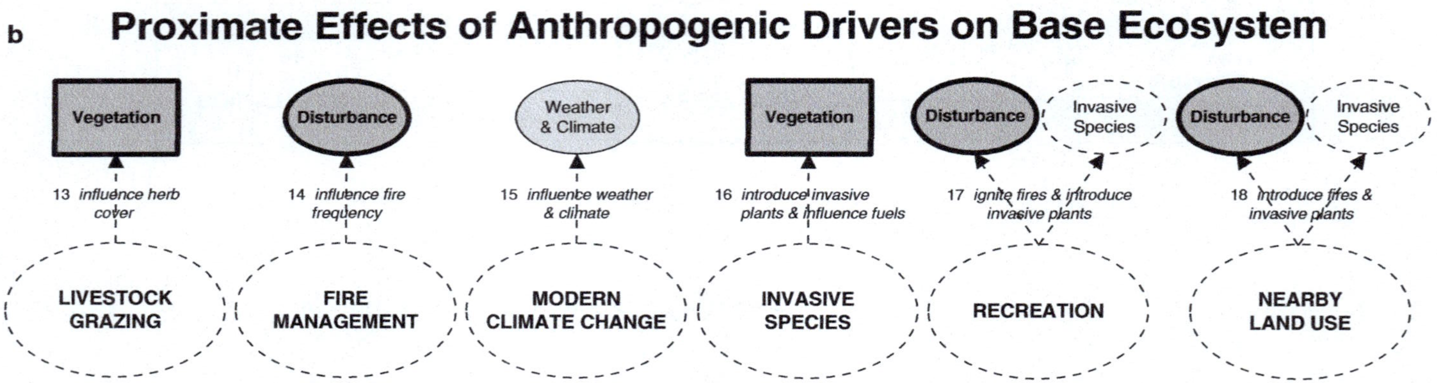

Fig. 7.10 (**a**, **b**) Ecosystem-characterization model for Gambel Oak Shrubland. In the base model (**a**), biotic components are shown as *rectangles* and drivers as *solid ovals*. Text within them identifies major features in *upper case* and other important features in *lower case*. Disturbance and Vegetation are *highlighted*, because they are the basis for the vegetation-dynamics models (Fig. 7.11 and 7.12). In the anthropogenic supplement to the base model (**b**), drivers are shown as *dashed ovals*. Relationships in (**a**) and (**b**) are shown as *solid* and *dashed arrows*, respectively, and are described in Table 7.2

Table 7.2 Relationships in the ecosystem-characterization model of southwestern Gambel Oak Shrubland (Fig. 7.10a, b)

Relationships	Circa 1870	Present
1	Precipitation increases soil moisture and can cause soil erosion. Winter precipitation is critical for recharging soil moisture	Same, but climate change increases the severity and duration of drought
2	Precipitation and temperature largely determine fuel moisture and are important to plant vigor	Same, but climate change increases the severity and duration of drought
3	Lightning ignites fires, wind influences fire behavior, and consistently low precipitation initiates drought	Same, but climate change increases the severity and duration of drought
4	Topography influences weather and climate	Same
5	Fires can spread from other areas of the landscape, and fire patterns are influenced by elevation, topography, and position in landscape	Same
6	Infrequent high-severity fires kill shrub and tree stems in all height and diameter classes and stimulate sprouting. Drought increases fire size	Same, but climate change increases the frequency and severity of drought
7	Extensive shrub cover, herb layer, and leaf litter influence fires	Same
8	Fires expose soil surface and alter soil properties	Same
9	Water and minerals in soil affect plant vigor and growth, influencing vegetation structure	Same
10	Plants add organic matter to soil and stabilize soil	Same
11	Herbivory reduces plant cover	Same
12	Vegetation is habitat for animals	Same
13	Not applicable	Livestock grazing reduces herb cover
14	Not applicable	Fire suppression possibly reduces the frequency and spread of fires in some areas, but not in others
15	Not applicable	Modern climate change increases temperatures and the frequency and extent of drought
16	Not applicable	Invasive plants change species composition of herbaceous layer and have potential to change fire patterns
17	Not applicable	Recreationists ignite fires and introduce and spread invasive plants
18	Not applicable	Nearby land use that initiates fires and colonization by invasive plants can introduce fires and invasives into Gambel Oak Shrubland

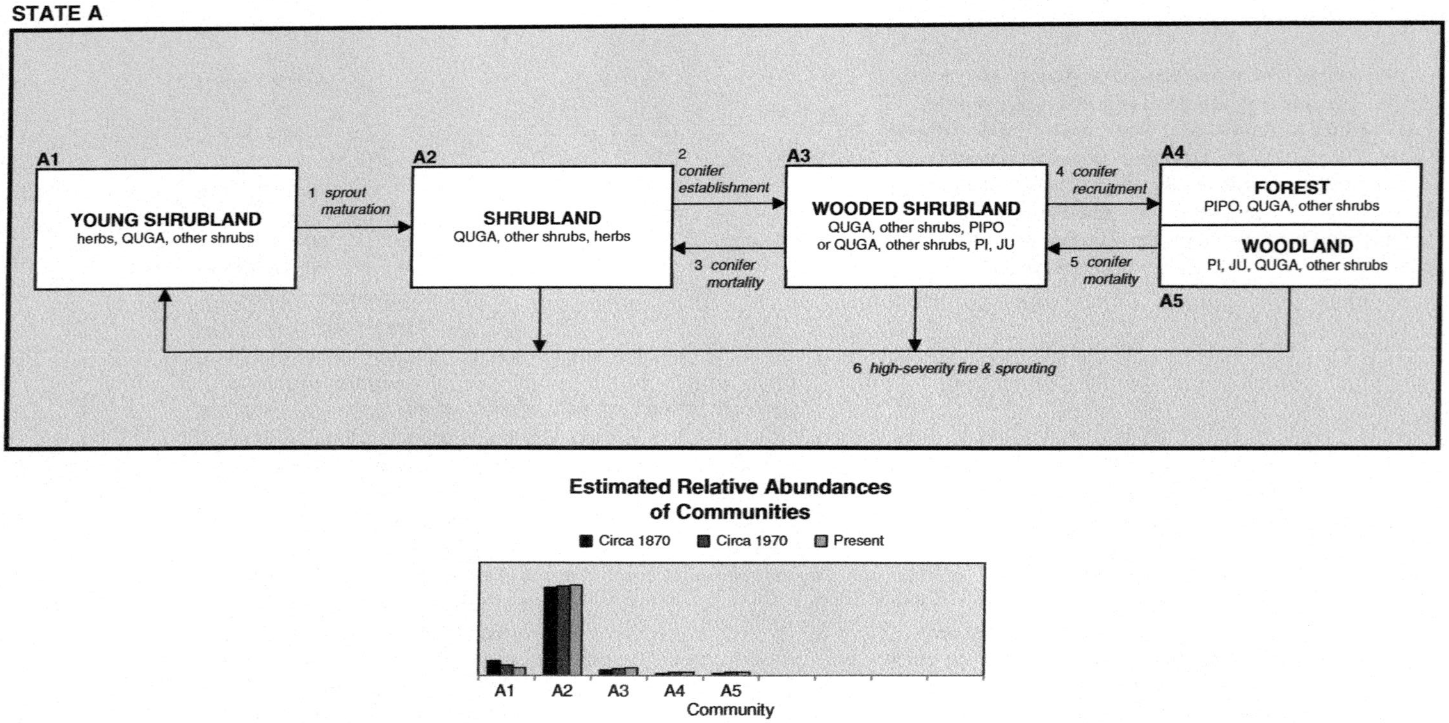

Fig. 7.11 Vegetation-dynamics model for southern Gambel Oak Shrubland. Model provides details on the vegetation-disturbance portion of the ecosystem-characterization model (Fig. 7.10a, b) by showing communities (*smaller, open rectangles*) within states (*larger, shaded rectangles*) and processes that form the communities. Text along *arrows* identifies pathways that are described in Table 7.3. *Bar graph* shows estimated relative shifts in abundances of the community types from circa 1870 (see Sect. 7.5.2). Species: *JU* junipers, *PI* pinyons, *PIPO* ponderosa pine, *QUGA* Gambel oak

Table 7.3 Relationships in the vegetation-dynamics model for southern Gambel Oak Shrubland (Fig. 7.11)

Relationships	Circa 1870	Present
1	Young sprouts mature, changing young shrubland into shrubland	Same
2	Invasion and establishment of ponderosa pine in upland sites and pinyon and/or juniper in lowland sites change shrubland into wooded shrubland	Same
3	Mortality of conifers changes wooded shrubland into shrubland	Same
4	Recruitment of ponderosa pine in upland sites changes wooded shrubland into forest. Alternatively, recruitment of pinyon and/or juniper in lowland sites changes wooded shrubland into woodland	Same
5	Mortality of conifers changes forest and woodland into wooded shrubland	Same
6	High-severity fire kills woody stems and promotes sprouting, forming young shrubland	Same

community. With invasion and establishment of ponderosa pine (upland sites) or pinyons and junipers (lowland sites), community *A2* forms *A3 Wooded Shrubland* with Gambel oak and other shrub species below an open overstory of these conifers. Community *A3* can revert to *A2* with mortality of the conifers. Community *A3* also can change into either community *A4 Forest* (upland sites) or *A5* Woodland (lowland sites) with additional recruitment of the conifers. These two communities are dominated by an overstory of conifers above an understory of Gambel oak and other shrubs. High-severity fire returns communities *A2*, *A3*, *A4*, and *A5* to *A1*.

7.5.2.2 Northern Gambel Oak Shrubland

The vegetation-dynamics model for northern Gambel Oak Shrubland also has one state, but three communities (Fig. 7.12, Table 7.4). All communities occurred historically. Community *A1 Young Shrubland* is formed by post-fire sprouting and is dominated by Gambel oak, bigtooth maple, and other shrubs. As sprouts mature and, where present, bigtooth maple infills, community *A1* changes into community *A2 Shrubland* dominated by the same species. This has been the most common community. With recruitment of white fir and/or Douglas-fir, community *A2* forms *A3 Wooded Shrubland* with the same shrub species below an open overstory of bigtooth maple and these conifers. Community *A3* can revert to A2 with mortality of the conifers. High-severity fire returns communities *A2* and *A3* to *A1*.

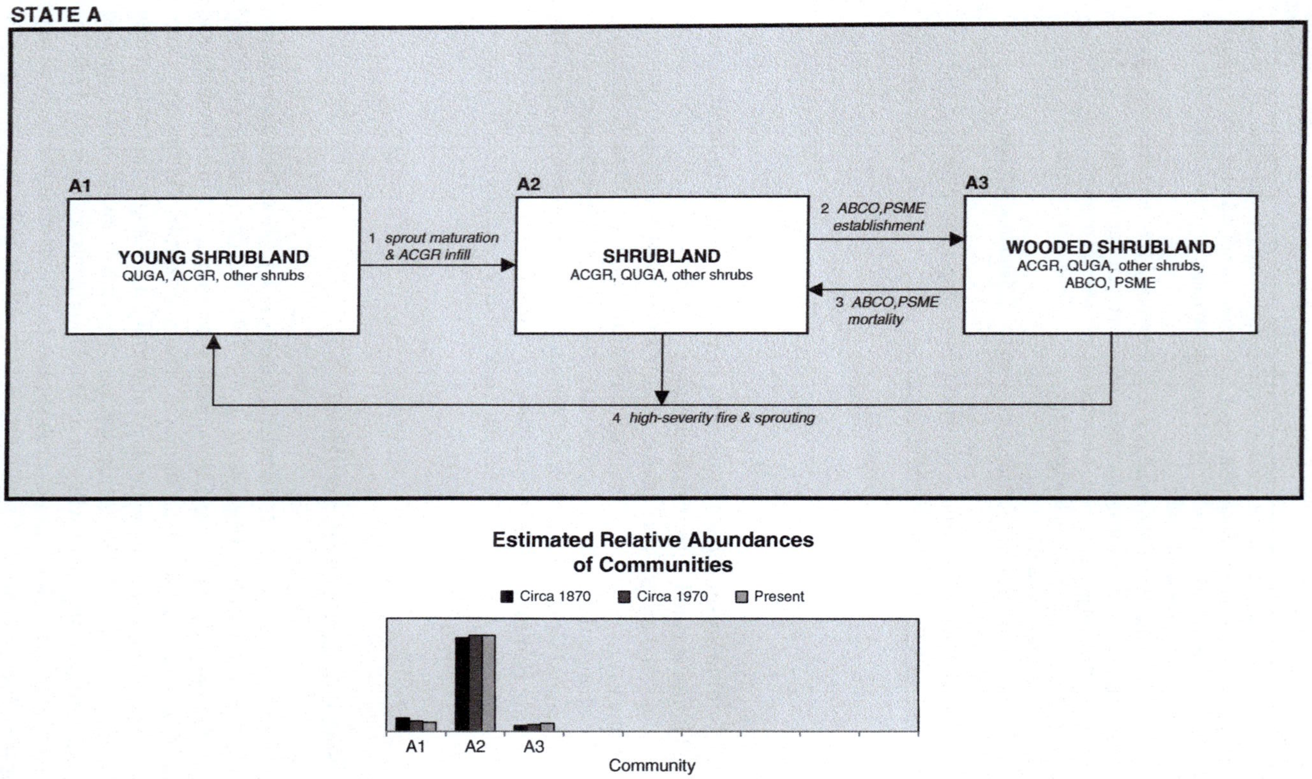

Fig. 7.12 Vegetation-dynamics model for northern Gambel Oak Shrubland. Model provides details on the vegetation-disturbance portion of the ecosystem-characterization model (Fig. 7.10a, b) by showing communities (*smaller, open rectangles*) within states (*larger, shaded rectangles*) and processes that form the communities. Text along *arrows* identifies pathways that are described in Table 7.4. *Bar graph* shows estimated relative shifts in abundances of the community types from circa 1870 (see Sect. 7.5.2). Species: *ABCO* white fir, *ACGR* bigtooth maple, *PSME* Douglas-fir, *QUGA* Gambel oak

Table 7.4 Relationships in the vegetation-dynamics model for northern Gambel Oak Shrubland (Fig. 7.12)

Relationships	Circa 1870	Present
1	Young sprouts mature and, where present, bigtooth maple infills, changing young shrubland into shrubland	Same
2	Invasion and establishment of white fir and/or Douglas-fir change shrubland into wooded shrubland	Same
3	Mortality of white fir and Douglas-fir changes wooded shrubland into shrubland	Same
4	High-severity fire kills woody stems and promotes sprouting, forming young shrubland	Same

7.5.3 *Mechanistic Model*

Both vegetation-dynamics models are explained by the same mechanistic model (Fig. 7.13). It has six biotic components on the right side of the figure (including three aspects of fuels), two drivers on the left side, and six anthropogenic factors at the bottom. In general, *Herbs, Shrubs & Trees*, and *Precipitation & Temperature* affect the fuel characteristics. *Fuel Moisture*, *Fuel Continuity*, and *Fuel Type & Loading* influence *Fire*, which affects characteristics of *Shrubs & Trees*, such as cover and species composition. *Shrubs & Trees* influence cover of *Herbs*, and both of these biotic components determine *Community Type* (of the three/five appearing in the vegetation-dynamics models). *Modern Climate Change* influences *Precipitation & Temperature. Fire Management* and fire ignitions caused by *Nearby Land Use* and *Recreation* can affect *Fire*. *Nearby Land Use* and *Recreation* are also sources of *Invasive Species*, and they, along with *Livestock Grazing*, affect *Herbs* cover.

7.6 Conclusions and Challenges

Southwestern Gambel Oak Shrubland is the least-researched type of vegetation covered in this book. Therefore, there are relatively few conclusions and many challenges for researchers and land managers. Research is needed on ecological and floristic differences across the range of southwestern Gambel Oak Shrubland. Such studies will add clarity to interpreting variation in this vegetation and provide context for researchers and land managers. Additional research on fire regimes would elucidate the role, if any, of low-severity fires and clarify what appear to be large differences in fire turnover times between southwestern Colorado and more northern sites. More information is also needed on anthropogenic disturbances, especially impacts of fire management, modern climate change, and invasive species. Successional patterns need to be clarified, as do changes in stand structure and composition since Euro-American settlement.

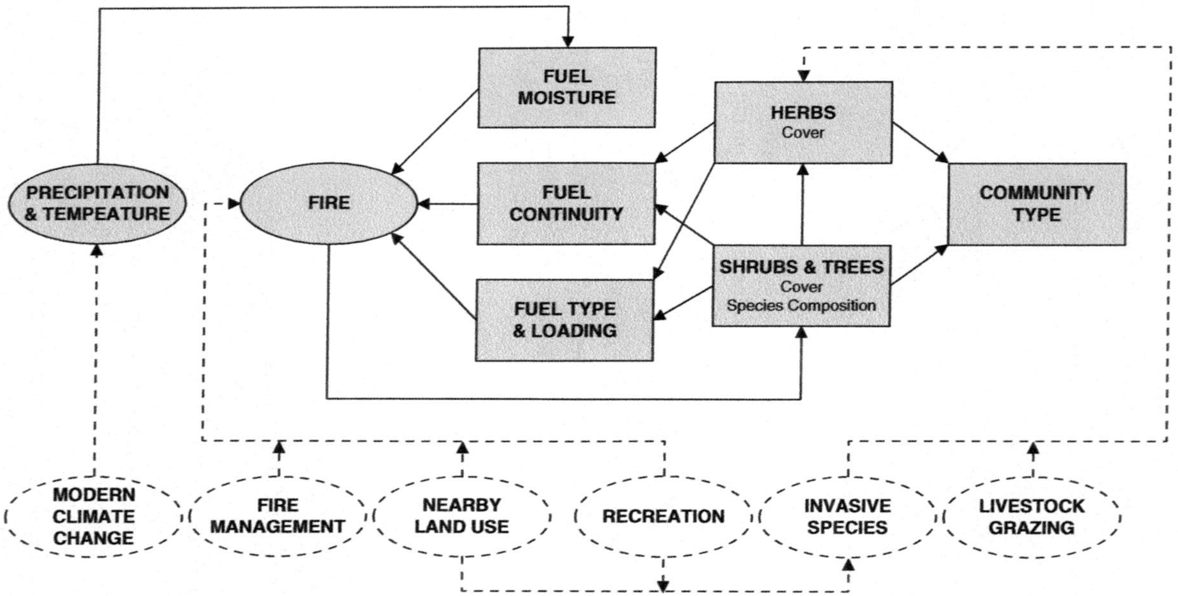

Fig. 7.13 Mechanistic model for Gambel Oak Shrubland. Model provides details on the mechanisms involved in the vegetation-dynamics models (Figs. 7.11 and 7.12). *Rectangles* are biotic components (including fuels), *ovals* are drivers, and *dashed ovals* are anthropogenic drivers

References

Bolton HE (1950) Pageant in the wilderness: the story of the Escalante Expedition to the Interior Basin, 1776, including the diary and itinerary of Father Escalante translated and annotated. Utah State Historical Society, Salt Lake City, Utah. Also published in Utah Hist Q 18(1,2,3,4):1–265. http://utah.ptfs.com/awweb/pdfopenerpage?sid=97BD5618D2DE879424E231AD31CFEBFA&did=34101&fl=%2Fpublications%2Farchive%2Frm000426.pdf#toolbar=0. Accessed 16 May 2012

Bradley AF, Noste NV, Fischer WC (1992) Fire ecology of forests and woodlands in Utah. U.S. Department of Agriculture, Forest Service General Technical Report INT-287, Intermountain Research Station, Ogden, UT

Brown DE (1994) Great Basin montane scrubland. In: Brown DE (ed) Biotic communities: southwestern United States and northwestern Mexico. University of Utah Press, Salt Lake City

Brown HE (1958) Gambel oak in west-central Colorado. Ecology 39:317–327

Christensen EM (1949) The ecology and geographic distribution of oak brush (*Quercus gambelii*) in Utah. Thesis, University of Utah, Salt Lake City

Christensen EM (1950) Distributional observations of oak brush (*Quercus gambelii* Nutt.) in Utah. Proc Utah Acad Sci Arts Lett 27:22–25

Christensen EM (1955) Ecological notes on the mountain brush in Utah. Proc Utah Acad Sci Arts Lett 32:107–111

Christensen EM (1957) Photographic history of the mountain brush on "Y" Mountain, central Utah. Proc Utah Acad Sci Arts Lett 34:154–155

Cranshaw W, Leatherman D, Kondratieff B (1994) Insects that feed on Colorado trees and shrubs. Colorado State University Cooperative Extension Bulletin 506A, Fort Collins, CO

Dick-Peddie WA (1993) New Mexico vegetation: past, present, and future. University of New Mexico Press, Albuquerque

Engle DM, Bonham CD, Bartel LE (1983) Ecological characteristics and control of Gambel oak. J Range Manag 36:363–365

Erdman JA (1970) Pinyon-juniper succession after natural fires on residual soils of Mesa Verde, Colorado. Brigh Young Univ Sci Bull Biol Ser 11(2):1–26

Evans FR (1936) A comparative study of the vegetation of a grazed and an ungrazed canyon of the Wasatch Range. Thesis, University of Utah, Salt Lake City

Ex SA, DeRose RJ, Long JN (2011) Stand development and population dynamics of curlleaf mountain mahogany (*Cercocarpus ledifolius* Nutt.) woodlands in Utah's Bear River Mountains. West J Appl For 26:183–188

Floyd ME (1982) The interaction of piñon pine and Gambel oak in plant succession near Dolores, Colorado. Southwest Nat 27:143–147

Floyd ML, Romme WH, Hanna DD (2000) Fire history and vegetation pattern in Mesa Verde National Park, Colorado, USA Ecol Appl 10:1666–1680

Floyd ML, Hanna DD, Salamacha G (2001) Post-fire treatment of noxious weeds in Mesa Verde National Park, Colorado. In: van Riper C III, Thomas KA, Stuart MA (eds) Proceedings of the fifth biennial conference of research on the Colorado Plateau. U.S. Geological Survey/FRESC report series USGSFRESC/COPL/2001/24

Floyd ML, Hanna D, Romme WH, Crews TE (2006) Predicting and mitigating weed invasions to restore natural post-fire succession in Mesa Verde National Park, Colorado, USA. Int J Wildland Fire 15:247–259

Forsling CL, Storm EV (1929) The utilization of browse forage as summer range for cattle in southwestern Utah. U.S. Department of Agriculture Circular 62, Washington, DC

Fulé PZ, Covington WW, Moore MM, Heinlein TA, Waltz AEM (2002) Natural variability in forests of the Grand Canyon, USA. J Biogeogr 29:31–47

Harper KT, Wagstaff FJ, Kunzler LM (1985) Biology and management of the Gambel oak vegetative type: a literature review. U.S. Department of Agriculture, Forest Service General Technical Report INT-179, Intermountain Forest and Range Experiment Station, Ogden, UT

Hayward CL (1948) Biotic communities of the Wasatch chaparral, Utah. Ecol Monogr 18:473–506

Hendricks DM (1985) Arizona soils. University of Arizona Press, Tucson

Integrated Taxonomic Information System (2012) http://www.itis.gov. Accessed 25 Aug 2012

Kunzler LM, Harper KT (1980) Recovery of gambel oak after fire in central Utah. Gt Basin Nat 40:127–130

Kunzler LM, Harper KT, Kunzler DB (1981) Compositional similarity within the oakbrush type in central and northern Utah. Gt Basin Nat 41:147–153

MacMahon JA (1988) Vegetation of Utah. In: Albee BJ, Schultz LM, Goodrich S (eds) Atlas of the vascular plants of Utah. Utah Museum of Natural History Occasional Publication No. 7, Salt Lake City, UT

Madany MH, West NE (1983) Livestock grazing – fire regime interactions within montane forests of Zion National Park, Utah. Ecology 64:661–667

Madany MH, West NE (1984) Vegetation of two relict mesas in Zion National Park. J Range Manag 37:456–461

Neilson RP, Wullstein LH (1983) Biogeography of two southwest American oaks in relation to atmospheric dynamics. J Biogeogr 10:275–297

Nixon ES (1967) A comparative study of the mountain brush vegetation in Utah. Gt Basin Nat 27:59–66

Omi PN, Emrick RL (1980) Fire and resource management in Mesa Verde National Park. Final report for Contract CS-1200-9-B015 between the U.S. National Park Service and Colorado State University

Paulson DD, Baker WL (2006) The nature of southwestern Colorado: recognizing human legacies and restoring natural places. University Press of Colorado, Boulder

Poreda SF, Wullstein LH (1994) Vegetation recovery following fire in an oakbrush vegetation mosaic. Gt Basin Nat 54:380–383

Price R, Evans RB (1937) Climate of the west front of the Wasatch Plateau in central Utah. Mon Weather Rev 65:291–301

Prior-Magee JS, Boykin KG, Bradford DF, Kepner WG, Lowry JH, Schrupp DL, Thomas KA, Thompson BC (2007) Ecoregional gap analysis of the southwestern United States: the Southwest regional gap analysis project final report. http://fws-nmcfwru.nmsu.edu/swregap/report/SWReGAP%20Final%20Report.pdf. Accessed 11 July 2012

Reynolds, RVR (1911) Grazing and floods: a study of conditions in the Manti National Forest, Utah. U.S. Department of Agriculture, Forest Service Bulletin 91, Washington, DC

Rogers GG (1982) Then and now: a photographic history of vegetation changes in the central Great Basin desert. University of Utah Press, Salt Lake City

Romme WH, Barry PJ, Hanna DD, Floyd ML, White S (2006) A wildfire hazard assessment and map for La Plata County, Colorado, USA. Fire Ecology 2:7–30, http://fireecology.org/docs/Journal/pdf/Volume02/Issue01/007.pdf. Accessed 21 May 2012

Romme WH, Redders JS, Floyd-Hanna L, Hanna DD (2009) Other vegetation types. In: Romme WH, Floyd ML, Hanna D (eds) Historical range of variability and current landscape condition analysis: south central highlands section, southwestern Colorado & northwestern New Mexico. Colorado Forest Restoration Institute, Colorado State University and Region 2, U.S Forest Service, Fort Collins

Spence JR, Romme WH, Floyd-Hanna L, Rowlands PG (1995) A preliminary vegetation classification for the Colorado Plateau. In: van Riper C III (ed) Proceedings of the second biennial conference on research in Colorado Plateau National Parks, 25–28 Oct 1993. National Park Service Transactions and proceedings series NPS/NRNAU/NRTP-95/11, Flagstaff

Steinhoff HW (1981) Gambel oak ecology. Colorado State University district office report, Durango, CO

Sudworth GB (1900) The White River Plateau and Battlement Mesa Forest Reserves. Department of the Interior, U.S. Geological Survey, Extract from the Twentieth Annual Report of the Survey, 1898–1899, Part V, Forest Reserves, Government Printing Office, Washington, DC

Swetnam TW, Betancourt JL (1998) Mesoscale disturbance and ecological response to decadal climatic variability in the American Southwest. J Clim 11:3128–3147
Tew RK (1967) Soil moisture depletion by Gambel oak in central Utah. U.S. Department of Agriculture, Forest Service Research Note INT-74, Intermountain Forest & Range Experiment Station, Ogden, UT
Tiedemann AR, Clary WP, Barbour RJ (1987) Underground systems of Gambel oak (*Quercus gambelii*) in central Utah. Am J Bot 74:1065–1071
U.S. Forest Service (2011) *Agrilus quercicola*. U.S. Department of Agriculture, Forest Service, Rocky Mountain Region, Forest Health Protection. http://www.fs.usda.gov/Internet/FSE_DOCUMENTS/stelprdb5347769.pdf. Accessed 25 May 2012
Van Epps GA (1974) Control of Gambel oak with three herbicides. J Range Manag 27:297–301
Wadleigh LL, Parker C, Smith B (1998) A fire frequency and comparative fuel load analysis in Gambel oak of northern Utah. In: Pruden TL, Brennan LA (eds) Fire in ecosystem management: shifting the paradigm from suppression to prescription. Tall Timbers fire ecology conference proceedings, no. 20. Tall Timbers Research Station, Tallahassee
Western Regional Climate Center (2012) Colorado climate summaries. http://www.wrcc.dri.edu/summary/climsmco.html. Accessed 23 Aug 2012
Zier JL, Baker WL (2006) A century of vegetation change in the San Juan Mountains, Colorado: an analysis using repeat photography. For Ecol Manag 228:251–262

Chapter 8
Interior Chaparral Shrubland

Abstract Interior Chaparral Shrubland covers more area on the mountains and plateaus of the American Southwest than any other shrubland. It is dominated by broad-leaved, evergreen shrubs with dense, compact crowns and generally extensive, deep root systems. Interior Chaparral Shrubland overlaps the elevational range of Pinyon-Juniper vegetation, occurring below Ponderosa Pine Forest and above desert scrub or semi-desert grassland. The primary natural disturbance is infrequent, high-severity fire. Major anthropogenic disturbances are livestock grazing and fire management; others are modern climate change, invasive species, recreation, and nearby land use. Vegetation dynamics are dominated by rapid regrowth after fire. This is facilitated by sprouting from below-ground structures and by fire-stimulated germination of seeds in the seed bank. Some stands are successional, particularly in the transition with Ponderosa Pine Forest. Historical conditions are poorly known, but Interior Chaparral Shrubland has been stable in its regional distribution. Herbaceous cover likely was reduced by intensive livestock grazing, and shrub cover increased by fire exclusion. The claim that shrubs of Interior Chaparral Shrubland expanded into other types of vegetation because of livestock grazing is unsupported. Vegetation dynamics are illustrated in a nested, three-tiered set of conceptual models. Key conclusions and challenges for researchers and land managers are summarized.

8.1 Introduction

Interior Chaparral Shrubland is the most extensive shrubland on the mountains and plateaus of the American Southwest (Figs. 8.1 and 8.2). It is dominated by 1–2.5 m (3–8 ft) tall, broad-leaved, sclerophyllous (i.e., hard-leaved), evergreen shrubs with dense, compact crowns. It overlaps the elevational range of Pinyon-Juniper vegetation (*Pinus-Juniperus*), occurring below Ponderosa Pine Forest (*Pinus ponderosa*) and above desert scrub or semi-desert grassland. Interior Chaparral Shrubland is also known as Arizona chaparral and Mogollon chaparral.

J.L. Vankat, *Vegetation Dynamics on the Mountains and Plateaus of the American Southwest*, Plant and Vegetation 8, DOI 10.1007/978-94-007-6149-0_8,

Fig. 8.1 Interior Chaparral Shrubland in foreground in Prescott National Forest, central Arizona. Note contrast with Ponderosa Pine Forest on hillside in background (Photograph by author)

Fig. 8.2 Interior Chaparral Shrubland on hillsides in foreground and background in Prescott National Forest, central Arizona. Ponderosa Pine Forest occurs in drainage at *left center* and on ridgetop at *upper right* (Photograph by author)

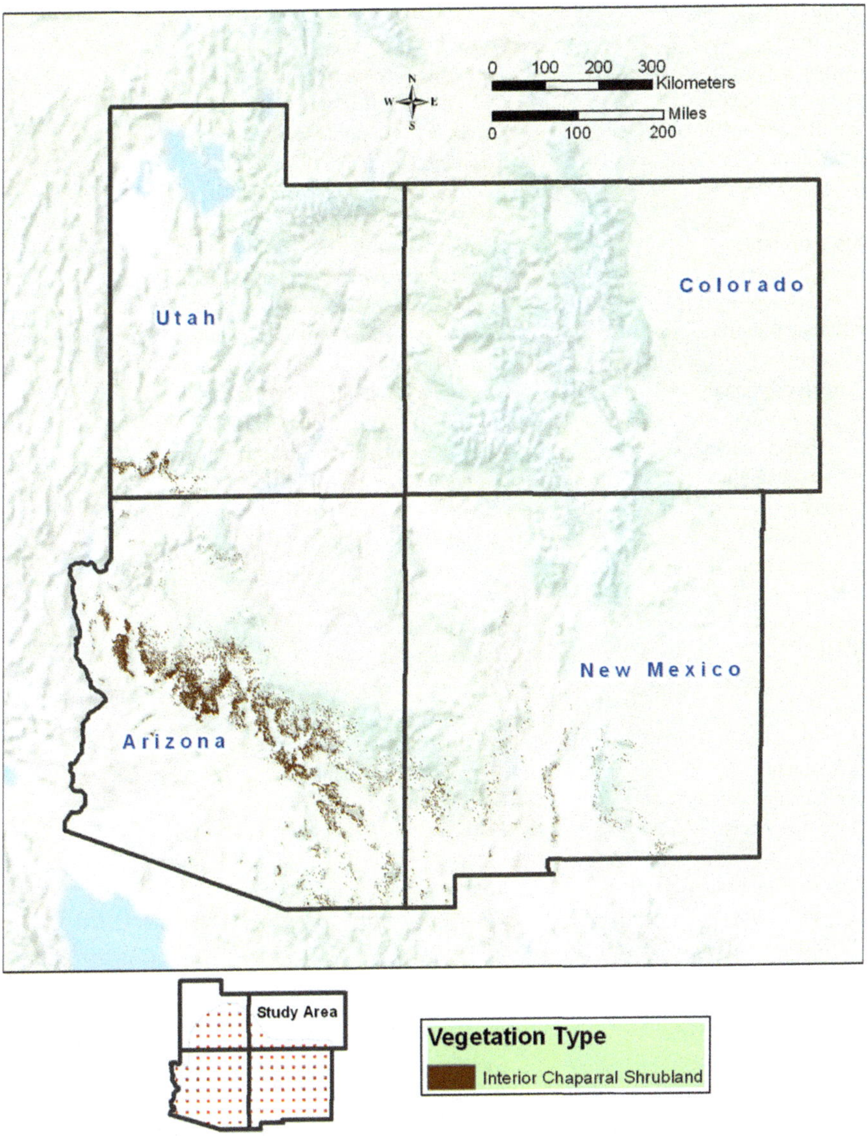

Fig. 8.3 Distribution of Interior Chaparral Shrubland in the American Southwest. The map shows all of the states of Arizona, New Mexico, Colorado, and Utah, and the American Southwest is illustrated in *red* on the small map (Source of data: U.S. Geological Survey National Gap Analysis Program 2005 Southwest Regional Gap Analysis Project, RS/GIS Laboratory, College of Natural Resources, Utah State University, Logan, Utah. Map prepared by Monica Swihart)

Interior Chaparral Shrubland covers approximately 10,800 km^2 (4,200 mi^2), which is 1.4 % of the American Southwest (Fig. 8.3; calculations based on Prior-Magee et al. 2007). There are other upland shrublands in the region (cf. Dick-Peddie 1993), but only Gambel Oak Shrubland (*Quercus gambelii*) is also common and is

Table 8.1 Common and scientific names of species in this chapter. Primary source: Integrated Taxonomic Information System (2012)

Plants	
Agave	*Agave* L.
Alligator juniper	*Juniperus deppeana* Steud.
Arizona white oak	*Quercus arizonica* Sarg.
Birchleaf mountain mahogany	*Cercocarpus betuloides* Nutt.
Blackfoot	*Melampodium longicorme* A. Gray
Bluestem	*Andropogon* L., *Bothriochloa* Kuntze, and *Schizachyrium* (Michx.) Nash
Boer lovegrass	*Eragrostis chloromelas* Steud.
Cane bluestem	*Bothriochloa barbinodis* (Lag.) Herder
Catclaw acacia	*Senegalia greggii* (A. Gray) Britton & Rose
Catclaw mimosa	*Mimosa aculeaticarpa* var. *biuncifera* (Benth.) Barneby
Cholla	*Opuntia* P. Mill.
Crowfoot grama	*Bouteloua eriopoda* (Torr.) Torr.
Curly mesquite	*Hilaria belangeri* (Steud.) Nash
Deerbrush ceanothus	*Ceanothus integerrimus* Hook. & Arn.
Desert ceanothus	*Ceanothus greggii* A. Gray
Dropseed	*Sporobolus* R. Br. and *Blepharoneuron* Nash
Emory oak	*Quercus emoryi* Torr.
Fendler's ceanothus	*Ceanothus fendleri* A. Gray
Fir	*Abies* P. Mill.
Gambel oak	*Quercus gambelii* Nutt.
Grama	*Bouteloua* Lag.
Hollyleaf buckthorn	*Rhamnus crocea* Nutt.
Juniper	*Juniperus* L.
Lehmann lovegrass	*Eragrostis lehmanniana* Nees
Manzanita	*Arctostaphylos* Adans.
Mountain mahogany	*Cercocarpus montanus* Raf.
Pine bunchgrass	*Festuca arizonica* Vasey
Pinyon	*Pinus* L.
Pointleaf manzanita	*Arctostaphylos pungens* Kunth
Ponderosa pine	*Pinus ponderosa* Douglas ex P. Lawson & C. Lawson
Prickly pear	*Opuntia* P. Mill.
Pringle's manzanita	*Arctostaphylos pringlei* Parry
Red brome	*Bromus rubens* L.
Seepwillow baccharis	*Baccharis salicifolia* (Ruiz & Pav.) Pers.
Shrub live oak	*Quercus turbinella* Greene
Skunkbush	*Rhus trilobata* Nutt.
Spruce	*Picea* A. Dietr.
Stork's bill	*Erodium cicutarium* (L.) L'Hér. Ex Aiton
Sugar sumac	*Rhus ovata* S. Watson
Threeawn	*Aristida* L.
Weeping lovegrass	*Eragrostis curvula* (Schrad.) Nees
Wright's silktassel	*Garrya wrightii* Torr.
Yellowleaf silktassel	*Garrya flavescens* S. Wats.
Yucca	*Yucca* L.
Animals	
Cattle	*Bos taurus* Linnaeus, 1758
Deer	*Odocoileus* Rafinesque, 1832
Goat	*Capra hircus* Linneaus, 1758
Sheep	*Ovis aries* Linnaeus, 1758

included in this book. Interior Chaparral Shrubland is most extensive in Arizona, where it occurs in a discontinuous band from the southeastern to the west-central portion of the state. It also occurs in extreme southwestern Utah and in small, scattered areas in southwest, south-central, and central New Mexico. It is absent from Colorado. Interior Chaparral Shrubland has little geographic overlap with Gambel Oak Shrubland.

The term chaparral comes from the Spanish word *chaparro*, which refers to short, evergreen oaks (*Quercus* spp.). Chaparral vegetation similar to Interior Chaparral Shrubland is better known from California. It also occurs in parts of northern Mexico. All areas of chaparral are dominated by shrubs, most of which have the broad-sclerophyll growth form. Some shrub species occur in two or more of these regions. Despite similarities in growth forms and overlap of species, the various regions of chaparral vegetation have climates that differ in timing of precipitation. California chaparral, the most widespread type of chaparral, occurs in California and the adjacent Mexican state of Baja California, where precipitation is concentrated in winter and summers are typically dry. Chaparral vegetation elsewhere in northern Mexico includes northeastern Mexico where winters are usually dry and precipitation is concentrated in summer. Interior Chaparral Shrubland of the American Southwest has peaks of precipitation in both winter and summer (Sect. 1.2.2).

In addition, the name "Petran chaparral" has been used for Gambel Oak Shrubland since the early twentieth century, a time when "chaparral" was generically applied to scrub and thicket vegetation in the western United States (Clements 1920). Today, the name "Petran chaparral" incorrectly implies substantial similarity with the chaparral vegetation of California, Mexico, and the American Southwest, even though the dominant shrub species in Gambel Oak Shrubland are deciduous, not evergreen (Sect. 7.1). Continued use of this name is confusing and inapplicable.

Interior Chaparral Shrubland occurs on a wide variety of sites, and therefore stands differ in structure and species composition. Stand structure ranges from moderately open to closed, with shrub cover typically 40–80 % and openings (interspaces) between shrubs (Figs. 8.4 and 8.5). Interior Chaparral Shrubland has more woody species than any other type of vegetation covered in this book. Ninety-nine shrub species (including approximately a dozen succulents and semi-succulents) are listed in Knipe et al. (1979). About 15 of the shrub species are widespread, and most stands are dominated by one to three species. Species composition and shrub cover depend on elevation, time since fire, slope aspect and inclination, soil depth, and soil water holding capacity (cf. DeBano et al. 1999). There is little information on vegetation classification, but see Carmichael et al. (1978), who reviewed previous classifications and described eight chaparral associations in the Mazatzal Mountains of central Arizona.

Shrub live oak (*Quercus turbinella*) dominates most stands and usually is present in others. Other common evergreen shrubs include pointleaf and Pringle's manzanitas (*Arctostaphylos pungens*, *A. pringlei*), desert ceanothus (*Ceanothus greggii*), Wright's and yellowleaf silktassels (*Garrya wrightii*, *G. flavescens*), hollyleaf buckthorn (*Rhamnus crocea*), and sugar sumac (*Rhus ovata*). Partially evergreen shrubs include birchleaf mountain mahogany (*Cercocarpus betuloides*) and mountain mahogany

Fig. 8.4 Open stand of Interior Chaparral Shrubland in central Arizona (Photograph by Jessa Fisher)

Fig. 8.5 Dense stand of Interior Chaparral Shrubland in Prescott National Forest, central Arizona (Photograph by author)

Fig. 8.6 Extensive root system of shrub live oak in Three Bar Wildlife Area, Tonto National Forest, central Arizona. Above-ground height of shrub is approximately 1.8 m (6 ft) (Photograph courtesy of Fort Valley Experimental Forest Historic Images, U.S. Forest Service)

(*C. montanus*). Fully deciduous shrubs are also present, and especially common are skunkbush (*Rhus trilobata*) and catclaw mimosa (*Mimosa aculeaticarpa* var. *biuncifera*). Other growth forms include stem succulents such as prickly pear and cholla (both are *Opuntia* spp.) and semisucculents such as agave (*Agave* spp.) and yucca (*Yucca* spp.). Moreover, some stands have scattered ponderosa pine, pinyons, or junipers. Herb cover is inversely proportional to shrub cover. Grasses are generally more common than forbs and include threeawn (*Aristida* spp.), grama (*Bouteloua* spp.), and cane bluestem (*Bothriochloa barbinodis*). Forbs are abundant only after fire. A species list is in Knipe et al. (1979); also see Hibbert et al. (1974), Cable (1975), and Keeley et al. (2012).

Shrubs of Interior Chaparral Shrubland can have extensive, deep root systems. Observations of roots of shrub live oak in quarries and mines document depths of at least 9 m (30 ft; Saunier and Wagle 1967). Excavation of a mature shrub live oak in central Arizona (Fig. 8.6) revealed a large root crown from which a branched tap root extended to bedrock at 6.4 m (21 ft; Davis 1977; Davis and Pase 1977). This branched root had a horizontal spread of 5 m (16 ft). In addition, the shrub had a dense network of fine lateral roots throughout the upper 0.3 m (1 ft) of soil. Laterals extended nearly 7 m (23 ft) downslope and 3 m (11 ft) upslope before turning downward. By having both deep and shallow roots, shrub live oak is well-adapted to accessing soil water at various depths. Below-ground structures are also important in that most shrub species sprout from them following fire (Sect. 8.3.1).

Compared to adjacent vegetation such as Ponderosa Pine Forest and Pinyon-Juniper vegetation, little research has been done on Interior Chaparral Shrubland. Early studies emphasized grazing and its relationship to erosion. Beginning in the 1950s, most research focused on conversion of Interior Chaparral Shrubland to grassland for the purposes of increasing forage, increasing water yield for use at lower elevations, and reducing fire hazards (Hibbert et al. 1974). Applied research topics have also included biomass conversion to liquid fuels (Davis et al. 1984). Applied research dramatically declined in the 1970s because of environmental concerns (Sect. 8.2.6.1), and little basic research has been published. The situation contrasts with California, where there has been extensive applied and basic research on its chaparral vegetation (cf. Keeley 2000; Keeley and Davis 2007).

The paucity of basic research is surprising. Although Interior Chaparral Shrubland has less economic value than southwestern forests and grasslands, it is extensive in Arizona. Moreover, research on Interior Chaparral Shrubland has potential to reflect on the origins of western North American vegetation and the evolution of ecosystems. Paleoecological research has indicated that Interior Chaparral Shrubland and chaparral vegetation in northeastern Mexico are more similar to paleo-chaparral vegetation than is the chaparral of California and Baja California (cf. Axelrod 1975; Ackerly 2009). Also, chaparral in California has been a focus of research addressing the hypothesis that similar selection pressures in similar environments produce similar ecosystems, even in different regions of the world. This research has examined evidence of ecosystem convergence among broad-sclerophyll shrublands occurring in winter-wet, summer-dry climates in California, Chile, the Mediterranean Basin, South Africa, and Australia (e.g., Mooney 1977; Cody and Mooney 1978). However, research on Interior Chaparral Shrubland in Arizona has shown that chaparral species are not uniquely adapted to a winter-wet, summer-dry "Mediterranean-type" climate, but are more generically adapted to seasonal drought (Vankat 1989; see also Valiente-Banuet et al. 1998).

With the scarcity of basic research, many accounts of Interior Chaparral Shrubland have included research results from California chaparral although there are differences in species, stand structure, climate, etc. (see Keeley et al. 2012 for consideration of some of the similarities and differences). This chapter focuses on research findings from Interior Chaparral Shrubland.

8.2 Drivers

The primary drivers of Interior Chaparral Shrubland are landscape, climate, soil, animals, natural disturbance, and anthropogenic disturbance. Each driver is important in vegetation dynamics, and therefore all are incorporated in the conceptual models (Sect. 8.5).

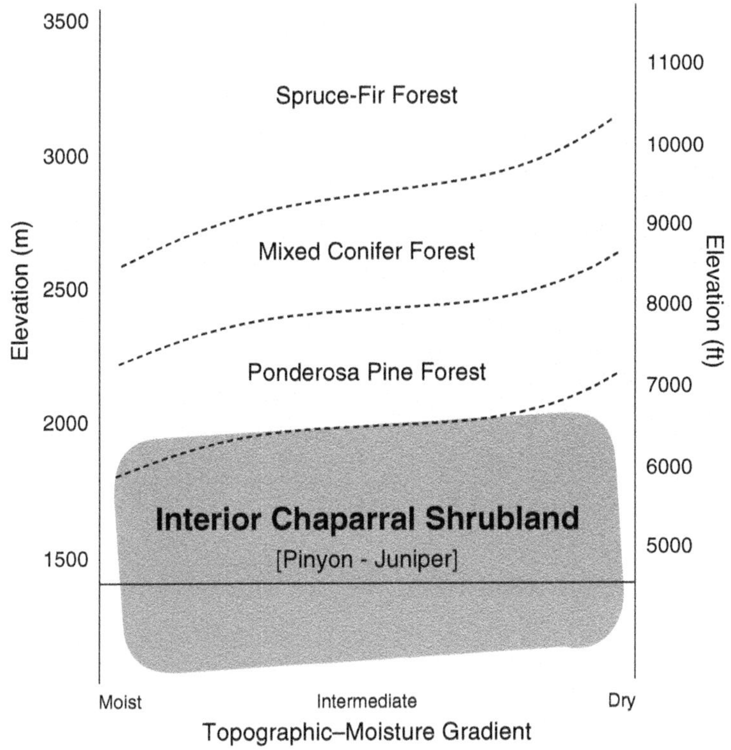

Fig. 8.7 Ecological distribution of Interior Chaparral Shrubland on the mountains and plateaus of the American Southwest along gradients in elevation and topographic-moisture. Interior Chaparral Shrubland (*shaded area*) is superimposed on Pinyon-Juniper vegetation. The elevation of Interior Chaparral Shrubland is typical of central Arizona. Elevations immediately below those shown have desert shrublands and semi-desert grasslands

8.2.1 Landscape

The typical elevational range of Interior Chaparral Shrubland in central Arizona is approximately 1,100–2,000 m (3,600–6,600 ft; Fig. 8.7). It is influenced by slope exposure, soils, and local climate (Hibbert et al. 1974; Carmichael et al. 1978). Interior Chaparral Shrubland can intergrade with Ponderosa Pine Forest at high elevation (Fig. 8.8), especially where the shrubland is a stage in forest succession (Sects. 4.3.2 and 4.5.2.3). In addition, there is much elevational overlap with Pinyon-Juniper vegetation, and these two vegetation types – along with semi-desert grassland – can occur in a patchy mosaic (Huebner and Vankat 2003). Where species of adjacent vegetation types intergrade, vegetation type is determined by the relative abundances of shrubs and trees.

Fig. 8.8 Scattered ponderosa pines in Interior Chaparral Shrubland illustrate transition with Ponderosa Pine Forest to right in Prescott National Forest, central Arizona (Photograph by author)

The range of Interior Chaparral Shrubland includes areas of highly variable topography such as foothills and mountain slopes. Therefore, topography plays a key role in the variation of stand structure and composition. Stands at higher elevation and on steeper sites generally have more closed structure, and shrub cover usually decreases with elevation and on flatter slopes. These decreases in cover are possibly related to fires spreading and burning less thoroughly and thereby causing less widespread sprouting, a phenomenon that also occurs in Gambel Oak Shrubland (Sect. 7.2.1). Influences of topography on species composition are complex, in part because there are so many species, but mountain mahogany generally decreases and shrub live oak and catclaw mimosa generally increase with decreased elevation.

8.2.2 Climate

The climate of Interior Chaparral Shrubland is characterized by warm springs and falls, warm to hot summers, and cool to cold winters. Mean annual temperatures are 10–18 °C (50–65 °F), and mean monthly temperatures range from <4 °C (40 °F) in January to >27 °C (80 °F) in July (Hibbert et al. 1974). Mean annual precipitation is 38–64 cm (15–25 in.; Cable 1975) and is positively correlated with elevation. Somewhat more than half falls in November-April. Snowfall averages 10–64 cm (4–25 in.), depending on elevation and local site conditions (Cable 1975). May and

June are typically very dry and windy (Bolander 1982). Most of the rest of the precipitation falls in spatially and temporally scattered monsoonal storms in July and August. Annual precipitation is highly variable, with the driest years receiving about half the mean and the wettest years receiving about twice the mean (Hibbert et al. 1974). Lightning is an important component of the climate, because it can ignite fires.

Modern climate change is described in Sect. 1.6.3 and is treated as an anthropogenic disturbance in Sect. 8.2.6.3.

8.2.3 Soil

Interior Chaparral Shrubland occurs on soils formed from a variety of parent materials of igneous, sedimentary, and metamorphic origins (Cable 1975). Granite is the most common parent material (Hibbert et al. 1974). Soils are typically poorly developed, yet deep. They tend to be coarsely textured (which facilitates the infiltration of water), ranging from cobbly and gravelly loamy sands to gravelly loams (Hibbert et al. 1974). There is little organic matter (Pase and Brown 1994). The A horizon is shallow, the B horizon is typically absent, and the C horizon makes up most of the soil (Hibbert et al. 1974). Soils appear to determine differences among Interior Chaparral Shrubland, grassland, and vegetation with trees (U.S. Forest Service 1970). Research on the vegetation mosaic near the city of Prescott in central Arizona indicated that stands of Interior Chaparral Shrubland are separated from Pinyon-Juniper Woodland and grassland in part by coarser soils (and steeper slopes; Huebner and Vankat 2003).

8.2.4 Animals

No animals, including insects, have been shown to affect vegetation dynamics of Interior Chaparral Shrubland. The most common large ungulates are deer (*Odocoileus* spp.), but their population densities are low (U.S. Forest Service 1970).

8.2.5 Natural Disturbance – Fire

Fire is the only natural disturbance to meaningfully impact Interior Chaparral Shrubland. Research findings are sparse, but the importance of fire is illustrated by observations of its frequent occurrence and its behavior (Fig. 8.9a, b). Central Arizona's Tonto and Prescott National Forests reported an average of 50 fires per year in Interior Chaparral Shrubland over an 11-year period (Brown and Boster 1974). Human- and lightning-caused fires were combined in that report, but most

Fig. 8.9 (**a, b**) Fire and soon after fire in Interior Chaparral Shrubland in central Arizona. *Reddish-brown* color at *top* of hill (**b**) is fire retardant dropped by aerial fire-fighting tanker (Fig. 1.31) (Photographs by Kari Greer/U.S. Forest Service)

fires in Interior Chaparral Shrubland are caused by lightning (Bolander 1982). In general, stands of Interior Chaparral Shrubland either burn intensely or do not burn, i.e., there is little or no gradation in fire intensity and severity (Lindenmuth and Davis 1973). Therefore, most fires are crown fires, regardless of whether they are wildfires or prescribed fires. Fires generally cover large areas.

Other evidence of the importance of fire is that Interior Chaparral Shrubland is considered a "fire type", in that its species are well-adapted to fire. Species regenerate after fire by sprouting from below-ground structures such as root crowns (Fig. 8.10) or by fire-stimulated germination of seeds in the seed bank. All but 4 of the 31 shrub species listed by Carmichael et al. (1978) for Interior Chaparral Shrubland are thusly adapted to fire. Some species are also combustible by having volatile oils, as well as abundant dead stems as individuals age (Fig. 8.11).

The mean fire-return interval in Interior Chaparral Shrubland has been estimated as 50–100 years (Pase and Brown 1994). This interval is not documented, but the finding that stands often do not burn supports a long interval. For example, prescribed burning is challenging, as it is possible only when fire conditions are extreme (Pieper and Wittie 1990). The generic fire-return interval of 50–100 years possibly masks wide variations in local return intervals (Brooks et al. 2007).

Numerous factors influence fire in Interior Chaparral Shrubland (e.g., Lindenmuth and Davis 1973; Brown and Boster 1974; Davis and Dieterich 1976; Pase and Granfelt 1977; Bolander 1982; Pieper and Wittie 1990). These include weather conditions prior to and at the time of the fire, such as temperature, humidity, precipitation, wind, and of course lightning. May and June are critical periods for fires, because of dry conditions, strong winds, and the possibility of lightning unaccompanied by precipitation. The previous year's precipitation also can be important because it affects herbaceous growth that cures and provides highly ignitable, fine fuels (Fig. 8.11).

Stand structure also influences fire. Open stands generally have more herbaceous growth, which is important in igniting and carrying fire between shrubs. Ignitions are reportedly less common in denser stands, but fires are higher in severity because of greater combustible material, especially dead material in the crowns of shrubs. In addition to the amount of fuel, fire is influenced by fuel moisture, fuel type (herbaceous vs. woody), and the shrub species present (species with volatile compounds enhance fire intensity).

Fire size is influenced by the above factors, as well as by topography. Fires move more rapidly on steep slopes and can be stopped by topographic barriers.

The paucity of data on the historical fire regime of Interior Chaparral Shrubland is due in part to the absence of fire scars of the type that enables dating past fires in forests (Sect. 1.2.5.1). Stems of shrubs of Interior Chaparral Shrubland are small and thin-barked. Therefore, they don't scar but instead are killed by fire. Also, there are no reports of datable fire scars on trees scattered within stands. These trees possibly do not survive the high-severity fires typical of Interior Chaparral Shrubland. Fire-scarred trees mentioned in early twentieth century reports such as Leopold (1924) were snags charred by a stand-replacing fire and not living or dead trees scarred by multiple fires.

Fig. 8.10 Post-fire sprouting of shrubs. Photograph taken four months after fire in Prescott National Forest, central Arizona. *Bright green* sprouts are mostly sugar sumac, and *dull green* sprouts are mostly shrub live oak (Photograph by author)

Fig. 8.11 *Gray*, dead stems of shrub live oak throughout much of the photograph and of subshrubs and herbs in the surface layer in the foreground facilitate the spread and increase the intensity of fires (Photograph by U.S. Forest Service)

As an alternative to examining fire scars, fire historians possibly need to use stem ages, as has been done in Gambel Oak Shrubland (Sect. 7.2.5.1). The only such study in Interior Chaparral Shrubland concluded that mean fire-return intervals were 30–40 years in Prescott National Forest (Sneed et al. 2002). The study examined two high-elevation sites, at least one of which was transitional with Ponderosa Pine Forest. Dates of past fires were not determined by the absence of stems predating fire, but instead were inferred from periods of relatively low abundance of stems. Alternative explanations for temporal variation in stem abundance, such as climate variation, were not pursued. Stems of Gambel oak, Arizona white oak (*Quercus arizonica*), and Emory oak (*Q. emoryi*) were analyzed (stems of shrub live oak, pointleaf manzanita, mountain mahogany, and skunkbush did not evidence known dates of fires on protocol-development sites; but see Welsh 1985). If the return interval of 30–40 years is accurate, it likely applies to low-severity fires in high-elevation stands associated with Ponderosa Pine Forest and does not reflect high-severity fires that appear to characterize most stands of Interior Chaparral Shrubland.

Study of the fire history of adjacent vegetation has provided further insight. Fire scars in an 87-ha (215-acre) stand of Ponderosa Pine Forest surrounded by Interior Chaparral Shrubland in Prescott National Forest indicated the forest had a mean fire-return interval of 1.5 years (all fires) during 1770–1870 (Dieterich and Hibbert 1990). The researchers assumed that many of these fires spread to the perimeter of the forest stand and burned into Interior Chaparral Shrubland where stands were burnable (it takes at least 20 years for burned Interior Chaparral Shrubland to recover before it can carry fire again). The inferred result was that the surrounding Interior Chaparral Shrubland had been a mosaic of uneven-aged stands. This mosaic contrasted with the landscape of uniformly old-age stands that developed with fire exclusion by approximately 1920 and continued to the time of the study. Sneed et al. (2002) also inferred that their Interior Chaparral Shrubland sites had been mosaics of uneven-aged stands.

Additional research on the historical fire regime is needed. Further examination of maximum ages of shrubs across a landscape has potential. This approach has been used with post-fire cohorts of quaking aspen in Spruce-Fir (*Picea-Abies*) and Mixed Conifer Forests (Sects. 2.2.5.1 and 3.2.5.1, respectively) and, as mentioned above, of Gambel oak in Gambel Oak Shrubland. However, it will be challenging in Interior Chaparral Shrubland because many shrub species, including shrub live oak, desert ceanothus, birchleaf mountain mahogany, skunkbush, and catclaw acacia (*Senegalia greggii*), continuously recruit and lose stems (Welsh 1985; Sneed et al. 2002). Pointleaf manzanita, even though it is a non-sprouting species, similarly recruits and loses stems as it spreads by layering and older stems in the center of shrubs die (Fig. 8.12; cf. Pond 1971). It is unknown how long stems of shrubs of Interior Chaparral Shrubland persist, but some have lasted for nearly five decades (Pond 1971). An analogous but potentially less complex method would be to determine if the ages of trees in drainages approximate the date of the previous high-severity fire (Pase and Johnson 1968). Another possible approach would be to date fires from Holocene charcoal deposits in alluvial sediments, as done by Jenkins et al. (2011) in Ponderosa Pine Forest (Sect. 4.2.5.1).

Fig. 8.12 Manzanita shrubs expand by layering, followed by the death of center stems. This pointleaf manzanita in Pinyon-Juniper vegetation in Coconino National Forest, north-central Arizona, extends across the *bottom* half of the photograph. The shrub was cut in half and shows an arc of living stems around dead stems in the center of the shrub (Photograph by author)

8.2.6 Anthropogenic Disturbance

Land use by Native Americans in the American Southwest was concentrated in Pinyon-Juniper vegetation (Sects. 1.5.1.1 and 5.2.6), but Interior Chaparral Shrubland was likely used for hunting and possibly other purposes, particularly where it occurred in a landscape mosaic with Pinyon-Juniper vegetation. The two Euro-American anthropogenic factors affecting Interior Chaparral Shrubland that have received most attention are livestock grazing and fire management. Yet there are few data on them and on the other anthropogenic disturbances: modern climate change, invasive species, recreation, and nearby land use. Nevertheless, each of these can be important in vegetation dynamics, and therefore all are incorporated in the conceptual models (Sect. 8.5).

8.2.6.1 Livestock Grazing

Livestock grazing appears to be the anthropogenic disturbance that has had the greatest impact on Interior Chaparral Shrubland. The Euro-American history of the land in central Arizona that became Tonto National Forest is probably typical for much of the area of Interior Chaparral Shrubland. The federal government established military posts in central Arizona soon after the Civil War ended in 1865 (Croxen 1926). That brought soldiers, packers, traders, and prospectors to the area. Word spread about extensive grass cover in the region, and livestock were brought into the area as early as 1874, first cattle (*Bos taurus*) and later both sheep (*Ovis aries*) and goats (*Capra hircus*).

Interviews with early stockmen and their descendants indicated that livestock grazing changed the landscape by greatly reducing grass cover and increasing brush (small trees and shrubs; Croxen 1926). In many areas with Interior Chaparral Shrubland, the weather was suited to year-around grazing, which resulted in greater impacts, especially on perennial grasses (Bolander 1982). Grazing was especially intensive from 1880 to 1920 (Pase and Brown 1994).

It is difficult to ascertain the relative impacts of this early livestock grazing on different vegetation types in this region: Subalpine-Montane Grassland, Ponderosa Pine Forest, Pinyon-Juniper vegetation, Interior Chaparral Shrubland, semi-desert grassland, and desert scrub. For example, Leopold's (1924) description of decreasing grass and increasing brush in central and southeastern Arizona has been interpreted as related to Interior Chaparral Shrubland (e.g., Pase and Brown 1994; Brooks et al. 2007). However, modern ecological knowledge indicates that he likely observed post-fire succession in Pinyon-Juniper persistent woodland (Sect. 5.3.2).

Concern over reduced grazing capacity of the land led to attempts beginning in the 1950s to convert stands of Interior Chaparral Shrubland into grassland (additional reasons for conversion included increased water yield and reduced fire hazard; Sect. 8.1). Conversion included using mechanical, fire, chemical, and biological (especially domestic goats) means to remove shrub cover, followed by reseeding with grasses that included non-native species (Sect. 8.2.6.4). Attempts at conversion dramatically declined in the early 1970s, following concerns about the use of herbicides and the desirability of conversion (McClaran and Brady 1994; Ffolliott et al. 2003).

Livestock in Interior Chaparral Shrubland primarily affect herbaceous cover in shrub interspaces (Pase and Brown 1994). Cattle and sheep do not browse shrub live oak, except for post-fire sprouts (U.S. Forest Service 1970), and do not affect recovery of shrubs following fire (Pond and Cable 1960). Other, little-browsed species include skunkbush, sugar sumac, birchleaf mountain mahogany, and Emory oak; shrubs that are browsed include mountain mahogany, Wright silktassel, hollyleaf buckthorn, desert ceanothus, and deerbrush ceanothus (*Ceanothus integerrimus*; Pond and Cable 1960; U.S. Forest Service 1970).

Livestock grazing likely affected the historical fire regime of Interior Chaparral Shrubland, at least to the extent that historical fires had been carried by herbaceous fuels (Sect. 8.2.5) and that livestock grazing increased shrub densities. Human attempts to convert stands of Interior Chaparral Shrubland into grassland to increase forage (see above) also affected the fire regime. In fact, this was part of the justification for conversion (Sect. 8.1).

8.2.6.2 Fire Management

The effects of active fire exclusion are under-studied, but possibly include greater shrub cover (Bolander 1982) and tree encroachment, especially at the interface with Ponderosa Pine Forest (Pase and Brown 1994; Brooks et al. 2007). Conversely, ponderosa pine is reduced if not eliminated for lengthy periods where fire suppression has been followed by intensive wildfires in this ecotone (Dickey 1982). Other effects of fire exclusion included increased landscape homogeneity (Dieterich and Hibbert 1990; see Sect. 8.2.5) and increased dead fuels.

Prescribed burning has been used in Interior Chaparral Shrubland to temporarily increase forage for livestock and game and to create a mosaic of different-aged stands to reduce the risk of large wildfires (Davis 1989). As indicated in Sect. 8.2.5, prescribed burning is difficult, because stands either do not burn or burn intensely.

8.2.6.3 Modern Climate Change

Climate warming and drying likely impact fire regimes and species composition of Interior Chaparral Shrubland, but no studies addressing these issues have been completed.

8.2.6.4 Invasive Species

Little information is available on invasive species in Interior Chaparral Shrubland. Invasive plants include red brome (*Bromus rubens*) and stork's bill (*Erodium cicutarium*). Non-native annual grasses have been said to increase fire frequency in southwestern Utah such that Interior Chaparral Shrubland does not have sufficient time to recover between fires (Brooks et al. 2007). Non-native plants such as Boer, Lehmann, and weeping lovegrasses (*Eragrostis chloromelas*, *E. lehmanniana*, and *E. curvula*, respectively) have been seeded in attempts to convert stands to grasslands (Carmichael et al. 1978); see Sect. 8.2.6.1.

8.2.6.5 Recreation

The impacts of recreation activities such as camping, hiking, hunting, and mountain biking have not been studied for Interior Chaparral Shrubland. Nevertheless, they likely include introduction of invasive species, fire ignitions, and local erosion. Recreation is expected to continue to increase as nearby cities such as Phoenix, Arizona grow in population.

8.2.6.6 Nearby Land Use

The impacts of nearby land use are also poorly studied. The introduction of fire from adjacent Ponderosa Pine Forest has been inferred (Dieterich and Hibbert 1990; see Sect. 8.2.5). In addition, nearby lands have invasive species that can spread into Interior Chaparral Shrubland. Both of these likely impacts are expected to increase as human development expands near and in Interior Chaparral Shrubland.

8.3 Processes

Because fire is the only important natural disturbance (Sect. 8.2.5), the primary vegetation dynamics of Interior Chaparral Shrubland are what follows fire. Some authorities have characterized post-fire changes as succession (e.g., Cable 1975). However, plant communities that follow fire consist largely of the same species present before fire. They regrow quickly, and there is little or no sequential replacement of species. Therefore, herein this process is considered rapid regrowth, not succession (see Sects. 6.3.1 and 7.3.1 for description of rapid regrowth in Subalpine-Montane Grassland and Gambel Oak Shrubland, respectively). Rapid regrowth plays a key role in the conceptual models (Sect. 8.5). Succession also occurs, but appears limited to transition zones, especially transitions with Ponderosa Pine Forest.

8.3.1 Rapid Regrowth

Most shrub species of Interior Chaparral Shrubland regrow by sprouting from below-ground structures that have buds and carbohydrate storage undamaged by fire (Carmichael et al. 1978; Sect. 8.2.5; Fig. 8.10). These species include shrub live oak, skunkbush, Wright's silktassel, and hollyleaf buckthorn. Some perennial grasses and forbs also sprout. Most sprouting species also can regenerate from seed. Non-sprouting species such as manzanitas (*Arctostaphylos* spp.) and weak-sprouting species such as desert ceanothus are prolific producers of seeds that build up in the seed bank and are stimulated to germinate by fire, resulting in rapid regrowth following fire.

The overall pattern of rapid regrowth is shown in Fig. 8.13a–f. Native grasses can recover 1 year after fire, and forbs peak in the second and third years and then rapidly decline (Pase and Pond 1964; Cable 1975). Grasses peak in years 5–7 (Cable 1975). Shrubs can sprout in the same year as fire (Fig. 8.13b) and most prolifically in the first 2 years (Cable 1957, 1975). Sprouts return to pre-fire densities in 5 years (Cable 1975; Fig. 8.13d). Seedlings of sprouting shrubs can be present the year after fire, depending on post-fire weather (Pase 1969), and seedlings of non-sprouting species can be abundant within 5 years (Pase and Pond 1964). Shrub cover can

Fig. 8.13 (**a–f**) Rapid regrown of Interior Chaparral Shrubland following fire in Three Bar Wildlife Area, Tonto National Forest, central Arizona. Photographs from approximately the same point with two overlapping camera angles show the watershed (**a**) before fire, (**b**) a few weeks after fire, and (**c**) one, (**d**) five, (**e**) ten, and (**f**) 15 years after fire. Initial sprouting of shrubs occurred within weeks after the fire, and shrub cover returned to pre-fire levels within 10 years. Some slopes in the background had additional management treatments that are especially apparent in (**f**) (Photographs by U.S. Forest Service)

Fig. 8.13 (continued)

approach that of adjacent unburned areas in 7–10 years (Pase and Pond 1964; Hibbert et al. 1974; Fig. 8.13e). Soon thereafter, seeding shrubs such as manzanitas and desert ceanothus mature sexually and begin to produce seeds that build up in the soil seed bank, generally before the stand can carry fire again (U.S. Forest Service 1970).

Fig. 8.13 (continued)

Study of 40 sites burned by six fires in southeastern Arizona and southwestern New Mexico determined that the primary shrub species one to 2 years after fire were sprouters shrub live oak, skunkbush, and seepwillow baccharis (*Baccharis salicifolia*) and obligate seeders pointleaf manzanita, desert ceanothus, and Fendler's ceanothus (Keeley et al. 2012). The herbaceous cover was dominated by annuals in both years (and there were large floristic differences in the herbaceous layer between spring and late summer/fall). Exotic species accounted for 8 % of the flora in the second year.

Without fire, stands can be quite stable. As described in Sect. 8.2.5, resprouting shrubs such as shrub live oak, skunkbush, catclaw acacia, and birchleaf mountain mahogany continually produce new stems, and some non-sprouting species such as manzanitas persist by layering. A study of tagged individuals from several shrub species indicated little change over 47 years, other than plants becoming larger (Pond 1971). Individuals of shrub live oak were especially persistent, and species with individuals that had not persisted generally had reproduced near the locations of the tagged plants.

8.3.2 Succession

The overall successional status of stands of Interior Chaparral Shrubland has received much attention (e.g., U.S. Forest Service 1970; Cable 1975; Carmichael et al. 1978; Pase and Brown 1994). Some stands show no evidence of substantial, directional changes in species composition and therefore are interpreted to be late-successional, "climax", or – with dependence on disturbance by fire – "disclimax". The presence of scattered ponderosa pines, pinyons, or junipers in some stands can be interpreted as tree invasion in succession. Alternatively, invasion is possibly driven by shifts in climate and disturbance regimes, but this has not been researched.

Succession does occur at the transition with Ponderosa Pine Forest. Disturbance is followed by rapid regrowth of shrubs, especially manzanita and Fendler's ceanothus (*Ceanothus fendleri*; cf. Dickey 1982). As manzanitas expand by layering and their centers senesce in 20–25 years, ponderosa pine seedlings invade. The growth of ponderosa pine into the overstory is paralleled by decreases in manzanita. Ponderosa pine can continue to reproduce and increasingly dominate stands (see Sect. 4.3.2). At lower elevations, the apparent invading trees are pinyons and junipers. Successional replacement by them is unstudied, but if it occurs, it would document that some stands of Interior Chaparral Shrubland are a seral stage in the post-disturbance recovery of Pinyon-Juniper persistent woodland (Sect. 5.3.2).

8.4 Historical Changes

8.4.1 Overstory

8.4.1.1 Pre-Euro-American Settlement

There are no known historical descriptions of the structure and composition of Interior Chaparral Shrubland from near the time of Euro-American settlement. Shrublands lack the economic value that motivated many early descriptions of

Fig. 8.14 Historical photograph showing Interior Chaparral Shrubland in the background in 1891 along Lynx Creek, east of Prescott, Arizona (Photograph courtesy of Sharlot Hall Museum Library and Archives, Prescott, Arizona)

forests and grasslands. The closest to a historical description appears in a summary of interviews of early ranchers in the area of Tonto National Forest a half-century after Euro-American settlement. It confirms that brushlands, apparently including Interior Chaparral Shrubland, occurred on sites where currently present:

> All the men interviewed state that there was little brush in the country at the time stock was first brought in… The little that there was, was only on some of the mountains and some of the slopes. (Croxen 1926)

Photographic evidence of historical conditions (e.g., Fig. 8.14) dates to the 1870s. Unfortunately, many early photographs were intended to document human land uses such as mining and smelting. Trees likely had been removed from some of these sites for fuel and construction materials, including mine supports. Many early photographs of more remote sites (Fig. 8.15) also give a biased perspective, having been taken by U.S. Forest Service personnel to document the presence of trees in Interior Chaparral Shrubland. However, regardless of the biases of early photographers, stands shown in historical photographs are remarkably similar in appearance to stands visible today in landscapes with Interior Chaparral Shrubland.

Alternatives to determining pre-Euro-American conditions include study of relict areas, reconstruction of past conditions using living and dead woody stems, and inference.

Fig. 8.15 Historical photograph of Interior Chaparral Shrubland in 1909 in Cameron Creek Watershed, Gila National Forest, west-central New Mexico. Shrubs are primarily shrub live oak, and scattered, darker trees on hillside are alligator juniper (*Juniperus deppeana*) (Photograph by G.A. Pearson, courtesy of Fort Valley Experimental Forest Historic Images, U.S. Forest Service)

Although relict areas can provide quantitative insight into historical conditions, the representativeness of the data is questionable. For example, do extreme sites – such as steep slopes that are likely to have undisturbed stands of Interior Chaparral Shrubland – adequately reflect overall historical conditions? See the discussion of limitations of data from southwestern relict sites in Sect. 5.4.1.1.

Woody stems, snags, and logs have been used to reconstruct historical conditions for forests (e.g., Sect. 4.4.1.1). Implementing this approach for stands of Interior Chaparral Shrubland could yield dates of on-going stem recruitment and possibly the date of stand origin (Sect. 8.2.5), but is otherwise unlikely to provide insight into historical stand structure and composition.

Inference can be used to elucidate historical conditions in Interior Chaparral Shrubland. The rapid regrowth of shrubs after disturbance and the relative stability of stands thereafter (Sect. 8.3.1) suggest pre-Euro-American shrub composition and structure were similar to today.

In conclusion, historical photographs and inference suggest that historical conditions in Interior Chaparral Shrubland are similar to conditions present today, with the exception of changes resulting from livestock grazing and fire exclusion (Sects. 8.2.6.1 and 8.2.6.2, respectively).

8.4.1.2 Post-Euro-American Settlement

Euro-American settlement of landscapes with Interior Chaparral Shrubland is also best known from the region that became Tonto National Forest (see previous section). Croxen (1926) outlined the area's Euro-American history and interviewed early settlers, who had brought livestock to the area beginning in 1874. The ranchers stated that the range became fully stocked around 1890 and peaked around 1900 at levels 15–20 times the stocking present at the time of the interviews a quarter of a century later. They witnessed changes in the landscape (all quotations are from Croxen 1926):

> …Blackfoot and Crowfoot Grama grass … touched ones [sic] stirrups when riding through it, where no grama grass grows at present. (Florance A. Packard; blackfoot may be *Melampodium longicorme* and crowfoot grama is likely *Bouteloua eriopoda*)

> There were perennial grasses on the mesas along Tonto Creek where only brush grows at the present time. (Florance A. Packard)

> …nearly all the north slope of Mt. Ord was a Pine Bunch grass country. At present this is one of the brushiest pieces of range on the Tonto, as anyone will agree who has been unfortunate enough to have come in contact with it. (Chub Watkins and Fred W. Croxen; pine bunchgrass is likely *Festuca arizonica*)

Such accounts in Croxen (1926) have been interpreted as documenting that grazing changed Interior Chaparral Shrubland by reducing the herbaceous layer, increasing shrub cover, and causing shrubs of Interior Chaparral Shrubland to invade other vegetation types (e.g., Cable 1975; Paulsen 1975; U.S. Forest Service 1975). Is this interpretation valid?

There is little doubt that the herbaceous layer of Interior Chaparral Shrubland would have been impacted by the intensive livestock grazing, at least to the degree that Interior Chaparral Shrubland was grazed. Stands on steep slopes and stands with dense shrub cover likely were little grazed because of poor accessibility.

The propositions that Interior Chaparral Shrubland increased in shrub cover and invaded other types of vegetation are less tenable. First, there appears to be no evidence other than Croxen (1926) to support these changes in Interior Chaparral Shrubland. Another source sometimes used is Leopold (1924; see Sect. 8.2.6.1), who wrote about increases in brush in and near Tonto, Prescott, Coronado, and Gila National Forests in Arizona and New Mexico. However, as described in Sect. 8.2.6.1, he likely described post-fire successional changes in Pinyon-Juniper persistent woodlands (Sect. 5.3.2), not increased shrub cover in Interior Chaparral Shrubland or invasion of other vegetation types.

Second, evidence for increased cover and invasion by shrubs of Interior Chaparral Shrubland in Croxen (1926) is weak. Although the early ranchers clearly saw increased brush, it is unclear it was in Interior Chaparral Shrubland. Shrubs/brush also occurred in all other types of vegetation in the region: Ponderosa Pine Forest, Pinyon-Juniper vegetation, desert scrub, and even in semi-desert grassland. Although a few ranchers mentioned specific sites of increased brush, accurate recollection of conditions at specific sites in a complex landscape observed several decades earlier is problematic (author, personal observation).

Third, the implication in Croxen (1926) that increases in brush in Interior Chaparral Shrubland were due to overgrazing possibly was not accurate. Other causes are likely, including climate change and disturbances such as tree cutting. In fact, there appears to be no scientific evidence that livestock grazing in Interior Chaparral Shrubland directly increases shrub cover. Instead, evidence suggests that increases in shrub cover have been due to fire exclusion (cf. Huebner et al. 1999).

Although the question of whether shrubs of Interior Chaparral Shrubland invaded other vegetation types has received much discussion (cf. Saunier and Wagle 1967), there has been little research, and no scientific evidence supports it. Patches of shrubs of Interior Chaparral Shrubland in grassland are not signs of invasion, but rather reflect unique habitat (Saunier and Wagle 1967). Repeat ground photography showed that stands of Interior Chaparral Shrubland did not spread in an area of grassland-juniper woodland ecotone in central Arizona from 1916 to 1977 (Johnsen and Elson 1979). Moreover, repeat aerial photography documented that Interior Chaparral Shrubland has not invaded grassland or Pinyon-Juniper vegetation in a landscape mosaic in central Arizona from 1940 to 1989 (Huebner and Vankat 2003).

Nevertheless, the early ranchers apparently witnessed increases in brush cover both where shrubs had been present and where they had been absent (Croxen 1926). Regardless of whether the increase of brush involved species of Interior Chaparral Shrubland or not, there is an alternative hypothesis to increases being due to grazing. It is that anthropogenic burning shortly before and during the time of the early ranchers reduced the extent and cover of shrubs, and the ranchers witnessed post-fire recovery of the vegetation. There were multiple possible sources of anthropogenic burning at that time: the military and the Apache Tribe possibly used fire in warfare (cf. Seklecki et al. 1996), prospectors who soon followed the military set fires to expose mineral deposits (cf. Dieterich and Hibbert 1990), and livestock herders set fires with the intent of increasing forage the following spring (William Craig, as quoted in Croxen 1926). The regrowth of shrubs following burning (Sect. 8.3.1) was interpreted by the ranchers (and subsequent ecologists) as resulting from intensive livestock grazing.

Another change that followed Euro-American settlement and affected Interior Chaparral Shrubland was the exclusion of fire. As in Ponderosa Pine Forest (Sects. 4.2.6.1 and 4.2.6.2), fire exclusion possibly began when historical grazing removed fine fuels and later continued as part of early fire management (Sects. 8.2.6.1 and 8.2.6.2, respectively). The resultant decrease in fire frequency led to increased shrub cover within stands and to greater homogeneity among stands across Interior Chaparral Shrubland landscapes (Bolander 1982; Dieterich and Hibbert 1990). Increases in shrub density due to fire exclusion have continued, as documented for 1940–1989 in Prescott National Forest (Huebner et al. 1999).

In addition, Interior Chaparral Shrubland – or a stage in forest succession resembling it in structure and composition – appears to have expanded with disturbance in areas transitional with Ponderosa Pine Forest (U.S. Forest Service 1975; Dickey 1982) and possibly with Pinyon-Juniper vegetation. Changes do not represent invasion by Interior Chaparral Shrubland, but instead the loss of tree canopies with anthropogenic burning or clear-cutting of trees for timber and charcoal.

In conclusion, Interior Chaparral Shrubland changed with Euro-American settlement. Shrub cover and stand ages have increased with fire exclusion. Landscapes of Interior Chaparral Shrubland have become more homogeneous, especially in terms of stand ages, structures, and fuels. The spatial arrangement of stands of Interior Chaparral Shrubland in landscape mosaics with Pinyon-Juniper vegetation and grassland remains unchanged, i.e., shrubs of Interior Chaparral Shrubland have not invaded other vegetation types.

8.4.2 Understory

Little information is available on the herbaceous layer of Interior Chaparral Shrubland prior to Euro-American settlement. Grasses said to have been more abundant include gramas, threeawns, dropseeds (*Sporobolus* spp. and *Blepharoneuron* spp.), bluestems (*Andropogon* spp., *Bothriochloa* spp., and *Schizachyrium* spp.), and curly-mesquite (*Hilaria belangeri*; Paulsen 1975). Where intensive livestock grazing occurred, it reduced herbaceous cover and changed species composition by reducing palatable herbs. Reductions in herbaceous cover apparently ranged from minimal on inaccessible sites to more significant (but undetermined) on accessible sites with open shrub cover. Herbaceous cover also likely decreased with fire exclusion because of increases in shrub cover. Another change is the presence of non-native invasive herbs (Sect. 8.2.6.4).

8.5 Conceptual Models

The following nested, three-tiered set of conceptual models is based on my interpretation of best-available information on Interior Chaparral Shrubland of the American Southwest. The models summarize and synthesize material on drivers (Sect. 8.2) and processes (Sect. 8.3) of vegetation dynamics, and they account for historical changes (Sect. 8.4). Their format is explained in Sect. 1.7. The models can be used to facilitate understanding of the past and the present. They also can provide insight into effects of future changes in the environment and outcomes of alternative land-management decisions. Moreover, they can guide the development of quantitative models.

8.5.1 Ecosystem-Characterization Model

The ecosystem-characterization model for Interior Chaparral Shrubland emphasizes *Vegetation* (a biotic component) and *Disturbance* (a driver; Fig. 8.16a, Table 8.2). Key aspects of *Vegetation* are structure, fuel, sprouting, and fire-stimulated seed germination, and these affect *Disturbance*. The only major *Disturbance* is fire,

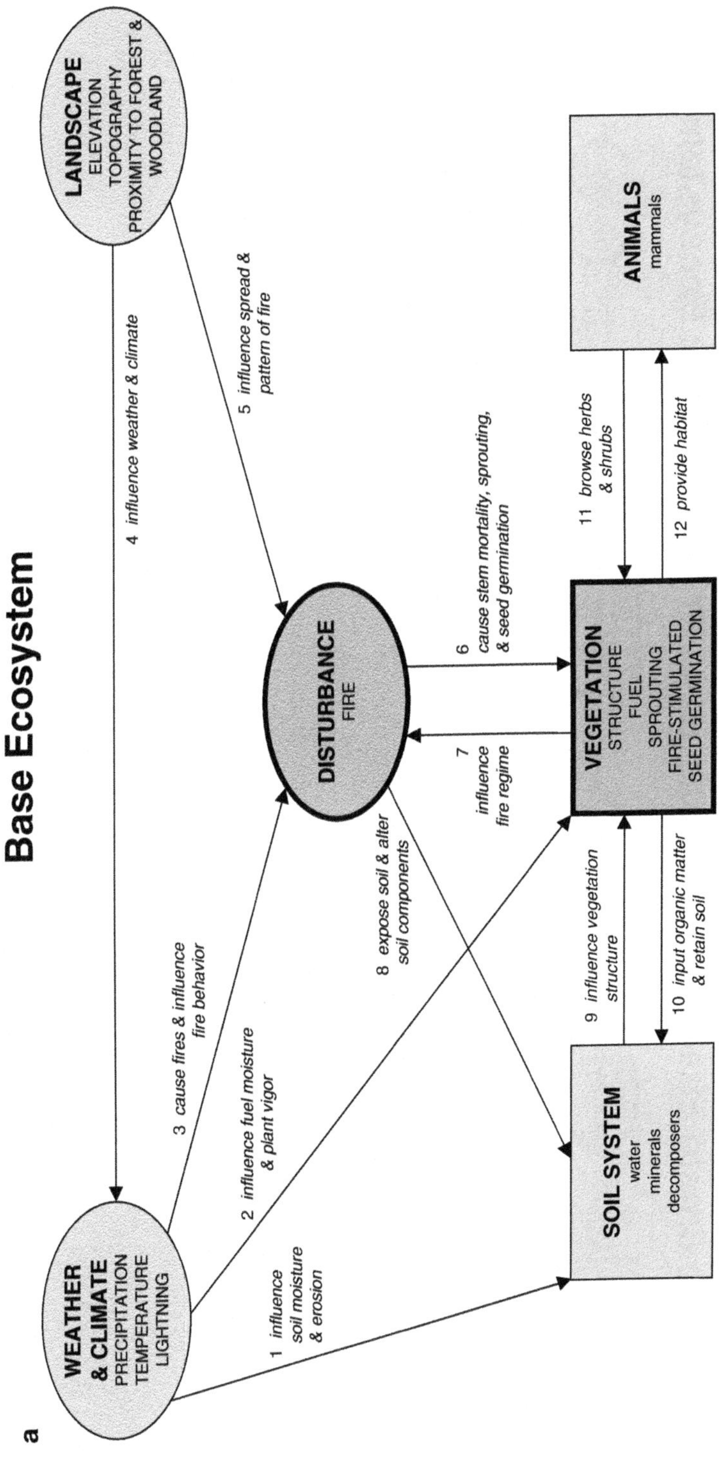
a
Base Ecosystem
LANDSCAPE
ELEVATION
TOPOGRAPHY
PROXIMITY TO FOREST &
WOODLAND
WEATHER
& CLIMATE
PRECIPITATION
TEMPERATURE
LIGHTNING
DISTURBANCE
FIRE
VEGETATION
STRUCTURE
FUEL
SPROUTING
FIRE-STIMULATED
SEED GERMINATION
SOIL SYSTEM
water
minerals
decomposers
ANIMALS
mammals
1 influence soil moisture & erosion
2 influence fuel moisture & plant vigor
3 cause fires & influence fire behavior
4 influence weather & climate
5 influence spread & pattern of fire
6 cause stem mortality, sprouting, & seed germination
7 influence fire regime
8 expose soil & alter soil components
9 influence vegetation structure
10 input organic matter & retain soil
11 browse herbs & shrubs
12 provide habitat

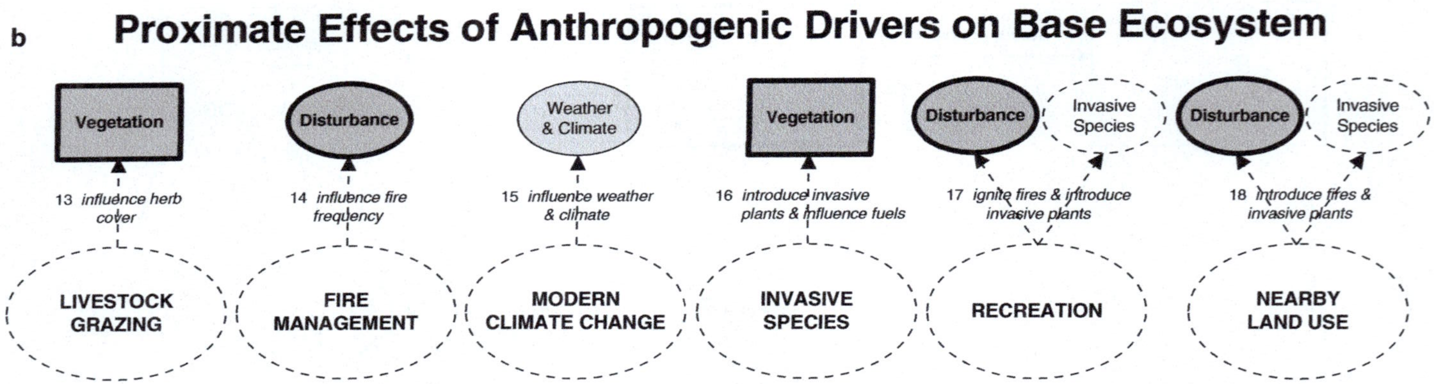

Fig. 8.16 (**a**, **b**) Ecosystem-characterization model for Interior Chaparral Shrubland. In the base model (**a**), biotic components are shown as *rectangles* and drivers as *solid ovals*. Text within them identifies major features in *upper case* and other important features in *lower case*. Disturbance and Vegetation are *highlighted*, because they are the basis for the vegetation-dynamics model (Fig. 8.17). In the anthropogenic supplement to the base model (**b**), drivers are shown as *dashed ovals*. Relationships in (**a**) and (**b**) are shown as *solid* and *dashed arrows*, respectively, and are described in Table 8.2

Table 8.2 Relationships in the ecosystem-characterization model of Interior Chaparral Shrubland (Fig. 8.16a, b)

Relationships	Circa 1870	Present
1	Precipitation increases soil moisture and can cause soil erosion. Winter precipitation is likely important for recharging soil moisture	Same, but climate change increases the severity and duration of drought
2	Precipitation and temperature largely determine fuel moisture and are important to plant vigor	Same, but climate change increases the severity and duration of drought
3	Lightning ignites fires and wind influences fire behavior	Same
4	Topography influences weather and climate	Same
5	Fires can spread from other areas of the landscape (especially from Ponderosa Pine Forest), and fire patterns are influenced by elevation and topography	Same
6	Infrequent high-severity fires kill shrub stems and stimulate sprouting and seed germination	Same
7	Shrub and herb cover influence fires	Same
8	Fires expose soil surface and alter soil properties	Same
9	Water and minerals in soil affect plant vigor and growth, influencing vegetation structure	Same
10	Plants add organic matter to soil and stabilize soil	Same
11	Herbivory reduces plant cover	Same
12	Vegetation is habitat for animals	Same
13	Not applicable	Livestock grazing reduces herb cover
14	Not applicable	Fire suppression possibly reduces the frequency and spread of fires in some areas
15	Not applicable	Modern climate change increases temperatures and the frequency and extent of drought
16	Not applicable	Invasive plants change species composition of herbaceous layer and have potential to change fire patterns
17	Not applicable	Recreationists ignite fires and introduce and spread invasive plants
18	Not applicable	Nearby land use that initiates fires and colonization by invasive plants can introduce fires and invasives into Interior Chaparral Shrubland

which causes stem mortality followed by sprouting and seed germination. The two other biotic components are *Soil System* and *Animals*. A second driver is *Weather & Climate*, which causes fires and influences fire behavior and fuel and soil moisture. The third driver is *Landscape*, with its primary features being elevation, topography, and proximity to Ponderosa Pine Forest and possibly Pinyon-Juniper persistent woodland. It influences weather and climate, as well as spread and pattern of fire. The model also includes six anthropogenic drivers (Fig. 8.16b, Table 8.2): *Livestock Grazing*, *Fire Management*, *Modern Climate Change*, *Invasive Species*, *Recreation*, and *Nearby Land Use*.

8.5.2 Vegetation-Dynamics Model

The vegetation-dynamics model for Interior Chaparral Shrubland is accompanied by a bar graph to show estimated relative shifts in abundances of the community types from circa 1870 (prior to widespread Euro-American land use) to circa 1970 (end of strict fire exclusion policy) to the present. The estimates are based on my understanding of vegetation dynamics and should be interpreted as showing relative differences and not interpreted quantitatively (hence the absence of a scale for abundance). Although not shown on the graph, relative abundances shifted prior to circa 1870 as a result of variation in climate and disturbance. Localized differences also have been and are present, but are beyond the scope of this generalized bar graph.

The vegetation-dynamics model has one state with five communities (Fig. 8.17, Table 8.3). All communities occurred historically. Community *A1 Young Shrubland* is formed by post-fire sprouting and fire-stimulated seed germination and is dominated by herbs and sprouts and seedlings of shrub live oak and other shrub species. As sprouts and seedlings mature, community *A1* changes into community *A2 Shrubland* dominated by shrub live oak, other shrub species, and herbs. Historically, this has been the most common community. With invasion and establishment of ponderosa pine at upland elevations or pinyons and/or junipers at lowland elevations, community *A2* forms *A3 Wooded Shrubland* with shrub live oak and other shrub species below an open overstory of these conifers. Community *A3* can revert to *A2* with mortality of the conifers. Community *A3* also can change into community *A4 Forest* (uplands) or possibly *A5 Woodland* (lowlands) with recruitment of conifers. Both of these communities are dominated by an overstory of the conifers above an understory of shrub live oak and other shrub species. High-severity fire followed by sprouting and fire-stimulated seed germination returns communities *A2*, *A3*, *A4*, and *A5* to *A1*.

8.5.3 Mechanistic Model

The mechanistic model has six biotic components on the right side of the figure (including three aspects of fuels), two drivers on the left side, and six anthropogenic factors at the bottom (Fig. 8.18). In general, *Herbs, Shrubs & Trees*, and *Precipitation*

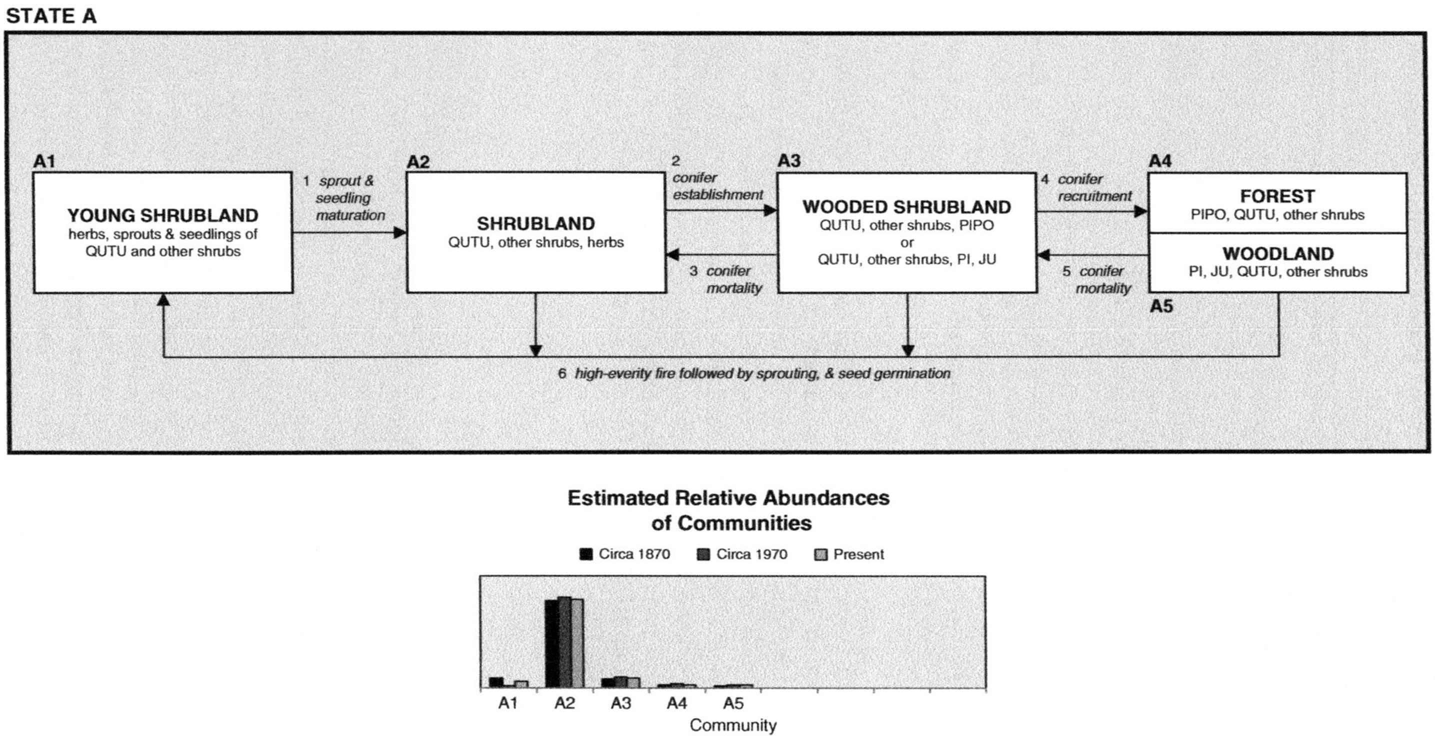

Fig. 8.17 Vegetation-dynamics model for Interior Chaparral Shrubland. Model provides details on the vegetation-disturbance portion of the ecosystem-characterization model (Fig. 8.16a, b) by showing communities (*smaller*, *open rectangles*) within states (*larger*, *shaded rectangles*), processes that form the communities, and transitions between states. Text along *arrows* identifies pathways that are described in Table 8.3. *Bar graph* shows estimated relative shifts in abundances of the community types from circa 1870 (see Sect. 8.5.2). Species: *JU* junipers, *PI* pinyons, *PIPO* ponderosa pine, *QUTU* shrub live oak

Table 8.3 Relationships in the vegetation-dynamics model for Interior Chaparral Shrubland (Fig. 8.17)

Relationships	Circa 1870	Present
1	Young sprouts and seedlings mature, changing young shrubland into shrubland	Same
2	Invasion and establishment of ponderosa pine in uplands and pinyon and/or juniper in lowlands changes shrubland into wooded shrubland	Same
3	Mortality of ponderosa pine, pinyon, and juniper changes wooded shrubland into shrubland	Same
4	Recruitment of ponderosa pine in uplands changes wooded shrubland into forest. Possible recruitment of pinyon or juniper in lowlands changes wooded shrubland into woodland	Same
5	Mortality of ponderosa pine in uplands changes forest into wooded shrubland. Possible mortality of pinyon and juniper in lowlands changes woodland into wooded shrubland	Same
6	High-severity fire kills woody stems and promotes sprouting and seed germination, forming young shrubland	Same

& Temperature affect the fuel characteristics. *Fuel Moisture*, *Fuel Continuity*, and *Fuel Type & Loading* influence *Fire*, which affects characteristics of *Shrubs & Trees*, especially cover and species composition. *Shrubs & Trees* influence the cover of *Herbs*, and both *Shrubs & Trees* and *Herbs* determine *Community Type* (of the five appearing in the vegetation-dynamics model). *Modern Climate Change* influences *Precipitation & Temperature. Fire Management* and fire ignitions caused by *Nearby Land Use* and *Recreation* affect *Fire*. *Nearby Land Use* and *Recreation* are also sources of *Invasive Species*, and they, along with *Livestock Grazing*, affect the cover of *Herbs*.

8.6 Conclusions and Challenges

The relative lack of basic research on Interior Chaparral Shrubland leaves many challenges for researchers and for land managers. Fire is the only important natural disturbance, and more information on the fire regime is needed, particularly on fire-return intervals and turnover times (rotations), as well as the historical landscape mosaic of different-aged stands. The effects of livestock grazing are well-studied, but more information is needed on the effects of fire exclusion, climate change, invasives (especially in relation to fire), recreation, and nearby land use. These are major needs of land managers. The process of rapid regrowth of Interior Chaparral Shrubland after fire is well-known, providing important insight for land managers, but more information is needed on succession at the transition with Ponderosa Pine Forest and at the transition with Pinyon-Juniper vegetation. The conclusion that present-day conditions in Interior Chaparral Shrubland are likely similar to conditions before Euro-American settlement – a conclusion of great importance to land managers – needs additional study.

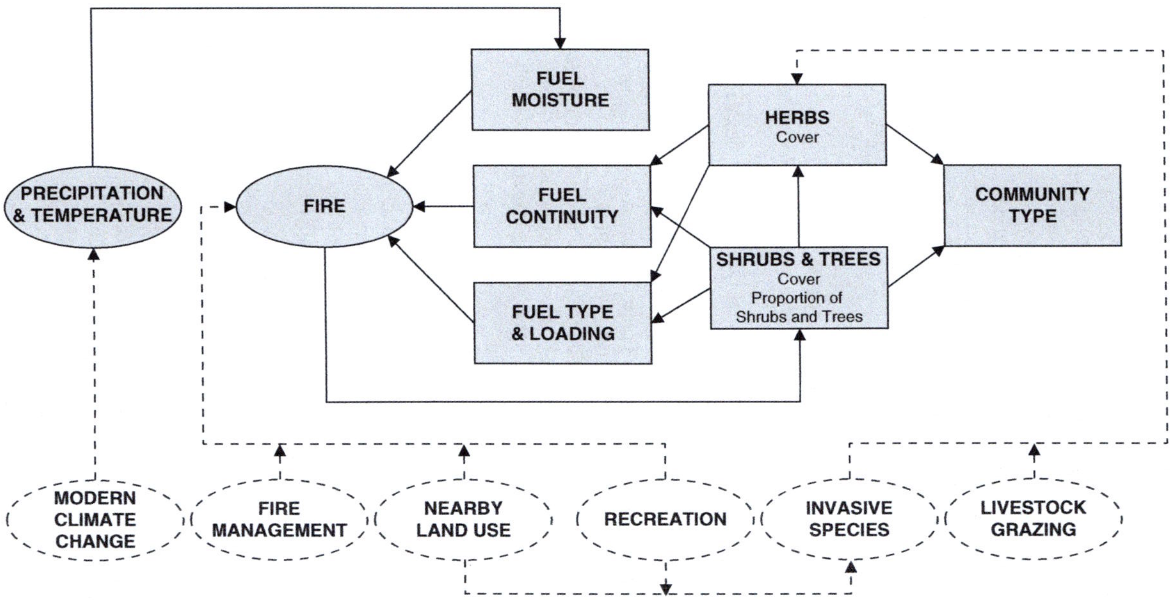

Fig. 8.18 Mechanistic model for Interior Chaparral Shrubland. Model provides details on the mechanisms involved in the vegetation-dynamics model (Fig. 8.17). *Rectangles* are biotic components (including fuels), *ovals* are drivers, and *dashed ovals* are anthropogenic drivers

References

Ackerly DD (2009) Evolution, origin and age of lineages in the Californian and Mediterranean floras. J Biogeogr 36:1221–1233

Axelrod DI (1975) Evolution and biogeography of Madrean-Tethyan sclerophyll vegetation. Ann Mo Bot Gard 62:280–334

Bolander DH (1982) Chaparral in Arizona. In: Conrad CE, Oechel WC (tech coords) Proceedings of the symposium on dynamics and management of Mediterranean-type ecosystems, San Diego, CA, 22–26 June 1982. U.S. Department of Agriculture, Forest Service, Pacific Southwest Forest and Range Experiment Station, Berkeley, CA

Brooks ML, Esque TC, Duck T (2007) Creosotebush, blackbrush, and interior chaparral shrublands. In: Hood SM, Miller M (eds) Fire ecology and management of the major ecosystems of southern Utah. U.S. Department of Agriculture, Forest Service General Technical Report RMRS-GTR-202. Rocky Mountain Research Station, Fort Collins, CO

Brown TC, Boster RS (1974) Effects of chaparral-to-grass conversion on wildfire suppression costs. U.S. Department of Agriculture, Forest Service Research Paper RM-119. Rocky Mountain Forest and Range Experiment Station, Fort Collins, CO

Cable DR (1957) Recovery of chaparral following burning and seeding in central Arizona. U.S. Department of Agriculture, Forest Service Research Note 28. Rocky Mountain Forest and Range Experiment Station, Fort Collins, CO

Cable DR (1975) Range management in the chaparral type and its ecological basis: the status of our knowledge. U.S. Department of Agriculture, Forest Service Research Paper RM-155. Rocky Mountain Forest and Range Experiment Station, Fort Collins, CO

Carmichael RS, Knipe OD, Pase CP, Brady WW (1978) Arizona chaparral: plant associations and ecology. U.S. Department of Agriculture, Forest Service Research paper RM-202. Rocky Mountain Forest and Range Experiment Station, Fort Collins, CO

Clements FE (1920) Plant indicators: the relation of plant communities to process and practice. Publication 290. Carnegie Institution of Washington, Washington, DC

Cody ML, Mooney HA (1978) Convergence versus nonconvergence in Mediterranean-climate ecosystems. Annu Rev Ecol Syst 9:265–321

Croxen FW (1926) History of grazing on Tonto. Unpublished manuscript of presentation at the Tonto grazing conference, Phoenix, Arizona, 4–5 Nov 1926

Davis EA (1977) Root system of shrub live oak in relation to water yield by chaparral. In: Proceedings of the 1977 meetings of the Arizona Section of the American Water Resources Association and the Hydrology Section of the Arizona Academy of Science, Las Vegas, Nevada, 15–16 Apr 1977. Hydrol Water Resour Ariz Southwest 7:241–248

Davis EA (1989) Prescribed fire in Arizona chaparral: effects on stream water quality. For Ecol Manage 26:189–206

Davis EA, Pase CP (1977) Root system of shrub live oak: implications for water yield in Arizona chaparral. J Soil Water Conserv 32:174–180

Davis EA, Kuester JL, Bagby MO (1984) Biomass conversion to liquid fuels: potential of some Arizona chaparral brush and tree species. Nature 307:726–728

Davis JR, Dieterich JH (1976) Predicting rate of fire spread (ROS) in Arizona oak chaparral: field workbook. U.S. Department of Agriculture, Forest Service General Technical Report RM-24. Rocky Mountain Forest and Range Experiment Station, Fort Collins, CO

DeBano LF, Baker MB Jr, Overby ST (1999) Managing chaparral in Yavapai County. In: Baker MB Jr (compiler) History of watershed research in the central Arizona highlands. U.S. Department of Agriculture, Forest Service General Technical Report RMRS-GTR-29 Rocky Mountain Research Station, Fort Collins, CO

Dickey AM (1982) Vegetational dynamics and fire succession in chaparral-ponderosa pine transition zone. Dissertation, Arizona State University, Tempe

Dick-Peddie WA (1993) New Mexico vegetation: past, present, and future. University of New Mexico Press, Albuquerque

Dieterich JH, Hibbert AR (1990) Fire history in a small ponderosa pine stand surrounded by chaparral. In: Krammes JS (tech coord) Effects of fire management of southwestern natural resources. U.S. Department of Agriculture, Forest Service General Technical Report RM-191. Rocky Mountain Forest and Range Experiment Station, Fort Collins, CO

Ffolliott PF, Baker MB Jr, DeBano LF (2003) Arizona watershed management program. J Ariz-Nev Acad Sci 35:5–10

Hibbert AR, Davis EA, Scholl DG (1974) Chaparral conversion potential in Arizona. Part I: Water yield response and effects on other resources. U.S. Department of Agriculture, Forest Service Research Paper RM-126. Rocky Mountain Forest and Range Experiment Station, Fort Collins, CO

Huebner CD, Vankat JL (2003) The importance of environment vs. disturbance in the vegetation mosaic of central Arizona. J Veg Sci 14:25–34

Huebner CD, Vankat JL, Renwick WH (1999) Change in the vegetation mosaic of central Arizona USA between 1940 and 1989. Plant Ecol 144:83–91

Integrated Taxonomic Information System (2012). http://www.itis.gov. Accessed 30 Aug 2012

Jenkins SE, Sieg CH, Anderson DE, Kaufman DS, Pearthree PA (2011) Late Holocene geomorphic record of fire in ponderosa pine and mixed-conifer forests, Kendrick Mountain, northern Arizona, USA. Int J Wildland Fire 20:125–141

Johnsen TN Jr, Elson JW (1979) Sixty years of change on a central Arizona grassland-juniper woodland ecotone. U.S. Department of Agriculture, Agricultural Research, Western Region, Science and Education Administration Agricultural Reviews and Manuals ARM-W-7, Oakland, CA

Keeley JE (2000) Chaparral. In: Barbour MG, Billings WD (eds) North American terrestrial vegetation, 2nd edn. Cambridge University Press, New York

Keeley JE, Davis FW (2007) Chaparral. In: Barbour MG, Keeler-Wolf T, Schoenherr AA (eds) Terrestrial vegetation of California, 3rd edn. University of California Press, Berkeley/Los Angeles

Keeley JE, Fotheringham CJ, Rundel PW (2012) Postfire chaparral regeneration under Mediterranean and non-Mediterranean climates. Madroño 59:109–127

Knipe OD, Pase CP, Carmichael RS (1979) Plants of the Arizona chaparral. U.S. Department of Agriculture, Forest Service General Technical Report RM-64. Rocky Mountain Forest and Range Experiment Station, Fort Collins, CO

Leopold A (1924) Grass, brush, timber, and fire in southern Arizona. J For 22:1–10

Lindenmuth AW Jr, Davis JR (1973) Predicting fire spread in Arizona's oak chaparral. U.S. Department of Agriculture, Forest Service Research Paper RM-101. Rocky Mountain Forest and Range Experiment Station, Fort Collins, CO. http://www.fs.fed.us/rm/pubs_rm/rm_rp101.pdf. Accessed 5 July 2012

McClaran MP, Brady WW (1994) Arizona's diverse vegetation and contributions to plant ecology. Rangelands 16:208–217

Mooney HA (ed) (1977) Convergent evolution in Chile and California: Mediterranean climate ecosystems. Dowden, Hutchinson & Ross, Stroudsburg

Pase CP (1969) Survival of *Quercus turbinella* and *Q. emoryi* seedlings in an Arizona chaparral community. Southwest Nat 14:149–156

Pase CP, Brown DE (1994) Interior chaparral. In: Brown DE (ed) Biotic communities: southwestern United States and northwestern Mexico. University of Utah Press, Salt Lake City

Pase CP, Granfelt CE (tech coords) (1977) The use of fire on Arizona rangelands. Arizona Interagency Range Committee Publication No. 4. Arizona Interagency Range Committee, Arizona

Pase CP, Johnson RR (1968) Flora and vegetation of the Sierra Ancha Experimental Forest, Arizona. U.S. Department of Agriculture, Forest Service Research Paper RM-41. Rocky Mountain Forest and Range Experiment Station, Fort Collins, CO

Pase CP, Pond FW (1964) Vegetation changes following the Mingus Mountain burn. U.S. Department of Agriculture, Forest Service Research Note RM-18. Rocky Mountain Forest and Range Experiment Station, Fort Collins, CO

Paulsen HA Jr (1975) Range management in the central and southern Rocky Mountains: a summary of the status of our knowledge by range ecosystems. U.S. Department of Agriculture, Forest Service Research Paper RM-154. Rocky Mountain Forest and Range Experiment Station, Fort Collins, CO

Pieper RD, Wittie RD (1990) Fire effects in southwestern chaparral and pinyon-juniper vegetation. In: Krammes JS (tech coord) Effects of fire management of southwestern natural resources. U.S. Department of Agriculture, Forest Service General Technical Report RM-191. Rocky Mountain Forest and Range Experiment Station, Fort Collins, CO

Pond FW (1971) Chaparral: 47 years later. U.S. Department of Agriculture, Forest Service Research Paper RM-69. Rocky Mountain Forest and Range Experiment Station, Fort Collins, CO

Pond FW, Cable DR (1960) Effect of heat treatment on sprout production of some shrubs of the chaparral in central Arizona. J Range Manage 13:313–317

Prior-Magee JS, Boykin KG, Bradford DF, Kepner WG, Lowry JH, Schrupp DL, Thomas KA, Thompson BC (2007) Ecoregional gap analysis of the southwestern United States: the Southwest regional gap analysis project final report. http://fws-nmcfwru.nmsu.edu/swregap/report/SWReGAP%20Final%20Report.pdf. Accessed 11 July 2012

Saunier RE, Wagle RF (1967) Factors affecting the distribution of shrub live oak (*Quercus turbinella* Greene). Ecology 48:35–41

Seklecki MT, Grissino-Mayer GT, Swetnam TW (1996) Fire history and the possible role of Apache-set fires in the Chiricahua Mountains of southeastern Arizona. In: Ffolliott PF, DeBano LF, Baker MB, Gottfried GJ, Solis-Garza G, Edminster CB, Neary DG, Allen LS, Hamre RH (tech coords) Effects of fire on Madrean Province ecosystems – a symposium proceedings, Tucson, Arizona, 11–15 Mar 1996. U.S. Department of Agriculture, Forest Service General Technical Report RM-GTR-289. Rocky Mountain Forest and Range Experiment Station, Fort Collins, CO

Sneed P, Floyd-Hanna L, Hanna D (2002) Prescott Basin fire history project. Unpublished final report to Prescott National Forest, Arizona

U.S. Forest Service (1970) Chaparral vegetation type. U.S. Department of Agriculture, Forest Service, Southwestern Region, Division of Range and Wildlife

U.S. Forest Service (1975) Chaparral: characteristics and management in Arizona. U.S. Department of Agriculture, Forest Service, Southwestern Region, Albuquerque, NM

Valiente-Banuet A, Flores-Hernández N, Verdú M, Dávila P (1998) The chaparral vegetation in Mexico under nonmediterranean climate: the convergence and Madrean-Tethyan hypotheses reconsidered. Am J Bot 85:1398–1408

Vankat JL (1989) Water stress in chaparral shrubs in summer-rain versus summer-drought climates – whither the Mediterranean-type climate paradigm? In: Keeley SC (ed) The California chaparral: paradigms re-examined. Publication no. 34, Science Series, Natural History Museum of Los Angeles County, Los Angeles

Welsh L (1985) The demographic structure of Arizona chaparral. Honors thesis, Occidental College, Los Angeles

Brief Professional Biography of John L. Vankat

John L. Vankat received an A.B. degree in biology from Carleton College, Northfield, Minnesota and M.A. and Ph. D. degrees in botany from the University of California, Davis. He taught and researched plant ecology in the Department of Botany, Miami University, Oxford, Ohio for 31 years. John's interest in education led to his textbook *The Natural Vegetation of North America: An Introduction*. He completed 49 master's and Ph.D. students, directing field research projects from California to Pennsylvania and from Alaska to Belize. John helped organize and was elected the first person to chair the Vegetation Section of the Ecological Society of America. Additionally, he helped organize and later chaired the North American Section of the International Association for Vegetation Science. After retiring from Miami University, John moved to Flagstaff, Arizona, where he accepted a 4-year position as ecologist with Grand Canyon National Park to do research and bring research to bear on management issues. At present, John is Professor Emeritus from Miami University and both Senior Research Ecologist with the Merriam-Powell Center for Environmental Research and Adjunct Professor in the School of Forestry at Northern Arizona University in Flagstaff. He continues to research forests, woodlands, and shrublands in the American Southwest, focusing on vegetation dynamics, and continues his interest in using science to enhance management of natural areas. John has published in 20 peer-reviewed journals, including *Ecology*, *Forest Ecology and Management*, *Journal of Vegetation Science*, *Journal of Biogeography*, *Conservation Biology*, *Ecological Modelling*, and *Landscape Ecology*. He has presented invited lectures at 37 universities in 10 countries.

J.L. Vankat, *Vegetation Dynamics on the Mountains and Plateaus of the American Southwest*, Plant and Vegetation 8, DOI 10.1007/978-94-007-6149-0,

Author Index

A
Abella, S.R., 201, 212, 214, 226, 227, 230, 232, 234
Abolt, R.A.P, 83, 84, 91, 92, 103, 137, 143, 149, 162
Abruzzi, W.S, 42, 48
Ackerly, D.D, 411
Adams, H.D., 91, 208, 283, 285, 290
Alban, D.H., 101, 158
Alcoze, T., 40, 41, 233, 288
Alexander, B.G., 127, 128, 163, 194, 234
Alexander, M.E., 33, 147, 207
Alexander, R.R., 76, 80, 87, 88, 90, 94
Allen, C.D., 22, 25–27, 30, 33, 39–42, 47, 51, 81, 135, 137, 139, 141, 142, 159, 198–200, 204, 205, 208, 222, 231, 276, 279, 282, 283, 338, 342–344, 349, 352, 357, 358
Alward, R.D., 51
Anderson, R.S., 25, 38, 84, 88, 103
Andrews, S.R., 207
Antos, J.A., 350, 351
Aoki, C.F., 126, 137, 140–143, 158
Aplet, G.H., 94, 95
Arbaugh, M.J., 51
Arnold, J.F., 233, 234, 279, 297
Arundel, T.R., 228
Axelrod, D.I., 411

B
Bailey, J.D., 292, 293, 302, 305
Bailey, L.R., 40
Bailey, V.O., 222
Baisan, C.H., 25–27, 30, 35, 41, 46, 47, 81, 137, 139, 141, 199, 200, 279
Baker, F.S., 140
Baker, W.L., 27, 41, 52, 53, 87, 88, 90, 91, 152, 155, 201, 203, 204, 227, 230, 278, 279, 281, 285–288, 290, 292, 296, 297, 349, 351, 352, 357, 358, 383, 389
Bakker, J.D., 209
Balling, R.C., 50
Barger, N.N., 285, 287, 296, 302
Barlein, P.J., 38
Barnett, T.P., 50
Barney, M.A., 297
Barrett, S.W., 40
Barrows, J.S., 25–27, 44, 81, 137, 199, 203, 278
Barton, A.M., 219, 234
Bartos, D.L., 137, 149, 154, 162, 210
Bastian, H.V., 162
Bates, S.T., 75, 127, 193, 271
Battaglia, M.A., 82, 88, 92, 95, 145, 149, 162, 205, 209, 210, 214, 233, 235
Baxter, J.O., 40
Beale, E.F., 222, 223, 233, 354
Bebi, P., 86, 90, 91
Bell, D.M., 230
Belnap, J., 272
Belsky, A.J., 48
Bennett, D.D., 205
Benninger-Truax, M., 93, 150, 151
Bentz, B., 283
Beschta, R.L., 135, 198
Bestelmeyer, B.T., 57
Betancourt, J.L., 23, 25, 34, 37, 38, 41, 200, 282, 283, 285, 296, 297, 381
Beymer, R.J., 288, 300
Bigio, E., 137, 139, 142
Bigler, C., 91

J.L. Vankat, *Vegetation Dynamics on the Mountains and Plateaus of the American Southwest*, Plant and Vegetation 8, DOI 10.1007/978-94-007-6149-0,

Binkley, D., 48, 51, 88, 147
Biondi, F., 24, 228, 231
Blackford, D., 88
Böhm, M., 51
Bolander, D.H., 414, 416, 420, 421, 430
Bolton, H.E., 388
Bowman, C.B., 150, 212
Box, G.E.P., 54
Bradley, A.F., 81, 126, 127, 151–152, 200, 217, 219, 234, 344, 350, 357, 382, 387, 389
Breece, C.R., 33, 205, 207
Breshears, D.D., 32, 34, 35, 51, 205, 208, 276, 282, 283, 285, 290, 292, 297
Briske, D.D., 57
Brockway, D.G., 307
Brookes, M.H., 145
Brooks, M.I., 52, 416, 420, 421
Brown, D.E., 4, 6, 37, 40, 76, 80, 132, 198, 233, 234, 276, 277, 338, 339, 342–344, 372, 376, 379, 380, 390, 414, 416, 420, 421, 426
Brown, D.P., 24
Brown, H.E., 376, 379, 380, 383–385, 387
Brown, M., 19
Brown, P.M., 26, 101, 132, 137, 139, 140, 200, 216, 217, 285
Brown, T.C., 414, 416
Brugge, D., 40
Burwell, T., 41

C
Cable, D.R., 410, 413, 414, 420, 422, 426, 429
Carmichael, R.S., 408, 412, 416, 421, 422, 426
Cartledge, T.R., 286
Cayan, D.R., 50, 51
Chambers, C.L., 80, 104, 105, 133, 198, 339
Chambers, J.C., 276, 292
Chapin, F.S., 55
Chornesky, E.A., 52
Christensen, E.M., 300, 304, 306, 307, 376, 379, 380, 383, 388–390
Clary, W.P., 198, 231, 236, 307
Clements, F.E., 408
Cocke, A.E., 80, 92, 101, 102, 135, 157–159, 198
Cody, M.L., 411
Cole, K., 38
Cole, K.L., 38, 290
Conklin, D.A., 145, 147, 285
Coop, J.D., 338, 344, 347
Cooper, C.F., 48, 201, 216, 217, 233, 235
Covington, W.W., 194, 199, 204, 226, 228–232, 234, 235
Cranshaw, W., 380
Crawford, J.A., 83, 150, 162, 164, 214, 235
Crimmins, M.A., 26
Crocker-Bedford, D.C., 232
Cronk, Q.C.B., 52
Croxen, F.W., 420, 427, 429, 430

D
Dahms, C.W., 32–34, 40, 44, 88, 90, 92, 102, 103, 128, 197, 198, 204, 231, 232
D'Arrigo, R.D., 23
Das, T., 50
Davis, E.A., 410, 411, 421
Davis, J.R., 416
Davis, O.K., 4, 38, 296
de Paula Vasconcelos, L.A., 51
DeBano, L.F., 24, 408
deBuys, W., 42
DeByle, N.V., 76, 128, 143, 149, 152, 210
DeRose, R.J., 90, 94, 97
Despain, D.W., 279
DeVelice, R.L., 76, 80, 104, 128, 163, 194, 234
Dickey, A.M., 421, 426, 430
Dick-Peddie, W.A., 4, 152, 272, 274, 297, 338, 339, 344, 347, 374, 406
Dickson, B.G., 25, 27
Dieterich, J.H., 47, 128, 137, 139, 142, 147, 159, 416, 418, 421, 422, 430
Diffenbaugh, N.S., 51
Dixon, H., 88
Douglas. L.P., 205
DuBois, C., 233
Dugan, A.J., 27, 203, 204, 212, 216, 217
Dukes, J.S., 51
Dusenberry, W.H., 288
Dutton, C.E., 99, 222, 223, 226, 228
Dwyer, D.D., 281
Dye, A.J., 80, 103–105
Dyer, J.M., 338, 349, 352, 357, 358
Dymerski, A.D., 88, 90, 95, 97

E
Eager, T., 282
Ehleringer, J.R., 50
Elliott, G.P., 152
Ellison, L., 75, 79–81, 336, 339, 342–344, 352, 355–357
Elton, C., 52
Engle, D.M., 376
Erdman, J.A., 297, 384
Ernst, R., 286, 288
Evans, A.M., 128, 140, 141, 145, 147, 162, 197, 232
Evans, F.R., 383
Evans, R.B., 80, 133, 277, 379, 380

Evett, R.S., 22, 26, 27
Ex, S.A., 381

F

Fairweather, M.L., 88, 145, 147, 284, 285, 303
Farris, C.A., 27, 30
Feeney, S.R., 204, 232
Fellin, D.G., 144, 145
Fenn, M.E., 51, 52
Fenner, P., 292
Ffolliott, P.F., 4, 135, 231, 276, 277, 293, 420
Fisher, M.A., 93, 103, 104, 150, 163, 164
Fitzhugh, E.L., 194, 234
Fleischner, T.L., 357
Fletcher, R., 334, 338, 339, 349, 354, 357, 358
Floyd, M.E., 297, 377, 387
Floyd, M.L., 30, 279, 282, 283, 285, 287, 290, 291, 293, 376, 377, 381–385
Ford, P.L., 342, 343
Forsling, C.L., 383
Foxx, T.S., 234
Frechette, J.D., 30, 137, 142
Frey, B.R., 137
Fritts, H.C., 34
Fuchs, E.H., 293
Fulé, P.Z., 16, 27, 30, 41, 48, 49, 79–81, 85, 88, 91–93, 101–105, 126, 131, 132, 137, 139–143, 147, 150, 154, 157–160, 162–164, 200, 201, 210, 212, 219, 222, 226, 228–232, 234–236, 302, 389

G

Gaines, E.M., 217
Galloway, J.N., 52
Ganey, J.L., 147, 208
Getsinger, F.R., 88
Gill, R.A., 339
Gitlin, A.R., 283
Gordon, B.R., 293, 300, 304
Gordon, M.M., 300
Gosz, J.R., 101, 158, 227
Gottfried, G.J., 16, 127, 135, 143, 276, 277, 282, 286, 288, 291, 293
Grand Canyon National Park, 290
Gray, S.T., 296
Graybill, D.A., 51, 232
Greenamyre, H.H., 131, 163, 195
Griffis, K.L., 214, 235
Grissino-Mayer, H.D., 22, 41, 48, 81, 83, 137, 139–141, 200, 201, 229
Guenther, D., 288, 291, 302

H

Haire, S.L., 23, 219, 292
Hall, B.L., 22, 26, 27
Hanks, J.P., 152, 194, 210, 234
Hanley, D.P., 80, 85
Harper, K.T., 219, 277, 376, 377, 379, 380, 383–385, 387, 390
Harrington, M.G., 33, 199, 207, 229, 231
Harris, A.T., 287, 288, 296, 302
Harris, G.R, 204
Hasbargen, J., 38
Haskett, B., 39, 42
Hayward, C.L., 377, 380, 383
Hebertson, E.G., 25, 33, 90
Heinlein, T.A., 137, 139, 140, 157–159, 162
Hendricks, D.M., 24, 80, 135, 198, 380
Hereford, R., 23
Heyerdahl, E.K., 157, 226
Hibbert, A.R., 410–414, 418, 421, 422, 424, 430
Hicke, J.A., 91, 141
Higgins, R.W., 20
Hoerling, M., 50, 51
Hoffman, C., 207
Holden, Z.A., 141, 142, 200, 204
Holsinger, S.J., 222, 233
Hood, S.M., 127, 288, 291
Houghton, J.T., 50
Huebner, C.D., 412, 414, 430
Huffman, D.W., 76, 88, 128, 147, 203, 234, 279, 283, 296–298, 302, 305, 306, 338, 343, 344, 346, 347, 349, 352, 358
Huisinga, K.D., 150, 163, 164
Hurst, M., 104, 164

I

Ingersoll, E., 233
Iniguez, J.M., 139, 200, 201, 203, 219
Integrated Taxonomic Information System, 5, 75, 127, 193, 271, 336, 375, 407
Intergovernmental Panel on Climate Change, 50, 290
Ireland, K.B., 200
Ives, J.C., 222

J

Jacobs, B.F., 272, 276, 285, 288, 292, 293, 296, 303, 306
Jameson, D.A., 281, 288, 293, 297, 300, 302, 306, 307
Jenkins, M.J., 25, 32, 33, 90, 91, 95, 139, 141
Jenkins, S.E., 137, 141, 142, 203, 418
Jenny, H., 55

Johnsen, T.N., 276, 430
Jones, J.R., 76, 128, 143, 152

K
Karnosky, D.F., 92, 148, 212
Kaufmann, G.A., 232
Kaufmann, M.R., 199, 217
Kay, C.E., 26, 40
Kaye, M.W., 41
Kearns, H.S.J., 227, 283, 284, 303
Keeley, J.E., 199, 410, 411, 425
Kellogg, R.S., 223
Kerns, B.K., 233, 235
Kleinman, L.H., 154
Klemmedson, J.O., 24–25, 80, 132, 135, 198, 277, 339, 343
Knapp, A.K., 95
Knight, R.L., 53
Knipe, O.D., 408, 410
Koepke, D.F., 208, 285, 293
Kohler, T.A., 286, 296
Kolb, T.E., 91, 204, 208
Korb, J.E., 150, 163, 164, 204, 210, 214, 233, 235
Kuenzi, A.M., 214, 234
Kulakowski, D., 86, 87, 95
Kunzler, L.M., 377, 384, 385, 387
Kurzel, B.P., 154

L
Laing, L., 342, 343
Landis, A.G., 292, 293, 302, 305
Lang, D.K., 79, 83, 99, 101, 131, 155, 156, 163, 205, 216, 223, 226, 228
Laughlin, D.C., 49, 93, 103–105, 143, 150, 158, 163, 164, 214, 234, 236
Layser, E.F., 76, 128, 154, 194
Leaf, C.F., 80
Leiberg, J.B., 52, 75, 81–83, 99, 104, 155, 163, 354
Lenart, M., 50
Leopold, A., 297, 416, 420, 429
Lewis, D.B., 229
Lewis, M.E., 357
Lindenmuth, A.W., 416
Link, S.O., 52
Linnane, J.P., 144, 145
Logan, J.A., 51
Longwell, C.R., 223
Lowe, C.H., 4, 6, 16, 37, 75, 76, 81, 88, 95, 132
Lynch, A.M., 93, 136, 142, 144, 145, 162

M
MacMahon, J.A., 127, 376
Madany, M.H., 228, 302, 377, 382, 383, 388–390
Major, J., 55
Maker, H. J., 24, 25, 135, 198, 343
Mann, W.G., 44
Mantua, N.J., 24
Margolis, E.Q., 26, 30, 41, 81–83, 95, 103, 137, 139–142, 154, 162, 200, 201
Marshall, J.D., 51
Martens, S.N., 276
Mason, L.R., 277, 302
Mast, J.N., 128, 137, 139, 147, 159, 162, 197, 216, 217, 219, 222, 226, 228, 231, 232, 338
Mathiasen, R.L., 145
McClaran, M.P., 420
McDowell, N., 283
McGarigal, K., 23, 53, 219, 292
McGlone, C.M., 52, 212, 214
McHenry, D.E., 147, 339
Mead, P., 48
Meagher, G.S., 292
Meko, D.M., 22
Menakis, J.P., 52
Menzel, J.P., 226, 230, 231
Merkle, J., 82, 88, 104, 147, 162, 164, 338, 344, 347
Merola-Zwartjes, M., 349, 357
Merriam, C.H., 6, 127, 222, 223, 233
Milchunas, D.G., 92, 105, 149, 209, 210, 276–278, 286–288, 302, 307, 342, 344, 357
Miller, G., 343
Miller, J.M., 205
Miller, K.K., 205
Miller, M.E., 54, 277, 293, 307
Miller, P.R., 51, 205
Miller, R.F., 293, 296, 297
Miller, W.E., 101, 158
Mitchell, J., 48
Moir, W.H., 75, 76, 80, 81, 88, 95, 97, 103–105, 126–128, 133, 135, 136, 145, 147, 154, 159, 162, 163, 198, 199, 201, 208, 216, 219, 229, 233, 234, 272, 274, 338, 343
Möllhausen, B., 223
Mooney, H.A., 51, 411
Moore, M.M., 76, 88, 101, 128, 147, 158, 209, 227, 229, 230, 234, 235, 338, 343, 344, 346, 347, 349, 352, 358
Mueggler, W.F., 76, 104, 128, 149, 154, 163, 210

Mueller, R.C., 283, 285
Muldavin, E.H., 76, 104, 126, 128, 163, 194, 234
Munson, S.M., 290
Murdock, L., 38

N
Naiman, R.J., 42
National Park Service, 45, 51–53, 150, 212
National Weather Service, 23
Negrón, J.F., 205, 283, 285
Neilson, R.P., 376, 377, 387
Nelson, G.C., 13
Niering, W.A., 75, 76, 81, 88, 95, 135
Nixon, E.S., 383
Nowak, R.S., 285

O
O'Brien, R.A., 128
O'Dell, T.E., 54
Ogle, K., 83, 283
Omi, P.N., 382
Orcutt, J.D., 39
Ott, J.E., 291, 297
Oukrop, C.M., 154

P
Paine, T.D., 32
Pase, C.P., 76, 80, 198, 233, 234, 410, 414, 416, 418, 420–422, 424, 426
Patten, D.T., 75, 81, 88, 95, 97, 104, 164
Patton, D.R., 339, 344, 347, 357
Pauchard, A., 93, 150
Paulsen, H.A., 198, 276, 277, 307, 429, 431
Paulson, D.D., 351, 383
Paysen, T.E., 297
Pearson, G.A., 80, 133, 135, 143, 152, 194, 204, 217, 231, 338, 342
Pearson, K.M., 292
Peet, R.K., 25, 78, 343
Peterman, W., 283
Peters, M., 288
Petersen, K.L., 20, 21, 38
Peterson, C.S., 40
Pfister, R.D., 76, 79–81, 88, 95, 154
Phillips, A.M., 6
Pieper, R.D., 281, 282, 286, 288, 291, 307, 416
Pinchot, G., 41, 222
Plummer, F.G., 131, 155, 156, 195
Pollock, M.M., 223
Pond, F.W., 418, 420, 422, 424, 426
Poreda, S.F., 384, 385
Potter, L.D., 210, 288
Poulos, H.M., 281, 290
Powell, J.W., 46, 155, 222
Prevedel, D.A., 355, 357
Price, R., 80, 133, 277, 379, 380
Prior-Magee, J.S., 18, 72, 128, 192, 270, 336, 374, 406
Pyne, S.J., 26, 40, 42, 46

R
Raish, C., 49, 209, 288
Rambo, J.L., 357
Randall, J.M., 52
Rasmussen, D.I., 44, 147, 339
Rebertus, A.J., 95
Rehfeldt, G.E., 147
Rejmánek, M., 52
Reynolds, A.C., 41
Reynolds, H.G., 149, 210
Reynolds, R.V.R., 347, 383
Rietveld, W.J., 338
Roccaforte, J.P., 219, 222
Rogers, G.G., 304, 389
Rogers, P.C., 154
Rogers, T.J., 283
Rollins, M.G., 201
Romme, W.H., 41, 51, 76, 78, 82, 83, 88, 90, 91, 95, 103, 126–128, 140, 141, 143, 145, 152, 154, 210, 219, 226, 233, 234, 272, 274, 278, 281–288, 290, 293, 294, 296, 297, 303, 308, 339, 344, 347, 349, 350, 355, 357, 372, 377, 379, 382
Roos, C.I., 40, 41, 203, 208
Rothrock, J.T., 222, 354
Rowlands, P.G., 297, 300, 302, 306, 307
Rusby, H.H., 233, 354
Russo, J.P., 88
Ryerson, D.E., 144, 145

S
Sackett, S.S., 194, 204, 214, 229, 231, 233, 235
Sallach, B.K., 293
Salzer, M.W., 22, 37
Samuels, M.L., 296
Sánchez Meador, A.J., 216, 217, 226, 230
Sánchez-Martínez, G., 205
Santos, M.J., 282, 283
Saunier, R.E., 410, 430
Savage, M., 48, 216, 217, 219, 222, 229, 338
Sawyer, G.S., 44
Schlanger, S., 286

Schlegel, P.A., 39, 42
Schmid, J.M., 90, 94, 97, 205
Schmutz, E.M., 300, 302
Schoennagel, T., 87
Schreiber, K.V., 51
Schubert, G.H., 76, 128, 154, 194, 198
Scurlock, D., 40, 42, 214
Seager, R., 23, 50, 51
Seklecki, M.T., 41, 430
Shaw, H.G., 278, 293, 296, 300, 304, 307
Shaw, J.D., 282–285, 293, 297
Shepperd, W.D., 82, 88, 90, 92, 95, 145, 147, 149, 162, 205, 209, 210, 214, 233, 235
Shinneman, D.J., 278, 279, 281, 285–288, 290, 292, 296, 297
Sibold, J.S., 82
Sieg, C.H., 214, 291
Smith, E.A., 152
Smith, S.D., 51
Sneed, P., 418
Soulé, P.T., 290
Spence, J.R., 6, 19, 20, 50, 376
Springfield, H.W., 271, 276, 277
St. George, S., 26
Stanley, R.H., 300
Stein, S.J., 139
Steinhoff, H.W., 380
Stephens, S.L., 46, 212
Stoddard, M.T., 214
Stone, J.E., 232
Strom, B.A., 219, 222
Stromberg, J.C., 75, 81, 88, 95, 97, 104, 164
Sudworth, G.B., 388
Sullivan, A.P., 288
Swetnam, T.W., 23, 25–27, 30, 34, 35, 37, 41, 44, 46–48, 81, 83, 92, 128, 136, 137, 139–142, 144, 145, 149, 162, 197, 199, 200, 203, 229, 232, 279, 282, 283, 285, 286, 381

T

Takemoto, B.K., 51
Tausch, R.J., 288, 291, 293, 297
Taylor, K., 93, 151
Tew, R.K., 380
Thatcher, A.P., 277, 300
The Nature Conservancy, 57
Theobald, D.M., 54
Thomas, L.P., 54
Thompson, R.S., 38
Tiedemann, A.R., 376
Touchan, R., 48, 137, 139, 142, 200
Tress, J.A., 297

U

U.S. Forest Service, 30, 33, 90, 145, 204, 205, 282, 344, 350, 380, 414, 420, 424, 426, 429, 430

V

Vale, T.R., 26, 39–41, 286
Valiente-Banuet, A., 411
Van Devender, T.R., 38, 41, 296
Van Epps, G.A., 376
Van Pelt, N.S., 300, 302
Vankat, J.L., 16, 49, 52, 54, 76, 79, 82, 85, 92, 99, 102, 103, 126, 128, 131, 132, 136, 143, 147, 149, 154–156, 159, 161–163, 194, 219, 226, 228–232, 411, 412, 414, 430
Veblen, T.T., 81, 82, 86–88, 90, 91, 94, 95
Villanueva-Díaz, J., 127, 217
Vitousek, P.M., 52

W

Wadleigh, L.L., 54, 381, 383–385
Wagner, D.J., 52
Wallin, K.F., 228, 232
Warren, P.L., 336, 343
Weaver, H., 199, 200, 214, 229, 231, 235
Weckmann, L., 288
Wells, F.H., 93, 150
Welsh, L., 418
Weng, C., 20, 38, 86, 199
Westerling, A.L., 51, 279
Western Regional Climate Center, 21, 380
White, A.S., 194, 216, 217
White, C.S., 204
White, M.A., 16, 76, 79, 82, 92, 102, 103, 126, 128, 131, 132, 136, 143, 147, 149, 159, 162, 163
White, M.R., 339, 342, 349, 356, 357
White, P.S., 35
Whittaker, R.H., 135
Wildeman, G., 39, 40, 42
Williams, A.P., 35, 51
Williams, M.A., 203, 204, 227, 230
Wilson, J.L., 205, 283, 285
Witt, C., 282, 283, 285, 293
Wolf, J.J., 128, 137, 139, 147, 159, 162, 197, 232

Wolters, G.L., 344, 346, 347, 349
Wood, R.E, 208
Woolsey, T.S., 47, 48, 201, 204, 208, 216, 223, 226
Worrall, J.J., 88, 147
Wright, J.P., 42
Wu, R., 126, 130, 132, 137, 139–141, 143, 149, 152, 154, 216, 217
Wurtz, M.J., 40, 42
Wyman, W.D., 24, 300

Y
Yeager, L.E., 104
Youngblood, A.P., 76, 104, 126, 128, 163, 194, 234

Z
Zausen, G.L., 205, 232
Zier, J.L., 155, 349, 352, 357, 358, 389
Ziska, L.H., 291

Subject Index

A

Abies. *See* Fir
  A. concolor (*see* White fir)
  A. lasiocarpa var. *arizonica* (*see* Corkbark fir)
  A. lasiocarpa var. *lasiocarpa* (*see* Subalpine fir)
Acer grandidentatum. *See* Bigtooth maple
Acid precipitation, 52
Acoma Pueblo, 294
Agave, 407, 410
Agriculture, 39, 41, 54, 347
Agrilus quercicola. *See* Flathead borer
Air pollution, 48, 51, 53, 55, 58, 92, 117, 150, 179, 212, 252
Alligator juniper, 271, 272, 274, 407, 428
Amelanchier. *See* Serviceberry
  A. alnifolia (*see* Western serviceberry)
  A. utahensis (*see* Utah serviceberry)
Andropogon. *See* Bluestem
Annosus root rot, 127, 135, 144, 145, 193, 198, 207
Annual (plant), 105, 288, 291, 297, 307, 339, 355, 421, 425
Antelope bitterbrush, 271, 297, 375, 376
Antilocapra americana. *See* Pronghorn antelope
Apache, 41, 43, 44, 219, 345, 388, 430
Apache National Forest, 28, 156, 157, 339, 342, 349, 356, 357
Apache Pine, 193, 195, 230
Arceuthobium
  A. divaricatum (*see* Pinyon dwarf mistletoe)
  A. microcarpum (*see* Western spruce dwarf mistletoe)
  A. vaginatum spp. *cryptopodum* (*see* Ponderosa pine dwarf mistletoe)
Archips fumiferana. *See* Western spruce budworm
Arctostaphylos. *See* Manzanita
  A. pringlei (*see* Pringle's manzanita)
  A. pungens (*see* Pointleaf manzanita)
Aristida. *See* Threeawn
Arizona fescue, 336, 339, 343, 350, 351, 355
Arizona pine, 193, 195
Arizona Strip, 40
Arizona white oak, 193, 195, 407, 418
Armillaria. *See* Armillaria root rot
Armillaria root rot, 127, 135, 144, 193, 198, 207, 271, 277, 281, 283
Artemisia. *See* Sagebrush
  A. tridentata (*see* Big sagebrush)

B

Baccharis salicifolia. *See* Seepwillow baccharis
Bandelier National Monument, 49, 205
Bark beetle, 5, 32–35, 45, 75, 81, 88, 108, 109, 147, 193, 199, 205, 208, 240, 271, 277, 281–285, 295, 312, 322
Beaver, 5, 42
Biennial (plant), 105, 339
Big bluegrass, 271, 307
Big sagebrush, 271, 272, 297
Bigtooth maple, 375–377, 382, 384, 387–389, 396–398
Bill Williams Mountain, 222
Birchleaf mountain mahogany, 407, 408, 418, 420, 426
Bison bison. *See* Buffalo

J.L. Vankat, *Vegetation Dynamics on the Mountains and Plateaus of the American Southwest*, Plant and Vegetation 8, DOI 10.1007/978-94-007-6149-0,

Blackfoot, 407, 429
Black stain root disease, 271, 277, 281, 283, 284, 286, 303
Blepharoneuron. *See* Dropseed
Blowdown, 30, 32, 87–91, 94, 95, 108, 110, 112, 113, 115, 143, 144, 204
Bluebunch wheatgrass, 271, 307
Blue grama, 271, 272
Blue spruce, 5, 8, 75, 76, 93, 102, 124, 127, 128, 143, 155, 159
Bluestem, 407, 431
Bobcat, 5, 44
Boer lovegrass, 407, 421
Bos taurus. *See* Cattle
Bothriochloa. *See* Bluestem
  B. barbinodis (*see* Cane bluestem)
Bouteloua. *See* Grama
  B. eriopoda (*see* Crowfoot grama)
  B. gracilis (*see* Blue grama)
Bristlecone pine, 75, 79, 99, 101
Brome, 336, 347, 349, 351, 375, 390, 407, 421
Bromus. *See* Brome
  B. inermis (*see* Smooth brome)
  B. rubens (*see* Red brome)
  B. tectorum (*see* Cheatgrass)
Buffalo, 271, 287
Bunchgrass, 339, 350, 351, 354, 355, 357, 368, 407, 429

C
Caelifera. *See* Grasshopper
Canada thistle, 271, 291
Cane bluestem, 407, 410
Canis
  C. latrans (*see* Coyote)
  C. lupis (*see* Wolf)
Canyon de Chelly National Monument, 300
Capra hircus. *See* Goat
Capulin Volcano National Monument, 378
Carduus nutans. *See* Musk thistle
Carex. *See* Sedge
  C. foenea (*see* Dry spike sedge)
Castor canadensis. *See* Beaver
Catclaw acacia, 407, 418, 426
Catclaw mimosa, 407, 410, 413
Cattle, 5, 39, 40, 47, 336, 347, 348, 356, 357, 375, 383, 407, 420
Ceanothus
  C. fendleri (*see* Fendler's ceanothus)
  C. greggii (*see* Desert ceanothus)
  C. integerrimus (*see* Deerbrush ceanothus)
Cedar Mountain, 154
Cercocarpus. *See* Mountain mahogany
  C. betuloides (*see* Birchleaf mountain mahogany)
  C. ledifolius (*see* Curl-leaf mountain mahogany)
  C. montanus (*see* Mountain mahogany)
Cervus elaphus. *See* Elk
Chaco Canyon, 41
Chaparral. *See* Interior Chaparral Shrubland
Cheatgrass, 127, 150, 162, 164, 193, 214, 271, 291, 307, 309, 314–320, 375, 384
Chenopodium. *See* Goosefoot
Chihuahuan pine, 193, 195, 219
Chiricahua Mountains, 41, 223
Chokecherry, 233, 375, 376, 382, 388
Cholla, 407, 410
Church Mesa, 302
Chuska Mountains, 229
Cirsium arvense. *See* Canada thistle
Cliff fendlerbush, 375, 376, 379, 385
Cliff-rose, 375, 376
Climate
  change, 25, 38, 48, 50–52, 55, 58, 80, 91–93, 102, 108–110, 117, 130, 135, 148–150, 161, 165, 168, 169, 179, 198, 208, 212, 222, 231, 237, 240, 241, 252, 277, 286, 290, 297, 308, 312, 313, 315, 317, 318, 320, 322, 343, 347, 349, 354, 359, 362, 363, 365, 366, 380, 383, 384, 391, 394, 398, 414, 419, 421, 430, 434, 435, 437
  model, 50
  variation, 34, 35, 41, 58, 81, 91, 117, 147, 199, 208, 283, 285, 286, 293, 307, 322, 418
Coconino National Forest, 16, 43, 82, 136, 206–210, 218, 221, 225, 226, 419
Coconino Plateau, 3, 24, 203
Coloradia pandora. *See* Pandora moth
Colorado pinyon, 2, 5, 193, 195, 271, 272, 274, 279, 281, 285, 287, 290, 292, 297, 299, 302, 303, 305, 306, 375, 376
Colorado Plateau, 4, 6, 14, 38–40, 42, 50, 288
Common dandelion, 193, 214
Common mullein, 193, 214
Common salsify, 193, 214
Conceptual model (introduction), 4, 13, 35, 36, 48, 54, 56
Conversion, 54, 221, 222, 252, 282, 291, 411, 420
Corkbark fir, 5, 6, 75, 76, 78, 82, 93, 95, 97, 99, 101, 104, 110–115, 124, 127, 128, 130, 132, 143, 159, 160, 162, 164
Coronado, Francisco Vásquez de, 39

Coronado National Forest, 429
Coyote, 5, 44
Crowfoot grama, 407, 429
Curl-leaf mountain mahogany, 375, 376
Curly mesquite, 407, 431
Cygnus Mountain. *See* Mount Hope

D

Dactylis glomerata. *See* Orchardgrass
Dalmatian toadflax, 193, 214
Dandelion. *See* Common dandelion
Danthonia. *See* Oatgrass
Deer, 5, 25, 39, 41, 48, 75, 81, 88, 109, 127, 136, 146, 147, 152, 154, 168, 169, 193, 199, 208, 241, 271, 277, 336, 343, 346, 347, 351, 357, 375, 380, 407, 414
Deerbrush ceanothus, 407, 420
Defoliating insect, 32, 33, 151
Dendroctonus. *See* Bark beetle
  D. brevicomis (*see* Western pine beetle)
  D. ponderosae (*see* Mountain pine beetle)
  D. pseudotsugae (*see* Douglas-fir beetle)
  D. rufipennis (*see* Spruce beetle)
Deposition, 52, 86, 359, 362
Desert Archaic peoples, 39
Desert ceanothus, 407, 408, 418, 420, 422, 424, 425
Desert scrub, 404, 420, 429
Disturbance regime, 16, 27, 30, 51, 55, 81, 93, 117, 127, 158, 199, 272, 275, 278, 344, 387, 426
Dixie National Forest, 36, 289
Douglas-fir, 2, 5, 8, 9, 33, 75, 76, 93, 124, 127, 128, 143–145, 147, 151, 152, 155, 156, 159, 160, 162–164, 170, 173, 174, 177, 193, 194, 201, 336, 352, 375, 376, 386, 387, 396–398
Douglas-fir beetle, 127, 135, 144, 145
Douglas-fir tussock moth, 127, 136, 144, 145
Dropseed, 407, 431
Drought (introduction), 4, 23, 25, 33–35, 50, 51
Dryocoetes confusus. *See* Bark beetle
Dry spike sedge, 127, 162, 164
Dwarf mistletoe, 5, 32–34, 75, 88, 93, 127, 144, 145, 147, 193, 205, 207, 271, 281, 283, 285

E

Ecosystem-characterization model (introduction), 54–57
Ecosystem convergence, 411
Ecotone, 131, 205, 208, 276, 279, 338, 421, 430
Elatobium abietinum. *See* Spruce aphid
Elevation (introduction), 6, 9–12, 14, 16, 19–22, 24, 27, 35, 38, 39, 41, 42, 48, 53
Elk, 5, 25, 39, 41, 127, 136, 146, 147, 152, 154, 168, 169, 193, 199, 208, 241, 271, 277, 336, 344, 346, 347, 351, 356, 357, 375, 380
El Malpais National Monument, 229
El Niño-Southern Oscillation (ENSO), 22, 23, 34, 200, 277
Emissions, 50
Emory oak, 407, 418, 420
Enchanted Mesa, 294
Encroachment. *See* Tree encroachment
Endangered Species Act, 45
Engelmann spruce, 5, 6, 36, 75, 76, 78, 79, 81, 82, 86–88, 90, 91, 93–97, 99, 101–105, 108, 110–115, 124, 127, 128, 130, 143, 144, 151, 162, 164, 170, 173, 336, 352
ENSO. *See* El Niño-Southern Oscillation (ENSO)
Eragrostis
  E. chloromelas (*see* Boer lovegrass)
  E. curvula (*see* Weeping lovegrass)
  E. lehmanniana (*see* Lehmann lovegrass)
Erodium cicutarium. *See* Stork's bill
Euro-American (introduction), 39–42, 44–48, 52, 57, 58
Expansion, 35, 37, 58, 284, 292–297, 303, 306–309, 318, 322, 349, 354, 387, 389

F

Felis concolor. *See* Mountain lion
Fendlera rupicola. *See* Cliff fendlerbush
Fendler's ceanothus, 407, 425, 426
Fescue, 5, 11, 336, 339, 343, 344, 350, 351, 355, 357
Festuca. *See* fescue
  F. arizonica (*see* Arizona fescue and pine bunchgrass)
  F. thurberi (*see* Thurber fescue)
Fir, 2, 5, 75, 127, 193, 271, 336, 375, 407
Fire
  crown (introduction), 27–30, 35, 49
  exclusion (introduction), 4, 25, 27, 42, 44, 46, 48, 49, 57
  history, 30, 46, 47, 87, 93, 137, 149, 150, 201, 204, 229, 234, 278, 307, 381, 418
  individual fire intervals, 140, 141, 201
  management (introduction), 48, 49, 55, 58

Fire (*cont.*)
mixed severity, 27, 29, 30, 83–87, 103–105, 108, 117, 137, 141, 143, 147, 149, 150, 151, 159, 162, 168, 179, 201, 203, 204, 251
Native-American use, 40, 41, 199, 384
prescribed, 46, 47, 49, 52, 92, 109, 149, 150, 169, 204, 212, 214, 217, 231, 234, 241, 349, 366, 416, 421
regime (introduction), 27, 30, 38, 40, 41, 46, 47, 51
return interval, 92, 179, 278, 281, 344, 416, 418, 437
rotation (*see* Fire, Turnover)
scar, 30, 140, 416, 418
stand replacing, 27, 82, 83, 279, 291, 296, 297, 306, 307, 309, 312, 315, 317, 318, 320, 338, 365, 366, 390, 416
suppression, 42, 44, 45, 49, 117, 149, 159, 229, 278, 290, 349, 363, 381, 384, 394, 421, 434
surface (introduction), 27–30, 33, 47
turnover, 27, 84, 102, 203, 279, 381, 382, 398, 437
Fir engraver, 127, 135, 144
Fishlake National Forest, 13
Fishtail Mesa, 300, 306
Flagstaff, 228, 235, 282
Flathead borer, 375, 380
Forest Service. *See* U.S. Forest Service
Fort Valley Experimental Forest, 156, 157, 210, 211, 225, 287, 342, 348, 410, 428
Full Glacial, 38
Fungi, 32, 34, 75, 88, 105, 127, 147, 193, 198, 207, 208, 271, 272, 284

G

Gambel Oak Shrubland
northern Gambel Oak Shrubland, 377, 378, 391, 396–398
southern Gambel Oak Shrubland, 376, 378, 387, 391, 395, 396
Gap dynamics, 35, 37, 94, 97, 98, 110, 112, 113, 115, 143, 145, 151, 152, 155, 171, 173, 175, 177, 216, 217
Garrya
G. flavescens (*see* Yellowleaf silktassel)
G. wrightii (*see* Wright's silktassel)
Gila National Forest, 202, 213, 280, 345, 428, 429
Gila Wilderness, 83, 84, 141
Glen Canyon National Recreation Area, 279, 287, 290, 291
Goat, 5, 40, 407, 420
Goosefoot, 271, 288
Grama, 271, 272, 407, 410, 429, 431
Grand Canyon. *See* Grand Canyon National Park
Grand Canyon National Park
North Rim, 14, 15, 20, 24, 31, 80, 85, 98, 101–104, 117, 129, 130, 134, 139–142, 147, 150, 154, 155, 157, 158, 160–164, 179, 194, 201, 220, 228, 230, 234, 335, 343, 346, 351, 352, 358
South Rim, 15, 24, 47, 196, 203, 212, 222, 273, 276
Grand Staircase-Escalante National Monument, 283
Grasshoppers, 336, 344
Gray oak, 193, 195
Grazing (introduction), 41, 42, 44, 47–49, 54, 55, 58
Great Basin, 271, 293
Greatheart Mesa, 302
Greenhouse gas, 50, 51
Gus Pearson Natural Area, 228, 230, 231

H

Helianthus. *See* Sunflower
Herder, 46, 47, 430
Heterobasidion annosum. *See* Annosus root rot
Hilaria belangeri. *See* Curly mesquite
Hollyleaf buckthorn, 407, 408, 420, 422
Holocene, 25, 38, 84, 142, 203, 293, 418
Hunting, 39–41, 208, 286, 347, 383, 419, 421
Hydrology, 48

I

Ignacio, 233
Indian. *See* Native American
Indiangrass, 271, 288
Infill, 35, 37, 292–295, 296–297, 303, 304, 306–309, 315, 317, 318, 320, 322, 352, 354, 365, 366, 387, 396, 398
Insect (introduction), 27, 32, 33, 35, 41, 45, 51, 58
Interior Chaparral Shrubland, 6, 11, 12, 17, 37, 190, 197, 275, 375, 379, 385, 404–438
Intermediate wheatgrass, 336, 347, 349
Invasive (introduction) 4, 48, 49, 51, 52, 54, 55, 58
Ips. *See* Bark beetle and Ips engraver beetle
I. confusus (*see* Pinyon bark beetle)
Ips engraver beetle, 193, 205

J
Jemez Mountains, 22, 30, 135, 141, 142, 198, 204, 205, 222, 342–344, 349, 352, 358
Juniper (introduction), 2, 5, 6, 9, 10, 12, 15, 27, 37, 39, 41
Juniper mistletoe, 271, 285
Juniperus. *See* Juniper
  J. deppeana (*see* Alligator juniper)
  J. monosperma (*see* Oneseed juniper)
  J. occidentalis (*see* Western juniper)
  J. osteosperma (*see* Utah juniper)
  J. scopulorum (*see* Rocky Mountain juniper)

K
Kaibab National Forest, 29, 45, 287
Kaibab Plateau, 3, 20, 24, 44, 48, 49, 83, 88, 99, 131, 149, 151, 153, 155, 156, 163, 205, 216, 222, 223, 226, 228
Kendrick Mountain, 142, 203
Kentucky bluegrass, 193, 214, 336, 342, 347, 349, 351, 375, 383, 384, 390

L
Lactuca serriola. *See* Wild lettuce
Lambdina
  L. punctata (*see* Looper)
  L. vitraria (*see* Looper)
Landscape (introduction), 2, 4, 12–14, 16, 27, 30, 37, 39–41, 46, 49, 53, 55, 56, 58
Land use (introduction), 4, 13, 39–42, 46, 48, 52, 54, 55, 57, 58
Late Glacial, 38
Lehmann lovegrass, 407, 421
Leptographium wageneri var. *wageneri*. *See* Black stain root disease
Lightning, 22, 24–27, 38, 41, 44, 49, 80, 81, 87, 92, 108, 109, 135, 137, 149, 168, 169, 198, 199, 205, 212, 229, 240, 241, 278, 312, 344, 362, 380, 381, 394, 414, 416, 434
Lignotuber, 376
Limber pine, 5, 8, 75, 79, 124, 127, 143
Linaria dalmatica. *See* Dalmatian toadflax
Lincoln National Forest, 28, 43, 53, 144
Livestock (introduction), 39–42, 46–49, 54, 55, 58
Lodgepole pine, 75, 76
Logging, 4, 42–44, 46, 48, 54, 91, 148, 203, 209, 216, 288, 390
Looper, 375, 380
Lumber, 42, 156, 199, 216, 226
Lynx rufus. *See* Bobcat

M
Malacosoma californicum. *See* Western tent caterpillar
Manzanita, 193, 219, 407, 408, 418, 419, 422, 424–426
Mazatzal Mountains, 408
Mechanistic model (introduction), 54, 56, 57
Melampodium longicorme. *See* Blackfoot
Merriam, Clinton Hart, 6
Mesa Verde National Park, 33, 52, 279, 287, 290, 291, 293, 297, 380–384
Mescalero Apache Reservation, 43, 44
Mexican pinyon, 271, 272, 281
Mexico, 271, 408, 411
Mimosa aculeaticarpa var. *biuncifera*. *See* Catclaw mimosa
Mixed Conifer Forest
  dry-mesic Mixed Conifer Forest, 129–131, 140–143, 155, 157, 173–175, 339, 344, 390
  moist-mesic Mixed Conifer Forest, 129, 130, 156, 158, 165, 170, 171, 173, 339, 344
  Quaking aspen Mixed Conifer Forest, 130
Model (introduction), 4, 13, 35, 36, 48, 50, 54–57
Monsoon, 19, 21–24, 26, 27, 38, 76, 108, 137, 168, 198, 200, 240, 272, 276, 277, 414
Mosaic, 8, 10, 73, 75, 79, 85, 99, 127, 131, 132, 136, 143, 159, 162, 195, 197, 275, 345, 382, 412, 414, 418, 419, 421, 430, 431, 437
Mountain lion, 5, 44
Mountain mahogany, 5, 271, 272, 274, 297, 372, 375, 376, 379, 407, 408, 413, 418, 420, 426
Mountain muhly, 336, 339
Mountain pine beetle, 127, 135, 144, 145
Mountain snowberry, 375, 376, 385
Mount Hope, 301
Mt. Ord, 429
Muhlenbergia
  M. montana (*see* Mountain muhly)
  M. virescens (*see* Screwleaf muhly)
Mule deer, 5, 25, 127, 136, 147
Mullein. *See* Common mullein
Musk thistle, 271, 291

N
National Environmental Policy Act (NEPA), 45
National Forest, 26, 42, 44, 81, 137, 199, 208, 230, 278
National Forest Management Act, 45
National Park, 4, 45, 229

National Park Service, 3, 98, 290, 300
National Park System, 4, 42, 48, 52
Native American, 39–42, 47, 58, 91, 148, 199, 208, 286, 288, 296, 300, 347, 382–384, 419
Navajo, 40
Neodiprion. *See* Pine sawfly
NEPA. *See* National Environmental Policy Act (NEPA)
New Mexico locust, 193, 219, 222, 230, 232, 375, 376
Nitrogen, 51, 52, 81, 204, 232
No Man's Mesa, 300, 302
Northern pocket gopher, 336, 344
North Fork Cañon, 224

O
Oak. *See Individual species*
Oatgrass, 336, 339
Odocoileus. *See* Deer
O. hemionus (*see* Mule deer)
O. virginianus (*see* White-tailed deer)
Off-highway vehicle (OHV), 93, 151, 215, 292, 350
OHV. *See* Off-highway vehicle (OHV)
Old growth, 45, 76, 81, 87, 95, 97, 101, 110, 112, 113, 115, 142, 159, 171, 173, 175, 177, 210, 216, 237, 243, 244, 246, 247, 249, 252
Oneseed juniper, 271, 272, 281, 294, 299
Opuntia. *See* Cholla and prickly pear
Orchardgrass, 336, 347, 349
Oregon, 20, 21, 23, 214, 290
Orgyia pseudotsugata. *See* Douglas-fir tussock moth
Ovis aries. *See* Sheep
Ozone, 51, 52, 92, 148, 150, 212

P
Pacific Decadal Oscillation, 24
Pagosa Springs, 233
Paiute, 41
Pajarito Plateau, 16
Paleo Indians, 39
Palmer Drought Severity Index (PDSI), 25, 26
Pandora moth, 193, 205
Parent material, 24, 55, 80, 81, 135, 198, 234, 277, 380, 414
Pascopyrum smithii. *See* Western wheatgrass
PDSI. *See* Palmer Drought Severity Index (PDSI)
Perennial (plant), 233, 297, 339, 355, 357, 384, 420, 422, 429
Petran chaparral, 372, 408
Phleum pratense. *See* Timothy
Phoradendron juniperinum. *See* Juniper mistletoe
Picea. *See* Spruce
P. engelmannii (*see* Engelmann spruce)
P. pungens (*see* Blue spruce)
Pine. *See Individual species*
Pine bunchgrass, 407, 429
Pine sawfly, 193, 205
Pinus. *See* Pine and pinyon
P. aristata (*see* Bristlecone pine)
P. arizonica (*see* Arizona pine)
P. cembroides (*see* Mexican pinyon)
P. contorta var. *latifolia* (*see* Lodgepole pine)
P. edulis (*see* Colorado pinyon)
P. engelmannii (*see* Apache pine)
P. flexilis (*see* Limber pine)
P. leiophylla (*see* Chihuahuan pine)
P. monophylla (*see* Singleleaf pinyon)
P. ponderosa (*see* Ponderosa pine)
P. strobiformis (*see* Southwestern white pine)
Pinyon. *See Individual species*
Pinyon bark beetle, 271, 277, 281–285, 322
Pinyon dwarf mistletoe, 271, 281, 283, 285
Pinyon-Juniper
persistent woodland, 272–274, 276, 278–282, 287, 290, 293, 296–298, 300, 303, 307–309, 312, 314, 315, 318, 321, 322, 420, 426, 429, 435
savanna, 2, 272, 274, 276, 281, 287, 288, 290, 293, 296, 300, 303, 304, 306–308, 312, 313, 318–320, 322, 323
wooded shrubland, 272–274, 276, 281, 290, 293, 296, 300, 303, 308, 309, 312, 316–318, 321, 322, 390, 396, 398, 435, 437
woodland, 2, 27, 41, 268, 274, 279, 290, 291, 297, 302–304, 414
Poa
P. pratensis (*see* Kentucky bluegrass)
P. secunda (*see* Big bluegrass)
Pointleaf manzanita, 407, 418, 419, 425
Pollution, 48, 51, 53, 55, 58, 92, 117, 150, 179, 212, 252
Ponderosa pine (introduction), 5, 6, 8, 9, 15, 27, 28, 30, 31, 34, 37, 38, 41–45, 47–49, 51, 75, 124, 190, 268, 334, 375, 404
Ponderosa pine dwarf mistletoe, 193, 205
Ponderosa Pine Forest

dry Ponderosa Pine Forest, 194–196, 219, 220, 231, 244, 247–249
mesic Ponderosa Pine Forest, 76, 130, 194–196, 202, 206, 209, 225, 244–247
moist Ponderosa Pine Forest, 130, 131, 194, 219, 220, 231, 232, 237, 242, 243
Populus tremuloides. *See* Quaking aspen
Precipitation (introduction), 19–21, 23–27, 38, 50–52
Predator, 42, 44, 48, 90
Prescott, 233, 414, 427
Prescott National Forest, 17, 206, 405, 409, 413, 414, 417, 418, 429, 430
Prickly pear, 407, 410
Pringle's manzanita, 407, 408
Pronghorn antelope, 271, 277
Prunus virginiana. *See* Chokecherry
Pseudoroegneria spicata. *See* Bluebunch wheatgrass
Pseudotsuga menziesii. *See* Douglas-fir
Purshia
P. mexicana var. *stansburyana* (*see* Cliff-rose)
P. tridentata (*see* Antelope bitterbrush)
Pygmy conifer woodland. *See* Pinyon-Juniper

Q

Quaking aspen (introduction), 2, 5, 6, 8, 33, 35–37, 41
Quercus. *See* Oak
Q. arizonica (*see* Arizona white oak)
Q. emoryi (*see* Emory oak)
Q. gambelii (*see* Gambel oak)
Q. grisea (*see* Gray oak)
Q. turbinella (*see* Shrub live oak)
Q. X *undulata* (*see* Wavyleaf oak)

R

Railroad, 42, 43, 286
Rapid regrowth, 35, 37, 350–352, 368, 385, 387, 422, 426, 428, 437
Recreation, 4, 48, 49, 53, 55, 58, 91, 93, 94, 109, 110, 113, 115, 117, 148, 150, 165, 169, 179, 208, 215, 237, 241, 251, 252, 279, 286, 291, 292, 300, 308, 313, 318, 322, 347, 349, 350, 359, 363, 366, 383, 385, 391, 394, 398, 419, 421, 434, 435, 437
Recruitment, 48, 88, 147, 152, 162, 197, 216, 217, 285, 296, 297, 308, 309, 312, 315, 317, 318, 320, 362, 366, 389, 396, 418, 428, 435, 437
Red brome, 407, 421
Regeneration, 35–37, 51, 76, 92, 95, 97, 102, 108, 109, 113, 115, 137, 143–145, 147, 149, 152, 154, 159, 160, 162, 168, 169, 172, 173, 176, 177, 199, 201, 204, 208, 210, 211, 216–218, 225, 229, 231–233, 237, 240, 241, 243, 244, 246, 247, 249, 251, 252, 284, 286–288, 292, 293, 296, 297, 309, 317, 350, 351, 416, 422
Relict area, 158, 228–230, 233, 277, 287, 300, 302–306, 354, 382, 383, 388–390, 427, 428
Repeat photography, 154, 210, 304, 389
Rhamnus crocea. *See* Hollyleaf buckthorn
Rhizome, 350, 351, 357, 376
Rhus
R. ovata See Sugar sumac)
R. trilobata See Skunkbush)
Rio Grande National Forest, 53
Roads, 42, 53, 54, 93, 150, 151, 215, 292, 300
Robinia neomexicana. *See* New Mexico locust
Rocky Mountain juniper, 375, 384, 390
Rocky Mountains, 14, 76, 78, 80–84, 86–88, 90, 91, 93, 94, 117, 131, 142, 150, 198, 339, 372
Root-decay fungi, 32, 34, 198, 207
Ruidoso, 215

S

Sagebrush, 271, 272, 274, 295, 297, 304, 306, 307, 375, 376, 385
Salsify. *See* Common salsify
San Francisco Mountain. *See* San Francisco Peaks
San Francisco Peaks, 49, 78, 80, 82, 83, 96, 99, 101, 102, 104, 135, 136, 142, 146, 155, 157–159, 163, 198, 215, 222, 223, 225, 233, 335, 340, 353–355
San Francisco volcanic field, 305
Sangre de Cristo Mountains, 83, 141, 142, 162
San Juan Mountains, 3, 41, 73, 83, 85, 95, 125, 139–143, 152, 155, 158, 200, 212, 233, 352, 358, 389
San Juan National Forest, 53, 148, 233
Santa Fe National Forest, 29, 32, 191
Santa Rita Mountains, 19
Savanna, 2, 228, 272, 274, 276, 281, 287, 288, 290, 293, 296, 300, 303, 304, 306–308, 312, 313, 318–320, 322, 323
Schizachyrium. *See* Bluestem
Scolytus ventralis. *See* Bark beetle and fir engraver
Screwleaf muhly, 336, 339

Sedge, 75, 105, 127, 162, 164, 336, 339, 350, 355, 357, 359
Sedona, 269
Seedling, 48, 52, 97, 99, 143, 151, 156, 162, 201, 204, 210, 216–219, 226, 231, 276, 283, 287, 288, 292, 296, 297, 299, 338, 344, 354, 422, 426, 435, 437
Seepwillow baccharis, 407, 425
Semi-desert grassland, 12, 268, 275, 404, 412, 420, 429
Senegalia greggii. *See* Catclaw acacia
Serviceberry, 5, 12, 271, 297, 375, 376, 379, 385
Sheep, 5, 39, 40, 42, 44, 47, 48, 193, 209, 287, 336, 347, 348, 352, 356, 357, 375, 383, 407, 420
Shiva Temple, 300
Shrubland, 2, 6, 11, 12, 17, 37, 177, 190, 197, 219, 221, 222, 244, 247, 251, 252, 268, 272–276, 281, 290, 293, 295, 296, 300, 303, 304, 306, 308, 309, 312, 316–318, 321, 322, 372–399, 404, 406, 408–416, 418–422, 426–438
Shrub live oak, 5, 12, 407, 408, 410, 413, 417, 418, 420, 422, 425, 426, 428, 435, 436
Sierra Blanca Creek, 224
Singleleaf pinyon, 271, 272, 285
Sitgreaves National Forest, 28, 339, 342, 349, 356, 357
Skunkbush, 375, 385, 407, 410, 418, 420, 422, 425, 426
Sky island, 14, 19
Smooth brome, 336, 347, 349, 351
Snow, 20, 21, 24, 80, 82, 87, 90, 135, 137, 141, 198, 276, 342, 380, 413
Snowberry, 375, 376, 379, 385
Sod, 338, 339, 347, 355, 357
Soil (introduction), 4, 13, 24, 48, 55
Sorghastrum nutans. *See* Indiangrass
Southwestern white pine, 5, 8, 124, 127, 145, 152, 193, 336, 352
Spanish, 39, 42, 286, 288, 300, 357, 408
Sporobolus. *See* Dropseed
Spruce, 2, 5, 6, 41, 75, 88, 99, 102, 103, 127, 132, 134, 144, 155, 159, 160, 336, 352
Spruce aphid, 75, 93, 108, 109, 117
Spruce beetle, 75, 81, 86, 88–91, 94–97, 117, 127, 135, 144
Spruce-Fir Forest
  lower-elevation Spruce-Fir Forest, 72, 75–77, 85–87, 91, 92, 95, 99, 102, 103, 109, 110, 113–115, 130
  Quaking aspen Spruce-Fir Forest, 76, 78, 83, 85, 92, 95–97, 99, 103–105, 110, 113, 115, 117, 418
  upper-elevation Spruce-Fir Forest, 76, 77, 78, 82, 86, 87, 91, 92, 103, 110–112
Spy Mesa, 300
Stork's bill, 407, 421
Subalpine fir, 5, 6, 75, 76, 78, 81, 87, 88, 91, 93, 95, 97, 102–105, 110–114, 124, 127, 128, 130, 132, 143, 144, 151, 159–162, 164, 336, 352
Subalpine-Montane Grassland
  mesic-dry Subalpine-Montane Grassland, 339, 340, 342–346, 348, 349, 351, 353, 357, 358, 359, 364, 365
  moist Subalpine-Montane Grassland, 338, 339, 340, 342, 344, 346, 349, 355, 357, 359, 364, 365
Succession (introduction), 4, 35–37
Sudden Aspen Decline, 88, 147, 148, 208
Sugarite Canyon State Park, 17
Sugar sumac, 407, 408, 417, 420
Sulfate, 51, 52
Sunflower, 271, 288
Symphoricarpos. *See* Snowberry
  S. oreophilus (*see* Mountain snowberry)

T

Taraxacum officinale. *See* Common dandelion
Tecalote Creek, 301
Temperature (introduction), 19, 21, 22, 33, 35, 38, 50, 51
Tertiary Period, 37
Thinning, 35, 37, 97, 108, 112, 115, 143, 145, 151, 152, 162, 168, 171, 175, 177, 201, 204, 211, 214, 216–218, 222, 227, 229, 232–237, 240, 243, 244, 246, 247, 249, 281, 290, 292, 293, 297, 303, 304, 308, 309, 312, 315, 317, 318, 320, 322, 347, 387
Thinopyrum intermedium. *See* Intermediate wheatgrass
Thomomys talpoides. *See* Northern pocket gopher
Threeawn, 407, 410, 431
Thurber fescue, 336, 339, 343, 344, 350, 351, 355, 357
Timothy, 336, 347, 349
Tonto Creek, 429
Tonto National Forest, 221, 410, 414, 419, 423, 427, 429
Topography (introduction), 8, 14, 16, 18–20, 24, 27, 38, 55
Tornado, 30, 204–206
Tragopogon dubius. *See* Common salsify
Trails, 53, 54, 93, 150, 151, 215

Tree encroachment, 35, 37, 349, 350, 352–354, 357–359, 368, 421
Truxton, 304
Tusayan National Forest, 226

U
Uncompahgre Plateau, 279
U.S. Forest Service, 28–30, 33, 43–46, 53, 82, 84, 90, 94, 138, 144–146, 148, 156, 157, 202, 204–206, 209, 211, 213, 215, 221, 225, 280, 282, 287, 289, 344, 345, 348, 350, 380, 410, 414, 415, 417, 420, 423, 424, 426–430
U.S. Geological Survey, 7, 8, 18, 74, 126, 192, 225, 270, 294, 305, 337, 374, 406
Utah juniper, 2, 5, 193, 195, 271, 272, 290, 292, 299, 302, 305–307, 375, 376
Utah serviceberry, 271, 297, 375, 376, 379, 385

V
Valles Caldera National Preserve, 338, 346
Vegetation-Dynamics Model (introduction), 54–57
Verbascum thapsus. *See* Common mullein

W
Wasatch Mountains, 388
Wasatch Plateau, 16, 336, 343, 347, 355
Wavyleaf oak, 193, 195, 219
Weeping lovegrass, 407, 421
Western balsam bark beetle, 75, 81, 88
Western juniper, 271, 290
Western pine beetle, 193, 205
Western serviceberry, 375, 376
Western spruce budworm, 5, 33, 127, 135, 144–147, 168
Western spruce dwarf mistletoe, 75, 88, 93
Western tent caterpillar, 5, 33
Western wheatgrass, 271, 307, 375, 383
White fir, 5, 8, 9, 33, 75, 76, 124, 127, 128, 134, 143–145, 147, 149, 151, 152, 155, 159–162, 164, 174, 177, 193, 194, 197, 201, 226, 227, 232, 237, 242, 336, 352, 375, 376, 384, 387, 396–398
White Mountains, 84, 94, 142, 150, 155, 163, 352, 358
White-tailed deer, 5, 25, 127, 136, 147
Wide Rock Butte, 300
Wild lettuce, 193, 214
Wind, 25, 30, 32, 46, 75, 81, 86–88, 91, 94, 95, 105, 108, 113, 117, 143, 168, 179, 199, 204, 206, 240, 277–279, 281, 312, 343, 344, 362, 394, 414, 416, 434
Witches broom, 34, 145, 207, 285
Wolf, 5
Woodland (introduction), 2, 6, 27, 35, 41
Wright's silktassel, 407, 408, 422

Y
Yellowleaf silktassel, 407, 408
Yucca, 407, 410

Z
Zion National Park, 53, 228, 302, 382, 383, 388–390
Zuni Mountains, 222

The manufacturer's authorised representative in the EU is Springer Nature Customer Service Centre GmbH, Europaplatz 3, 69115 Heidelberg, Germany. If you have any concerns regarding our products, please contact ProductSafety@springernature.com

Printed and bound by CPI Group (UK) Ltd, Croydon, CR0 4YY
30/09/2026
02225445-0014